Industrial Microbiology

Industrial Microbiology

SECOND EDITION

ARVIND H. PATAL M.Sc; MT (ASCP)
Adjunct Professor of Microbiology
Bergen Community College
Paramus, New Jersey

TRINITY
PRESS
(An Imprint of Laxmi Publications Pvt. Ltd.)
An ISO 9001:2015 Company
BENGALURU • CHENNAI • GUWAHATI • HYDERABAD • JALANDHAR
KOCHI • KOLKATA • LUCKNOW • MUMBAI • RANCHI
NEW DELHI

INDUSTRIAL MICROBIOLOGY

Printed and bound in India

First Published : 1985, *Reprinted* : 1996-2008 (Twelve times), 2010, 2011,
Second Edition : 2012, 2016, *Reprint* : 2017, 2019, 2022, *Edition* : 2023

ISBN : 978-93-85750-26-7

Branches

Bengaluru	080-26 75 69 30
Chennai	044-24 34 47 26
Guwahati	0361-254 36 69
Hyderabad	040-27 55 53 83
Jalandhar	0181-222 12 72
Kochi	0484-405 13 03
Kolkata	033-40 04 77 79
Lucknow	0522-430 36 13
Ranchi	0651-224 24 64

PUBLISHED IN INDIA BY

TRINITY PRESS

(An Imprint of Laxmi Publications Pvt. Ltd.)
An ISO 9001:2015 Company
113, GOLDEN HOUSE, GURUDWARA ROAD,
DARYAGANJ, NEW DELHI-110002, INDIA
Telephone : 91-11-4353 2500, 4353 2501
www.laxmipublications.com
info@laxmipublications.com

C—00466/022/08
Printed at: Ajit Printing Press, Delhi.

To

My Wife Minaxi

and

Son Ravi

Preface

The overwhelming response to my First Edition, evoked among the university students and teachers, has prompted me to come out with the Second Edition. I hope that it will also be received with the same warmth.

There is a dearth of authentic books on the subject, and the few that are available in the market are written by foreign authors. The reader would be gratified to note that this is the first book written by an Indian author on the subject.

Biotechnology is a fast developing science, and I believe there is a wide scope for its further development, especially in an agricultural country like ours. If this book proves helpful in some way, I would feel my efforts fully rewarded.

I have tried to make this book as comprehensive as possible. Hence, our university students would find all the topics dealt with at length in the book. I am also sure industrial microbiologists, engaged in the production and processing of fermentation products, would find this book useful in their respective fields.

The book has been amply illustrated, with the use of flow-sheet diagrams and other drawings, to make the scientific principles crystal-clear.

I have made sincere efforts to make the book up-to-date by embracing all the relevant aspects of the subject. I cordially invite constructive suggestions from students and positive criticism from teachers.

The entire text has been revised and updated, and a substantial amount of new material has been added. Seven new chapters have been added.

I hope this book will be warmly welcomed by the community of microbiology and other related sciences.

Arvind H. Patel

New Jersey, U.S.A.
January 1, 2011

Preface to the First Edition

The overwhelming response that my earlier publication, **A Text-book of Microbiology,** has prompted me to publish this book with the hope that it will also be received with the same warmth as its forerunner.

There is a dearth of authentic books on the subject, and the few that are available in the market are written by foreign authors. The reader would be gratified to note that this book is the novel attempt on the subject, by an Indian author.

I have tried to make this book as comprehensive as possible. Hence our university students would find all the topics, listed in the syllabi, dealt with at length in the book. I am also sure industrial microbiologists engaged in the production and processing of microbial products would also find this book useful in their respective fields.

The book has been amply illustrated to make the principles discussed in the book crystal clear. The flow-sheets would further help the reader in understand the topics dealt within the book.

Biotechnology is a fast developing science, and I think there is a wide scope for its further development, especially in an agricultural country like ours. If this book proves helpful, in someway, in advancing the cause of biotechnology, I would feel my efforts fully rewarded.

I am particularly thankful to Professor H.M. Pandya of Mahudha college for going through the press-copy of the manuscript and providing invaluable suggestions. I am also grateful to all those who have helped me, knowingly or otherwise, in my endeavour to bring out this publication in time and to my entire satisfaction.

Nadiad, India
June 4, 1984

Arvind H. Patel

Acknowledgments

First, my sincere thanks go to all the teachers, who have used **Industrial Microbiology,** for their wonderful support and encouragement. I wish to thank my always delightful, and ever-so-patient students for their insight, humour and hard work. I am grateful to the holders of copyright who have given me permission to reproduce diagrams and tables.

While working on the second edition of **Industrial Microbiology,** I was fortunate to have the assistance of Faizan Nawab, Jr. commissioning editor, Macmillan Publishers India Ltd. Thank you, Faizan, for your commitment to the pedagogy in **Industrial Microbiology.** Your cooperation in the editorial process and your patience with me is commendable.

Last, I am indebted to the people who have contributed in various ways in bringing the edition and taking it into so many classrooms.

Important Notes

Considerable care has been taken in the choice and compilation of the 'recipes' included in this book, also in the wording of the working instructions and directions for manufacture. '**Industrial Microbiology**' by A. H. Patel, is offered for sale on the understanding that neither the author, publisher nor selling agents can be held responsible, or liable for any errors, loss, or damage arising from the working of any of these recipes and their attendant instructions and directions.

Contents

1 Introduction

Industrial microbiology is an important branch of microbiology dealing with those areas of microbiology which involve economic aspects, where valuable products are prepared from cheaper, and often disposable, substrates. Therefore, it has become possible for the industrial microbiologist to compete with the industrial chemist. For example, fermentative production costs of all antibiotics, except one or two, are appreciably less than the synthetic production costs of the same.

There were problems with certain industries regarding the disposal of some by-products, (e.g. cheese whey from the cheese industry). These problems have been solved by the fermentation industries, (e.g. acetic acid production by the use of whey). Here, conversion of the substrate into a desired product is affected by the microbial activity, (i.e. bioconversion). In some cases, mass production of microbes is required, (i.e. biomass production).

The microbes useful to the industrial microbiologist include viruses, bacteria, actinomycetes, yeasts, moulds and algae. Special strains are developed for fermentation industries. Actually, microbes are present in natural sources, such as soil and water. For economizing the fermentation processes, production strains are developed from natural isolates. This is accomplished through sequential genetic selection.

Industrial microbiology can be well appreciated by exposing only the useful applied aspects of many different branches of microbiology. In other words, selected areas of many different branches of microbiology engaged, directly or indirectly, in the human services constitute industrial microbiology. This is illustrated in Figure 1.1.

The industrial microbiologist is always dependent upon a biochemical engineer and vice versa. Apart from the cultural responsibilities, the remaining bioparameters of the fermentation process are to be controlled by the biochemical engineer. Thus, economization of a fermentation process requires both a biochemical engineer and a microbiologist. Recently, computer systems have been employed in fermentation operations. This has helped the industrious research microbiologist and the biochemical engineer. With this application of computer systems, it has become very easy to optimize the different bioparameters. Advanced biotechnology and computerisation of fermentation operations have brightened the future of industrial fermentations.

The development of any fermentation process requires constant experimental work and patience. No book can provide complete details about fermentation processes, since such details are trade secrets. Thus, microbiological industries differ from chemical and related industries. In other words, experimentation is a must in establishing new fermentation plants.

Many fermentation products are commercially produced today. Table 1.1 lists the major classes of fermentation products:

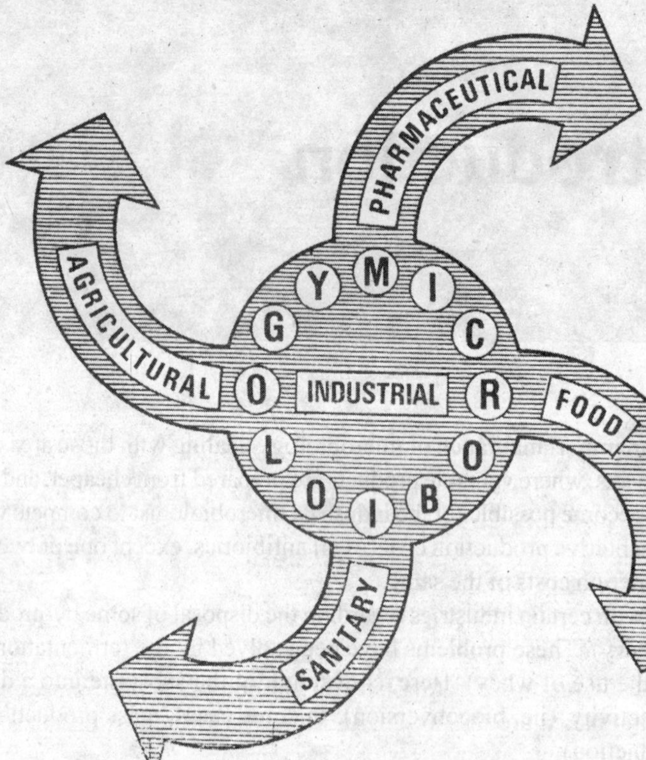

Fig. 1.1 *This shows the area of industrial microbiology. It is scattered among many different branches of the discipline.*

Table 1.1 *Major classes of fermentation products.*

Antibiotics
Beverages
Enzymes
Foods
Organic acids
Fats
Solvents
Bacterial insecticides
Vitamins and growth factors
Nucleosides and Nucleotides
Amino acids
Steroids
Miscellaneous

Industrial applications of micro-organisms tend to fall into two categories:

(i) Those which utilize a pure culture or pure cultures and involve large scale manufacturing techniques and

(ii) Those involving the use of naturally occurring mixtures of micro-organisms under conditions that may be rather primitive, as a means of producing a desired change in some products of industrial value, (e.g. sauerkraut).

2 History of Development

1 INTRODUCTION

We can conveniently consider the history of the development of industrial microbiology in terms of three periods of time:

(a) A period of ignorance (pre-1800),

(b) A period of discovery (1800-1900), and

(c) A period of industrial development (post-1900).

Many of man's domestic processes dating back to antiquity depended on fermentative changes. For example, bread was first baked around 4000 B.C. Moreover, wine was produced from grapes and the brewing of beer had the elements of a professional occupation.

Hindsight would associate the end of the period of ignorance with the development of the microscope and the discovery of living micro-organisms during the eighteenth century.

The industrialization of fermentation processes had already been started by the beginning of the twentieth century.

2 THE ERA OF THE DISCOVERY OF MICROBES

Roger Bacon in 1267 described a lens, and Salvino D'Armati (d. 1317) was said to have been the inventor of spectacles. However, the microscope did not come into existence until nearly three centuries later. At some time during the period 1591–1608 Zaccharias Jensen, while working with his father (a maker of spectacles), placed two lenses together to make a crude, compound microscope. This instrument had no provision for focussing. However, a compound microscope made by Galileo Galilei in 1610, which he called the *occiale*, had the tube mounted so that it could be focussed. Under it, according to his own description, a water flea looked as large as a hen. The name microscope was first proposed by Fabri in 1625. In 1650 Huygens evolved the eyepiece lens system which bears his name, and which is essentially the same as that being used in most modern microscopes.

Hooke, the great British microscopist, in *Micrographia* (1665) first applied the term *cell* to the box-like structures he observed in the thin sections of a cork when they were examined by means of a compound microscope.

With these developments the stage was set for revealing of the world of micro-organisms. However, until Antony van Leeuwenhoek began to squint through his simple microscope, the existence of micro-organisms was largely a matter of philosophical postulation.

It was about 1670 when Antony van Leeuwenhoek (1632–1723) started his hobby of making simple microscopes. He was a Dutch merchant living in Delft, Holland. During his lifetime he made more than 250 microscopes. These consisted of home-ground lenses mounted in brass and silver. Thus, Antony van Leeuwenhoek was able to obtain magnifications up to 160 diameters (some commentaries say 270 diameters). His descriptions of protozoa were so accurate that many of the forms were easily recognized.

For him, nothing was too sacred for observation. He turned his microscope on a great variety of substances. Leeuwenhoek carefully recorded his observations in a series of letters to the British Royal Society. In one of the letters dated September 7, 1674, he described the 'very little animalcules' which we recognize as free-living protozoa. On October 9, 1676, he wrote:

> "In the year 1675, I discovered living creatures in rain water which had stood but a few days in a new earthen pot, glazed blue within. This invited me to view this water with great attention, especially those little animals appearing to me ten thousand times less than those . . . which may be perceived in the water with the naked eye."

Leeuwenhoek described his little animals in great details, and there is no doubt that he observed bacteria, fungi and many forms of protozoa. For example, he reported that on June 16, 1675, while examining well water into which he had soaked peppercorns the day before:

> "I discovered, in a tiny drop of water, incredibly many very little animalcules, and these of diverse sorts and sizes. They moved with bendings, as an eel, always swimming with its head in front, and never tail first, yet these animalcules swam backwards as well as forwards, though their motion was very slow."

In addition to the diversity of this microbial world, Leeuwenhoek emphasized its incredible abundance. For example, in one letter describing for the first time, the characteristic bacteria of the human mouth, he wrote:

> "I have had several gentlewomen in my house, who were keen on seeing the little eels in vinegar; but some of them were so disgusted at the spectacle, that they vowed they'd never use vinegar again. But, what if one should tell such people in future that there are more animals living in the scum on the teeth in a man's mouth, than there are men in a whole kingdom?"

For almost 50 years, until his death in 1723, Leeuwenhoek transmitted his discoveries to the Royal Society in the form of a long series of letters written in Dutch. These letters were largely translated and published in English in the *proceedings of the Royal Society*. Thus, his discoveries were quickly and widely disseminated.

While Leeuwenhoek may not have been the first to see bacteria under the microscope, he was certainly the first to describe these forms accurately and was probably, the first to recognize the value of the microscope as a laboratory instrument.

3 PASTEUR AND FERMENTATION

In the middle of the last century, fermentation was considered to be a chemical process. The chemist Liebig described the process as depending on a chemical ferment, an alterable substance that decomposes and in doing so excites chemical change, that is, fermentation in a ground substance. The disintegration of a ferment was thus believed to cause a disturbance among the molecules that brought about the changes associated with fermentation. According to this view, the primary cause of fermentation was a chemical instability. Other chemists of the period, including Berzelius (1779–1848) and Bertholet (1827–1907) held similar views concerning alcoholic fermentation. In 1837 Schwann had described alcoholic fermentation as a process dependent on a yeast, but none of the leading chemists of the time considered seriously the possibility of a living organism as the causative agent of fermentation.

Pasteur pursued Schwann's idea further and became a pioneer in experimental studies on fermentation. He confirmed the observation made by Schwann and showed that a living organism, a yeast, is required for the chemical change that transforms sugar into alcohol and carbonic acid. In 1857, Pasteur discovered that a different kind of organism was associated with the process by which *sugars were broken down to lactic acid*. These observations led Pasteur to a hypothesis that micro-organisms of some kind are essential for each kind of fermentation. To obtain evidence concerning this hypothesis, Pasteur isolated micro-organisms associated with fermentation and raised them in nutrient solutions. These were inoculated into the appropriate natural media from which the natural organisms had been removed and fermentation was produced experimentally. A temperature of 30 to 50 degrees centigrade was found to be the optimum for the process.

Louis Pasteur

Pasteur followed carefully the cycle of fermentation. He repeatedly removed the organisms from fermenting cultures, transferred them to other sterile media, and found that fermentation always resulted when the proper organisms were present and the appropriate conditions were arranged. At the end of the experiment, Pasteur identified the materials that had been produced. He then described the chemical changes which required organisms that were involved in the transformation of sugar to lactic acid.

Different organisms were observed under the microscope and their characteristics were carefully described. Lactic acid organisms were found to resemble in some ways to those associated with alcoholic fermentation. More recent studies have shown that the two kinds of organisms are not as closely related as Pasteur supposed. The lactic acid organism is a bacterium rather than a yeast. Distinctly different kinds of organisms were thus shown to be responsible for different kinds of fermentation. When brewer's yeast was added to a standard medium containing sugar, alcoholic fermentation would proceed; whereas when lactic acid organisms were added to a similar medium, lactic acid was the product.

During the next 20 years (1857–77), Pasteur studied a number of other fermentable materials and became an authority in the field. In 1858, he discovered that the organism associated with fermentation in ammonium tartrate was a mould. Yeasts, bacteria, and moulds had now been identified with fermentation, Pasteur's memoir on alcoholic fermentation, written in 1860, remains to this day a classic on the subject.

Pasteur (1861) investigated butyric acid fermentation and made another important discovery that fermentation can proceed in the absence of oxygen. The rod-shaped organism associated with butyric fermentation was considered by Pasteur to be animal in character and was named vibrio. While examining a drop of fluid culture containing butyric vibrio under the microscope, Pasteur observed that the organisms in the margin of the drop where they came in contact with the air were not active, but in the centre they were actively *motile*. He developed the hypothesis that oxygen had an inhibiting effect on the growth and activity of these organisms. To test the hypothesis, Pasteur passed a stream of oxygen through an active butyric fermenting culture and observed that all action was inhibited. Two words, *aerobic* and *anaerobic*, were used respectively to distinguish between the two kinds of organisms. Also, in 1861 Pasteur published his research on acetic acid fermentation. He showed that this fermentation was accomplished by organisms of the genus *Mycoderma*, which he studied in detail in 1862. A full scale study of vinegar preparation was undertaken and the results were published in 1864 and 1868.

Pasteur was then asked by his former professor, J.B. Dumas, and personally requested by Napoleon III to study the problem of souring wine that threatened the important French wine industry. He pursued the investigation with vigour and enthusiasm, and in 1866 published an impressive memoir on wine. This 264-page treatise dealt with the so-called diseases of wine, which Pasteur described as being due to foreign organisms that invaded the wine and altered its chemical and physical properties. Pasteur showed that undesirable organisms could be removed by partially sterilizing, at a temperature below boiling point, the juice from which wine was to be made. This was not sufficient to destroy the valuable properties of the juice. Other organisms capable of producing desirable qualities could then be introduced from pure cultures. The process, now called pasteurization, was later adapted to milk for the purpose of removing most of the natural organisms and all of the pathogens such as the tuberculosis, brucellosis, and typhoid organisms.

Pasteur returned in 1871 to his investigation of fermentation, this time devoting his attention to the fermentation of beer. The results were published in 1876 in a detailed, a nearly 400-page paper which not only discussed the results of investigations on beer, but also summarized Pasteur's general knowledge substantiating his concept that the fermentation process was based on living organisms. The proposed explanation of fermentation was again challenged, this time by Liebig, who had *developed a modified version of his older theory* in which enzymes were postulated to be responsible for fermentation. No enzymes could be demonstrated, however, and Pasteur's view that living organisms were responsible for the process eventually won support.

4 THE ERA OF THE DISCOVERY OF ANTIBIOTICS

By the beginning of the twentieth century, the German scientists named Emmerich and Low isolated an antimicrobial substance from bacteria. This substance was called pyocyanase. It was able to destroy some disease-producing micro-organisms. Thus, the first antibiotic was discovered. Unfortunately,

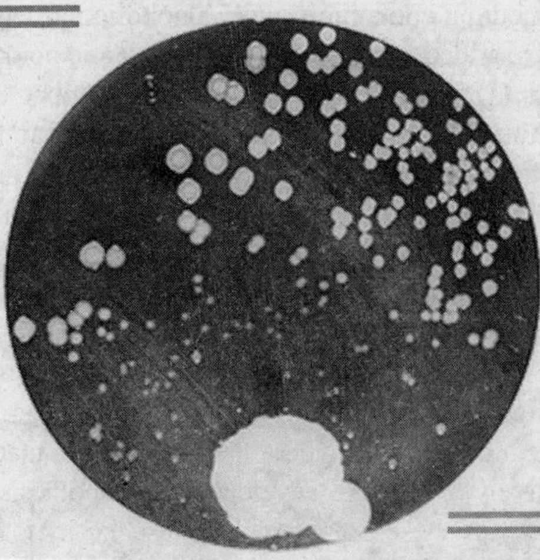

Fig. 2.1a *Top: The original culture plate on which the observation of the action of penicillin was made. Below: Sir Alexander Fleming in his laboratory in 1954.*

techniques were not available to guarantee the quality of each of the batches and the project was abandoned as it was thought to be too unreliable.

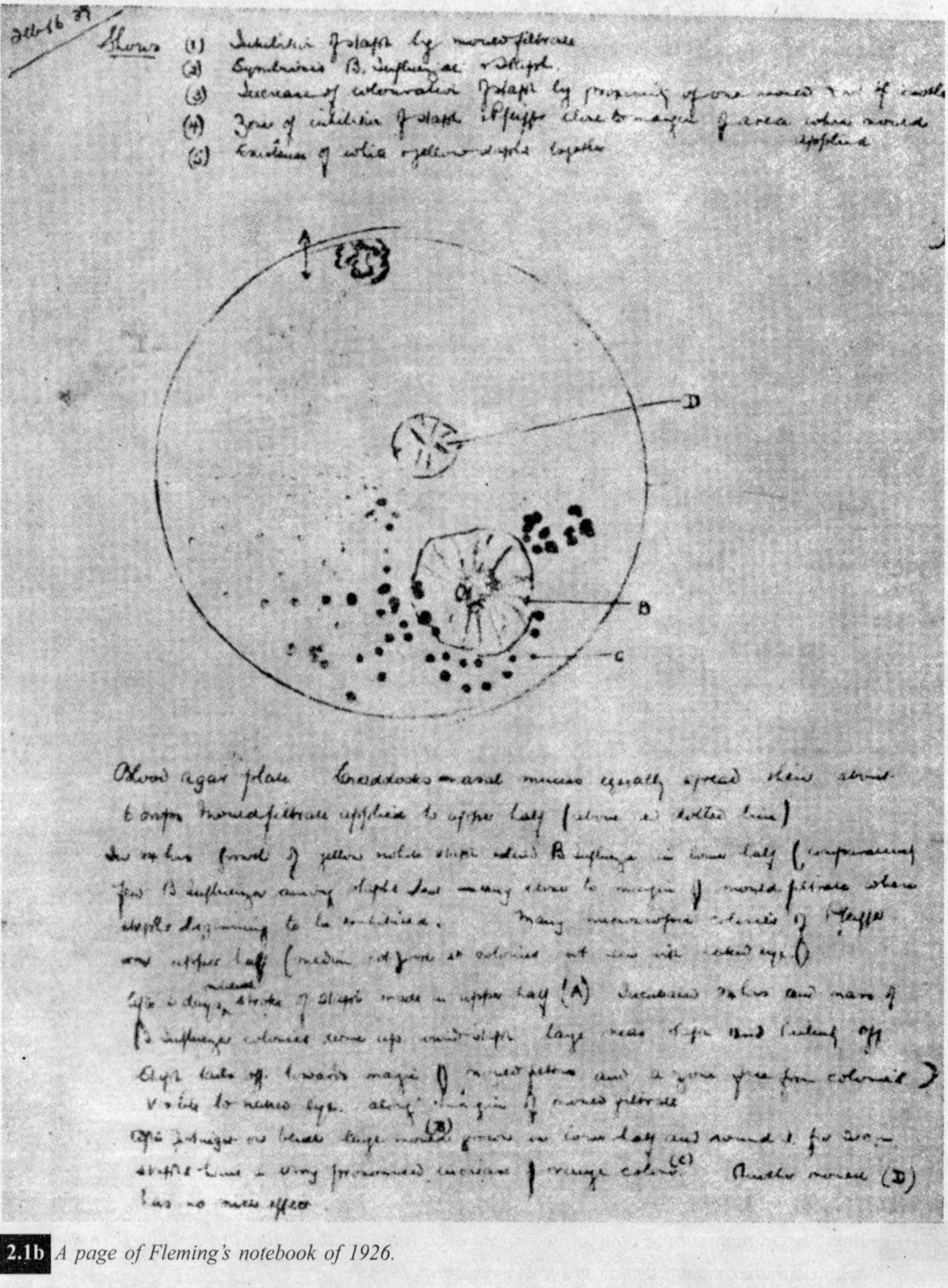

Fig. 2.1b *A page of Fleming's notebook of 1926.*

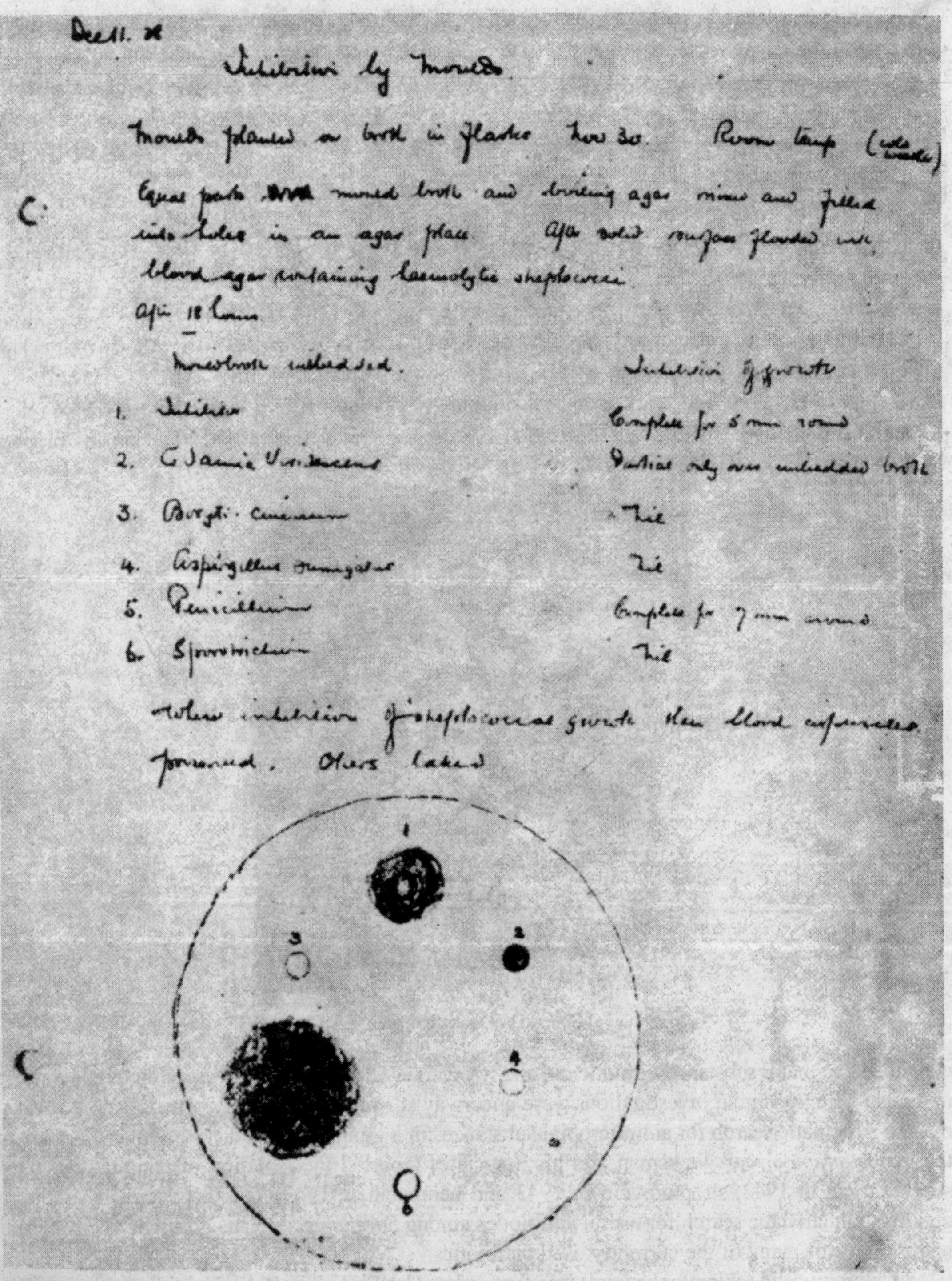

Fig. 2.1c *Top: Fleming's original notebook of 1926.*
Below: The fungus Penicillium notatum.

Thirty years later, the real progress in the field of antibiotics was initiated after the chance discovery of penicillin. Sir Alexander Fleming (1881–1955), British bacteriologist, observed that on a plate culture of bacteria which had become contaminated by a mould, bacterial growth in the vicinity of the mould colony was inhibited (Figures 2.1a, 2.1b and 2.1c). He carried on his experiments at St. Mary's Hospital, London, in 1928. His results were published in the *Journal of Experimental Pathology* in 1929. Since the mould was a strain of *Penicillium* (*P. notatum*), he named the antibacterial substance, penicillin. For lack of time and money, Fleming did not pursue the investigation further, but simply shelved his discovery in the hope that some day it would be investigated further. A decade later his hope was realized when two other Englishmen, Chain and Florey engaged themselves in furthering Fleming's investigations. Thus, an antibiotic preparation was developed for human use. The problems inherent in the large-scale, submerged, aerobic culture of micro-organisms were quickly solved, with the result that commercial production in large, mechanically agitated and aerated fermentors began in the U.S.A. in 1942. This led quickly to a similar development in the United Kingdom.

At about the same time, Dubos, working at the Rockefeller Institute described a series of microbial products to which he gave the name, tyrothricin. Further investigation showed that his preparation

Selman A. Waksman in 1964

contained two major substances; gramicidin and tyrocidine. These were found useful in treating infections in human-beings. Similar investigations were underway at Rutgers University, where Dr. S.A. Waksman made a systematic search for antimicrobial substances in a group of soil-inhabiting micro-organisms of the genus *Streptomyces*. Waksman and his associates reported the isolation of actinomycin in 1940, streptothricin in 1942, streptomycin in 1943 and neomycin in 1949. The discovery of streptomycin greatly stimulated the search for useful antibiotics among *Streptomyces*. This group of micro-organisms is the source of many of the currently used antibiotics.

Chloromycetin (chloramphenicol) was first isolated by Ehrlich in 1947. The following year, Duggar described aureomycin. This even provided fresh opportunities for the search and the discovery of an important class of antibiotics, called the tetracyclines. Following these discoveries, the pace of new discoveries accelerated. The years of discovery of the most important antibiotics are shown in Table 2.1.

Table 2.1 *Years of discovery of important antibiotics*

Name of the antibiotic	*Year of discovery*
Penicillin	1929
Tyrothricin {gramicidin, tyrocidine}	1939
Griseofulvin	1939
Streptomycin	1943
Bacitracin	1945
Chloramphenicol	1947
Polymyxin	1947
Chlortetracycline	1948
Cephalosporin C, N, P	1948
Neomycin	1949
Oxytetracycline	1950
Nystatin	1950
Erythromycin	1952
Novobiocin	1955
Kanamycin	1957
Fusidic Acid	1960
Ampicillin	1961
Cephalothin	1962
Lincomycin	1962
Gentamycin	1963
Carbenicillin	1964
Cephalexin	1967
Clindamycin	1968

5 A CENTURY OF GROWTH OF INDUSTRIAL FERMENTATIONS

In the last fifteen years, there has been a tremendous development in the field of industrial fermentations. This development is both qualitative and quantitative. In other words, industrial fermentations have been expanded both in number of products produced on a commercial scale and in fermentation capacity. Thus, the number of products has more than doubled, and the volume has increased about five–fold. There are more than 500 companies using fermentation processes in production operations.

The fermentation industries in the United States (and in most of the industrialized world) date back to about 1880 when a small-scale lactic acid production plant was started in Avery, Massachusetts. As shown in Table 2.2, during the initial phase of development (1880–1910) fermentation products had immediate application in foods (lactic acid and bakers' yeast), as a raw material for chemical processing (ethanol) and in the textile industry (amylases for desizing of textiles). Certain practices were based on experience rather than science, although the fermentation industries prospered. Thus, the element of 'practical witch-craft' limited the development of the science.

Table 2.2 *A century of growth of industrial fermentations*

Year of commerciali-zation	*Fine chemicals*	*Enzymes*	*Therapeutic substances*	*Microbial transformations*	*Other*
1880–1990	lactic acid				bakers' yeast
1900–1910	ethanol glycerol	amylases			
1910–1920	acetone *n*-butanol	invertase			
1920–1930	citric acid				
1930–1940	gluconic acid	proteinases	riboflavin	sorbose acyloins	
1940–1950	itaconic acid 2-keto-D-gluconic acid	cellulases pectinases	vitamin B_{12} penicillin G bacitracin streptomycin 7-chlortetracycline tyrothricin gramicidin polymyxin neomycin		
1950–1960	kojic acid glutamic acid lysine	glucose oxidase catalase	amphotericin B colistin cycloheximide cycloserine erythromycin griseofulvin kanamycin novobiocin nystatin oleandomycin penicillin V semisynthetic penicillins oxytetracycline polymyxins tetracycline demeclocycline variotin viomycin	steroid oxidations	gibberellins dextran single-cell protein
1960–1970	valine	glucose isomerase glucamylase lipases lactase	amphomycin blasticidin S cactinomycin candicidin cephalosporins dactinomycin fusidic acid gentamicins	dihydroxy-acetone	xanthan 5′-nucleosides bioinsecticide

Table 2.2. (continued)

Year of commerciali-zation	*Fine chemicals*	*Enzymes*	*Therapeutic substances*	*Microbial transformations*	*Other*
			hygromycin B		
			kitasatomycin		
			lincomycin		
			mikamycins		
			monensin		
			pimaricin		
			polyoxins		
			pristinamycins		
			ribostamycin		
			rifamycins		
			siomycin		
			spectinomycin		
			thiostrepton		
			tylosin		
			vancomycin		
1970–1977	comenic acid	microbial enzyme as rennet replacement	adriamycin	sterol cleavage	ribose
1980		melibiase	avoparcin	xylitol	zearalenol
		dextranase	bambermycin	aspartic acid	
			bicyclomycin	malic acid	
			bleomycin		
			candidin		
			capreomycin		
			daunorubicin		
			enduracidin		
			fortimicin		
			josamycin		
			lasalocid		
			lividamycin		
			macarbomycin		
			mepartracin		
			medicamycin		
			mocimycin		
			myxin		
			quebemycin		
			sagamicin		
			salinomycin		
			sisomicin		
			tetranactin		
			thiopeptin		
			tobramycin		
			validamycin A		

The wartime needs for acetone in 1914–18 stimulated some practical research. This led to the establishment of a substantial scientific base for the development of fermentation industries from 1920–1940. The Weizmann process (acetone-butanol fermentation) required the combined efforts of biochemists, engineers, bacteriologists and chemists to bring this rather complicated fermentation from the laboratory scale to the production scale, and emphasized the need for sterile media, pure culture technique and understanding of bacterial physiology. This experience underscored the value of a team approach in solving fermentation problems.

Development of chemical industries in the United States after the First World War also had a stimulating effect on the fermentation industries, as efforts were initiated to develop processes that were new and different from those already in operation in Europe. The early success of the mycological (mould) processes for production of citric acid, gluconic acid, and the continued success of the acetone-butanol fermentation process convinced some chemical producers that microbial processes could be dependable and economically successful.

The interest in vitamins to supplement human and animal diets stimulated the search for a fermentation process for riboflavin, thiamine, ascorbic acid, and other therapeutically useful compounds. This enthusiasm increased when the first reports on penicillin suggested that this microbial product could have tremendous impact on the therapy of infectious diseases.

The 1945–60 period saw a further exploitation of fermentation technology for the production of antibiotics, enzymes, and additional organic acids having a great deal of commercial potential. Some of the more remarkable practical developments included: microbial processes for glutamic acid and lysine; commercial production of the anti-inflammatory agents including cortisone, hydrocortisone, prednisolone and triamcinolone, based on selective microbial oxidation of the 1, 2-cyclopentenophenanthrene nucleus; and large-scale preparation of yeast and other organisms as protein sources in human and animal diets.

The list of new fermentations activated during the 1960–80 period includes about 50 antibiotics, and several enzymes, polysaccharides, flavour-enhancing purine nucleosides, microbial insecticides and fragrances. More than half of the protein-coagulating proteases used in manufacturing various kinds of cheese in the United States in 1980 are mould enzymes. These have replaced and supplemented the limited amount of calf-stomach-derived rennet for this purpose. For example, marzyme is an enzyme produced by *Mucor miehei*. Nowadays many cheese industries have started using it for milk-coagulation.

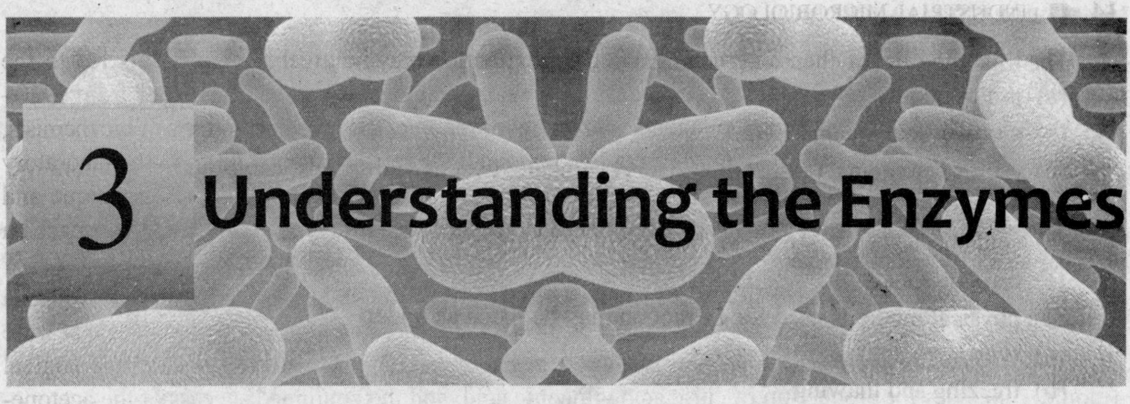

3 Understanding the Enzymes

An enzyme may be defined as a thermolabile organic catalyst liberated by a living cell and can function independently of the cell.

In a sense, the enzymes may be regarded as the working part of the cell. There can be no life without enzymes. An enzyme acts by catalysis, i.e. the enzyme increases the velocity of a chemical reaction without itself being permanently changed, it may be recovered in a functional state at the end of the reaction.

A pure enzyme preparation consists of a large number of protein molecules each with an identical amino acid sequence. Enzymes are denatured by heat, precipitated by ethanol or high concentrations of inorganic salts like ammonium sulphate, and do not dialyze through semipermeable membranes, (i.e. proteinic as all enzymes are proteins but all proteins are not enzymes).

Some enzymes consist entirely of protein, (e.g. pepsin and trypsin). Many enzymes consist of a protein linked with a low molecular weight organic molecule known as a coenzyme. Enzymes of this type are called conjugate enzymes. This can be represented as under:

Apoenzyme	+	Coenzyme	⟶	Holoenzyme
Inactive		Inactive		Active
High molecular weight		Low molecular weight		
Non-dialyzable		Dialyzable		

In some cases the nonprotein part of an enzyme may be a metal ion, (e.g. iron in the enzyme catalase). It may be firmly combined to the protein or loosely linked and easily dissociable, depending upon the particular enzyme. Such metal ions (Mg^{2+}, Fe^{2+}, Zn^{2+}, Mn^{2+}, etc.) are considered as inorganic coenzymes or cofactors. Sometimes both a cofactor and a coenzyme (organic) are needed before enzyme becomes active.

1 EXTRACTION, ISOLATION AND PURIFICATION

All enzymes are proteinic in nature and, therefore, can easily be damaged when exposed to drastic conditions, namely excess of heat, acidity, or alkalinity. These produce chemical or physical changes in enzymes leading to insolubilization or the other alterations, which impair enzymic activity. The problem is how to extract, isolate and purify enzymes in their active forms, preferably just the desired one, from the complex mixture of enzymes and other components of the cell, including non-enzymic structural protein.

Extraction

The term **extraction** will be applied to define the techniques used for the disruption of micro-organisms. Effective bacterial disintegration can be brought about, either by chemical, or physical methods.

The following list of chemical methods used for extraction may be given:

(a) use of alkali
(b) use of lysozyme and EDTA
(c) use of detergents, (e.g. sodium lauryl sulphate, tweens and tritons, cetyl-diethyl ammonium bromide)
(d) cold-shock
(e) osmotic shock

The physical methods used for extraction may be listed as under:

(a) sonication
(b) freezing and thawing
(c) solid shear
(d) liquid shear
(e) agitation with abrasives (grinding)

The most commonly used method for fractionation is '**grinding**', a physical method. This technique was developed by Booth and Green (1938). This was unsatisfactory, as both *E. coli* and *B. subtilis* required two hours' treatment and *Sarcina lutea* four hours for enzyme release.

Detailed study about the disruption of bacteria by agitation in the presence of glass beads was made by Curran and Evans (1942), and King and Alexander (1948). King and Alexander (1948) found the following conditions as optimum during mechanical disintegration of bacteria.

(a) beads of 0.13 – 0.26 mm. diameter
(b) 300 – 500 strokes per minute
(c) an amplitude of $2\frac{1}{2}$ inches

Mickle (1948) developed an apparatus, which has found wide application in the disintegration of micro-organisms. It comprises a glass cup with indentations suspended on the end of a variable length arm. Into the glass cup is placed 10 ml. of bacterial suspension, (10–20 mg. of protein/ml.) together with an equal volume of Ballatini beads (0.1–0.2 mm. in diameter). The cups are then stirred at a frequency of 50 cycles/second. Overheating can be avoided by working in a cold-room. Similar technique was described by Shockman, Kolb and Toennies (1957), which could cope with up to 6 g. of material per run.

There are various factors affecting the degree of disruption of bacteria, and they may be given as under:

(a) the rate of stirring
(b) the concentration of micro-organisms
(c) the concentration and size of beads
(d) the contact time

Fractionation and subsequent separation of sub-cellular structures by ultra-centrifugation is considered as an important step in isolation and purification of enzymes. Fractionation is carried out by addition of glass beads with the cells. The tube containing glass beads and cells is subjected for centrifugation (20,000 revolutions/minute). Collision occurs between glass beads and cells and this disintegrates the cell-wall. This method is known as **mechanical disintegration**. Also lysozyme may be used for disruption of the cell-wall. For separation of sub-cellular structures, the tube is subjected to ultra-centrifugation at very high speed. As a result, three fractions are produced in the form of different bands as follows:

(a) Structures which are large enough to be observed with a light microscope by sedimenting at 10,000 X g for 15 min;

(b) Submicroscopic particles which sediment at 100,000 X g for 60 to 120 min; and

(c) The soluble fraction which contains substances that remain in solution in the supernate that is separated from the submicroscopic particles.

The bulk of the respiratory enzymes of bacteria are recovered in fraction (a). This fraction consists of cytoplasmic membrane and a network of internal membranes in the forms of vesicles and tubules in section. These envelopes are free of all cytoplasmic material and ribosomes. The intracytoplasmic membranes are the probable locus of the respiratory enzymes of the cell. The proteolytic enzymes appear to be about equally divided between cytoplasm and cytoplasmic membrane.

The ribosomes are found in fraction (b). They are rich in RNA and phospholipids and do not contain respiratory enzymes. Thus, crude extract of enzyme is separated from other unwanted cellular material. This crude extract can be refined to obtain enzyme in pure form. It is necessary to keep near 0°C at all times to avoid cellular organelles disruption and destruction of enzyme activities.

Because of the sensitivity of enzymes to extremes of pH they should be extracted from the disrupted cells with cold buffer solutions at pH near neutrality. The crude mixture obtained in this way must be separated into material active and inactive with respect to the enzymic activity under investigation. Obviously a biochemical technique for the estimation of the amount of desired enzyme present must be developed first. This estimation technique is referred to as an '*enzyme assay*' and it generally involves the measurement of the rate of appearance of the products of that enzymic reaction by chemical and physical procedures.

Isolation and Purification

The terms **isolation** and **purification** will be used to define the methods involved in obtaining a pure, or specified product.

The important contaminating material is nucleic acid. This can be removed, either by precipitation and final centrifugation, or nuclease treatment. Bovine pancreatic RNAase and DNAase can be used to hydrolyze bacterial nucleic acids (Davidson, 1965).

Concentration may be made by different methods as under:

(a) precipitation, (e.g. ammonium sulphate)

(b) ultrafiltration

(c) freeze drying

(d) gel chromatography

(e) ion-exchange chromatography

(f) affinity column chromatography

Many separation techniques are available for isolation of particular enzymes. For example, the '*selective precipitation*' of the required enzyme can often be achieved using ammonium sulphate. It is necessary to bring the final concentration of this highly soluble salt to whatever level has been observed to give the best yield of that enzyme in the precipitate of protein material resulting from this '**salting-out**' technique. Further purification of the enzyme may be achieved by a variety of methods. The well-known '**affinity column chromatography**' technique is used for separating the desired enzyme from contaminating enzymes and other proteins. In this technique, column is filled with a suitable adsorbent (activated charcoal or alumina) or resin as shown in the Figure 3.1. The crude enzyme mixture is applied from the top of column and elution is done.

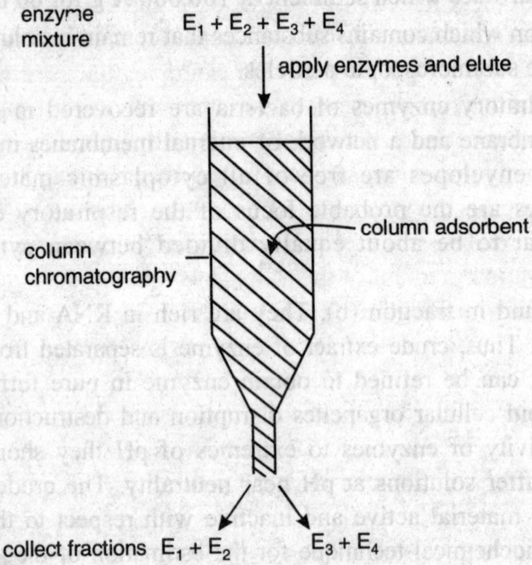

Fig. 3.1 *Affinity column chromatography used to remove impurities.*

The purified enzyme is collected as a particular fraction from the eluted column, and may even be induced to form crystals although it may still be only about 50 per cent pure.

Unfortunately, criteria of purity of an enzyme are often difficult to establish. One such criterion is that a pure enzyme is homogeneous upon '*electrophoresis*' so that a single band (or zone) is observed. This technique relies on the fact that the same characteristic distance is travelled by all the molecules of a pure enzyme upon application of an electrical field to an aqueous solution of that enzyme. An enzyme carries a characteristic overall positive or negative charge, at a particular pH of its aqueous solution, due to the amino-acid composition of that enzyme. This can be shown in Figure 3.2.

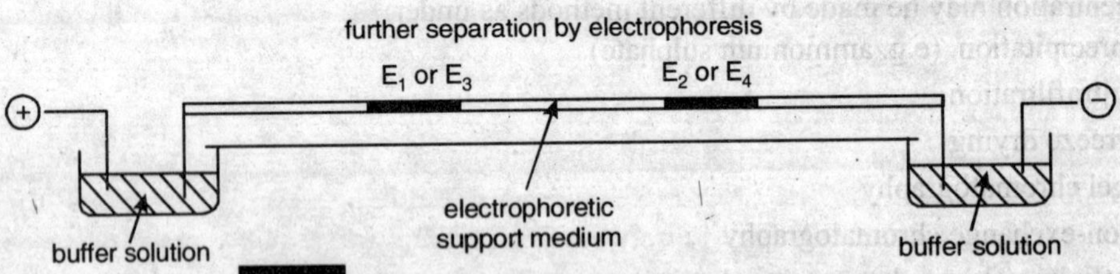

Fig. 3.2 *Electrophoresis used for further separation.*

The test for purity include:

(a) comparison of enzymic activity of purified enzyme with standard enzyme under uniform conditions

(b) ascending paper chromatography

(c) descending paper chromatography

(d) circular paper chromatography
(e) thin layer chromatography

Many enzymes are isolated from brewers' and bakers' yeast because of the availability and cheapness of large quantities of these materials of the brewing and baking industries. One such enzyme, called alcohol dehydrogenase, catalyses the formation of ethanol from acetaldehyde. The yeast must be disrupted before extraction of enzymes is attempted. For example, the yeast may be ground, after which it is possible to extract alcohol dehydrogenase with a dilute solution of disodium hydrogen phosphate (pH slightly alkaline) at room temperature. An initial precipitation with ice–cold acetone is followed by further purification by the use of ammonium sulphate to yield the required alcohol dehydrogenase. Any residual ammonium sulphate in this enzyme preparation can be removed by dialysis. As the enzyme is protein, it is held within the thin dialysis sac used and ammonium sulphate moves out into the surrounding water because of the concentration gradient of this salt. Thus ammonium sulphate moves outside from inside of the dialysis sac as shown in Figure 3.3.

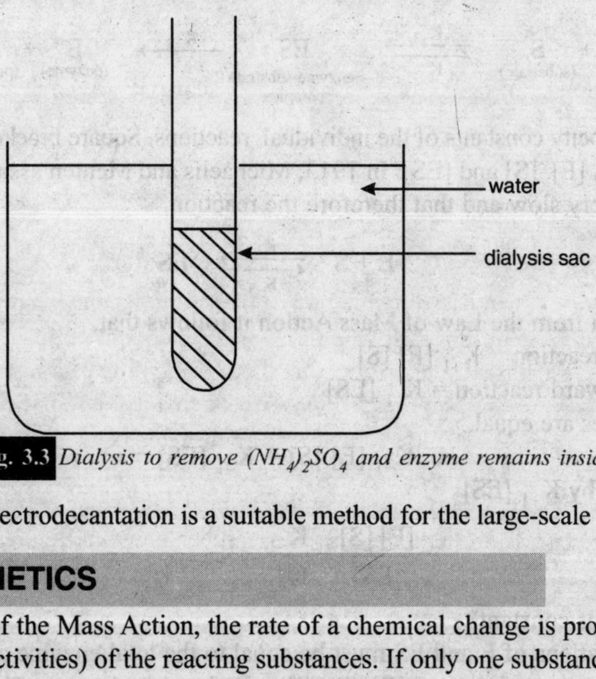

Fig. 3.3 *Dialysis to remove $(NH_4)_2SO_4$ and enzyme remains inside.*

Multimembrane electrodecantation is a suitable method for the large-scale purification of enzymes.

2 ENZYME KINETICS

According to the Law of the Mass Action, the rate of a chemical change is proportional to the product of the concentrations (activities) of the reacting substances. If only one substance is reacting, the rate of reaction is proportional to the concentration of this substance. This is a first-order reaction. Many hydrolytic reactions appear to be first-order reactions and, therefore, have velocities proportional to the concentration of the material undergoing hydrolysis. During the inversion of sucrose by H^+ a molecule of water reacts for every molecule of sucrose hydrolyzed, but since the concentration of water is great compared to that of sugar, little overall change in water concentration accompanies a large change in sugar concentration, and the reaction rate is governed by the concentration of sugar present:

$$V \propto [a]$$

This equation states that the velocity, V, of the reaction is proportional to the concentration of reacting substance, [a]. By introducing a velocity constant, K, the speed of reaction can be equated to K times [a]:

$$V = K [a]$$

After the reaction has proceeded for time t, the original concentration has changed and there remains [a – x] as the concentration of reacting substance. The velocity of the reaction at time t must then be governed by the amount of material equal to [a – x] and this velocity can be expressed in the differential equation

$$V = \frac{dx}{dt} = K[a - x]$$

where K = a velocity constant for the specific reaction, x = the amount of substance changed in time, t, a – x = the concentration of original material remaining at time t and V = the reaction velocity at time t.

3 DERIVATION OF THE MICHAELIS–MENTEN EQUATION

A simple enzyme reaction may be represented by:

$$\underset{\text{(enzyme)}}{E} + \underset{\text{(substrate)}}{S} \underset{K_{-1}}{\overset{K_{+1}}{\rightleftharpoons}} \underset{\substack{\text{enzyme-substrate}\\ \text{complex}}}{ES} \xrightarrow{K_{+2}} \underset{\text{(enzyme)}}{E} + \underset{\text{(product)}}{P}$$

where the K's are the velocity constants of the individual reactions. Square brackets are used to represent molar concentrations, i.e. [E], [S] and [ES]. In 1913, Michaelis and Menten assumed that the formation of E + P from ES was very slow and that therefore the reaction,

$$E + S \underset{K_{-1}}{\overset{K_{+1}}{\rightleftharpoons}} ES$$

was in equilibrium. Then from the Law of Mass Action it follows that,
The rate of the forward reaction = K_{+1} [E] [S]
and the rate of the backward reaction = K_{-1} [ES]
At equilibrium these rates are equal,

$$\text{i.e. } K_{+1} [E] [S] = K_{-1} [ES] \qquad \text{...(a)}$$

Divide both sides of (a) by K_{+1} [ES]

$$\frac{[E] [S]}{[ES]} = \frac{K_{-1}}{K_{+1}} = K_m \qquad \text{...(b)}$$

where K_m is the Michaelis constant*
The sum of the concentrations of E and ES must be equal to the total enzyme concentration, $[E_0]$

$$[E] + [ES] = [E_0] \qquad \text{...(c)}$$

Substract [ES] from both sides of (c),

$$[E] = [E_0] - [ES] \qquad \text{...(d)}$$

We can now substitute this value for [E] in (b),

$$K_m = \frac{([E_0] - [ES])[S]}{[ES]} \qquad \text{...(e)}$$

* In this derivation the Michaelis constant is equal to the dissociation constant of ES. This has been proved to be correct for some enzymes. Fortunately, a slightly more complex theory can account for the most exceptions.

Expansion of the bracket and multiplication of both sides of (e) by [ES] gives,

$$K_m [ES] = [E_0] [S] - [ES] [S] \quad \text{...(f)}$$

Collect the terms that contain [ES]

$$[ES] (K_m + [S]) = [E_0] [S] \quad \text{...(g)}$$

Divide both sides of (g) by $(K_m + [S])$

$$[ES] = \frac{[E_0][S]}{K_m + [S]} \quad \text{...(h)}$$

Michaelis and Menten made the assumption that the slowest, or velocity-determining step, was the breakdown of the enzyme-substrate complex. Then it followed from the Law of Mass Action that the velocity, v, of the overall reaction was,

$$v = K_{+2} [ES] \quad \text{...(i)}$$

Substitute the value of ES from (h) in (i),

$$v = \frac{K_{+2}[E_0][S]}{K_m + [S]} \quad \text{...(j)}$$

If the enzymes were completely saturated by substrate, v would increase to a maximal velocity, V, c.f. (i),

$$V = K_{+2} [E_0] \quad \text{...(k)}$$

Substitute (k) in (j),

$$v = \frac{V[S]}{K_m + [S]} \quad \text{...(l)}$$

This is the Michaelis-Menten equation.

The equilibrium assumption used by Michaelis and Menten is not always valid but this same equation, although with a somewhat different meaning for K_m, accounts for the relationship found between velocity and substrate concentration for many enzymes. In fact K_m is always equal to the substrate concentration that gives half maximum velocity, but its significance in terms of velocity constants is dependent upon the mechanism of the reaction under consideration. In practice it is often necessary to consider enzymes which catalyze reactions involving more than one substrate—these too can be fitted into the above scheme, for most purposes.

4 FACTORS AFFECTING ENZYME ACTIVITY

Among the factors affecting the activity of an enzyme are the following:

(1) nature and concentration of enzyme
(2) nature and concentration of substrate
(3) time
(4) pH
(5) temperature
(6) ultraviolet light
(7) inhibitors
(8) activators
(9) allosteric enzymes
(10) ionic strength

(1) Nature and concentration of enzyme

Enzymes exhibit high degree of specificity, i.e. specific enzymes can act only on their own substrates (urea can be acted upon by urease).

The rate of an enzyme-catalysed reaction is directly proportional to the concentration of enzyme present as shown in Figure 3.4. The validity of this linear relationship is absolutely essential to the accurate measurement of enzyme activity. The relationship only holds true under certain rigidly defined conditions. The pH and temperature of the system must be constant and the substrate must be present in excess. If the substrate is not present in excess there will be significant variations in the substrate concentration during the course of the measurement and the extent of the variations will be dependent on the enzyme concentration. In general, the concentration of enzyme and substrate are adjusted so that less than 1 per cent of the substrate is used up during the measurement, the sensitivity of modern equipment allowing the accurate detection of such relatively small changes.

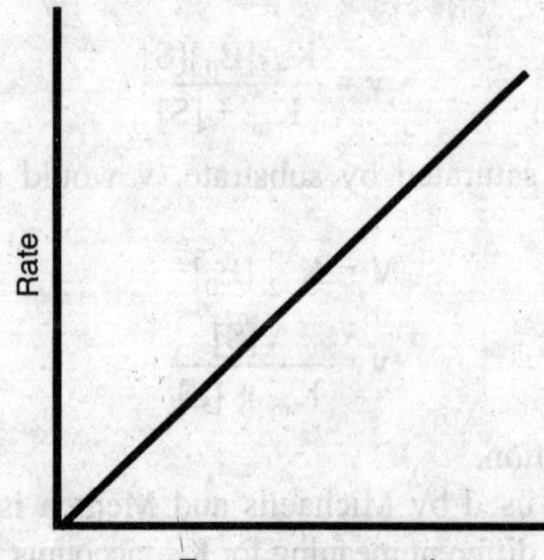

Fig. 3.4 *The effect of enzyme concentration on the rate of an enzyme reaction.*

(2) Nature and concentration of substrate

The nature of substrate is equally important as in the case of an enzyme. Only particular substrates are attacked by particular enzymes.

If the initial rate of an enzyme reaction is measured at a series of substrate concentrations, for most enzymes the results obtained take the form shown in Figure 3.5. At low substrate concentrations, the rate is directly proportional to the substrate concentration, while at high substrate concentrations the rate is constant. This characteristic graph is a rectangular hyperbola which has the general equation:

$$y = \frac{ax}{x + b}$$

where a is the maximum value of y reached and b is the value of x when y is half the maximum value, i.e. when $y = \frac{1}{2}$ a. In the specific case of the variation of rate with substrate concentration the constants a and b are given the symbols V and K_m respectively and their relationship to the graph is shown. If v is the rate at any substrate concentration, S, then they are related by the equation:

$$v = \frac{V.S}{K_m + S}$$

This is the so-called Michaelis-Menten equation and over the whole range of substrate concentrations represents graphs of the type shown in Figure 3.5. V represents the maximum rate attained by the reaction at high substrate concentrations and K_m, the Michaelis constant, is the substrate concentration when the rate is half the maximum attainable rate. Thus V and K_m are experimentally determinable constants and, provided the enzymes is studied under controlled conditions, for example constant defined pH and temperature, they are characteristic and constant for a particular enzyme.

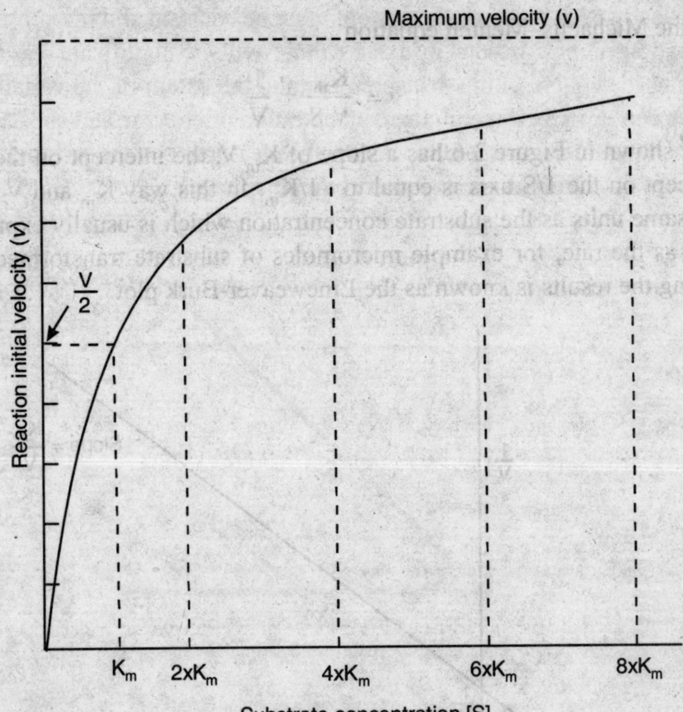

Fig. 3.5 *The reaction velocity against increasing substrate concentration.*

When v is equal to half the maximum velocity then

$$\frac{V}{2} = \frac{V[S]}{K_m + [S]}$$

cancelling out V, and cross multiplying

$$K_m + [S] = 2[S]$$

$$K_m = [S]$$

K_m thus has the units of molarity, and is equal to the substrate concentration required to reach an initial velocity of half the maximum velocity at that particular enzyme concentration, pH and temperature. The value of K_m obtained is independent of enzyme concentration, but is dependent on the pH and temperature of the experiment. Thus K_m is a characteristic of an enzyme, at defined pH and temperature. Typically one finds values of K_m in the range 10^{-2} to 10^{-5} moles/litre for various enzymes. This represents an enormous difference between enzymes in their ability to work near their maximum velocities despite a low concentration of the required substrate being present. Enzymes with low K_m, e.g. 10^{-5} moles/litre, can work rapidly even at this low substrate concentration. This difference is thought to provide an important control mechanism in the cell, especially in the presence of more than one enzyme competing for one substrate.

The determination of V and K_m

Although it is possible to obtain the value of V and K_m directly from the graph of rate against substrate concentration, it is more usual to plot the reciprocal of the rate, 1/v against the reciprocal of the substrate concentration, 1/S.

Rearranging the Michaelis–Menten equation

$$1/v = \frac{K_m}{V.S} + \frac{1}{V}$$

The reciprocal plot shown in Figure 3.6 has a slope of K_m/V, the intercept on the 1/v axis is equal to –1/V and the intercept on the 1/S axis is equal to $-1/K_m$. In this way K_m and V can be conveniently found. K_m has the same units as the substrate concentration which is usually expressed as molarity. V has the same units as the rate, for example micromoles of substrate transformed per unit time. This method of expressing the results is known as the Lineweaver-Burk plot.

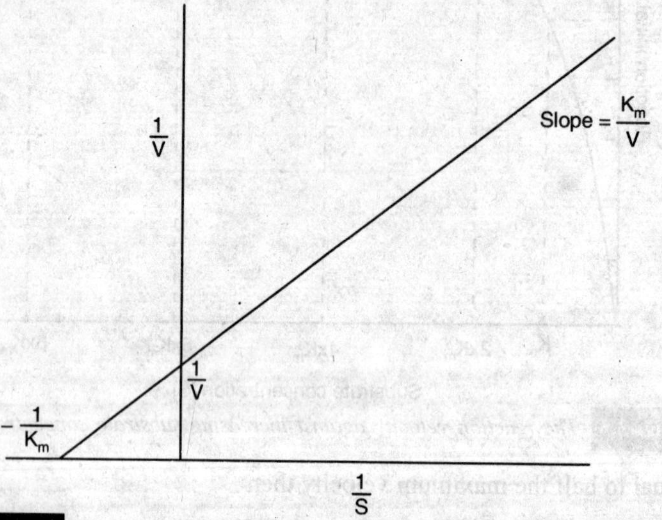

Fig. 3.6 *The Lineweaver–Burk plot for the determination of K_m and V.*

Since the maximum rate V, is equal to K_{+2} . E, if the maximum rates of two reactions are compared at equal enzyme concentrations, the relative values of the rate of breakdown of the enzyme–substrate complex can be found.

(3) Time

The usual variation of the extent of reaction with time is shown in Figure 3.7. There are two distinct regions in the graph. In the initial period of time the amount of substrate which has been transformed is directly proportional to the length of time. After this initial period the rate of reaction starts to decrease and the amount of reaction is no longer directly proportional to the time. Provided the substrate is present in excess the explanation of this phenomenon is the progressive loss of enzyme activity after a period of time. This may be due to the effect of heat on the tertiary structure of the enzyme or to the formation of some product or side-product of the reaction which inhibits the enzyme. In the initial investigation of any enzyme reaction it is important to establish the period over which the reaction varies linearly with time. This period can be as short as a minute while some enzymes have a linear variation over a period of days. For meaningful results the initial rate of reaction should be measured. In Figure 3.7, the initial rate of reaction is given by the slope of the line A. If, however, the reaction had only been studied by measuring the substrate transformed after time C, then the apparent rate of reaction, given by the slope of line B, would have been considerably lower.

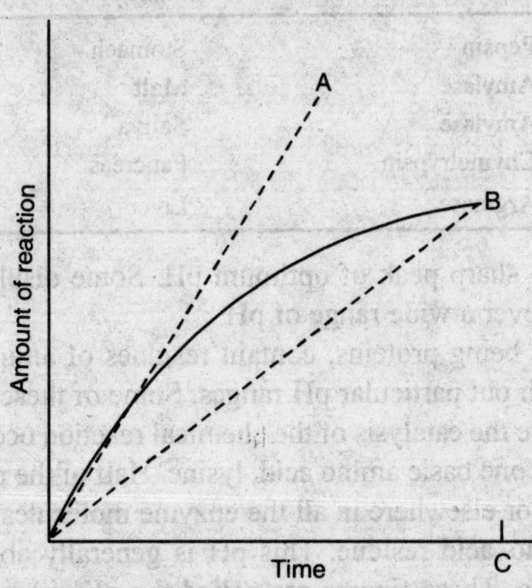

Fig. 3.7 *The time–course of an enzyme reaction, showing the difference between initial and final rate.*

(4) pH

An enzyme is only active as a catalyst within a relatively narrow range of pH and usually the enzyme is most active at a certain pH value, called the *optimum pH*. Figure 3.8 shows a typical relationship between activity and pH for an enzyme catalyzed reaction. The bell-shaped curve reaches its peak at the

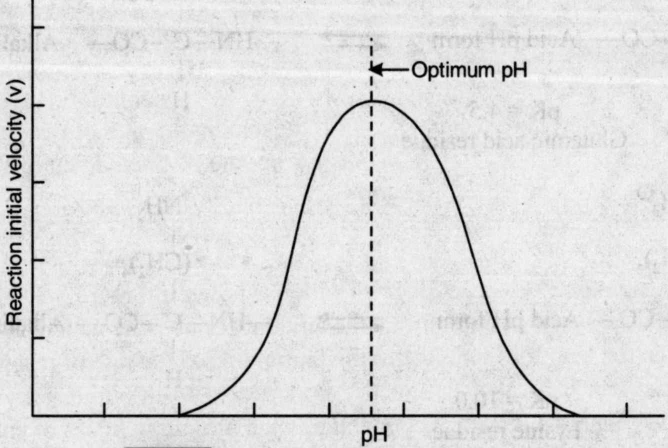

Fig. 3.8 *The effect of pH on reaction rate.*

optimum pH. For most enzymes this optimum pH is near neutrality (pH 5 to pH 9) but there are some exceptional enzymes such as pepsin which act at extreme pH values. Chymotrypsin, however, performs its fastest at the mildly alkaline pH found in the intestine. Table 3.1 shows optimum pH value(s) for some enzymes.

Table 3.1 *Optimum pH value(s) for some enzymes.*

Sr. No.	*Enzyme*	*Source*	*Optimum pH*
1.	Pepsin	Stomach	2 and 3.3
2.	Amylase	Malt	5.2
3.	Amylase	Saliva	6.8
4.	Chymotrypsin	Pancreas	7 to 8
5.	Arginase	Liver	9.8

Not all enzymes show a sharp peak of optimum pH. Some display a plateau region where the reaction velocity is unaltered over a wide range of pH.

It is clear that enzymes, being proteins, contain residues of amino acid types which have side groups that are ionized through out particular pH ranges. Some of these residues will be present at the active site of the enzyme, where the catalysis of the chemical reaction occurs. Let us consider one acidic amino acid, glutamic acid, and one basic amino acid, lysine. Half of the residues of each of these amino acids present at the active site or elsewhere in all the enzyme molecules is ionized at any one time at a pH characteristic of that amino–acid residue. This pH is generally about 4.5 for glutamic acid and generally about 10.0 for lysine. These figures are called the pK value for glutamic acid and lysine, respectively, and refer to the midpoint of the equilibrium reaction between the ionized and unionized forms of each amino acid (Figure 3.9).

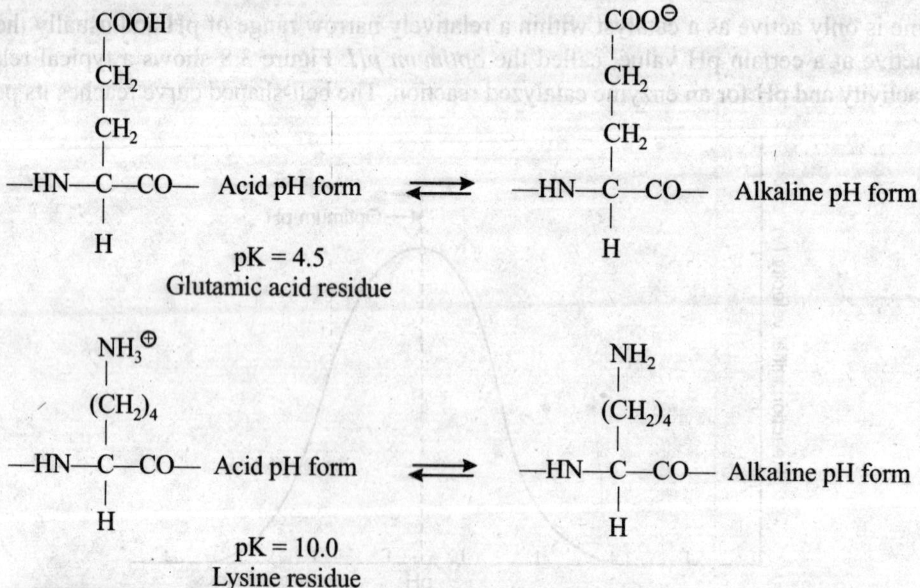

Fig. 3.9 *Ionized and unionized forms of glutamic acid and lysine residues in proteins.*

Let us suppose that one residue of each of these two amino acids is present at the active site of an enzyme molecule and that, for example, the charged form (ionized) of both amino acids must be present if that enzyme molecule is to function. Figure 3.10(a) shows that at least half the carboxyl and amino groups will be present in the form of $-COO^-$ and $-NH_3+$ over the pH range 4.5 to 10.0.

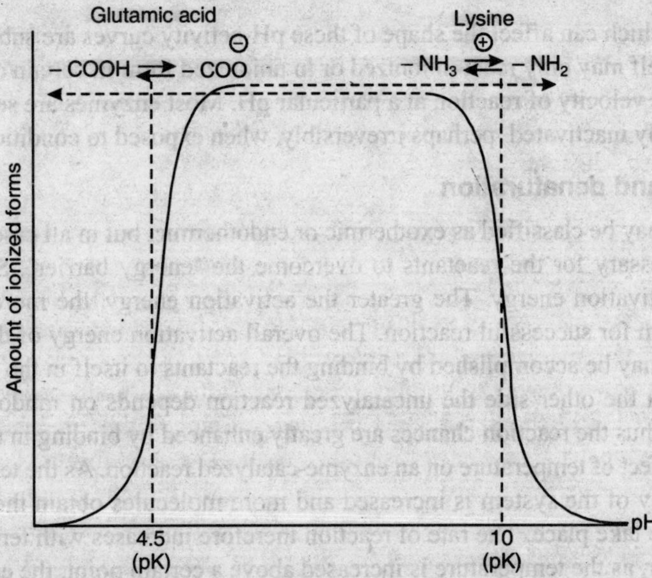

Fig. 3.10 (a) *At least half the carboxyl and amino groups in the form of $—COO^-$ and NH_3^+ over the pH range 4.5 to 10.*

Thus the probability of the presence of both $-COO^-$ and NH_3^- at the active site of any one enzyme molecule is at least 0.5 and rises to almost 1.0, (i.e. certainty) in a range of pH near neutrality (pH 6–8). We see from Figure 3.10(b) that the optimum pH for this system is indeed a wide plateau over a range of pH near neutrality.

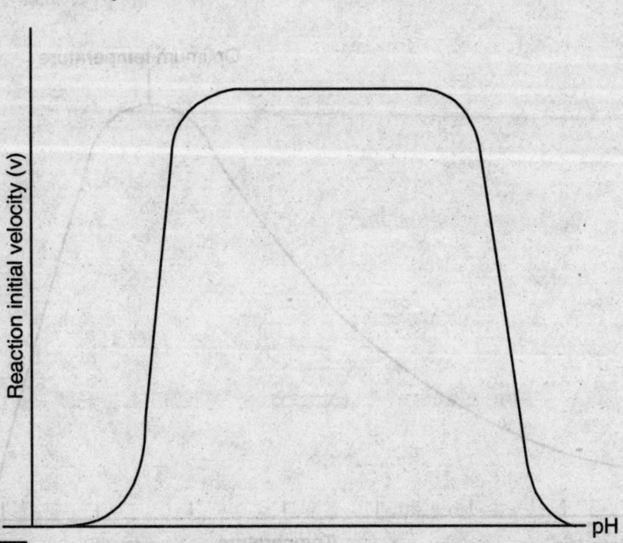

Fig. 3.10 (b) *Optimum pH is indeed a wide plateau over a range of pH near neutrality.*

It is clearly the pK value of each of the ionizable amino acids at the active site of the enzyme that can determine its response to pH changes. The bell-shaped curve usually observed may be due to amino acid residues being present at the active site which have pK values not greatly different from each other. In that case the pH range which satisfies the ionization requirements for enzyme activity would be quite narrow.

Other factors which can affect the shape of these pH-activity curves are substrate and extremes of pH. The substrate itself may only react in ionized or in unionized form in certain cases, and its pK value would then affect the velocity of reaction at a particular pH. Most enzymes are sensitive to extremes of pH and may be rapidly inactivated, perhaps irreversibly, when exposed to conditions of low or high pH.

(5) Temperature and denaturation

Chemical reactions may be classified as exothermic or endothermic, but in all cases, before reaction can take place, it is necessary for the reactants to overcome the 'energy barrier'. Such energy barrier is referred to as the activation energy. The greater the activation energy, the more heat which must be supplied to the system for successful reaction. The overall activation energy of the reaction is lowered by an enzyme. This may be accomplished by binding the reactants to itself in the correct orientation for reaction to occur. On the other side the uncatalyzed reaction depends on random collisions between reactant molecules. Thus the reaction chances are greatly enhanced by binding in the catalyzed reaction.

Consider the effect of temperature on an enzyme-catalyzed reaction. As the temperature is increased so the inherent energy of the system is increased and more molecules obtain the necessary activation energy for reaction to take place. The rate of reaction therefore increases with temperature as shown in Figure 3.11. However, as the temperature is increased above a certain point, the energy of the system is sufficient to cause the breakdown of hydrogen bonding and many other forces which hold the tertiary structure. Thus enzyme eventually becomes completely inactive. This process is referred to as denaturation. There comes a point where the increase of reaction due to temperature on the activation of molecules is equal to the decrease of reaction rate. At this temperature the activity is a maximum and this temperature is often known as the optimum temperature.

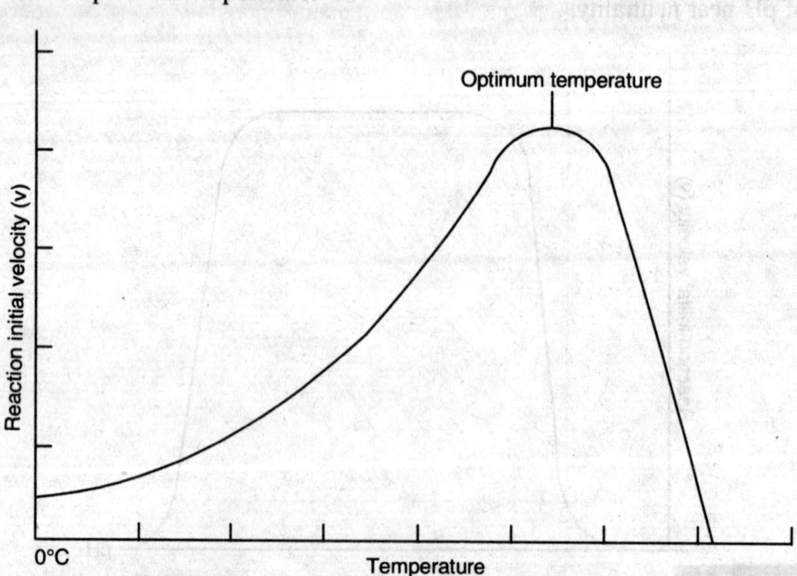

Fig. 3.11 *The effect of temperature on reaction rate.*

(6) Ultraviolet light

It either destroys or modifies the action of enzymes. The rate of destruction is practically independent of temperature but is affected by pH and other environmental conditions. Purified enzymes are more easily destroyed by ultraviolet light than the same enzymes in less purified preparations. Enzymes are most sensitive at a wavelength of 2650 A.

(7) Inhibitors

Inhibitors may be classified as irreversible and reversible. The use of group-specific reagents such as DFP, PCMB and FDNB is an example of irreversible inhibition. Once a covalent bond is formed with enzyme, it is relatively difficult to remove the group and restore the enzyme activity. In case of reversible inhibitors, the inhibitor is rarely covalently bound and may be easily removed with the restoration of enzyme activity by techniques such as dialysis, dilution or addition of substrate in excess. This may be written as under:

(a) E (Enzyme) + I (Inhibitor) $\longrightarrow$ EI

(b) E + I $\rightleftharpoons$ EI

EI → Enzyme-Inhibitor complex

Let us now consider reversible inhibitors in more detail. They can be divided into certain categories; we will mention two of these, competitive and non-competitive inhibitors. **Competitive inhibitors**, as the name implies, compete with substrate for the same binding site at the active site of the enzyme. Therefore, the degree of inhibition observed for a competitive inhibitor at a particular concentration is reduced by increasing the substrate concentration. Figure 3.12 (a) shows typical results obtained with a competitive inhibitor. When 1/v is plotted against 1/S at the various inhibitor concentrations, a series of straight lines is obtained which intersect on the 1/v axis. This point corresponds to the maximum velocity V. The lines intersect at various points on the 1/S axis and so the apparent value of the Michaelis constant changes in the presence of different concentrations of inhibitor. These results are to be expected from the theory of action of competitive inhibitors, since they affect the binding of substrate and hence K_m but, once the substrate is bound, there is no effect on the subsequent catalysis and hence on V.

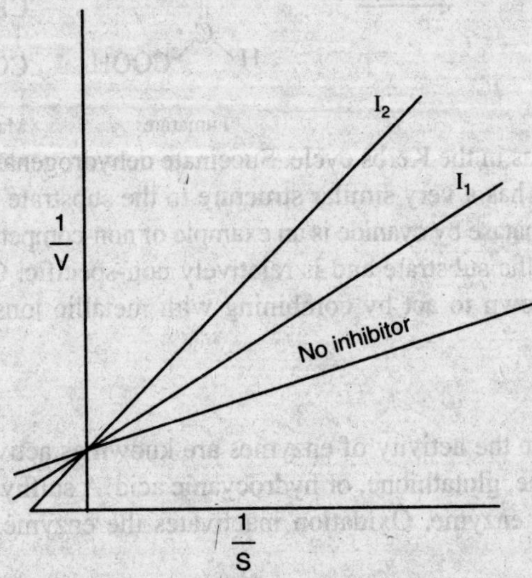

Fig. 3.12(a) *Reciprocal plots for a competitive inhibitor showing intersection on the 1/v axis.*

Figure 3.12 (b) shows comparable results for a non-competitive inhibitor. The lines intersect at the 1/S axis giving the same value of K_m in each case. This is also to be expected since the inhibitor does not interfere with the binding of substrate. The inhibitor acts by blocking catalytic groups and hence the value of V as obtained from the intercept on the 1/v axis is different in each case.

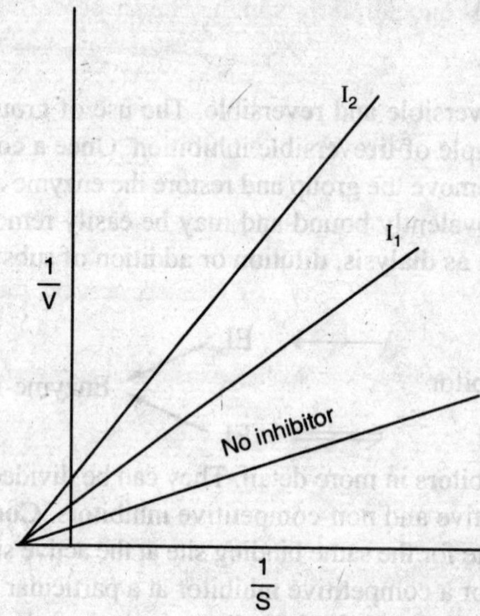

Fig. 3.12 (b) *Reciprocal plots for a non-competitive inhibitor showing intersection on the 1/S axis.*

One of the most widely quoted examples is the case of the enzyme, succinate dehydrogenase, which catalyzes the removal of two hydrogen atoms from succinate to yield fumarate as shown under:

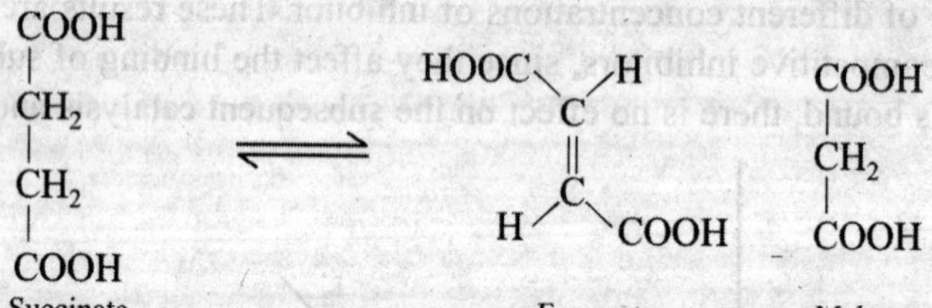

Succinate Fumarate Malonate

This is one of the reactions in the Kerbs cycle. Succinate dehydrogenase is competitively inhibited by malonate. Malonate indeed has a very similar structure to the substrate, succinate.

The inhibition of arylsulphatase by cyanide is an example of non-competitive inhibition. The inhibitor has no structural similarity to the substrate and is relatively non-specific. Cyanide is known to inhibit many enzymes and can be shown to act by combining with metallic ions which serve as prosthetic groups in many cases.

(8) Activators

Those substances which restore the activity of enzymes are known as activators. Activation of papain can be effected by H_2S, cysteine, glutathione, or hydrocyanic acid. A sulfhydryl group is believed to be an essential part of the active enzyme. Oxidation inactivates the enzyme, and reduction restores its activity.

(9) Allosteric enzymes

Certain enzymes do not show the typical Michaelis-Menten kinetics shown in Figure 3.5 but rather the variation of enzyme activity with substrate concentration is sigmoidal (Figure 3.13, control curve). It will be seen that such enzymes display low activity even at moderate substrate concentrations and only reach full activity when the substrate concentration is high. The shape of this curve is altered markedly

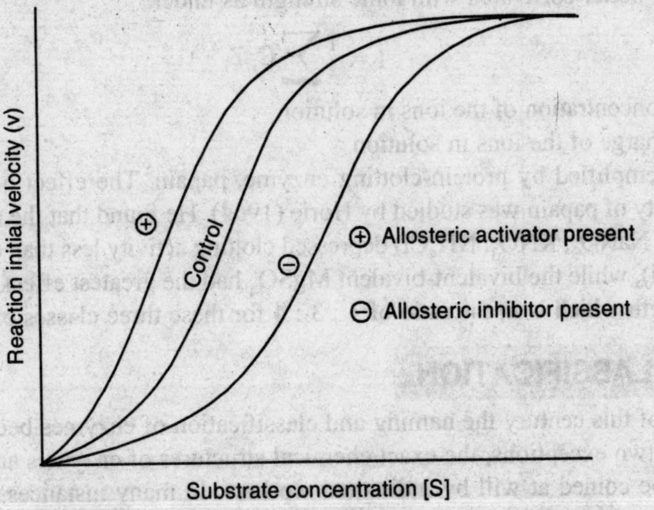

Fig. 3.13 *Michaelis plots for a typical allosteric enzyme.*

by the presence of allosteric effectors, either activators or inhibitors. In the presence of an activator, curve ⊕ is obtained which is similar to the normal Michaelis-Menten curve. Thus, with activator present, the enzyme is active even at fairly low substrate concentrations. On the other hand, in the presence of inhibitor (curve ⊖), the sigmoidicity of the curve is greatly increased and the enzyme requires even higher substrate concentrations for appreciable activity. It must be emphasized that in many cases these allosteric effectors bear no structural similarity to the substrates or products of the enzyme reaction. This is possible because an allosteric effector controls the velocity of the reaction by binding at its own allosteric site rather than the active site. In this way it is possible for an accumulation of some end product of a metabolic pathway to alter the activity of an enzyme acting many stages back in the formation of such a product.

This is due to the allosteric effector apparently either reversing or promoting the same conformational change as the substrate that interconverts active and inactive forms of these enzymes. These two types of enzymes are represented below:

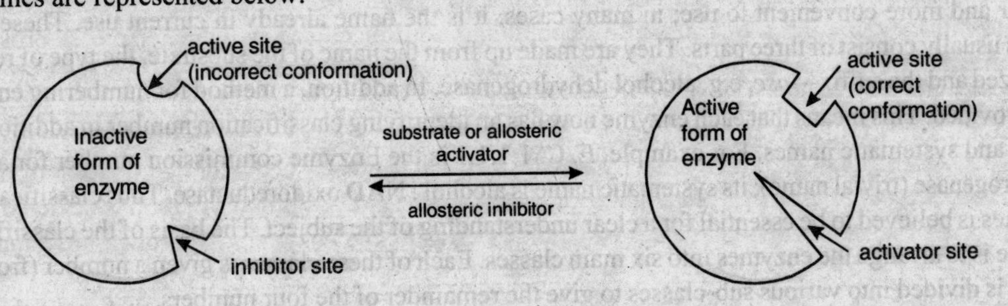

This shows how the allosteric effector combines with its enzyme at a site which is not the active site of the enzyme–thus the term "allosteric" refers to the "other site".

(10) Ionic strength

As the concentration of salts increase, one often finds a general decrease in enzyme activity. This characteristic may be better correlated with ionic strength as under:

$$I = \frac{1}{2}\sum C_i Z_i^2$$

where, C = Concentration of the ions in solution
Z = Charge of the ions in solution

This can be exemplified by protein-clotting enzyme, papain. The effect of various ions on the protein clotting activity of papain was studied by Horie (1964). He found that the addition of univalent-univalent salts (NaCl, $NaNO_3$, KNO_3, NH_4Cl) depressed clotting activity less than the univalent-bivalent salts (Na_2SO_4, NH_4Cl), while the bivalent-bivalent $MgSO_4$ had the greatest effect. The results correlate well with ionic strength which is in the ratio of 1 : 3 : 4 for these three classes of salts.

5 ENZYME CLASSIFICATION

During the first-half of this century the naming and classification of enzymes became very confusing. With possibly one or two exceptions, the exact chemical structures of enzymes are not known. Names of enzymes used to be coined at will by individual workers. In many instances, the same name was given to different enzymes or the same enzyme became known by several different names. Moreover, many of the names given conveyed little or no idea of the nature of the reactions catalyzed. By about 1955, it had become evident that the nomenclature of enzymes was in a chaotic state.

The situation was resolved by an International Commission on Enzymes established by the International Union of Biochemistry in 1955. Its recommendations were published in 1964, and they should be followed by every microbiologist working with enzymes.

Names of enzymes ending in '*ase*' should be used only for individual enzymes, e.g. succinate dehydrogenase. When it is desired to name a complex of several enzymes on the basis of the overall reaction catalyzed by it, the word *system* should be used, e.g. the succinate oxidase system, which catalyzes the oxidation of succinate by oxygen and involves succinate dehydrogenase, cytochrome oxidase, and several intermediate carriers. For classification, only individual enzymes, and not enzyme systems, are considered.

Enzymes are named and classified on the basis of what they do rather than on what they are. It is this specific property that distinguishes one enzyme from another. Each enzyme should be coded on a four-number system and two names: a working or trivial name and a systematic name. The trivial name is shorter and more convenient to use; in many cases, it is the name already in current use. These trivial names usually consist of three parts. They are made up from the name of the substrate, the type of reaction catalyzed and the suffix—*ase*, e.g. alcohol dehydrogenase. In addition, a method for numbering enzymes was provided. This means that each enzyme now has an identifying classification number in addition to its trivial and systematic names. For example, E. C. 1.1.1.1 is the Enzyme commission number for alcohol dehydrogenase (trivial name); its systematic name is alcohol : NAD oxidoreductase. Thus classification of enzymes is believed to be essential for a clear understanding of the subject. The basis of the classification scheme is to arrange the enzymes into six main classes. Each of these classes is given a number (from 1 to 6) and is divided into various sub-classes to give the remainder of the four numbers.

The new finding that many enzymes occur in different structural forms but possess identical (or nearly so) catalytic properties has created a new problem with respect to enzyme nomenclature. These enzymes are referred to as *isoenzymes*. Such enzyme molecules have essentially the same catalytic activity but different structures and chemical compositions. The present system for enzyme nomenclature and classification makes no provision for structural diversity with similar activity.

Enzymes have been placed in six main groups. Each group is further subdivided in different sub-groups as below:

1. oxidoreductases
2. transferases
3. hydrolases
4. lyases
5. isomerases
6. ligases (or synthetases)

1. Oxidoreductases

Oxidoreductases are those enzymes which bring about biological oxidation and reduction in a pair of compounds.

i.e. $$AH_2 + B \rightleftharpoons A + BH_2$$

In the above general reaction, AH_2 is a reduced form and B is an oxidized form of the compounds. On the left hand side, after action of oxidoreductase enzyme; end products are A and BH_2. Here, (i.e. on right hand side) A is oxidized whereas BH_2 is reduced. The substrate AH_2 (reduced) is oxidized and simultaneously B (oxidized) compound gets reduced on the right hand side in the reaction. The transfer of H_2 occurs by the activity of enzyme oxidoreductase. In such reactions, NAD and NADP coenzymes play an important role in bringing about oxidation-reduction reaction in a pair of compounds. For understanding this, specific reaction can be illustrated as under:

$$\underset{\substack{\text{Ethyl alcohol} \\ \downarrow \\ \text{(Red.)}}}{CH_3CH_2OH} + \underset{\substack{\text{Coenzyme} \\ \downarrow \\ \text{(Oxi.)}}}{NADP} \underset{\text{oxidoreductase}}{\overset{\text{Alcohol : NADP}}{\rightleftharpoons}} \underset{\substack{\text{Acetaldehyde} \\ \downarrow \\ \text{(Oxi.)}}}{CH_3CHO} + \underset{\substack{\text{Coenzyme} \\ \downarrow \\ \text{(Red.)}}}{NADPH_2}$$

In the above reaction, ethyl alcohol which is a reduced form of the substrate is oxidized by donating H_2-molecule to NADP (coenzyme present in oxidized form) and thus NADP itself gets converted into reduced form $NADPH_2$. Ethyl alcohol is converted to oxidized form, i.e. acetaldehyde. This oxidation-reduction reaction is catalyzed only in the presence of the enzyme known as alcohol : NADP oxidoreductase.

2. Transferases

Transferases are those enzymes which bring about transfer of a group other than hydrogen. The groups which are transferred can be listed as under: nitrogenous groups, phosphorus-containing groups, one carbon groups, sulphur-containing groups and acyl groups.

(i) Transferring nitrogenous groups:

$$\underset{\text{L-Alanine}}{CH_3 \cdot CHNH_2 \cdot COOH} + \underset{\text{2–Oxoglutarate}}{COOH \cdot CH_2 \cdot CH_2 \cdot CO \cdot COOH} \underset{\text{aminotransferase}}{\overset{\text{L–Alanine : 2–Oxoglutarate}}{\rightleftharpoons}}$$

$$\underset{\text{Pyruvic acid}}{CH_3 \cdot CO \cdot COOH} + \underset{\text{L–Glutamic acid}}{COOH \cdot CH_2 \cdot CH_2 \cdot CHNH_2 \cdot COOH}$$

(ii) Transferring phosphorus containing groups:

$$\text{D-Glucose} + \text{ATP} \xrightleftharpoons[\text{phosphotransferase}]{\text{ATP : D–hexose–6–}} \text{D–Glucose-6-phosphate} + \text{ADP}$$

In the above reaction, one phosphate group is transferred from energy rich compound, i.e. ATP to D-glucose at carbon number 6. Thus ATP gets converted into ADP and, D-glucose into D-glucose-6-phosphate. This reaction is catalyzed by ATP : D-hexose-6-phosphotransferase.

3. Hydrolases

Hydrolases are those enzymes which bring about hydrolyses of a substrate with the formation of two products, i.e. $AR + H_2O \rightleftharpoons AH + ROH$

This can be understood from the following illustrations:

(i)

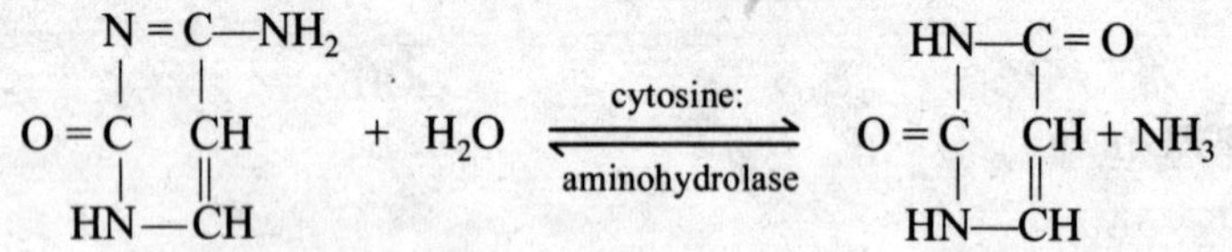

Cytosine Uracil

In the above reaction cytosine (cyclic amidine) is hydrolyzed forming uracil, with the liberation of NH_3. Here enzyme acts on C – N bond other than peptide bonds.

(ii) $$O{=}C\langle^{NH_2}_{NH_2} + H_2O \xrightleftharpoons[\text{hydrolase}]{\text{urea amido:}} 2NH_3 + CO_2$$

Urea

(iii) $$\underset{\text{Glycyl–glycine}}{CH_2NH_2 \cdot CONH \cdot CH_2 \cdot COOH} + H_2O \xrightleftharpoons[\text{hydrolase}]{\text{Glycyl–glycine :}} \underset{\text{Glycine}}{2CH_2NH_2COOH}$$

(iv) Peptide peptidohydrolases: e.g. Pepsin, Rennin, Trypsin and Papain.

$$\text{Native proteins} \xrightleftharpoons{\text{Papain}} \text{Polypeptides} + \text{dipeptides}$$

4. Lyases

Lyases are those enzymes which catalyze the addition of groups to double bonds or their formation.

(i) Carbon-oxygen lyases (Hydro-lyases)

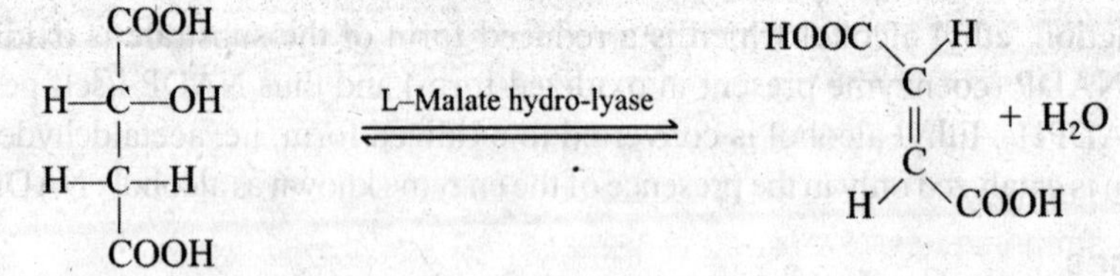

L-Malic acid Fumaric acid

Such enzymes are classified under sub-group known as carbon-oxygen lyases.

(ii) Carbon-nitrogen lyases (Ammonia-hydrolyases)

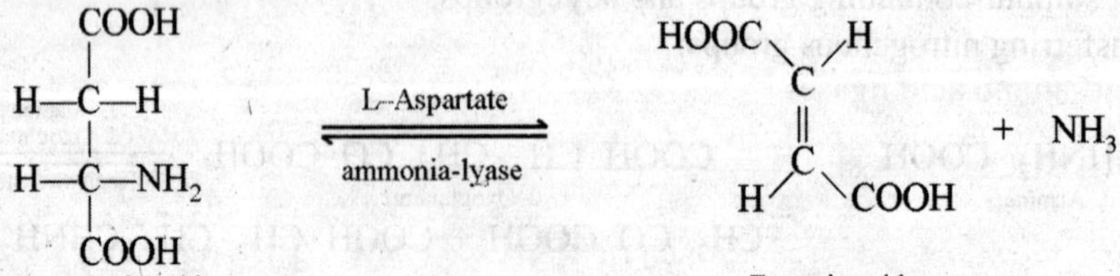

L-Aspartic acid Fumaric acid

(iii) Carbon-carbon lyases (Carboxy-lyases)

$$\underset{\text{Pyruvic acid}}{CH_3 \cdot CO \cdot COOH} \xrightleftharpoons[\text{carboxy–lyase}]{\text{2–Oxo–acid}} \underset{\text{Acetaldehyde}}{CH_3 \cdot CHO} + CO_2$$

5. Isomerases

Isomerases are those enzymes which bring about interconversion between the two isomers.

(i) Racemases and epimerases

(a) Acting on amino–acids

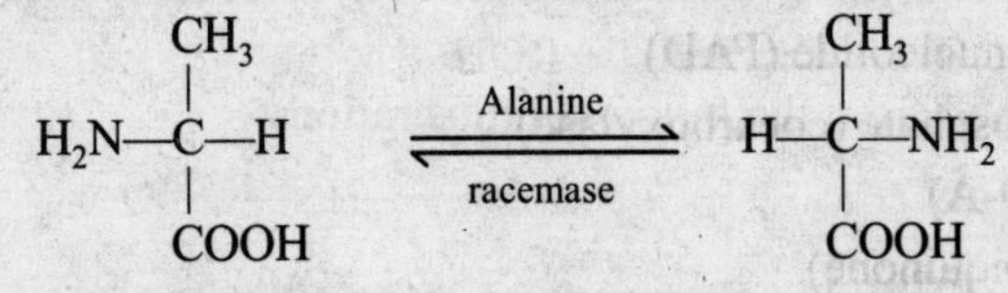

L–Alanine D–Alanine

(b) Acting on hydroxyacids

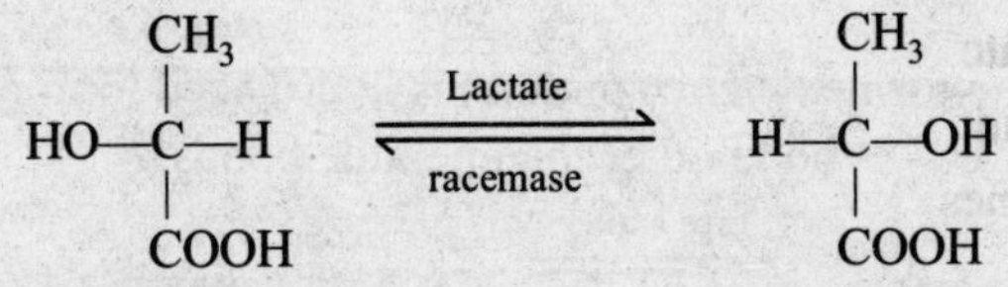

L–Lactic acid D–Lactic acid

(ii) Intramolecular oxidoreductases

(a) Interconverting aldoses and ketoses

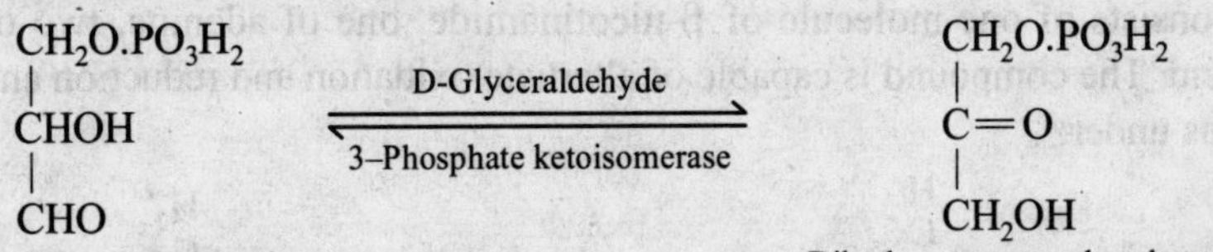

D–Glyceraldehyde–3–phosphate Dihydroxyacetone phosphate

(b) D–Glucose-6-phosphate $\xrightleftharpoons[\text{Ketoisomerase}]{\text{D-Glucose-6-phosphate}}$ D-Fructose-6-phosphate

6. Ligases or Synthetases

Ligases are those enzymes which bring about linking of two compounds with the breakdown of third energy giving compound. This can be illustrated by the following examples.

(i) Acid-ammonia ligases

$$\underset{\text{L-Aspartic acid}}{COOH \cdot CH_2CHNH_2COOH} + ATP + NH_3 \xrightleftharpoons[\text{ligase (ADP)}]{\text{L-Aspartate-ammonia :}}$$

$$\underset{\text{L-Aspargine}}{CONH_2 \cdot CH_2CHNH_2COOH} + H_3PO_4 + ADP$$

(ii) Acid-amino acid ligases

$$\underset{\text{2D–Alanine}}{2CH_3 \cdot CHNH_2COOH} + ATP \xrightleftharpoons[\text{ligase (ADP)}]{\text{D–Alanine : D–alanine}}$$

$$\underset{\text{D–Alanyl–D–alanine}}{CH_3CHNH_2 \cdot CONH \cdot CH(CH_3) \cdot COOH} + H_3PO_4 + ADP$$

6 COENZYMES

Coenzymes are catalytic agents produced by living cells. They are necessary in many enzymatic reactions. They are heat-stable, dialyzable through parchment membrane and have smaller molecular weights. Most of the coenzymes are synthesized from vitamins as precursors.

Coenzymes can be listed as under:

(1) nicotinamide adenine dinucleotide (NAD)
(2) nicotinamide adeninc dinucleotide phosphate (NADP)
(3) flavin mononucleotide (FMN)
(4) flavin adenine dinucleotide (FAD)
(5) thiamine pyrophosphate (cocarboxylase)
(6) coenzyme A (CO-A)
(7) coenzyme Q (Ubiquinone)
(8) adenosine diphosphate (ADP) and Adenosine triphosphate (ATP)
(9) lipoic acid
(10) pyridoxal phosphate
(11) tetrahydrofolic acid
(12) cobamide coenzymes

1. Nicotinamide adenine dinucleotide

It is the new name of diphosphopyridine nucleotide (DPN), also known as codehydrogenase I and coenzyme I.

NAD consists of one molecule of β-nicotinamide, one of adenine, two of pentose, and two of phosphoric acid. The compound is capable of alternate oxidation and reduction and functions as a carrier of hydrogen as under:

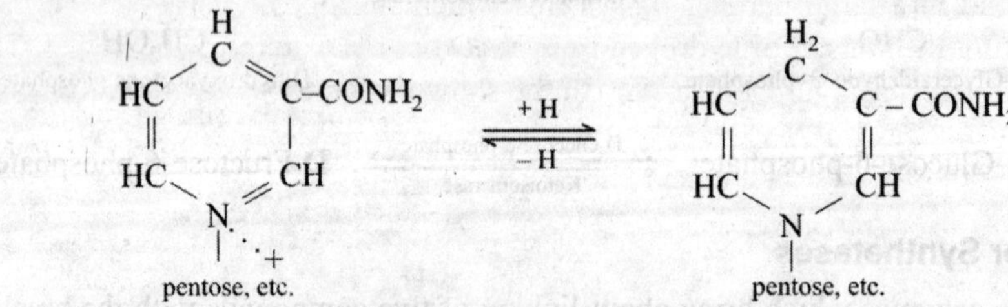

(Pyridine ring containing a strongly basic quaternary nitrogen atom) (Pyridine ring containing weakly basic tertiary nitrogen atom)

NAD is universally present in living cells. It is present in high concentration in baker's yeast. It functions as a coenzyme in many oxidoreductase reactions.

2. Nicotinamide adenine dinucleotide phosphate

It is the new name of triphosphopyridine nucleotide (TPN), also known as codehydrogenase II or coenzyme II. Like NAD, it is also widely distributed in living cells. It functions in the same manner as NAD and differs from it only in containing one more phosphate radical, which is joined to the D-ribose. The two coenzymes are generally not interchangeable. Some oxidoreductases require NAD; others must have NADP. Their function is to transport hydrogen from metabolites to other carriers, usually the flavin nucleotides, and become reoxidized again to accept more hydrogen.

This may be illustrated in the following series of reactions in which glucose becomes oxidized to gluconic acid.

$$\text{Glucose} + H_3PO_4 \rightleftharpoons \text{glucose-6-phosphate}$$

$$\text{Glucose-6-phosphate} + \text{NADP} \rightleftharpoons \text{gluconic acid-6-phosphate} + NADPH_2$$

$$NADPH_2 + \text{Flavin nucleotide (oxi.)} \rightleftharpoons \text{NADP} + \text{flavin nucleotide (red.)}$$

$$\text{Flavin nucleotide (red.)} + O_2 \rightleftharpoons \text{flavin nucleotide (oxi.)} + H_2O_2$$

3. Flavin mononucleotide

This coenzyme is the phosphate ester of riboflavin (vitamin B_2). It has the following structural formula:

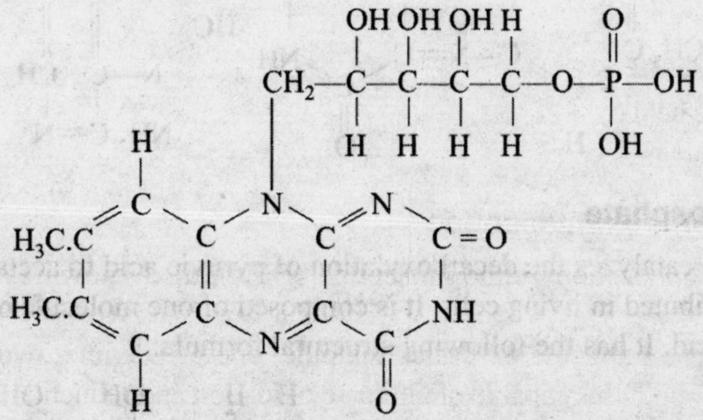

FMN is found in a large number of systems which function in carbohydrate metabolism and serves as an agent for the transfer of hydrogen from NAD or NADP to the cytochromes. The mechanism of FMN by which it acts as hydrogen carrier can be shown as under:

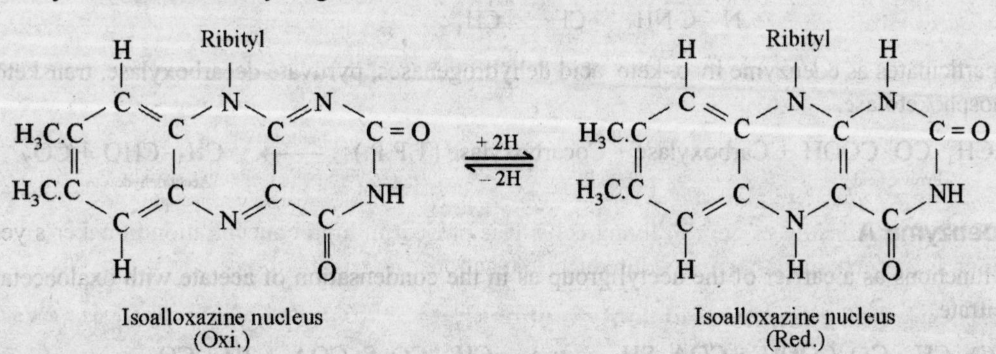

4. Flavin adenine dinucleotide (FAD)

Like FMN, FAD serves as an agent in the transfer of hydrogen from NAD or NADP to the cytochromes. The mechanism of hydrogen transfer is similar to that of FMN involving isoalloxazine nucleus. It has the following structural formula:

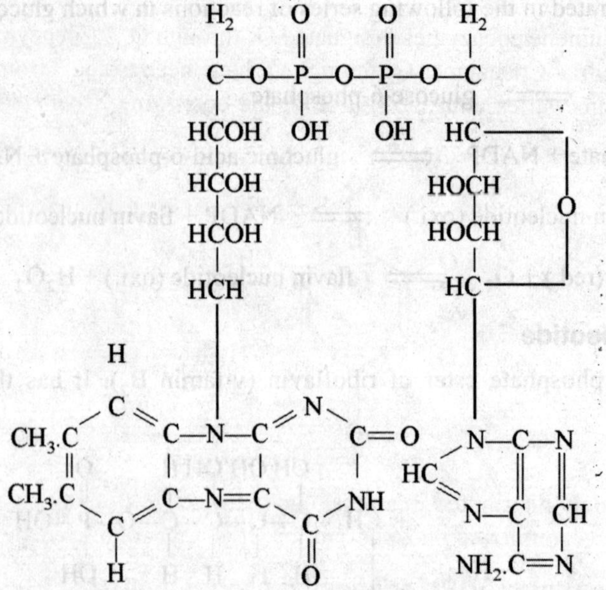

5. Thiamine pyrophosphate

The enzyme carboxylase catalyzes the decarboxylation of pyruvic acid to acetaldehyde and CO_2. This coenzyme is widely distributed in living cells. It is composed of one molecule of thiamine (vitamin B_1) and two of phosphoric acid. It has the following structural formula:

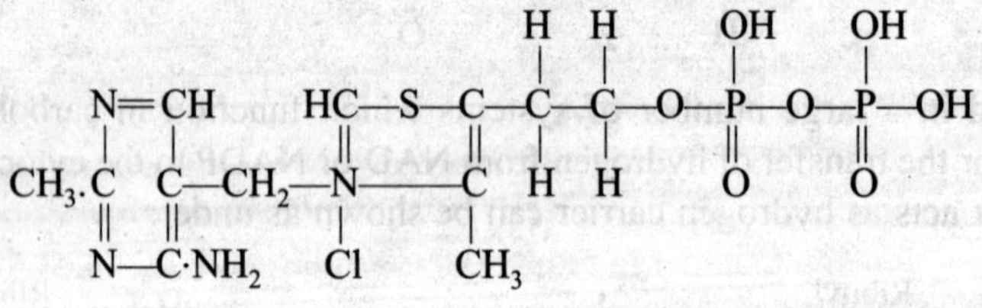

T.P.P. participates as coenzyme in α-keto acid dehydrogenases, pyruvate decarboxylase, transketolase and phosphoketolase.

$$\underset{\text{Pyruvic acid}}{CH_3 \cdot CO \cdot COOH} + \text{Carboxylase} + \text{Cocarboxylase (T.P.P.)} \longrightarrow \underset{\text{Acetaldehyde}}{CH_3 \cdot CHO} + CO_2$$

6. Coenzyme A

Co–A functions as a carrier of the acetyl group as in the condensation of acetate with oxaloacetate to form citrate.

$$\text{(i)}\quad \underset{\text{Pyruvic acid}}{CH_3 \cdot CO \cdot COOH} + COA \cdot SH \longrightarrow \underset{\text{Acetyl-coenzyme A}}{CH_3 \cdot CO \cdot S \cdot COA} + 2H + CO_2$$

$$\text{(ii)}\quad COOH \cdot CO \cdot CH_2 \cdot COOH + CH_3 \cdot CO \cdot S \cdot COA + H_2O \longrightarrow \begin{array}{c} CH_2 \cdot COOH \\ | \\ COH \cdot COOH \\ | \\ CH_2 \cdot COOH \end{array} + COA \cdot SH$$

7. Coenzyme Q

There exists a family of quinone coenzymes designated Q_6 through Q_{10}. Coenzyme Q_{10} was first isolated from beef heart mitochondria. Coenzymes Q_6 through Q_9 have been isolated from micro-organisms. The number of isoprenoid units varies with the source of the coenzyme. Coenzyme Q has the following structure:

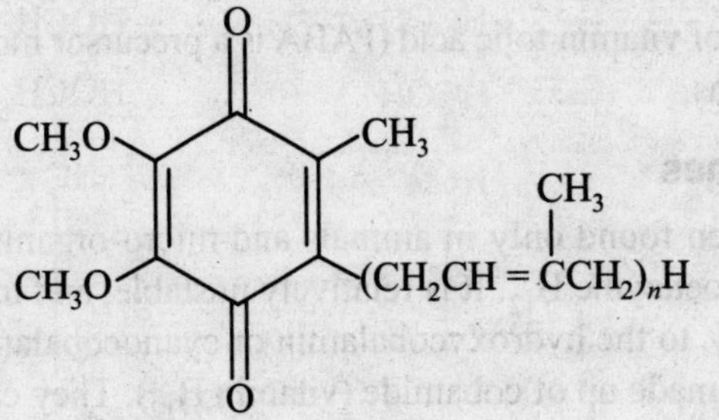

n = 6, 7, 8, 9 or 10 isoprenoid units.

Coenzyme Q is widely distributed in animal and plant cells. It functions as an electron carrier in citric acid cycle oxidations coupled to phosphorylation. It is a fat soluble coenzyme.

8. Adenosine diphosphate (ADP), adenosine triphosphate (ATP)

These are phosphate-transferring coenzymes. They are energy-rich compounds. They liberate a large amount of energy on breaking of high energy bonds (symbol, ~,). ATP is able to donate phosphate groups to a number of metabolic intermediates, thereby converting them to activated forms.

e.g. D-glucose + ATP $\rightleftharpoons$ D-glucose-6-phosphate + ADP
(Activated form of D-glucose)

Thus ATP and ADP play an important role in the bio-synthesis (see Chapter 4).

9. Lipoic acid

Liver and yeast are rich sources of lipoic acid. It exists in both oxidized and reduced forms due to the ability of the disulphide linkage to undergo reduction. Bound to protein, lipoic acid is released by acidic, basic, or proteolytic hydrolysis.

Lipoic acid is a cofactor of the multienzyme complexes: (i) Pyruvate dehydrogenase and (ii) α-Ketoglutarate dehydrogenase.

It has the following structural formula:

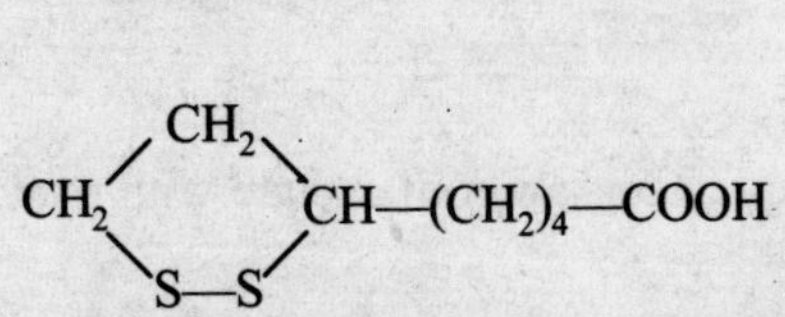

Lipoic acid (oxi.)

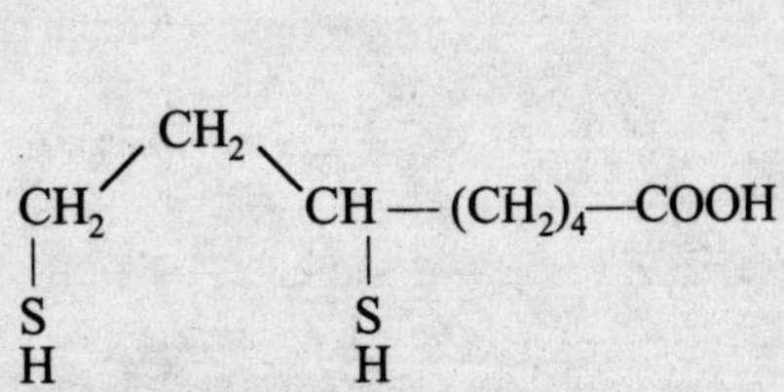

Lipoic acid (red.)

Lipoic acid plays an important role during condensation of active acetate with oxaloacetate to form citric acid.

10. Pyridoxal phosphate

It consists of pyridoxine (vitamin B_6). This coenzyme plays an important role in amino-acid metabolism, e.g. Transamination, Deamination and decarboxylation. Thus deficiency of amino acids can be fulfilled by such types of reactions.

11. Tetrahydrofolic acid

This coenzyme is consisted of vitamin folic acid (PABA is a precursor molecule). This coenzyme catalyzes transfer of one-carbon atoms.

12. Cobamide coenzymes

Vitamin B_{12}, which has been found only in animals and micro-organisms and not in plants, occurs as part of a coenzyme called coenzyme B_{12}. It is relatively unstable, and in the presence of light or cyanide is decomposed, respectively, to the hydroxycobalamin or cyanocobalamin form of the vitamin.

These coenzymes are made up of cobamide (vitamin B_{12}). They catalyze molecular rearrangement reactions.

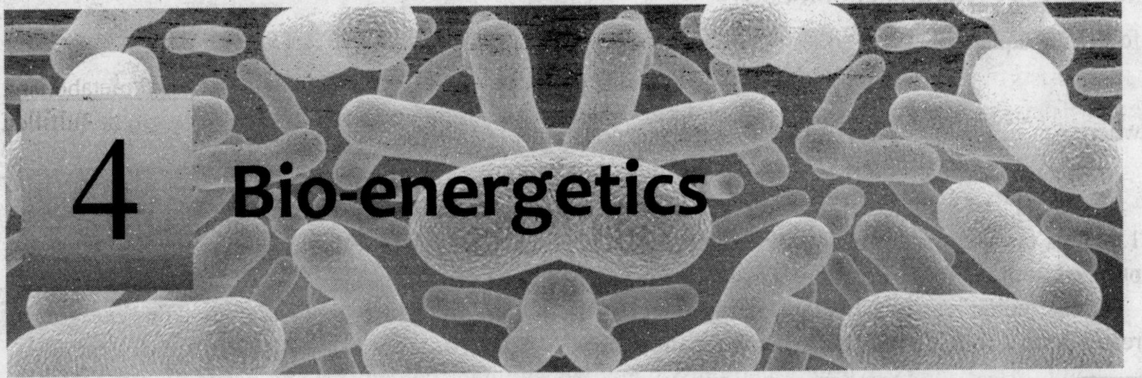

4 Bio-energetics

FUNDAMENTAL PRINCIPLES OF BIO–ENERGETICS

The metabolism of any organism may be subdivided into two categories, namely, catabolism and anabolism. Energy is generated during degradative pathways. An input of energy is required in many of the biosynthetic pathways. Therefore, the mobilization of *chemical energy* is an important part of the metabolism to drive biosynthetic reactions. The usual chemical form of energy is the highly reactive molecule, ATP. Many bacteria obtain energy by carrying out energy-yielding chemical reactions. Photosynthetic bacteria are capable of converting solar energy into chemical energy in the presence of light and photosynthetic pigments.

According to the Second law of thermodynamics,

$$\Delta H = \Delta G + T\Delta S$$

where ΔH = total energy

ΔG = free energy change

$T\Delta S$ = entropy increase

It is obvious that only part of the energy released in a chemical reaction is available. This chemical energy is the utilizable form of energy for the performance of work. Such energy is referred to as free energy (ΔG). The rest of it is lost as an increase of entropy ($T\Delta S$). Living systems are subject to this physical law.

If the ΔG of a reaction has a negative value, the reaction liberates energy. Such reactions are called exergonic or downhill reactions. If the ΔG of a reaction has positive value, the reaction requires energy. Such reactions are referred to as endergonic or uphill reactions. The numerical value of ΔG for a chemical reaction depends upon the concentration of the reactants and products in the reaction mixture. For purposes of comparisons between the energetics of various reactions, it is assumed that the concentration of all reactants is one molar in the steady state; this is referred to as standard concentration. Under conditions of standard concentration, the free energy change of a reaction is known by a special term, $\Delta G°$. In other words, under standard conditions, $\Delta G = \Delta G°$.

Moreover, the value of $\Delta G°$ is affected by temperature and pH. Hence one must make sure that the similar conditions of temperature and pH have been used while comparing energetics of two reactions by comparing their $\Delta G°$ values. $\Delta G°$ is related to the equilibrium constant, K_{eq}, of the reaction by the following equation:

$$\Delta G^\circ = -RT \ln K_{eq}$$

where R = the gas constant

T = absolute temperature

If ΔG° of a chemical reaction is a negative value, the equilibrium constant (K_{eq}) is greater than 1.0. This favours the formation of products. If it is positive, K_{eq} is less than 1.0 and reaction tends to proceed in the reverse direction.

Exergonic reactions liberate energy, which is used in driving endergonic reactions. Thus exergonic reactions are coupled with endergonic reactions.

1 KEY POSITION OF ATP IN METABOLISM

Adenosine triphosphate (ATP) plays an important role in cellular metabolism. It is required to drive biosynthetic reactions. It has the following chemical structure :

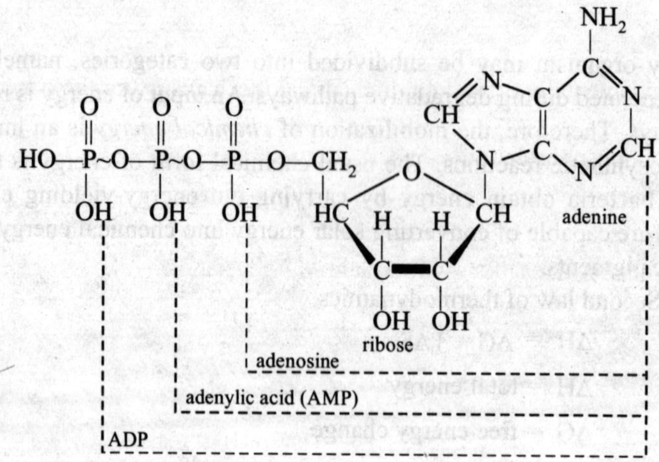

It is a derivative of adenosine monophosphate to which two additional phosphate groups are linked. The symbol, ~, suggests high-energy bonds. It can donate phosphate groups to the metabolic intermediates and thus convert them into activated forms. In other words, their standard free energy (ΔG°) is increased to a level, that permits the phosphorylated intermediates to take part in biosynthetic reactions.

Upon hydrolysis of the phosphate group of ATP, ADP and AMP; energy is liberated. ΔG° values of these compounds are different as shown under:

(i) 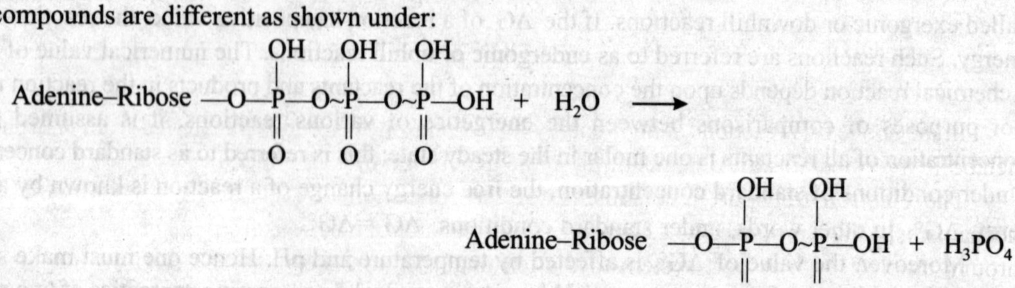

$\Delta G^\circ = 7.3$ kcal

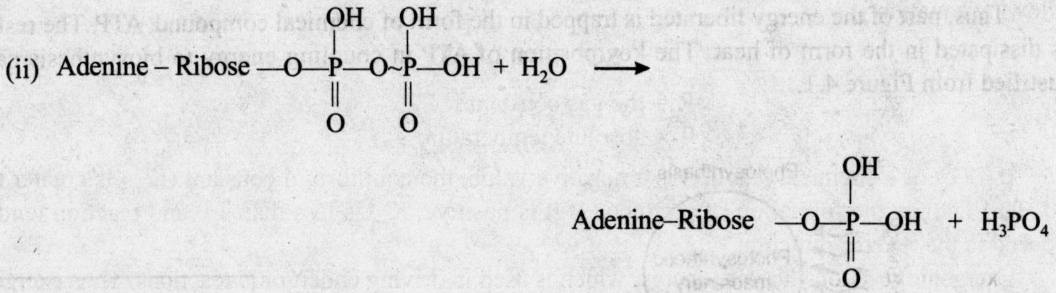

$\Delta G^\circ = 7.3$ kcal

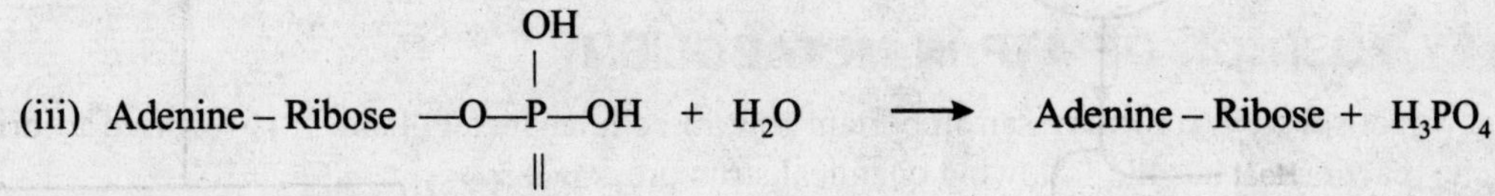

$\Delta G^\circ = 3.4$ kcal

The generation of ATP

There are three fundamentally different bio-chemical mechanisms for the generation of ATP. They are:

(i) substrate-level phosphorylation
(ii) electron transport
(iii) photosynthesis

The following two chemical reactions will explain ATP generation by substrate-level phosphorylation:

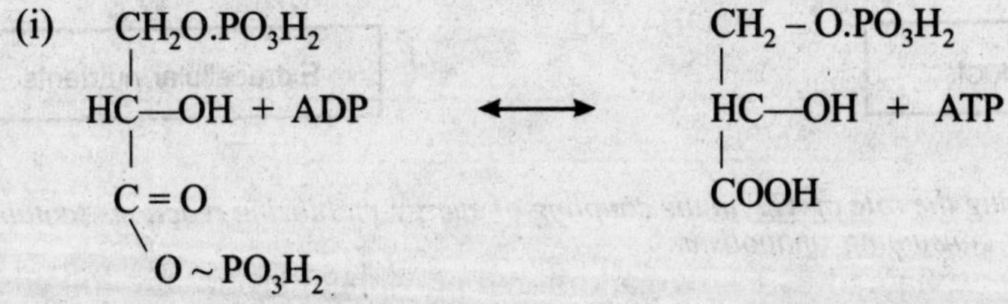

1, 3–Diphosphoglycerate 3–Phosphoglycerate

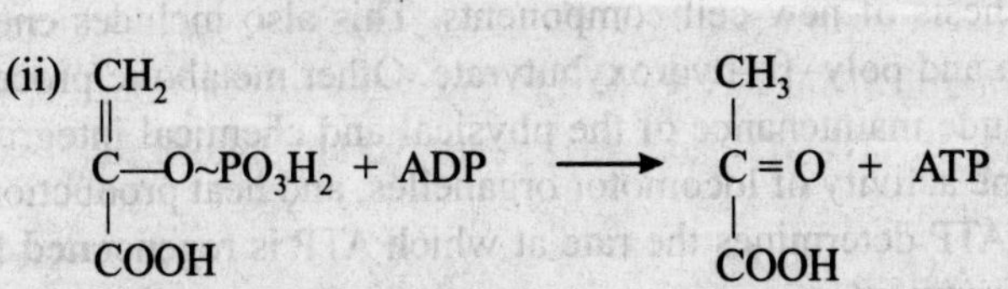

Phosphoenol pyruvate Pyruvate

In the above two reactions, a high-energy linked phosphate group is transferred from phosphorylated metabolic intermediates to ADP, resulting in the generation of ATP molecule.

The electron transport mechanisms for the formation of ATP occur in membrane. In such mechanisms, ATP is produced as a consequence of transfer of electrons from the primary electron donor through an electron transport system to the terminal electron acceptor. The electron transport-linked reactions of oxidative phosphorylation and photophosphorylation are included in this major category. These reactions will be dealt within the latter section of this chapter.

Thus, part of the energy liberated is trapped in the form of chemical compound, ATP. The rest of it is dissipated in the form of heat. The key position of ATP in coupling energy to biosynthesis can be justified from Figure 4.1.

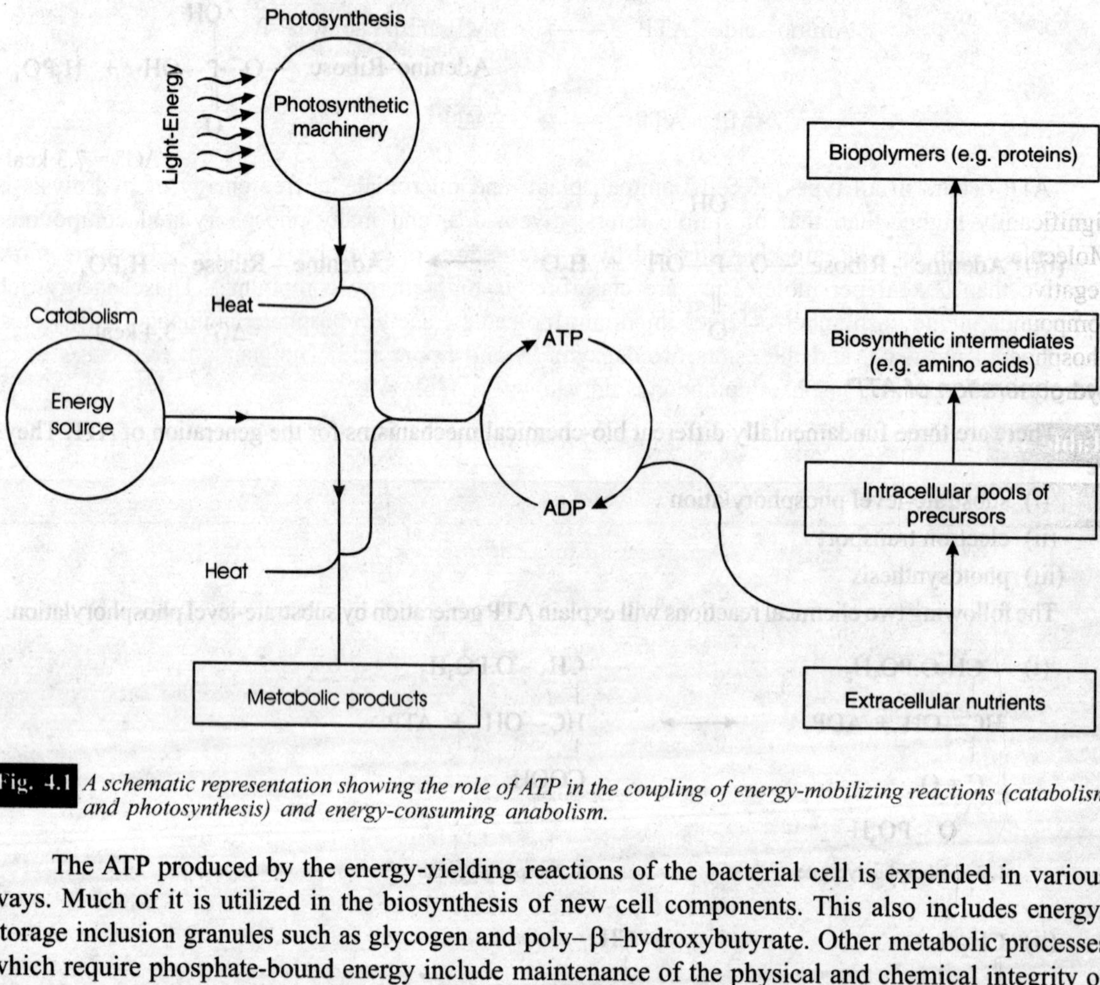

Fig. 4.1 *A schematic representation showing the role of ATP in the coupling of energy-mobilizing reactions (catabolism and photosynthesis) and energy-consuming anabolism.*

The ATP produced by the energy-yielding reactions of the bacterial cell is expended in various ways. Much of it is utilized in the biosynthesis of new cell components. This also includes energy-storage inclusion granules such as glycogen and poly–β–hydroxybutyrate. Other metabolic processes which require phosphate-bound energy include maintenance of the physical and chemical integrity of the cell, transport of solutes across membrane activity of locomotor organelles, and heat production. In general, the rate of utilization of energy in ATP determines the rate at which ATP is regenerated from ADP at the expense of energy from the environment.

In this way, ATP serves as a coupling agent. As the above Figure 4.1 shows, the phosphate bond energy of ATP intervenes at a number of different stages of the overall biosynthetic process. First, it requires to drive the entry of external nutrients into the cell. It is almost, always an active process that requires a net input of energy. An example is the activation of glucose:

$$\text{Glucose} + \text{ATP} \longrightarrow \text{glucose-6-phosphate} + \text{ADP}$$

Second, it is used to convert these nutrients to metabolic intermediates of low molecular weight. An example is the biosynthesis of L-tryptophan, L-phenylalanine and L-tyrosine from erythrose-4-phosphate and PEP.

Finally, it is utilized to achieve the polymerization of these biosynthetic intermediates into bio-polymers. An example is the polymerization of amino-acids to form proteins.

$$\text{Amino-acid} + \text{ATP} \longrightarrow \underset{\text{(activation of amino-acid)}}{\text{AMP-amino acid} + \text{P-P}}$$

$$\text{AMP} + \text{ATP} \longrightarrow 2\text{ADP}$$

ATP occurs in all types of cells: animal, plant, and microbial. Its free energy of hydrolysis is significantly higher than that of simple esters, glycosides, and many phosphorylated compounds. Molecules, such as ATP, are characterized by a free energy of hydrolysis at pH 7. They are more negative than 7 kcal per mole. They are classified as high-energy compounds. These energy-rich compounds include a number of other important molecules: acetyl phosphate, aminoacyl adenylates, phosphoenol pyruvate, and the esters of coenzyme A and lipoic acid. The standard free energies of hydrolysis of various phosphate compounds are shown in Table 4.1.

Table 4.1 *Standard free energy of hydrolysis of phosphorylated compounds.*

Sr. No.	*Energy-rich compound*	*$\Delta G°$. kcal*
1.	Phosphoenol pyruvate	–14.8
2.	1, 3–Diphosphoglycerate	–11.8
3.	Acetyl phosphate	–10.1
4.	Adenosine triphosphate (ATP)	–7.3
5.	Guanosine triphosphate (GTP)	–7.3
6.	Uridine triphosphate (UTP)	–7.3
7.	Cytidine triphosphate (CTP)	–7.3
8.	Glucose–1–phosphate	–5.0
9.	Fructose–6–phosphate	–3.8
10.	Glucose–6–phosphate	–3.3

2 ENERGETIC COUPLING THROUGH PYRIDINE NUCLEOTIDES

Like ATP, a net input of *reducing power* is necessary in many biosynthetic reactions. It is also derived from the degradative reactions. Pyridine nucleotides (NAD and NADP) participate in the transfer of reducing power form energy-producing metabolism to reductive biosynthetic step reactions. NAD has the following structure:

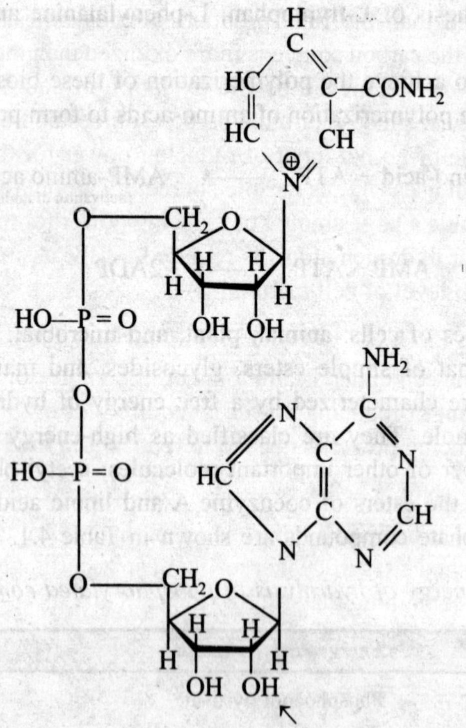

NADP has an identical structure except that it contains an additional phosphate group, linked to one of the ribose residues (indicated by the arrow). Both pyridine nucleotides can readily undergo reversible oxidation and reduction (refer coenzymes in Chapter 3). It may be symbolized as under:

(i) $NAD^+ + 2H \rightleftharpoons NADH + H^+$

(ii) $NADP^+ + 2H \rightleftharpoons NADPH + H^+$

Many enzymatic reactions that involve an oxidation or a reduction of pyridine nucleotides function specifically with either NAD or NADP. However, in terms of the overall metabolism of the cell, these two electron carriers are often equivalent, since a special enzyme, transhydrogenase, can mediate a reversible electron transfer between them. This enzyme is present in many organisms.

$$NADH_2 + NADP \rightleftharpoons NAD + NADPH_2$$

The reductive biosynthetic step reactions are mediated by a transfer of electrons from the reduced forms of the pyridine nucleotides, e.g.

$$\underset{\alpha\text{-Ketoglutaric acid}}{HOOC \cdot CH_2 \cdot CH_2 \cdot CO \cdot COOH} + NH_3 + NADH_2 \longrightarrow$$

$$\underset{\text{Glutamic acid}}{HOOC \cdot CH_2 \cdot CH_2 \cdot CHNH_2 \cdot COOH} + NAD + H_2O$$

The net input of reducing power is highly variable. This reflects the difference in oxidation state between the nutrients and the overall products of biosynthesis. If the compound being used as carbon source is as reduced as or more reduced than the final products of biosynthesis, its oxidation will generate the required reducing power. If the carbon source is more oxidized than the overall products of biosynthesis, however, net reducing power is achieved by the oxidation of additional carbon source molecules to CO_2. Chemoautotrophs that use glucose or other carbon sources on this oxidation level for the synthesis of cell material require little or no net input of reducing power for the assimilation of carbon. On the other hand, an organism that uses CO_2 as its sole source of carbon, (e.g. a plant or a chemoautotrophic bacterium) must have available a large amount of reducing power for the assimilation of carbon. Roughly speaking, a net input of four hydrogen atoms is required for the assimilation of each molecule of CO_2, to reduce it to the oxidation level of cell material:

$$CO_2 + 4H \longrightarrow (CH_2O) + H_2O$$

If nitrogen is furnished to the cell either as ammonia or as complex organic nitrogen compounds (amino acids, peptones), no net input of reducing power is required for nitrogen assimilation. If it is furnished as nitrate, considerable reducing power is needed to reduce it to the level of ammonia:

$$NO_3^- + 8H \longrightarrow NH_3 + 2H_2O + OH^-$$

The sulphur of cell material is principally present as sulphydryl groups, on the same oxidation level as sulfide. If sulfate is furnished as sulfur source, it must accordingly be reduced to sulfide prior to incorporation into sulphur-containing materials of the cell:

$$SO_4^{2-} + 8H \longrightarrow S^{2-} + 4H_2O$$

ENERGY–YIELDING METABOLISM

3 BIOLOGICAL OXIDATIONS

Several types of chemical reactions are involved in energy production, but oxidation-reduction is probably the commonest. Oxidative reactions occurring in the living cells are known as biological oxidations. Biological oxidation may be defined as the withdrawal of electrons from a substance with or without the addition of oxygen or elements analogous to oxygen, or as the withdrawal of electrons with or without the withdrawal of hydrogen or elements analogous to hydrogen. It is believed that oxygen acts as a hydrogen acceptor to form H_2O or H_2O_2. H_2O_2 is decomposed into H_2O and O_2 by the action of enzyme $H_2O_2 : H_2O_2$ oxidoreductase (catalase), i.e.

$$2H_2O_2 \xrightarrow{\text{catalase}} 2H_2O + O_2\uparrow$$

Oxygen, as such, does not enter the molecule of the substrate compounds, which are active hydrogen acceptors take the place of oxygen in various oxidations, being themselves reduced in the reactions. Bacterial oxidations and reductions are associated phenomena, and must be studied together.

Almost all oxidations in living cells can be explained on the basis of the removal of hydrogen, rather than of the addition of oxygen. Very few cases are known, in which oxygen is used directly as the oxidizing agent. Hydrogen acceptors are necessary for the reactions to take place. Oxidations that occur in this manner are called dehydrogenations. Exceptions include the oxidation of aldehydes and purine bases. In these instances, a preliminary hydration occurs prior to removal of hydrogen:

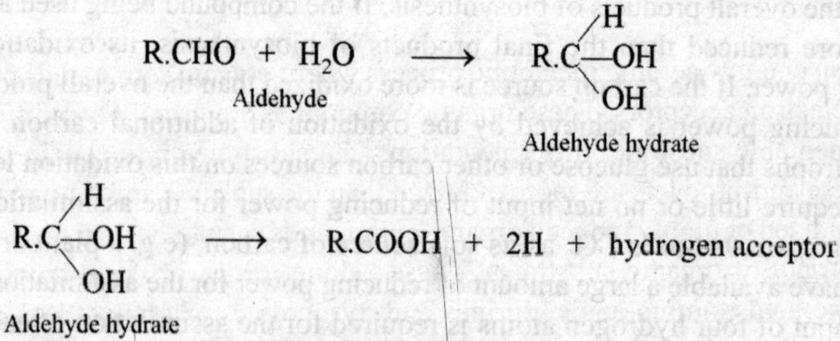

Bacteria are capable of reducing such stains as methylene blue, neutral red and litmus to their colourless forms. This can be understood by taking the case of methylene blue. Methylene blue acts as a hydrogen acceptor, and is reduced to the colourless form. If air or oxygen is bubbled through the medium containing the reduced compound, the blue colour is restored. The reaction may be represented as under:

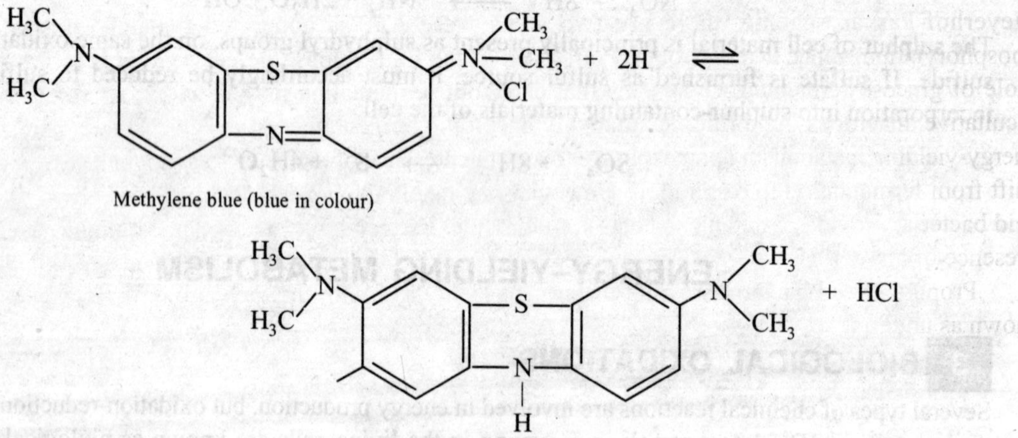

Methylene blue (leuco compound)

One atom of hydrogen is taken up by the double-bonded nitrogen, converting the stain into the reduced or colourless form. The reaction is freely reversible in both directions.

By means of the methylene blue technique, it is possible to demonstrate the presence of oxidoreductases, which are capable of oxidizing compounds by the removal of hydrogen. The speed of reduction of the stain is an indication of the rate of oxidation taking place. The reaction occurs in an anaerobic environment. The presence of air results in the reoxidation of the reduced methylene blue to the coloured form. A test based on this principle is the determination of the approximate bacterial count of milk by noting the methylene blue reduction time.

There are three principal modes of energy-yielding metabolism. They may be listed as under :

(A) fermentation

(B) respiration (aerobic and anaerobic)

(C) photosynthesis

(A) Fermentation

It may be defined as an energy-yielding metabolic process where organic compounds participate both as electron donors and electron acceptors. These organic compounds are usually two different metabolites obtained from a fermentable substrate. In fermentations, the substrate gives rise to a mixture of end-products, some of these end-products are more oxidized than the substrate and others are more reduced. Thus either highly oxidized or highly reduced organic compounds can not be degraded. Bacteria are able to ferment a wide variety of organic substrates. These belong to one of the following main classes: carbohydrates, organic acids, amino-acids, purines and pyrimidines. Carbohydrates are the principal substrates of fermentation. Practically ever carbohydrate or carbohydrate derivative can be used as a fermentable energy source by some organism.

Fermentable substrates are on roughly the same oxidation level, a cell material. They also usually serve as the principal source of carbon for biosynthesis. So the net biosynthetic requirement for reducing power during fermentative growth is negligible.

ATP is generated by substrate-level phosphorylation. Here, high-energy phosphate groups from phosphorylated intermediates are transferred to ADP, forming ATP molecules. In the case of Embden-Meyerhof-Parnas scheme, there are two steps were ATP is generated through substrate-level phosphorylation. Since 2 moles of ATP are used in the initial steps, the net ATP yield is 2 moles per mole of glucose fermented. Organisms that carry out fermentation may be either strict anaerobes or facultative anaerobes. As a rule, facultatively anaerobic fermentative organisms change their mode of energy-yielding metabolism upon exposure to air : the presence of molecular oxygen induces a metabolic shift from fermentation to respiration. However, one facultatively anaerobic bacterial group, the lactic acid bacteria, provides a notable exception to this rule. Fermentative metabolism continues even in the presence of air.

Propionic acid bacteria are able to ferment lactic acid with the generation of ATP. This can be shown as under:

$$3CH_3CHOHCOOH \xrightarrow{-6H} 3CH_3COCOOH$$

$$3CH_3COCOOH \xrightarrow[ADP + Pi \rightarrow ATP]{+6H} 2CH_3CH_2COOH + CH_3COOH + CO_2$$

Let us now discuss the fermentative metabolism in the case of clostridia. The most widespread type of amino acid fermentation among clostridia is the **Stickland reaction**. It is a coupled oxidation-reduction involving a pair of amino acids. One of the amino acids serves as electron donor and the other as electron acceptor. This can be understood by considering a fermentative metabolism of alanine and glycine as shown in Figure 4.2. In this reaction, alanine serves as the electron donor and glycine as the electron acceptor.

Clostridia can also ferment single amino-acids, (e.g. fermentation of threonine to give propionic acid). Certain clostridia and other anaerobic bacteria can obtain energy by the fermentation of ring compounds, such as uric acid, uracil, allantoin, and nicotinic acid.

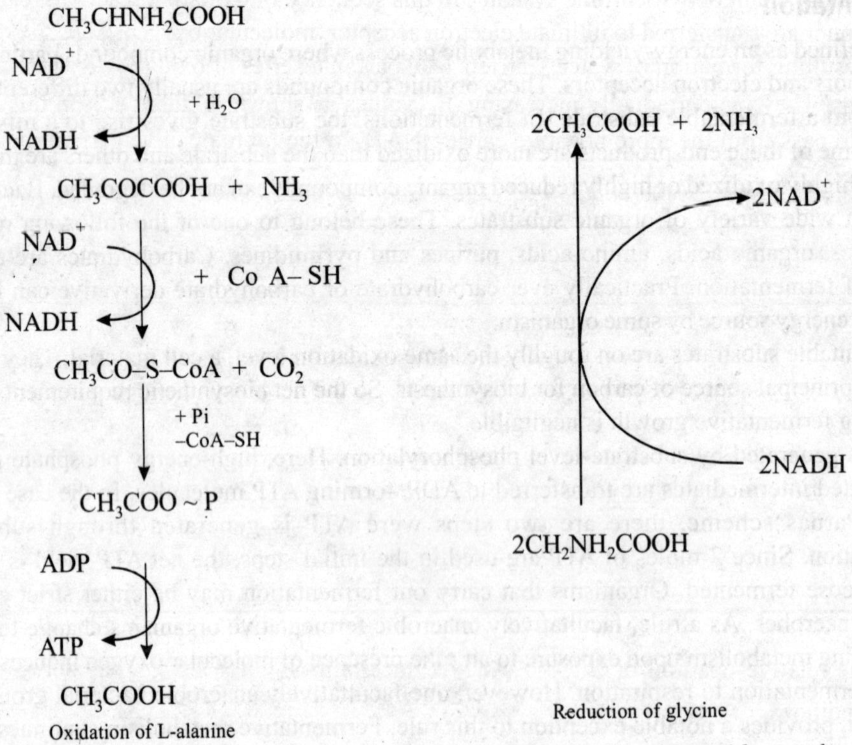

Fig. 4.2 *Stickland reaction illustrating fermentative metabolism which involves L-alanine (electron donor) and glycine (electron acceptor). There is net ATP generation.*

Fermentative processes always maintain a strict oxidation-reduction balance, in the sense that the average oxidation level of the end products is identical with that of the substrate fermented.

(B) Respiration

It is an energy-yielding metabolic process. In this process, either organic compounds or reduced inorganic compounds participate as electron donors, and molecular oxygen as the final electron acceptor. If final electron acceptor is molecular oxygen, respiratory process is termed as aerobic respiration. If an oxidized inorganic compound other than oxygen (sulphates, nitrates, or carbonates) acts as terminal electron acceptor, respiratory process is referred to as anaerobic respiration.

Aerobic Respiration

ATP generation occurs by oxidative phosphorylation. Energy is trapped in the form of ATP during the movement of electrons down the respiratory chain of carriers. In electron transport chain, electrons pass through a series of intermediate carriers, capable of undergoing freely reversible oxidation and reduction. The last carrier in the system reacts with molecular oxygen in a terminal oxidase-catalyzed reaction.

Electron Transport System

The series of carrier molecules that link the oxidation of an oxidizable substrate with the reduction of final molecular oxygen to water is called the **'electron transport system.'** It is also referred to as the 'respiratory chain or cytochrome system'. In this sequence of oxidation reactions, electrons from an electron donor are transferred to ultimate electron acceptor, molecular oxygen. Thus, oxygen is reduced, with the resulting formation of water. Carriers take part in a series of oxidation reactions in the transfer of electrons. They have gradually increasing E_0' values as shown in Table 4.2. In this way, electrons will tend to pass from the more negative carrier, NAD, to the more positive carrier above it.

Table 4.2 *E_0' values of the component O/R systems of the electron transport system*

Sr. No.	*O/R system*	*E_0' in volts at pH 7 and 25°C*
1.	$NAD^+/NADH + H^+$	–0.32
2.	Flavoprotein/flavoprotein–H_2	–0.03
3.	CoQ/CoQ – H_2	+ 0.10
4.	Cyt *b* – Fe^{3+}/cyt *b* – Fe^{2+}	+ 0.05
5.	Cyt c_1 – Fe^{3+}/cyt c_1 – Fe^{2+}	+ 0.21
6.	Cyt *c* – Fe^{3+}/cyt *c* – Fe^{2+}	+ 0.23
7.	Cyt *a* – Fe^{3+}/cyt *a* – Fe^{2+}	+ 0.29
8.	Cyt a_3 – Fe^{3+}/cyt a_3 – Fe^{2+}	+ 0.53
9.	Oxygen/water	+0.82

Note: The relative position of CoQ to cyt *b* is uncertain. E_0' for CoQ is measured under artificial conditions.

Components of ETS (see Figure 4.3)

The respiratory chain consists of enzymes having prosthetic groups or coenzymes. These can be regarded as the working parts of the enzymes, and in the case of the respiratory chain, each is in fact an O/R system. The spectrophotometry may be applied in distinguishing oxidized form of each prosthetic group or coenzyme from that of reduced form. This is possible because there is a difference in absorption spectrum between the oxidized form and reduced form of the coenzyme. The components may be summarized as under:

NAD and NADP: Certain enzymes known as dehydrogenases have NAD^+ or $NADP^+$ as their coenzyme, NAD^+ can exist in a reduced form, $NADH + H^+$, to form an O/R system:

$$NAD^+ + 2H^+ + 2e^- \rightleftharpoons NADH + H^+$$

Similarly, NADP can exist in a reduced state. These coenzymes have already been dealt with in Chapter 3.

FAD and FMN: Flavoproteins are enzymes that contain a yellow prosthetic group. The prosthetic group may be either flavin mononucleotide (FMN) or flavin adenine dinucleotide (FAD). These coenzymes are capable of undergoing reversible oxidation and reduction due to the presence of isoalloxazine moiety. The reduced forms of the coenzymes are $FADH_2$ and $FMNH_2$. The flavoproteins are a large class of enzymes, differing widely with respect to their E_0' values. They mediate many diverse oxidation-reduction reactions. Some are active in primary dehydrogenation of organic substrates. This may be exemplified by succinic dehydrogenase, mediating the oxidation of succinate to fumarate. Other flavoproteins accept electrons from reduced pyridine nucleotides formed in a primary substrate dehydrogenation.

The electron acceptors for the reoxidation of flavoproteins are also variable. Some reduced flavoproteins are auto-oxidizable by molecular oxygen, with the resulting formation of hydrogen peroxide:

$$\text{Reduced flavoprotein} + O_2 \longrightarrow \text{Oxidized flavoprotein} + H_2O_2$$

Hydrogen peroxide is a highly toxic substance. Catalase, present in most aerobic organisms, decomposes hydrogen peroxide into water and oxygen:

$$2H_2O_2 \xrightarrow{\text{catalase}} 2H_2O + O_2$$

The flavoproteins associated with the respiratory electron transport chain do not, however, transfer electrons directly to oxygen. Their reoxidation is mediated by subsequent carriers of the chain, either quinones or cytochromes.

UBIQUINONE: In the mitochondrial electron transport system, the major quinone is ubiquinone. A diversity of other quinones occur in bacterial electron transport systems. One of these, the natural napthoquinone K_9H occurs in *Mycobacterium phlei*. It has been shown to act between the flavoprotein and cytochrome *b*. When the quinone is separated from the mitochondria, the transfer of electrons from reduced substrates to molecular oxygen is prevented; the activity is restored when the quinone is incorporated back. It functions as an additional electron carrier as it is easily reduced and oxidized as shown under:

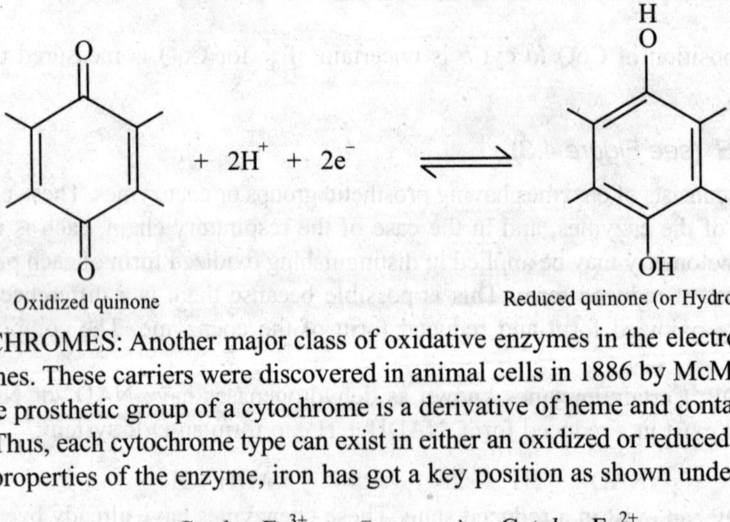

Oxidized quinone — Reduced quinone (or Hydroquinone)

CYTOCHROMES: Another major class of oxidative enzymes in the electron transport system is the cytochromes. These carriers were discovered in animal cells in 1886 by McMunn and rediscovered by Keilin. The prosthetic group of a cytochrome is a derivative of heme and contains a single iron atom (Figure 4.4). Thus, each cytochrome type can exist in either an oxidized or reduced form. In the oxidative or reductive properties of the enzyme, iron has got a key position as shown under:

$$\underset{\text{(Ferric)}}{\text{Cyt } b\text{—}Fe^{3+}} + e^- \rightleftharpoons \underset{\text{(Ferrous)}}{\text{Cyt } b\text{—}Fe^{2+}}$$

The cytochromes have characteristic absorption bands in the reduced state, and their spectral and functional properties allow the recognition of several different members of the class in electron transport systems. They are distinguished by a terminal letter, (e.g. cytochrome *c*). Cytochromes *a* and a_3 together are known as *cytochrome oxidase*. They also contain copper. But only cytochrome a_3 can react directly with O_2. It is easy to determine the order in which the cytochromes are reduced in the ETS.

If an oxidizable substrate of the TCA cycle, (i.e. malate) is added to a mitochondrial suspension, and the mixture observed in a spectrophotometer under anaerobic conditions, the absorption band of reduced cytochrome *b* appears first. Later on, the bands of cytochromes c_1, *c*, and *a* appear in that order.

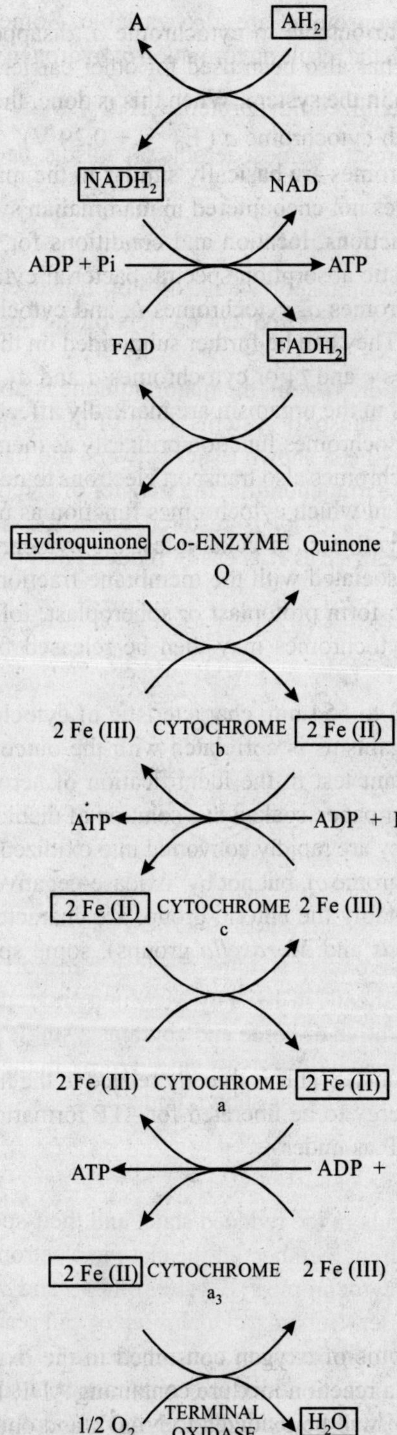

Fig. 4.3 *A schematic representation of a respiratory electron transport chain, linking the dehydrogenation of an oxidizable substrate, with the reduction of oxygen to H_2O.*

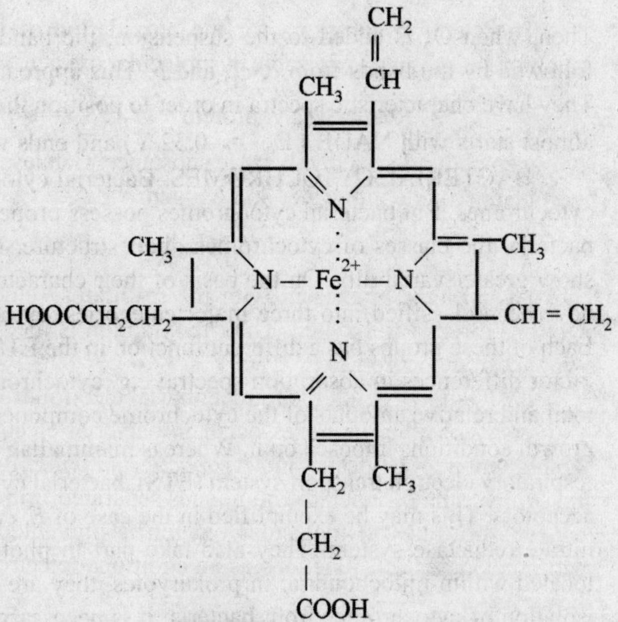

Fig. 4.4 *Structure of heme.*

Then, when O_2 is added to the suspension, the band attributable to cytochrome *a*, disappears first, followed by the bands from *c*, c_1, and *b*. This approach has also been used for other carriers in ETS. They have characteristic spectra in order to position them in the system. When this is done, the position almost starts with NADH (E_0' = –0.32 V) and ends with cytochrome *a* (E_0' = + 0.29 V).

BACTERIAL CYTOCHROMES: Bacterial cytochromes are basically similar to the mammalian cytochromes. But bacterial cytochromes possess properties not encountered in mammalian systems. In bacteria, the classes of cytochromes, their structure, functions, location and conditions for existence show greater variability. On the basis of their characteristic absorption spectra, bacterial cytochromes have been classified into three major categories: cytochromes *a*, cytochromes *b*, and cytochromes *c*. Each of these groups has a different function in the ETS. They can be further subdivided on the basis of minor differences in absorption spectra; e.g. cytochromes *c* and c_1 or cytochromes *a* and a_3. Both the total and relative amounts of the cytochrome components in the organism are markedly affected by the growth conditions imposed on it. Whereas mammalian cytochromes function primarily as members of a respiratory electron transport system (ETS), bacterial cytochromes also transport electrons to non-oxygen acceptors. This may be exemplified in the case of *E. coli* in which cytochromes function as part of the nitrate reductase system. They also take part in photosynthesis. In eukaryotes, the cytochromes are located within mitochondria; in prokaryotes, they are associated with the membrane fraction. For the isolation of cytochromes from bacteria, it is necessary to form protoplast or spheroplast, followed by osmotic shock to produce a membranous ghost. The cytochromes may then be released from their membrane environment by sonication.

A number of bacterial species show no peak at 552 to 554 nm, characteristic of cytochromes of the *c* type. The lack of a *c*-type cytochrome in such organisms is correlated with the outcome of an empirical procedure, the oxidase test. This is an important test in the identification of aerobes. It is carried out by placing a patch of bacteria on a piece of filter paper, soaked in a solution of dichlorophenol indophenol or N, N–dimethyl-para-phenylenediamine. They are rapidly converted into oxidized coloured products by 'oxidase-positive' species (containing a cytochrome *c*), but not by 'oxidase-negative' species (lacking a cytochrome *c*). Certain groups of aerobes, notably the enteric group, are characteristically oxidase negative. In other groups, (e.g. the *Pseudomonas* and *Moraxella* groups), some species are oxidase positive and others are oxidase negative.

Oxidative phosphorylation

A decline in free energy is associated with each electron transfer. This is directly related to the magnitude of the fall in electron pressure. This allows sufficient energy to be liberated for ATP formation. There are three sites where energy is trapped in the form of ATP as under:

(i) between NAD and FP
(ii) between cyt *b* and cyt *c*
(iii) between cyt *a* and cyt a_3

P : O Ratio

The ratio of the atoms of phosphate esterified to the atoms of oxygen consumed in the oxidation is referred to as P : O ratio. If NADPH is incorporated into a reaction mixture containing ADP, inorganic PO_4, Mg^{2+}, and animal or plant mitochondria, the NADH will be oxidized to NAD^+, and one atom of oxygen will be consumed. This occurs because mitochondria contain the intact electron transport chain. At the same time, formation of ATP from ADP and inorganic phosphate takes place. Under ideal conditions, between 2 and 3 moles of ATP will be synthesized per atom of oxygen used. It is believed that 3 moles of ATP are formed per mole of NADH oxidized or atom of oxygen consumed:

$$NADH + H^+ + \tfrac{1}{2}O_2 + 3ADP + 3H_3PO_4 \longrightarrow NAD^+ + 3ATP + 4H_2O$$

Such a reaction is also said to have a P : O ratio of 3.0. Since the oxidation of malate and isocitrate showed P : O ratios of 3.0, the phosphorylations associated with these substrates were assumed to occur after the NAD^+. A P : O ratio of 2.0 for succinate, similarly, indicated that one fewer phosphorylation step was involved during oxidation of the same.

In most bacteria, lower P : O ratio has been observed. In the case of *Mycobacterium phlei*, 3 moles of ATP are formed per mole of oxidized substrate by three distinct respiratory chains.

Sources of electrons for the ETS

Glucose is frequently used as nutritional source of energy. The following discussion will explain how glucose indirectly contributes electrons to the ETS.

In heterotrophs, a molecule of glucose (a six-carbon sugar) is dissimilated by a sequence of enzyme-catalyzed reactions, collectively called *glycolysis*. Thus, two molecules of pyruvic acid (a three-carbon compound) are formed (see Figure 5.1). The oxidation of glyceraldehyde-3-phosphate is accompanied with the removal of a pair of electrons. This pair of electrons may enter the ETS.

Pyruvic acid can further be oxidized according to the scheme in Figure 5.13. This results in the formation of acetyl-coenzyme A (active acetate) and it may enter the TCA cycle. The pair of electrons liberated can enter the ETS as shown in Table 4.3.

Table 4.3 *Synthesis of energy-rich phosphate during the oxidation of pyruvate by the Krebs cycle.*

Sr. No.	*Enzyme or process*	*Reaction*	*Energy-rich Phosphate produced*
1.	Pyruvic dehydrogenase	$Pyruvate + NAD^+ + CoASH \longrightarrow$ $Acetyl-CoA + NADH + H^+ + CO$	0
2.	Electron transport	$NADH + H^+ + \tfrac{1}{2}O_2 \longrightarrow NAD^+ + H_2O$	3
3.	Isocitric dehydrogenase	$Isocitrate + NAD^+ \longrightarrow$ $\alpha\text{-Ketoglutarate} + CO_2 + NADH + H^+$	0
4.	Electron transport	$NADH + H^+ + \tfrac{1}{2}O_2 \longrightarrow NAD^+ + H_2O$	3
5.	α-Ketoglutaric dehydrogenase	$\alpha-Ketoglutarate + NAD^+ + CoASH \longrightarrow$ $Succinyl-CoA + NADH + H^+ + CO_2$	0
6.	Electron transport	$NADH + H^+ + \tfrac{1}{2}O_2 \longrightarrow NAD^+ + H_2O$	3
7.	Succinic thiokinase	$Succinyl-CoA + GDP + H_3PO_4 \longrightarrow$ $Succinate + GTP + CoASH$	1
8.	Succinic dehydrogenase	$Succinate + FAD \longrightarrow Fumarate + FADH_2$	0
9.	Electron transport	$FADH_2 + \tfrac{1}{2}O_2 \longrightarrow FAD + H_2O$	2
10.	Malic dehydrogenase	$Malate + NAD^+ \longrightarrow Oxalacetate + NADH + H^+$	0
11.	Electron transport	$NADH + H^+ + \tfrac{1}{2}O_2 \longrightarrow NAD^+ + H_2O$	3
		Sum	15

In one turn of the TCA cycle, citric acid (a six-carbon compound) undergoes decarboxylation and oxidization to regenerate oxalacetate (a four-carbon molecule). Two carbon atoms are lost in the form of CO_2. In so doing, four pairs of electrons are removed from the intermediates of the cycle.

Thus, the above-mentioned three processes, together with the respiratory chain, can bring about the complete oxidation of glucose to carbon dioxide and water:

$$\underset{\text{Glucose}}{C_6H_{12}O_6} + 6O_2 \longrightarrow 6CO_2 + 6H_2O + \text{Energy}$$

Theoretically, 34 molecules of ATP per molecule of glucose are synthesized by the action of the ETS.

Glucose can also be oxidized by the hexose monophosphate shunt. There occurs the liberation of electron pairs, which may enter the respiratory chain. The steps showing liberation of electron pairs may be illustrated as under:

$$\text{(i) Glucose–6–phosphate} \xrightarrow{2H} \text{6–Phospho–glucono–lactone}$$

$$\text{(ii) 6–phospho–gluconic acid} \xrightarrow{CO_2\ \ 2H} \text{Ribulose–5–phosphate}$$

Glycerol as a component of fats can be metabolized by many micro-organisms. This is accomplished via its connection with the metabolism of carbohydrates:

$$\text{Glycerol} + \text{ATP} \xrightarrow[\text{Mg}^{2+}]{\text{glycerol kinase}} \text{ADP} + \alpha\text{-glycerolphosphate} + H_2O$$

$$\alpha - \text{Glycerolphosphate} + \text{NAD}^+ + H_2O \underset{\text{dehydrogenase}}{\overset{\text{glycerol phosphate}}{\rightleftharpoons}} \text{Dihydroxyacetone phosphate} + \text{NADH} + H_3O^+$$

The dihydroxyacetone phosphate formed would be metabolized by the mechanisms as shown in Figure 5.1. Fatty acids undergo oxidation with the successive removal of two-carbon fragments in the form of acetyl-CoA. The acetyl-CoA produced may then enter the Krebs cycle. At the same time, electrons are fed in the respiratory chain leading to oxidative phosphorylation.

Chemoautotrophs can use CO_2 as their sole source of carbon. CO_2 is reduced to glyceraldehyde–3–phosphate in order to take part in metabolic reactions. This cycle is called Calvin-Benson-Bassham cycle and is discussed in the latter part of this chapter. This reductive process needs a great deal of energy. This energy arises from the oxidation of inorganic compounds, (e.g. molecular hydrogen, ammonia, nitrite, thiosulfate, etc.). Electrons liberated from such oxidative reactions may enter a respiratory chain. Thus ATP formation takes place by oxidative phosphorylation. The genus *Nitrobacter* is a good example of electron transport phosphorylation in the chemoautotrophic bacteria because the E_0' of its O/R system is relatively high. As shown in Figure 4.5., flow of electrons occurs from cytochrome *c* to molecular oxygen. Therefore, chemoautotrophs are at a considerable energetic disadvantage as compared to the heterotrophs.

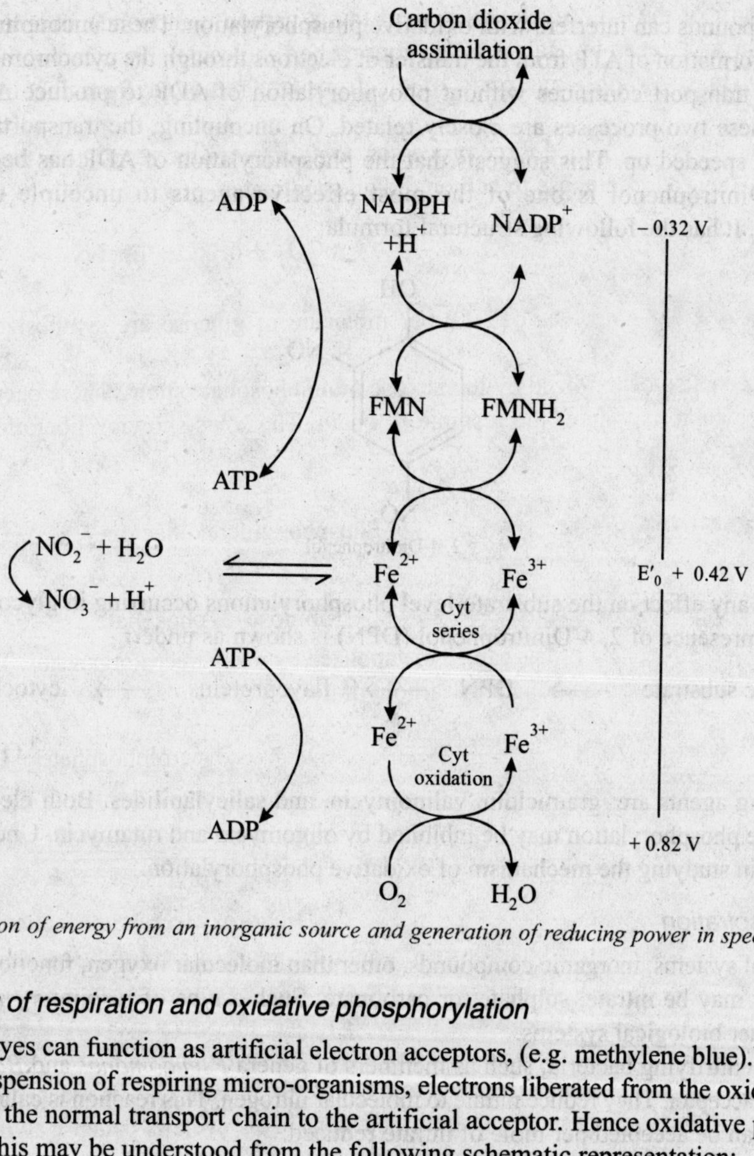

Fig. 4.5 *Assimilation of energy from an inorganic source and generation of reducing power in species of Nitrobacter.*

The uncoupling of respiration and oxidative phosphorylation

Some synthetic dyes can function as artificial electron acceptors, (e.g. methylene blue). When such dyes are added to a suspension of respiring micro-organisms, electrons liberated from the oxidizable substrate are shunted from the normal transport chain to the artificial acceptor. Hence oxidative phosphorylation also decreases. This may be understood from the following schematic representation:

$$MB + 2H \longrightarrow MBH_2$$

Reduced methylene blue is auto-oxidizable by molecular oxygen:

$$MBH_2 + O_2 \longrightarrow MB + H_2O_2$$

With this dye, the shunt occurs at the level of the flavoproteins. The incorporation of MB to a suspension of respiring cells, thus, creates an artificial bypass of the cytochrome system:

Substrate $\longrightarrow$ flavoproteins $\longrightarrow$ MB $\longrightarrow$ molecular oxygen $\longrightarrow$ hydrogen peroxide

Some other compounds can interfere with oxidative phosphorylation. These uncoupling agents are able to uncouple the formation of ATP from the transfer of electrons through the cytochrome system. In other words, electron transport continues without phosphorylation of ADP to produce ATP. In the intact mitochondria, these two processes are closely related. On uncoupling, the transportation of electrons may actually be speeded up. This suggests that the phosphorylation of ADP has been a rate-limiting process. 2, 4-Dinitrophenol is one of the most effective agents to uncouple respiratory-chain phosphorylation. It has the following structural formula:

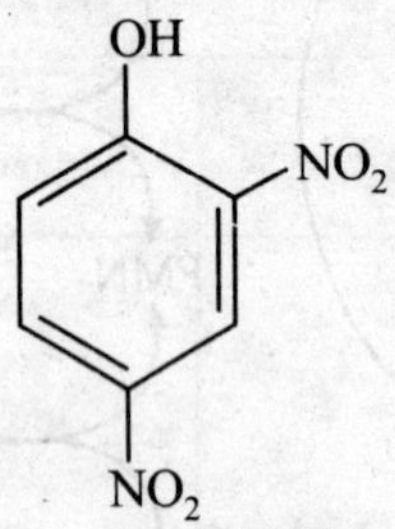

2, 4-Dinitrophenol

It does not have any effect on the substrate-level phosphorylations occurring in glycolysis. The path of electrons in the presence of 2, 4-Dinitrophenol (DPN) is shown as under:

$$\text{Oxidizable substrate} \longrightarrow \text{DPN} \longrightarrow \text{flavoproteins} \longrightarrow \text{cytochromes} \longrightarrow \tfrac{1}{2}O_2 \longrightarrow H_2O$$

Other uncoupling agents are: gramicidin, valinomycin, and salicylanilides. Both electron transport as well as oxidative phosphorylation may be inhibited by oligomycin and rutamycin. Uncouplers have also been employed in studying the mechanism of oxidative phosphorylation.

Anaerobic Respiration

In some bacterial systems, inorganic compounds, other than molecular oxygen, function as final electron acceptors. They may be nitrate, sulphate, or carbonate. Such a type of respiratory metabolism is not exhibited by other biological systems.

Nitrate: Denitrifying bacteria, such as members of genera *Pseudomonas* and *Bacillus*, use nitrate as final electron acceptor. They reduce nitrate to molecular nitrogen. This reaction is called **denitrification**. Five electrons can be accepted per mole of nitrate reduced:

$$2NO_3^- + 10e^- + 12H^+ \longrightarrow N_2 + 6H_2O$$

Many aerobic bacteria contain nitrate reductases. These enzymes can mediate the following reaction:

$$NO_3^- + 2e^- + 2H^+ \longrightarrow NO_2^- + H_2O$$

This ability, (i.e. reduction of nitrate to nitrite) of bacteria does not allow normal growth under anaerobiosis. For this, there are two following reasons:

(i) Large amount of nitrate must be reduced to oxidize a small amount of oxidizable substrate.

(ii) Nitrite is produced as a product of nitrate reduction. This nitrite exerts toxic effect on the growth.

Similarly, SO_4^{2-} and CO_3^{2-} can function as final electron acceptor in other bacterial systems.

Different groups of microbes are capable of utilizing different types of H donors and H acceptors. Either organic or inorganic compounds may serve as H donors. H acceptors may be molecular oxygen, inorganic compounds, or organic compounds. Altogether five patterns exist as shown in Table 4.4.

Table 4.4 *Summary of primary electron donors and ultimate electron acceptors in the various energy-yielding oxidation-reduction reactions used by micro-organisms.*

H donor	H acceptor		
	O_2 (Aerobic Respiration)	NO_3^-, SO_4^{2-}, CO_2 (Anaerobic Respiration)	Organic compounds (Fermentation)
Inorganic substances	I. Aerobic respiration of inorganic substances *e.g. Nitrosomonas* spp. $NH_3 \longrightarrow NO_2^-$ $O_2 \longrightarrow H_2O$	II. Anaerobic respiration of inorganic substances *e.g. Thiobacillus denitrificans* $S \longrightarrow SO_4^{2-}$ $NO_3^- \longrightarrow N_2$	None
Organic substances	III. Aerobic respiration of organic substances *e.g.* Many organisms $C_6H_{12}O_6 \longrightarrow CO_2$ $O_2 \longrightarrow H_2O$	IV. Anaerobic respiration of organic substances *e.g. Desulfovibrio* spp. $CH_3CHOHCOOH \longrightarrow CO_2$ $SO_4^{2-} \longrightarrow H_2S$	V. Fermentation of organic substances *e.g. Streptococcus* spp. $C_6H_{12}O_6 \xrightarrow{-4H} 2CH_3COCOOH \downarrow +4H$ $2CH_3CHOHCOOH$

(c) Photosynthesis

Since, most photosynthetic organisms use CO_2 as their principal carbon source, they require reducing power, Many of the photosynthetic organisms (plants, algae and blue-green bacteria) use water as final source of reductant. There occurs liberation of oxygen due to oxidation of water. This type of photosynthesis is referred to as **oxygenic photosynthesis**.

Certain photosynthetic prokaryotes (purple and green bacteria) use reduced inorganic compounds, (e.g. H_2S and H_2) as reductants. These organisms are not able to use water as final reductant and, therefore, it is not accompanied with the liberation of oxygen. This kind photosynthesis is called **anoxygenic photosynthesis**.

Here, mode of ATP formation is an independent reaction to molecular oxygen. Consequently, all photosyntheses can, in principle, take place under strictly anaerobic conditions. However, all organisms exhibiting oxygenic photosynthesis are aerobic organisms in the limited sense, that they must be able to tolerate oxygen. Organisms exhibiting anoxygenic photosynthesis are fundamentally anaerobes. The majority of these organisms are strict anaerobes; a few of them are facultative aerobes. In facultative aerobes, photophosphorylation is suppressed in the presence of oxygen. Under such circumstances, alternative respiratory generation of ATP occurs.

Some biological groups use light as a source of energy. This light energy is absorbed by the photosynthetic pigment system. Ejection of electrons takes place from such excited pigments. Electrons move through an electron transport chain, the concomitant movement of electrons leading to the formation of ATP. Thus light energy is converted into chemical energy, (i.e. ATP). This process is termed as **photophosphorylation**. The two types of phosphorylations are as under:

(i) cyclic photophosphorylation

(ii) non-cyclic photophosphorylation

This mode of energy-giving reaction is more or less similar to that of oxidative phosphorylation. Substrate-level phosphorylation does not occur in these organisms.

PLANT AND BACTERIAL PHOTOSYNTHESIS

Photosynthesis is a type of metabolism, in which light serves as an energy source for the reductive synthesis of cell materials from carbon dioxide. This unique metabolism is observed in green plants, algae and some bacteria.

All photosynthetic bacteria are included in a single order, Rhodospirillales, which is subdivided into four families as under:

(i) *Rhodospirillacęae* (non-sulphur purple bacteria)

(ii) *Chromatiaceae* (sulphur purple bacteria)

(iii) *Chlorobiaceae* (green sulphur non-filamentous bacteria)

(iv) *Chloroflexaceae* (green sulphur filamentous bacteria)

Each family contains several genera. In bacteria, the photosynthetic apparatus is essentially a part of the intracytoplasmic membrane system. It may assume different forms. Sometimes, it consists of a roughly spherical coloured particle. It is called chromatophore. It has a diameter of about 60 mm. (mol. wt. 10^7). It contains bacteriochlorophyll (BChl), carotenoids, phospholipids and all the enzymes responsible for bacterial photosynthesis. In eukaryotic organisms, it is contained within the chloroplast.

Until the first quarter of this century, it was believed that the mechanism of bacterial photosynthesis and green plant photosynthesis are similar; and that the lack of oxygen evolution in the former type was due to the immediate reaction of oxygen with hydrogen sulphide producing water and elemental sulphur (S^0)

$$H_2S + \frac{1}{2}O_2 \longrightarrow H_2O + S^0 \downarrow$$

But the work of American scientist C. B. Van Niel (Hopkin's Marine Station of Stanford University, California) with *chromatiaceae* proved it to be otherwise. He showed that H_2S is converted first into S^0. It is further converted into sulphate (SO_4--) under anaerobic condition. Whenever oxidizable sulphate compound is exhausted, there is an absolute cessation of photosynthesis, as shown below:

$$H_2S + 2CO_2 + 2H_2O \xrightarrow[\text{BChl}]{h\nu} H_2SO_4 + \underset{\text{Hexose sugar}}{2(CH_2O)}$$

Later, it was again demonstrated by Van Niel that in *Rhodospirillaceae*, organic compounds may act as reducing agent.

e.g.

$$CO_2 + \underset{\text{Iso-propanol}}{2CH_3 \cdot CHOH \cdot CH_3} \xrightarrow[\text{BChl}]{h\nu} (CH_2O) + H_2O + \underset{\text{Acetone}}{2CH_3 \cdot CO \cdot CH_3}$$

In 1944, Van Niel made a brilliant suggestion that fundamentally there is no difference between green plant photosynthesis and bacterial photosynthesis as shown under:

(i) Plant photosynthesis

$$CO_2 + 2H_2O \xrightarrow[\text{Chl, enzymes, etc.}]{h\nu} (CH_2O) + O_2 + H_2O$$

(ii) Bacterial photosynthesis

$$CO_2 + 2H_2A \xrightarrow[\text{BChl, enzymes, etc.}]{h\nu} (CH_2O) + 2A + H_2O$$

He made a unified equation of photosynthesis for both green plants and bacteria as mentioned below:

$$CO_2 + 2H_2A \xrightarrow[\text{Chl/BChl, enzymes, etc.}]{h\nu} (CH_2O) + 2A + H_2O$$

4 UNIFIED SCHEME OF PHOTOSYNTHESIS

It is now well-established that both in green plants and bacteria, photosynthesis involves two closely integrated but distinct phases of reactions :

(i) the photochemical phase (discovered by Robin Hill in 1937)

(ii) the biochemical phase

Photochemical phase

In this phase, packets of chemical energy in the form of adenosine triphosphate (ATP), and reductants in the form of reduced nicotinamide adenine dinucleotide (phosphate) [NAD(P)H] are formed in the presence of Chl or BChl and light. A photon (quantum or packet of light energy or hν) is absorbed by a molecule of light-harvesting Chl/BChl or of a carotenoid. The energy migrates to another Chl/BChl, called reaction centre (RC) Chl/BChl, causing it to eject an electron:

$$\text{Chl} + h\nu \longrightarrow \text{Chl}^*;\ \text{Chl}^* \longrightarrow \text{Chl}^+ + e^-$$

The electron is accepted by a ferroubiquinone (Fe-UQ) complex. Then it migrates to ferredoxin (Fd), while the oxidized Chl/BChl oxidizes the terminal cytochrome of an electron transport system (ETS):

$$\text{Chl}^+ + \text{Cyt}^{++} \longrightarrow \text{Chl} + \text{Cyt}^{+++}$$

The charged system, so formed, creates a large potential difference between its terminals. It is used to generate both ATP and NAD (P) H. The system operates in two different ways, which permits proportions of ATP and NAD (P) H to vary according to necessity. In cyclic photophosphorylation, flow of electrons is through a closed circuit (Figure 4.6). In non-cyclic system, electrons emitted by Chl/BChl

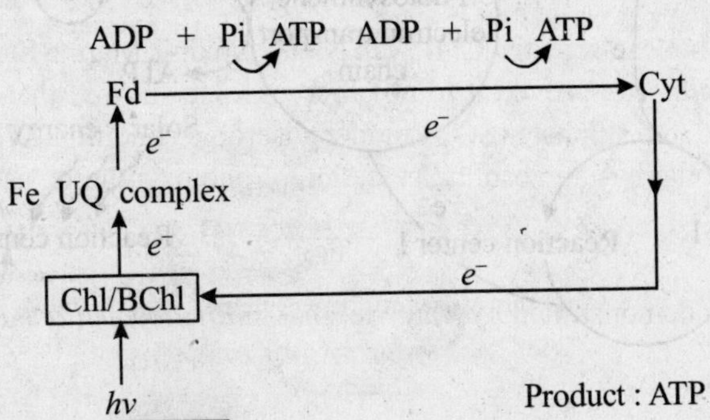

Fig. 4.6 *Cyclic photophosphorylation.*

are used for reduction of NAD $(P)^+$, and hence the system is coupled to substrate oxidation (Figure 4.7). In green plants and algae, this substrate is water, which is oxidized to supply electrons to Chl via Cyt. In bacteria, this is not water but another substrate, which is oxidized at a lower potential.

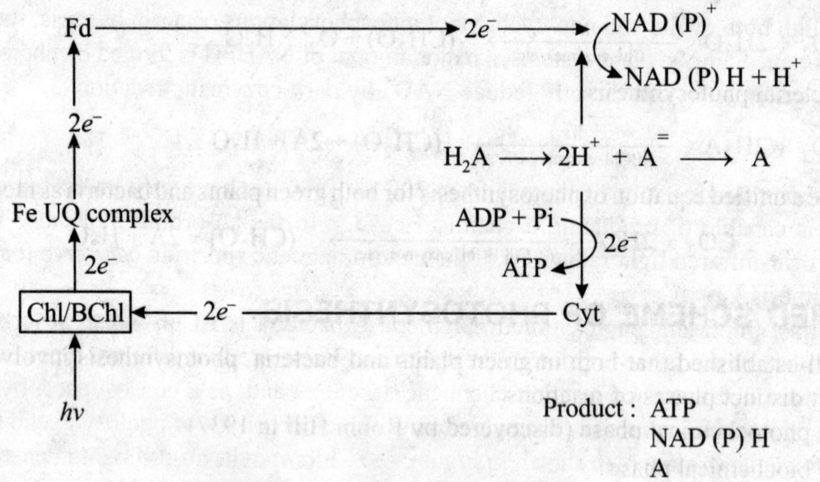

Fig. 4.7 *Non-cyclic photophosphorylation.*

The large potential difference required for the evolution of O_2 in plant photosynthesis is achieved by photoactivation of two types of reaction centres, RC I and RC II. They are connected in series (Figure 4.8). RC I reaches a low enough reduced potential to reduce Fe(–0.42V), and RC II reaches a high enough oxidized potential to oxidize water (+ 0.82V). The bacterial system is more primitive. It contains only one reaction centre (RC I); hence its oxidized potential is not so high. So it requires an electron donor with a potential, lower than that of water, such as H_2S or molecular hydrogen (H_2) or some simple organic compounds:

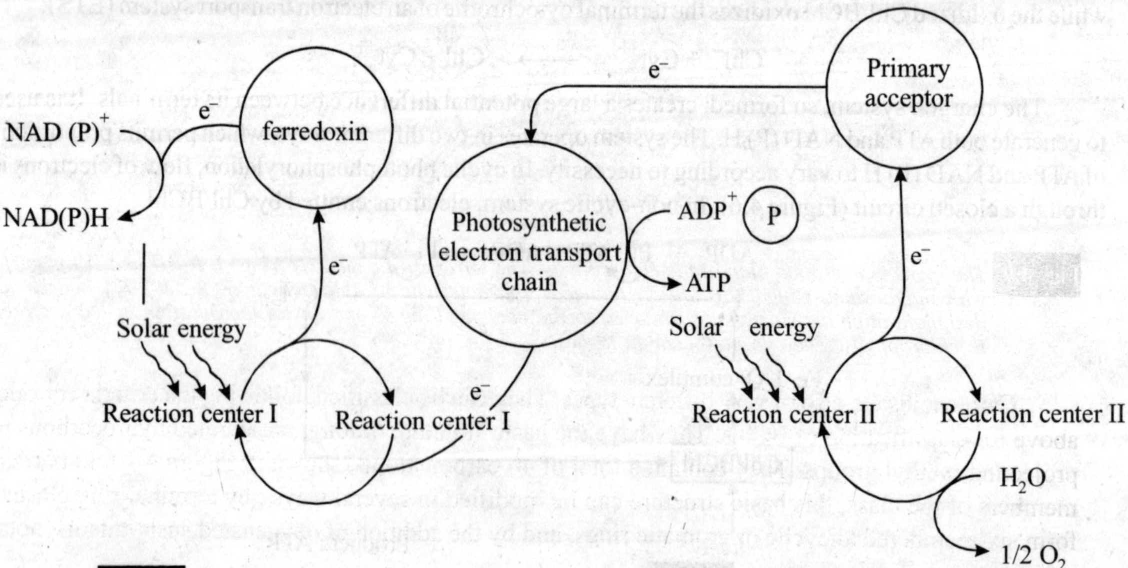

Fig. 4.8 *Schematic representation showing two reaction centres. They are connected in series.*

$$H_2S \xrightarrow[]{2Fe^{+++} \curvearrowright 2Fe^{++}} 2H^+ + S^0\downarrow$$

Although both cyclic and non-cyclic photophosphorylations occur in bacteria, the cyclic one is of greater importance. For, in plant systems, a major amount of NADPH is formed by photophosphorylation. In bacterial systems, H_2 can directly reduce NAD^+ by dark enzymatic reaction :

$$NAD^+ + H_2 \longrightarrow NADH + H^+$$

Radiations ranging in wavelengths from 200 to 1,200 nm have such an energy content that their absorption is capable of producing a chemical change in the absorbing molecule. Thus, ultraviolet, visible and near-infrared light radiations of the electromagnetic spectrum can serve for the performance of photosynthesis.

The most important photosynthetic pigment in bacteria is BChl, which absorbs radiant energy more effectively at the far-red region of the spectrum. There are five different types of BChl : a, b, c, d and e. They are distinguished by their chemical structures and, as a consequence, by their absorption maxima (λ_{max}) and Rf-values in chromatography. The chlorophylls of photosynthetic organisms have a common molecular ground plan shown in Figure 4.9. Chlorophylls absorb light intensely in two regions : the violet, around 400 nm; and the red or near-infrared, around 600 to 800 nm.

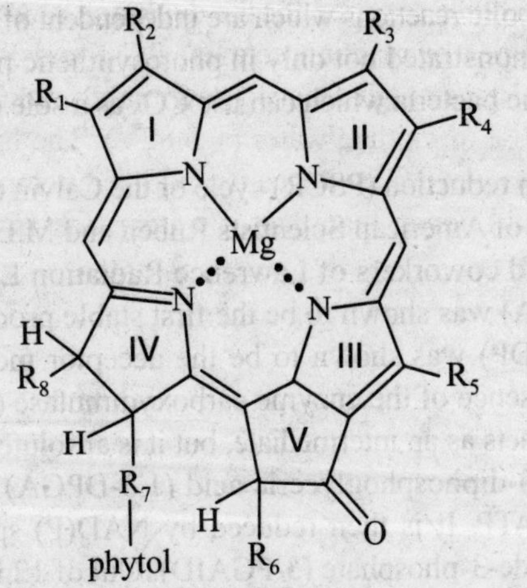

Fig. 4.9 *The molecular ground plan of the chlorophylls. The tetrapyrrolic nucleus (rings I, II, III and IV) is chelated with magnesium. Ring IV is shown reduced (characteristic of chlorophyll a). R_1, R_2, R_3, and so forth, designate aliphatic side chains attached to the nucleus. Ring V, and phytol (substitution of R_7) are present in chlorophylls. They do not occur in hemes.*

Carotenoids are also of very different types. They can be classified, following the criteria considered above for classification of BChls. They have the basic structure of long, unsaturated hydrocarbons with projecting methyl groups. They contain a total of 40 carbon atoms, shown in Figure 4.10. In particular members of the class, this basic structure can be modified in several ways : by terminal ring closure to form six-membered alicyclic or aromatic rings, and by the addition of oxygenated susbtitutents, notably

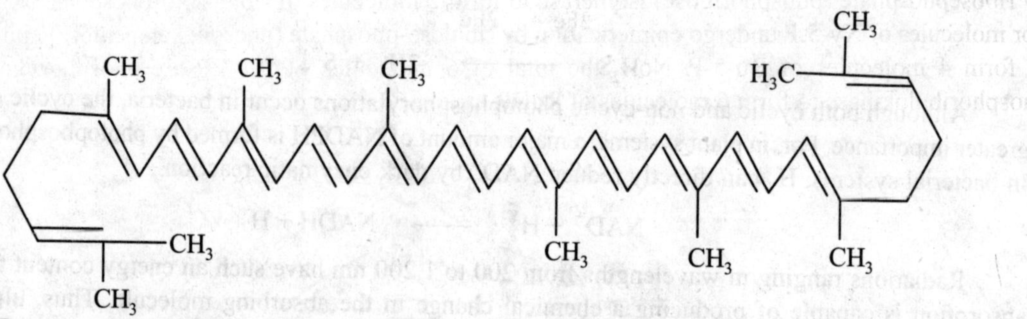

Fig 4.10 *The molecular ground plan of the carotenoids, illustrated by an open-chain carotenoid, which does not contain oxygen.*

hydroxyl, ethoxyl, or keto groups. Carotenoids have a single broad region of light absorption between 450 and 550 nm.

The phycobiliproteins are present in some groups of photosynthetic organisms. They are water-soluble chromoproteins. They contain linear tetrapyrroles and absorb light in a broad region from 550 to 650 nm.

Biochemical phase

This phase involves a set of anabolic reactions which are independent of light but dependent on temperature effects. The phase has been demonstrated not only in photosynthetic plants and bacteria but also in non-photosynthetic chemoautotrophic bacteria which can use CO_2 as a sole carbon source. Hence, the pathway is of very general use.

The photosynthetic carbon reduction (PSCR) cycle or the Calvin cycle (Figure 4.11) was established by using C^{14} in the laboratories of American Scientists Ruben and M.D. Kamen, University of California (1940-41) and of M. Calvin and coworkers of Lawrence Radiation Lab., California University (1954). 3-Phosphoglyceric acid (3-PGA) was shown to be the first stable product of carboxylation reaction and ribulose-1, 5-diphosphate (RuDP) was shown to be the acceptor molecule of CO_2. Six molecules of RuDP on carboxylation, in presence of the enzyme carboxydimutase (CD), give rise to 12 molecules of 3-PGA. A 6-carbon compound acts as an intermediate, but it is absolutely unstable. The 3-PGA molecules are then phosphorylated to 1, 3-diphosphoglyceric acid (1,3-DPGA) in the presence of phosphorylate kinase and at the expense of ATP. It is then reduced by NAD(P) specific glyceraldehyde phosphate dehydrogenase to glyceraldehyde-3-phosphate (3-PGAlD). Out of 12 molecules of the latter compound, only 5 are converted into dihydroxyacetone phosphate (DHAP), catalyzed by the enzyme triosephosphate isomerase. 3 molecules of DHAP are condensed with 3 molecules of 3-PGAld, in the presence of aldolase, to form 3 molecules of fructose-1, 6-diphosphate (Fr-1, 6-DP) which are then dephosphorylated by phosphatase to fructose-6-phosphate (Fr-6-P). One molecule of Fr-6-P is drained off as a pool. The remaining 2 molecules undergo a transketolase (TK) reaction with 2 molecules of 3-PGAld to give rise to 2 molecules of erythrose-4-phosphate (Er-4-P) and 2 molecules of xylulose-5-phosphate (Xy-5-P). 2 molecules of Er-4-P and 2 molecules of DHAP undergo an aldolase reaction to form 2 molecules of sedoheptulose-1, 7-diphosphate (Se-1, 7-DP) which are then dephosphorylated by phosphatase to form sedoheptulose-7-phosphate (Se-7-P). The latter undergo a TK-reaction with 2 molecules of 3-PGAld to form 2 molecules of Xy-5-P and 2 molecules of ribose-5-phosphate (R-5-P). R-5-P undergo isomerization by ribosephosphate (phosphoribose) isomerase to form 2 molecules of ribulose-5-phosphate (Ru-5-P). For molecules of Xy-5-P undergo epimerization by ribulose-phosphate (phosphoketopentose) epimerase to form 4 molecules of Ru-5-P. Now, the total of 6 molecules of Ru-5-P are phosphorylated by phosphoribulokinase to form 6 molecules of RuDP.

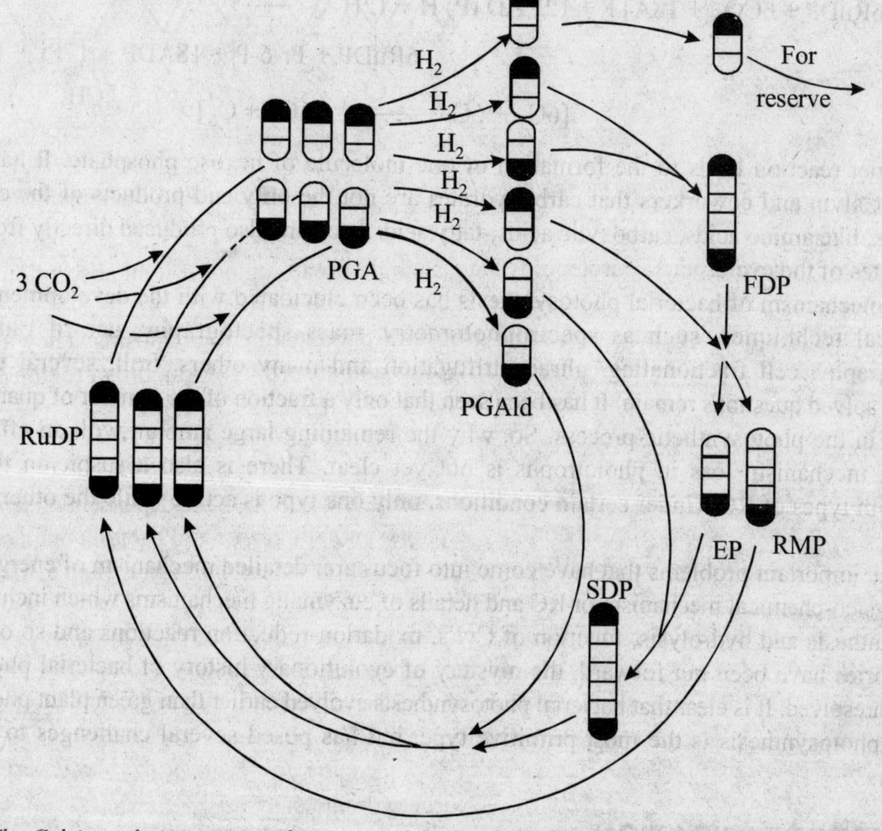

Fig. 4.11 *The Calvin cycle is represented in terms of the number of carbon atoms in substances which are taking part in the cycle. The blackened parts of the molecules represent carbon atoms to which phosphate groups are attached. Abbreviations : FDP. fructose diphosphate : EP. erythrose phosphate : RMP. ribose monophosphate and ; SDP. sedoheptulose diphosphate. It may be added that the regeneration of the sugar ribulose in its diphosphate form requires energy-rich phosphate bases from ATP. The ATP molecules are by-products of the main light reaction.*

So, the Calvin cycle can be summarized in 3 major phases:

1. Carboxylation phase:

$$6RuDP + 6CO_2 \longrightarrow 12\ 3\text{-}PGA$$

2. Reduction phase:

$$12\ 3\text{-}PGA + 12ATP + 12NAD(P)H + 12H^+ \longrightarrow 12\ 3\text{-}PGAld + 12ADP + 12Pi + 12NAD(P)^+$$

3. Regeneration phase:

$$12\ 3\text{-PGAld} + 6ATP \longrightarrow 6RuDP + 6ADP + 5Pi + Fr\text{-}6\text{-}P$$

These 3 phases can be further summarized by the equation:

$$6RuDP + 6CO_2 + 18ATP + 12NAD(P)H + 12H^+ \longrightarrow 6RuDP + Fr\text{-}6\text{-}P + 18ADP + 17Pi + 12NAD(P)^+$$

$$[6C_5 + 6C_1 \longrightarrow 6C_5 + C_6]$$

The net reaction leads to the formation of one molecule of hexose phosphate. It has also been shown by Calvin and coworkers that carbohydrates are not the only end-products of the cycle. Other compounds, like amino acids, carboxylic acids, fatty acids, etc., are also produced directly from different intermediates of the cycle.

The mechanism of bacterial photosynthesis has been elucidated with the development of several biochemical techniques, such as spectrophotometry, mass spectography, use of radioisotopes, chromatography, cell fractionating, ultracentrifugation and many others. Still, several unsolved or tentatively solved questions remain. It has been seen that only a fraction of the number of quanta absorbed is utilized in the photosynthetic process. So, why the remaining large amount with an efficient light-harvesting mechanism loss in phototrophs is not yet clear. There is also a suspicion that bacteria contain both types of RC. Under certain conditions, only one type is active while the other is idle or is inhibited.

Some important problems that have come into focus are: detailed mechanism of energy migration in RC, physico-chemical mechanism or RC and details of enzymatic mechanisms which include coupling of ATP synthesis and hydrolysis, function of Cyt's, oxidation-reduction reactions and so on. Although many theories have been put forward, the mystery of evolutionary history of bacterial photosynthesis remains unresolved. It is clear that bacterial photosynthesis evolved earlier than green plant photosynthesis. Bacterial photosynthesis is the most primitive type, but has posed several challenges to present-day researchers.

5 PHOTORESPIRATION

Photorespiration, as the name implies, is respiration or the evolution of carbon dioxide by green plants in light. The phenomenon has attracted much attention in recent years on two counts: (i) It appears to be distinct from ordinary respiration and biochemists are intrigued about the source of carbon dioxide evolved in the process and (ii) Photorespiration is of vital importance from the viewpoint of economics and human welfare. It is now known that respiration in light is several folds greater in magnitude than is the case in ordinary 'dark' respiration.

The discovery of photorespiration dates back to 1920 when the German biochemist, Otto Warburg, observed an inhibition of the photosynthetic rate in the green alga, *Chlorella* with increasing concentrations of oxygen in the ambient atmosphere. In honour of this discovery, later scientists called it the "Warburg effect" even though they did not fully realize its significance. Warburg himself did not elaborate upon this observation and the phenomenon would have been "lost" to science but for the fact that it kept cropping up in various kinds of studies on photosynthesis and respiration and several scientists began to suspect that light did indeed stimulate respiration, at least in some green plants.

A systematic study of the phenomenon began in the early fifties when J. P. Decker in the USA discovered photorespiration in a higher plant, namely tobacco. He used the sensitive infrared gas analyzer to measure the amount of carbon dioxide evolved.

Subsequent to Decker's studies, several other laboratories took up work on photorespiration. Ozbun and co-workers (1964) in the USA not only confirmed the earlier work but by a modification of the technique-employing the heavy isotope of O_2 namely $^{18}O_2$—they could observe high rates of respiration in light itself instead of relying upon the transient "post-illumination burst" as an indicator of photorespiration. They sealed the leaves of bean in a chamber, one end of which was connected to a regulated source of $^{18}O_2$ and the other to a mass spectrometer,—an instrument employed for the exact determination of the mass of an isotope. They observed that $^{18}O_2$ uptake almost doubled in light as compared to that in the dark. This was a clear-cut evidence for the existence of photorespiration.

In the light of these observations, it is now possible to explain the inhibition of photosynthesis by high concentrations of oxygen which had intrigued Warburg in the 1920s. High levels of oxygen stimulate photorespiration, a process in which gas exchange occurs in the reverse of that in photosynthesis.

Photorespiration differs from dark respiration in another aspect : whereas in dark respiration the substrate is generally considered to be a sugar, in photorespiration glycolic acid (a 2-C organic compound) serves as the substrate.

Some plants exhibit photorespiration and others not as shown in Table 4.5.

Table 4.5 *Some common examples of photorespiring and non-photorespiring plants.*

Photorespiring	*Non-photorespiring*
Wheat	Corn
Rice	Sugar-cane
Legumes	Sorghum
Potato	Sugar-beet
Tomato	Bajra

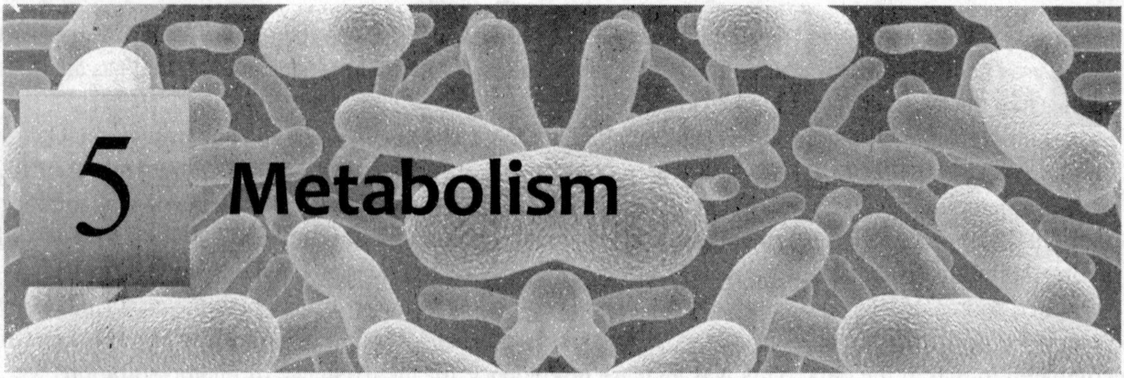

5 Metabolism

1 INTRODUCTION

Growth may be defined as the orderly increase in mass or number of all cellular components. It is the main function of any living cell. In order to grow, new components, (e.g. enzymes, chromosomes, protein-synthesizing systems, etc.) must be synthesized. These macromolecules are assembled from their respective sub-units, (e.g. proteins from amino acids, nucleic acids from nucleotides, etc.).

Condensation of sub-units requires an input of energy, usually in the form of ATP. Some biosynthetic reactions also require a net input of reducing power, usually in the form of $NADH_2$ or $NADPH_2$. Thus, both chemical energy and reducing power are generated through catabolism. In this sense, there is a fundamental unity underlying the superficial diversity of catabolism. This unity of bio-chemistry is a concept first emphasized by the microbiologist, A.J. Kluyver (1926).

Biosynthetic reactions, leading to the formation of end-products, (e.g. proteins, polysaccharides, etc.) show little variation from group to group among prokaryotes, and even between prokaryotes and eukaryotes.

In this chapter, we will focus our attention on three different metabolisms:

(A) metabolism of carbohydrates

(B) metabolism of nitrogenous compounds

(C) metabolism of nucleic acids

Each metabolism will be divided into two halves, i.e. (i) Catabolism and (ii) Anabolism (or Biosynthesis) and will be discussed at length for the purpose of simplification.

2 METHODOLOGY

Different techniques have been exploited to elucidate the biosynthetic pathways. Many facts have been traced in bacterial metabolism. The same nature of bacterial physiology is also seen in higher systems. The techniques can be considered more important than man's knowledge in developing bio-chemistry. Some important techniques which have found wide applications in bio-chemistry include:

(A) use of bio-chemical mutants

(B) isotope labelling

(C) different types of chromatographic techniques

(D) electrophoresis

These have been dealt with in Chapter 9.

(A) METABOLISM OF CARBOHYDRATES

Carbohydrates may be defined as substances containing carbon, hydrogen and oxygen, the hydrogen and oxygen being present in the same ratio as in water. Their general formula is, therefore, $C_m (H_2O)_n$. This definition is, however, not strictly accurate, since in a few carbohydrates the proportion of oxygen is lower than that required by the general formula.

The easiest way to understand their structure is to consider all carbohydrates as either hydroxy-aldehydes or hydroxy-ketones. The simpler carbohydrates are known as sugars. Those containing an aldehyde group, such as glyceraldehyde, are called aldoses.

CHO
|
CHOH
|
CH_2OH

glyceraldehyde

Those containing a keto group, such as dihydroxyacetone, are known as ketoses.

CH_2OH
|
C = O
|
CH_2OH

dihydroxyacetone

Glyceraldehyde and dihydroxyacetone are called trioses as they contain three carbon atoms. Tetroses contain four carbon atoms, pentoses contain five, and hexoses contain six carbon atoms. Such simple sugars are called **monosaccharides**.

When two monosaccharide units condense together with the removal of water, a *disaccharide* is formed. When three monosaccharide units condense together with the elimination of water, a **trisaccharide** is produced. Finally, a *polysaccharide* is that product formed by the condensation of a large number of monosaccharide units.

Monosaccharides are those sugars which cannot be further hydrolyzed. Glucose, fructose, galactose and mannose are the examples of monosaccharides.

Ribose and deoxyribose are also monosaccharides containing five carbon atoms.

Disaccharides are those sugars which give two monosugars upon hydrolysis, e.g.

(i) Sucrose $\xrightarrow{H_2O}$ glucose + fructose

(ii) Lactose $\xrightarrow{H_2O}$ glucose + galactose

(iii) Maltose $\xrightarrow{H_2O}$ glucose + glucose

Trisaccharides are those sugars which yield three monosugars upon hydrolysis, e.g.

Raffinose $\xrightarrow{H_2O}$ glucose + fructose + galactose

Polysaccharides are those sugars which give large numbers of monosaccharide units upon hydrolysis, e.g. glycogen, starch and cellulose. All these three are built up solely from glucose units.

The polysaccharides have high molecular weights. They are not usually soluble in water though some may form colloidal solutions.

3 CATABOLISM

Carbohydrates are major sources of energy for living organisms. Simple sugars such as sucrose, lactose, maltose, glucose, fructose and galactose are utilized by living organisms as a source of carbon and biological energy, usually in the form of ATP. Since glucose is the compound, formed from both starch and glycogen through catabolism, our discussion of carbohydrate catabolism should naturally begin with this monosaccharide.

The pathways of glucose catabolism leading to pyruvate formation

Certain degradative pathways are common to both fermentative and respiratory metabolism. There are three pathways of conversion of glucose to pyruvate:

(1) Embden-Meyerhof-Parnas pathway (or glycolysis)

(2) Pentose phosphate pathway (or hexose monophosphate shunt)

(3) Entner-Doudoroff pathway (also called the ED pathway)

The first two pathways occur in many organisms, belonging to both prokaryotes and eukaryotes. The latter is limited to certain groups of prokaryotes.

(1) Embden-Meyerhof-Parnas pathway (Figure 5.1) :

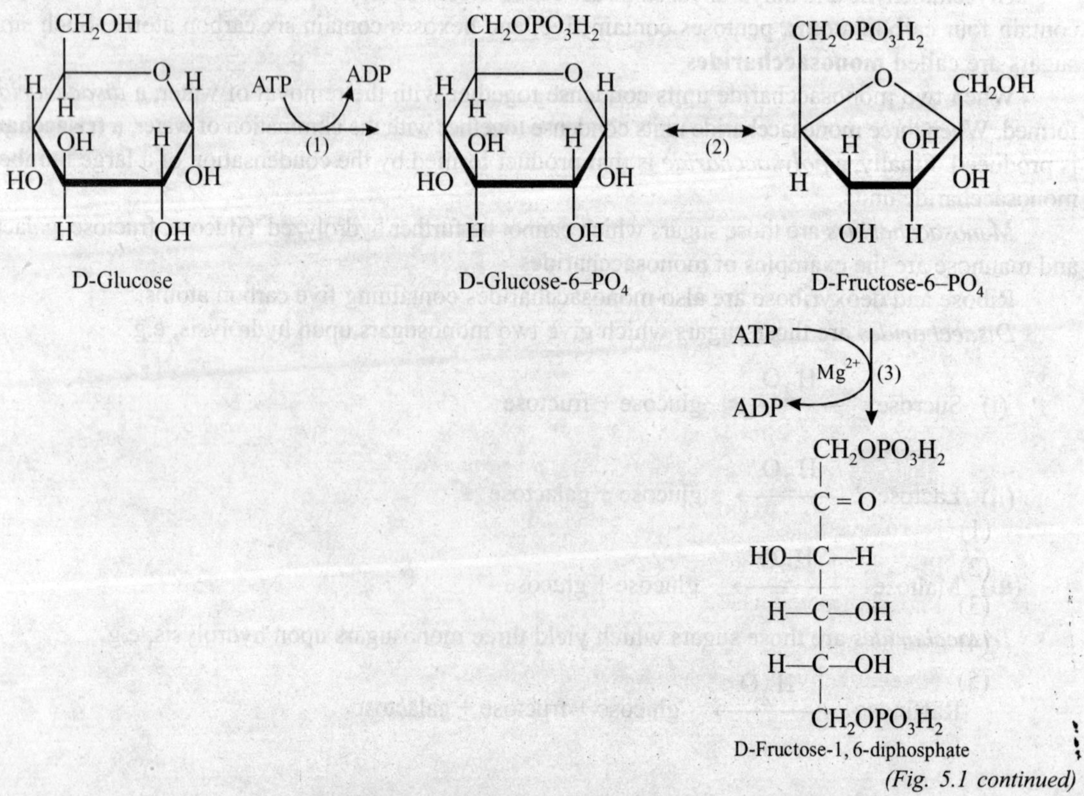

(Fig. 5.1 continued)

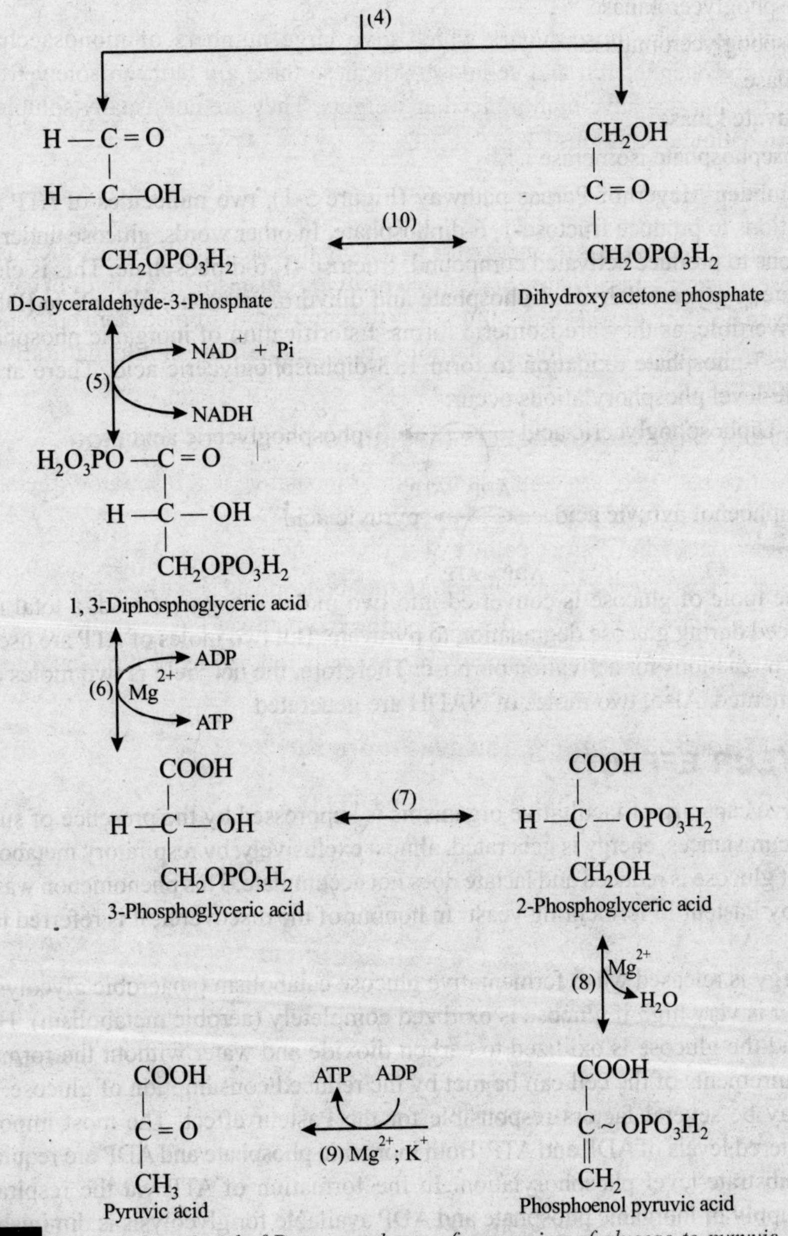

Fig. 5.1 *The Embden-Meyerhof-Parnas pathway of conversion of glucose to pyruvic acid.*

(1) Hexokinase
(2) Phosphoglucoisomerase
(3) Phosphofructokinase
(4) Aldolase
(5) Glyceraldehyde-3-phosphate dehydrogenase

(6) Phosphoglycerokinase
(7) Phosphoglyceromutase
(8) Enolase
(9) Pyruvate kinase
(10) Triosephosphate isomerase

In the Embden-Meyerhof-Parnas pathway (Figure 5-1), two molecules of ATP are consumed in the initial reactions to produce fructose-1, 6-diphosphate. In other words, glucose undergoes preliminary phosphorylations to produce activated compound, fructose -1, 6-diphosphate. This is cleaved to give the triose phosphates : glyceraldehyde-3-phosphate and dihydroxyacetone: phosphate (DHAP). These are freely interconvertible, as they are isomeric forms. Esterification of inorganic phosphate accompanies glyceraldehyde-3-phosphate oxidation to form 1, 3-diphosphoglyceric acid. There are the two stages where substrate-level phosphorylations occur:

(i) 1, 3–Diphosphoglyceric acid ⟶ 3-phosphoglyceric acid
ADP ATP

(ii) Phosphoenol pyruvic acid ⟶ pyruvic acid
ADP ATP

Since one mole of glucose is converted into two moles of pyruvic acid, a total of four moles of ATP are produced during glucose degradation to pyruvate. But two moles of ATP are used in preliminary phosphorylation reactions for activation purpose. Therefore, the net yield is two moles of ATP per mole of glucose fermented. Also, two moles of NADH are generated.

4 PASTEUR EFFECT

The fermentative capacity of facultative organisms is suppressed by the presence of sufficient oxygen. Under such circumstances, energy is generated, almost exclusively, by respiratory metabolism. Therefore, consumption of glucose is reduced and lactate does not accumulate. This phenomenon was recognized, for the first time, by Pasteur in fermenting yeast. In honour of the discoverer, it is referred to as the *Pasteur effect*.

Less energy is released when fermentative glucose catabolism (anaerobic glycolysis) is operative. But energy gain is very high if glucose is oxidized completely (aerobic metabolism). Hence, if oxygen is available, and the glucose is oxidized to carbon dioxide and water without the formation of lactate, the energy requirements of the cell can be met by the reduced consumption of glucose.

There may be several factors responsible for the Pasteur effect. The most important, probably, relates to the altered levels of ADP and ATP. Both inorganic phosphate and ADP are required in glycolytic pathway for substrate-level phosphorylation. In the formation of ATP via the respiratory pathways, however, the supply of inorganic phosphate and ADP available for glycolysis is diminished. Thus, ADP/ATP ratio is changed. The increased levels of ATP inhibit phosphofructokinase activity, thereby decreasing the flow of glucose into the glycolysis. This inhibition of enzyme activity by ATP is illustrative of regulation of an allosteric enzyme by a negative effector. The reaction, catalyzed by phosphofructokinase (formation of fructose-1,6-diphosphate from fructose–6-phosphate) is a critical control point. It is subject to strong metabolic regulation. It is susceptible to strong inhibition by high concentrations of ATP. It is activated by ADP or AMP. Such controls ensure that, whenever a plentiful supply of ATP is available (as when pyruvate undergoes aerobic metabolism via the TCA cycle), the glycolytic pathway will be essentially inhibited. The reverse is also true. When glycolysis is absolutely required for the synthesis of energy, i.e. when the ATP drops to low levels and ADP or AMP accumulates, glycolysis will be favoured.

Action of coliform organisms on carbohydrates

Coliform organisms ferment carbohydrates, (e.g. glucose) chiefly to acetic, formic, lactic, and succinic acids, ethanol, acetoin, 2, 3-butanediol, CO_2 and H_2.

The important products produced by *E. coli* and *Klebsiella pneumoniae*, through fermentative action on glucose, are as shown in Table 5.1.

Table 5.1 *End products produced by E. coli and Klebsiella pneumoniae as a result of glucose fermentation.*

E. coli	*Klebsiella pneumoniae*
Acetic acid	Acetic acid
Formic acid	Formic acid
Lactic acid	Lactic acid
Succinic acid	Succinic acid
Ethanol	Ethanol
CO_2	CO_2
H_2	H_2
	Acetoin
	2, 3-Butanediol

Fermentation reactions and identification of bacteria

There are two purposes behind incorporating carbohydrates and compounds of a similar nature in culture media :

(i) Carbohydrates serve as energy source.

(ii) Fermentation reactions help in the identification and classification of bacteria.

Fermentative ability of organisms is variable from one group to another. Some bacteria have the ability to attack many carbohydrates, others cannot ferment any carbohydrate. There are also some groups of organisms exhibiting fermentability between these two extremes. Moreover, some organisms ferment carbohydrates with the production of both acid and gas. Some organisms can produce acid only. Such reactions are very useful as a guideline in the identification and subsequent classification of bacteria.

(2) Pentose phosphate pathway (Figure 5.2) :

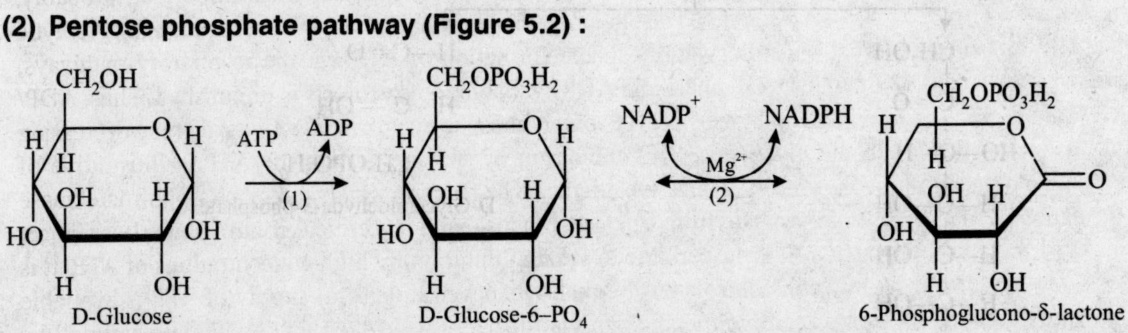

(Fig. 5.2 continued)

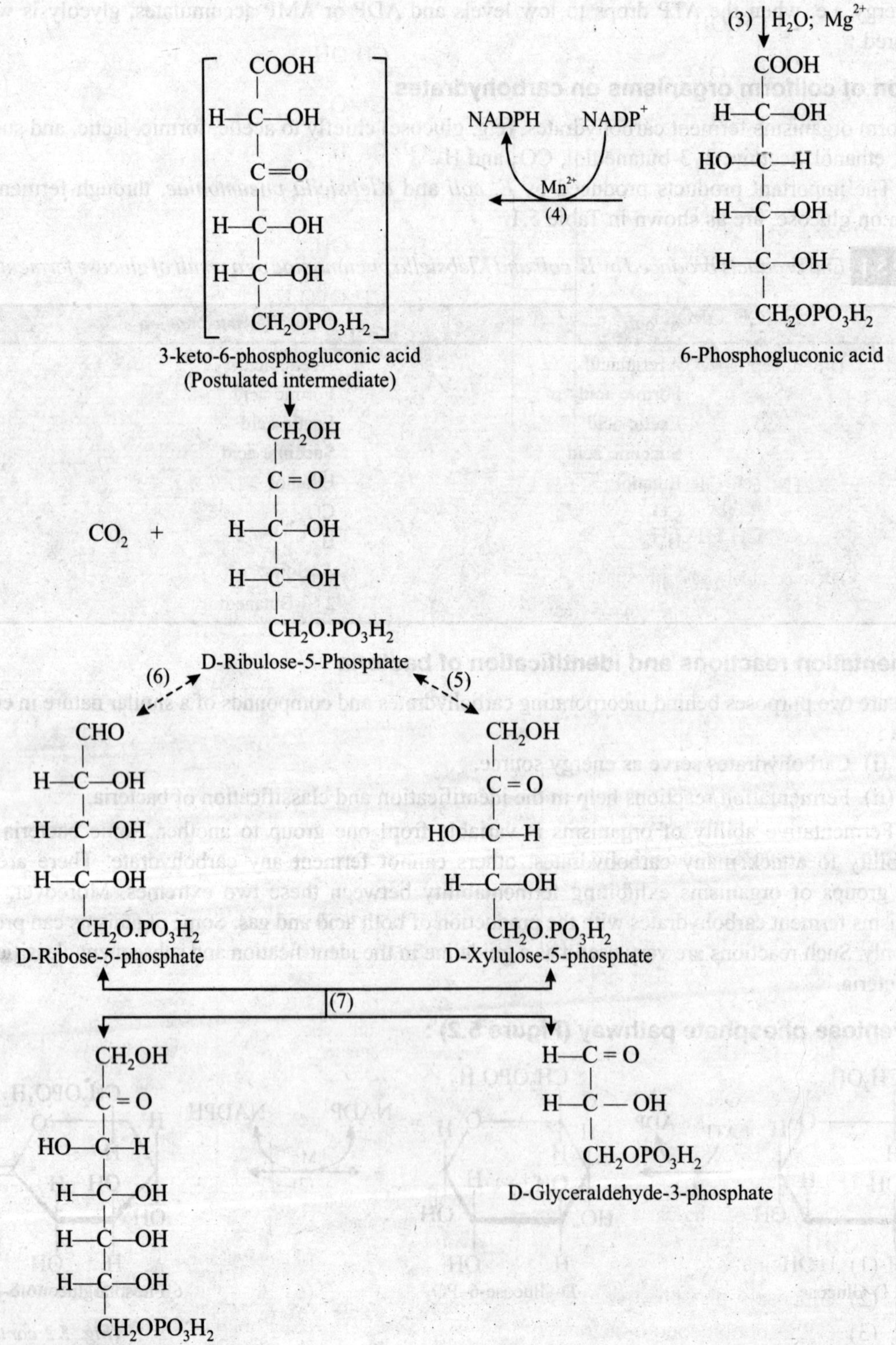

(Fig. 5.2 continued)

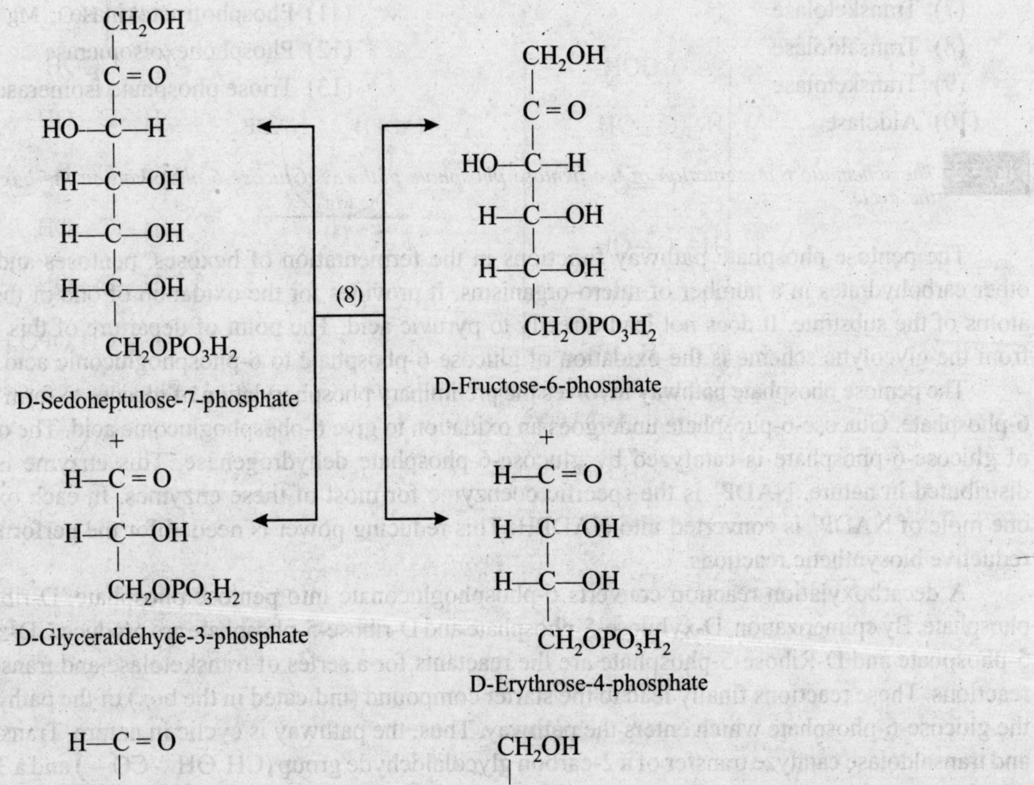

(1) Hexokinase
(2) Glucose-6-phosphate dehydrogenase
(3) 6-Phosphoglucono-δ-lactonase
(4) 6-Phosphogluconate dehydrogenase
(5) Phosphoketopentoisomerase
(6) Phosphoriboisomerase

(Fig. 5.2 continued)

(7) Transketolase
(8) Transaldolase
(9) Transketolase
(10) Aldolase
(11) Phosphofructokinase
(12) Phosphohexoisomerase
(13) Triose phosphate isomerase

Fig. 5.2 *The schematic representation of the pentose phosphate pathway. Glucose-6-phosphate in the box enters in the cycle.*

The pentose phosphate pathway functions in the fermentation of hexoses, pentoses and several other carbohydrates in a number of micro-organisms. It provides for the oxidation of one of the carbon atoms of the substrate. It does not lead directly to pyruvic acid. The point of departure of this pathway from the glycolytic scheme is the oxidation of glucose-6-phosphate to 6-phosphogluconic acid.

The pentose phosphate pathway involves the preliminary phosphorylation of glucose to form glucose-6-phosphate. Glucose-6-phosphate undergoes an oxidation to give 6-phosphogluconic acid. The oxidation of glucose-6-phosphate is catalyzed by glucose-6-phosphate dehydrogenase. This enzyme is widely distributed in nature. $NADP^+$ is the specific coenzyme for most of these enzymes. In each oxidation, one mole of $NADP^+$ is converted into NADPH. This reducing power is needed for the performance of reductive biosynthetic reactions.

A decarboxylation reaction converts 6-phosphogluconate into pentose phosphate, D-ribulose-5-phosphate. By epimerization, D-xylulose-5-phosphate and D-ribose-5-phosphate are produced. D-xylulose-5-phosphate and D-Ribose-5-phosphate are the reactants for a series of transketolase and transaldolase reactions. These reactions finally lead to the starter compound (indicated in the box) of the pathway. It is the glucose-6-phosphate which enters the pathway. Thus, the pathway is cyclic in nature. Transketolase and transaldolase catalyze transfer of a 2-carbon glycolaldehyde group (CH_2OH—CO—) and a 3-carbon dihydroxy-acetone group (CH_2OH—CO—CHOH—) respectively. The passage of six molecules through the cycle results in the complete oxidation of one molecule of glucose-6-phosphate to CO_2 and the reduction of six molecules of $NADP^+$ to NADPH.

This pathway serves the following functions in metabolism: (i) It produces the D-ribose-5-phosphate. This is required for the biosynthesis of nucleic acids and (ii) It produces much of the NADPH required in the various reductive biosynthetic reactions.

To understand this concept, the following examples may be considered. NADPH is specifically utilized in the synthesis of long chain fatty acids and steroids. This reducing agent is also used in the reduction of:

(i) glucose to sorbitol
(ii) dihydrofolic acid to tetrahydrofolic acid
(iii) glucuronic acid to L-gluconic acid

In addition, NADPH is utilized in the reductive carboxylation of pyruvate to malate by the malate enzyme. Finally, NADPH plays a unique role in hydroxylation reactions. These reactions are involved in the formation of unsaturated fatty acids.

(iii) Erythrose-4-phosphate is required in the first step of a biosynthetic pathway in micro-organisms and plants that lead to shikimic acid. Several amino acids are then obtained from shikimic acid.

When glucose is fermented through the pentose phosphate pathway, the net yield of ATP is half of that, characteristic of the glycolysis. This lower energy yield is characteristic of a pathway of dehydrogenation before cleavage.

(3) The Entner-Doudoroff pathway (Figure 5.3):

This route of glucose catabolism was first discovered in the bacterium *Pseudomonas saccharophila*. Some bacteria, (e.g. *Pseudomonas, Azotobacter* spp.) lack phosphofructokinase. They, therefore, cannot split glucose by the glycolytic pathway.

In this sequence, glucose is phosphorylated to form the first intermediate, glucose-6-phosphate. Glucose-6-phosphate undergoes oxidation to produce 6-phosphoglucono-δ-lactone. The lactone itself is unstable, and hydrolyzes spontaneously to phosphogluconate. This intermediary compound then undergoes a dehydration and rearrangement to form 2-keto-3-deoxy-6-phosphogluconic acid (KDPG). KDPG is cleaved to one molecule of pyruvate and one molecule of glyceraldehyde-3-phosphate. Glyceraldehyde-3-phosphate is further catabolized by enzymes of the EMP pathway. This results in the formation of a second molecule of pyruvate. The net yield of catabolism of one mole of glucose through the ED pathway is one mole of ATP and two moles of NADH.

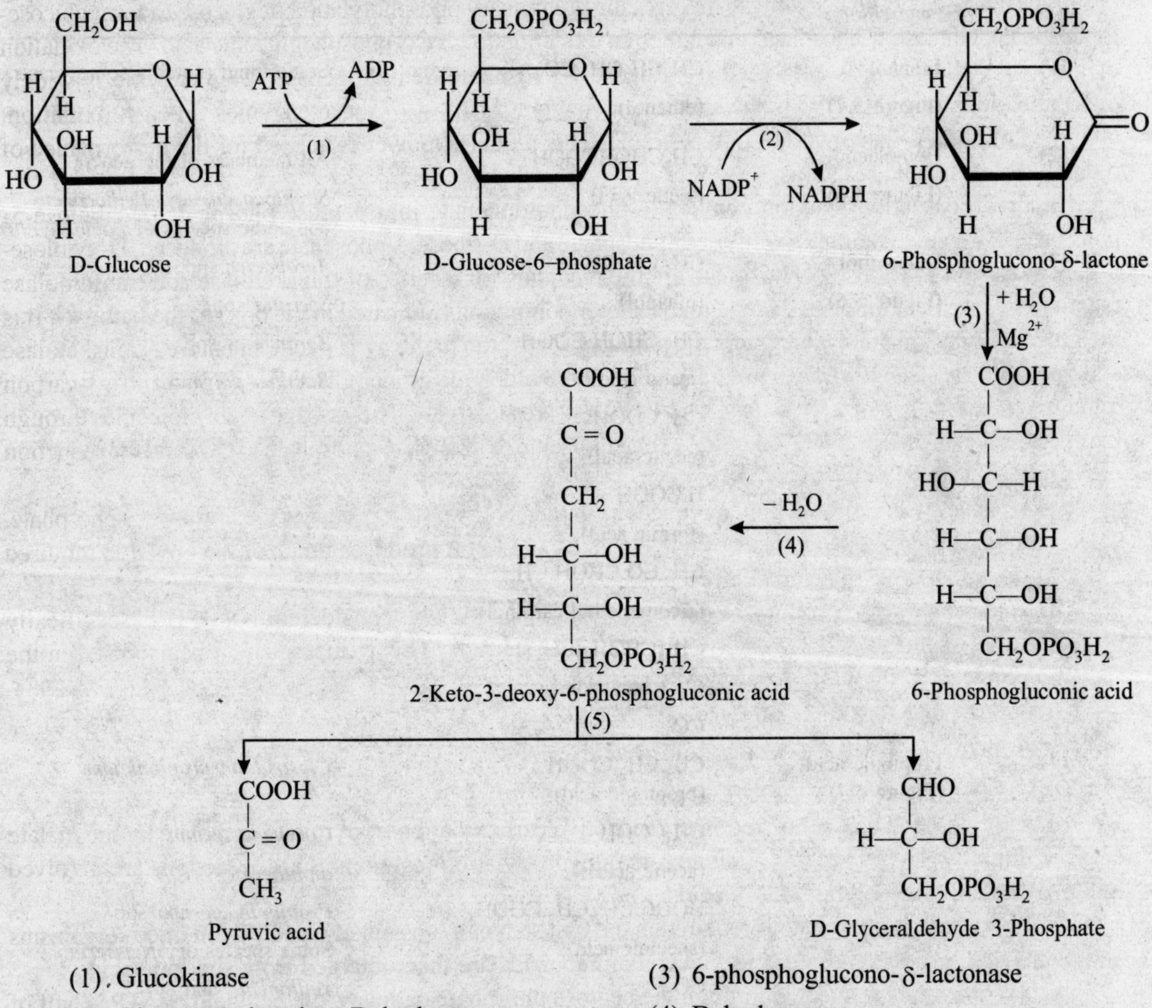

(1) Glucokinase

(2) Glucose-6-phosphate Dehydrogenase

(3) 6-phosphoglucono-δ-lactonase

(4) Dehydrase

(5) Aldolase

Fig. 5.3 *The Entner-Doudoroff pathway of conversion of glucose to pyruvic acid and glyceraldehyde-3-phosphate.*

Fate of pyruvate under anaerobic conditions

Pyruvate is considered as the key intermediary compound in the fermentative metabolism of all carbohydrates. Two molecules of $NADH_2$ formed during fermentation of glucose to pyruvate must be reoxidized in order to achieve a final oxidation reduction balance. $NADH_2$ is reoxidized to NAD by the reduction of pyruvate–either directly, as in lactic acid homofermentation, or indirectly, as in the others.

The different fermentations of this group are thus characterized by the reduction products formed from pyruvate. This reflects the differences in the enzymatic constitutions of different species. Microbial fermentations are of practical importance because they provide products of industrial value. They are also useful in the identification of microbial species. The major end-products of fermentation are as listed in Table 5.2.

Table 5.2 *Microbial sugar fermentations based on the Embden-Meyerhof-Parnas glycolysis*

Sr. No.	*Fermentation*	*End-products*	*Organisms*
(1)	Ethanol (Figure 5.4)	CH_3CH_2OH; CO_2 (ethanol)	Some fungi (notably some yeasts)
(2)	Homolactic (Figure 5.5)	$CH_3CHOHCOOH$ (lactic acid)	All members of the genera *Streptococcus* and *Pediococcus* and some species of *Lactobacillus*
(3)	Butanediol (Figure 5.6)	$CH_3.CH_2OH$ (ethanol); $CH_3.CHOH.COOH$ (lactic acid); $CH_3.COOH$ (acetic acid); $H.COOH$ (formic acid); $CH_3.CO.CHOH.CH_3$ (Acetylmethyl carbinol); $CH_3(CHOH)_2CH_3$ (2, 3 butanediol); CO_2.	*Aerobacter* spp., *Serratia* spp., *Aeromonas* spp., *Bacillus polymyxa*
(4)	Propionic acid (Figure 5.7)	$CH_3.CH_2.COOH$ (propionic acid); CH_3COOH (acetic acid); $HOOC.CH_2.CH_2.COOH$ (succinic acid); CO_2.	*Clostridium propionicum* *Corynebacterium diphtheriae* *Propionibacterium* spp. Some species of : *Neisseria*; *Veillonella*; and *Micromonospora*

(Table 5.2 continued)

Sr. No.	Fermentation	End-products	Organisms
(5)	Mixed acid (Figure 5.8)	$CH_3CHOHCOOH$ (lactic acid); $CH_3.COOH$ (acetic acid); HCOOH (formic acid); or CO_2 and H_2; $HOOC.CH_2.CH_2.COOH$ (succinic acid); CH_3CH_2OH (ethanol).	Many enteric bacteria (e.g. *Escherichia* spp.) *Salmonella* spp. *Shigella* spp. *Proteus* spp. *Yersinia* spp.
(6)	Butanol-butyrate (Figure 5.9)	$CH_3.CH_2.CH_2.CH_2OH$ (Butanol); $CH_3.CH_2.CH_2.COOH$ (Butyrate); $CH_3.CO.CH_3$ (acetone); $CH_3.CHOH.CH_3$ (isopropanol); $CH_3.COOH$ (acetic acid); $CH_3.CH_2OH$ (ethanol); H_2, CO_2.	*Zymosarcina maxima* *Butyribacterium* spp. Some species of : *Clostridium*, *Neisseria*

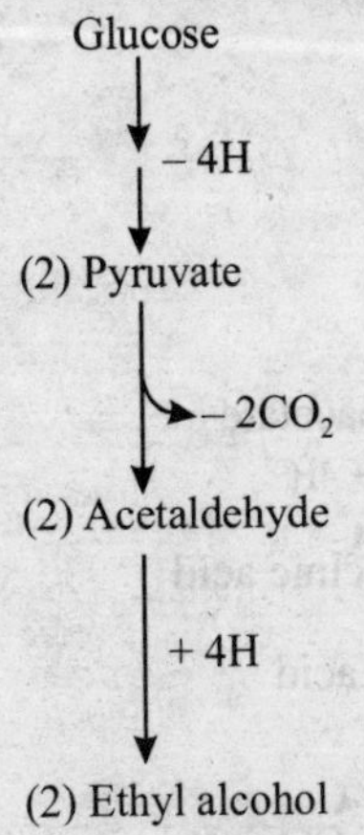

Fig. 5.4 *Ethyl alcohol fermentation.*

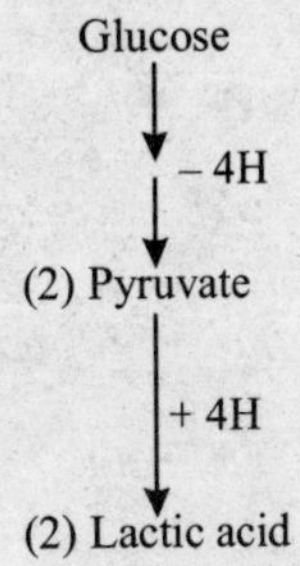

Fig. 5.5 *Lactic acid homofermentation.*

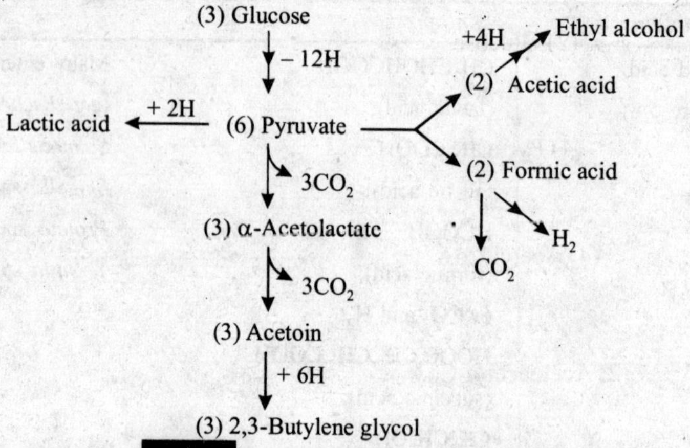

Fig. 5.6. Butylene glycol fermentation.

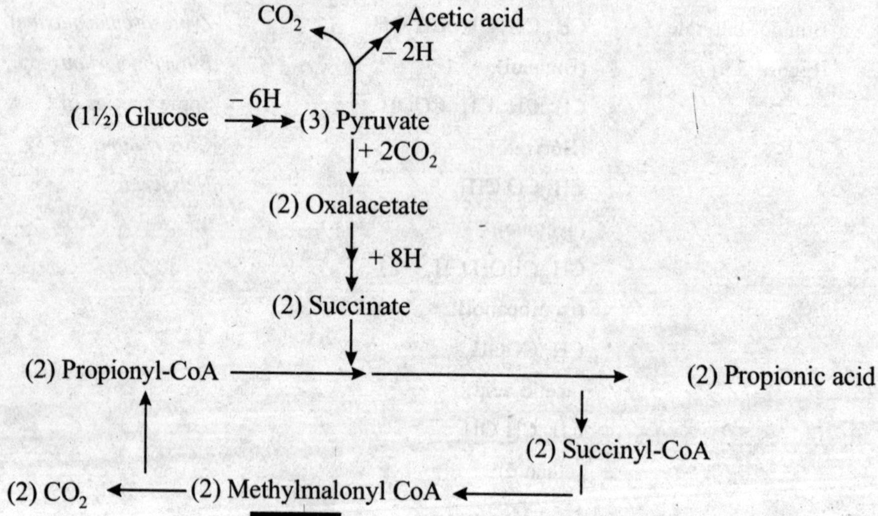

Fig. 5.7. Propionic acid fermentation.

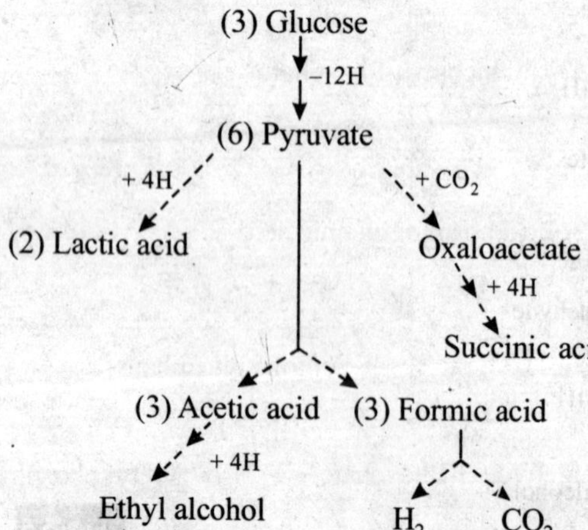

Fig. 5.8 *Mixed acid fermentation.*

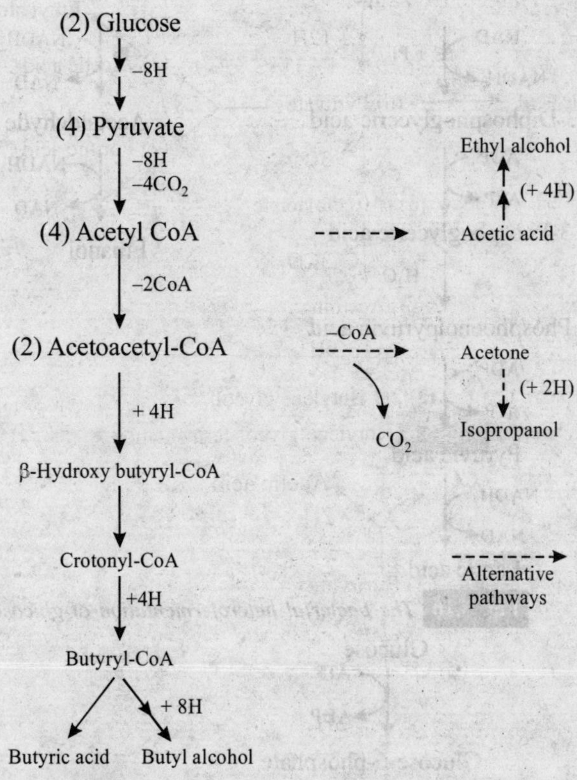

Fig. 5.9 *Butyl alcohol-butyric acid fermentation.*

There are two additional fermentations based on the pentose phosphate pathway. Their mechanisms are outlined in Figures 5.10 and 5.11.

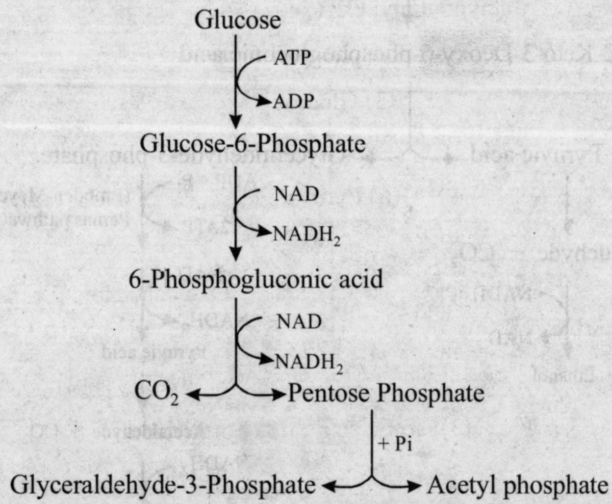

(Fig. 5.10 continued)

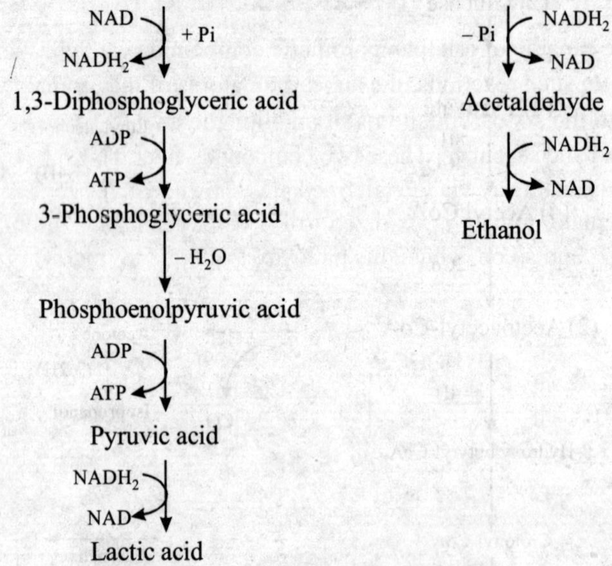

Fig. 5.10 *The bacterial heterofermentation of glucose.*

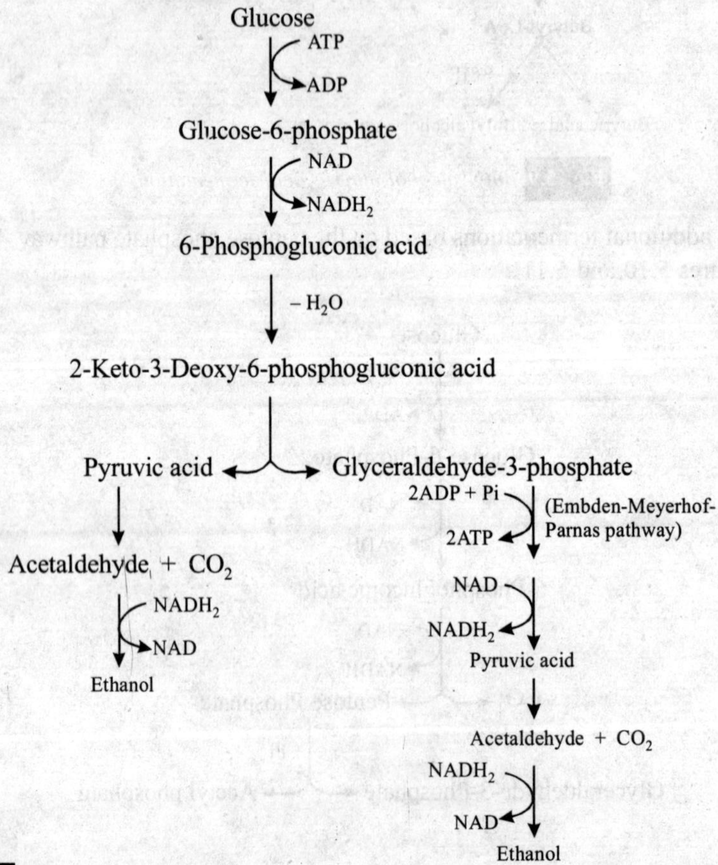

Fig. 5.11 *Mechanism of the alcoholic fermentation of glucose by the bacterium Zymomonas mobilis.*

5 THE β-KETOADIPATE PATHWAY (FIGURE 5.12)

Many micro-organisms are capable of oxidizing aromatic compounds via the β-ketoadipate pathway to acetyl-CoA and succinate. In such reactions, the six carbon atoms of the aromatic nucleus in the initial substrate are converted into the six carbon atoms of an aliphatic acid, β-ketoadipate. β-Ketoadipate is then cleaved to acetyl-S-CoA and succinate. These two compounds enter the Krebs cycle. Other oxidizable aromatic acids funnel into this pathway via the intermediates shown, e.g. benzene is oxidized to catechol, toluene is oxidized to benzoate, toluate is oxidized to P-hydroxybenzoate, shikimate and quinate are oxidized to protocatechuate, and so on. Thus, this pathway is unique to micro-organisms.

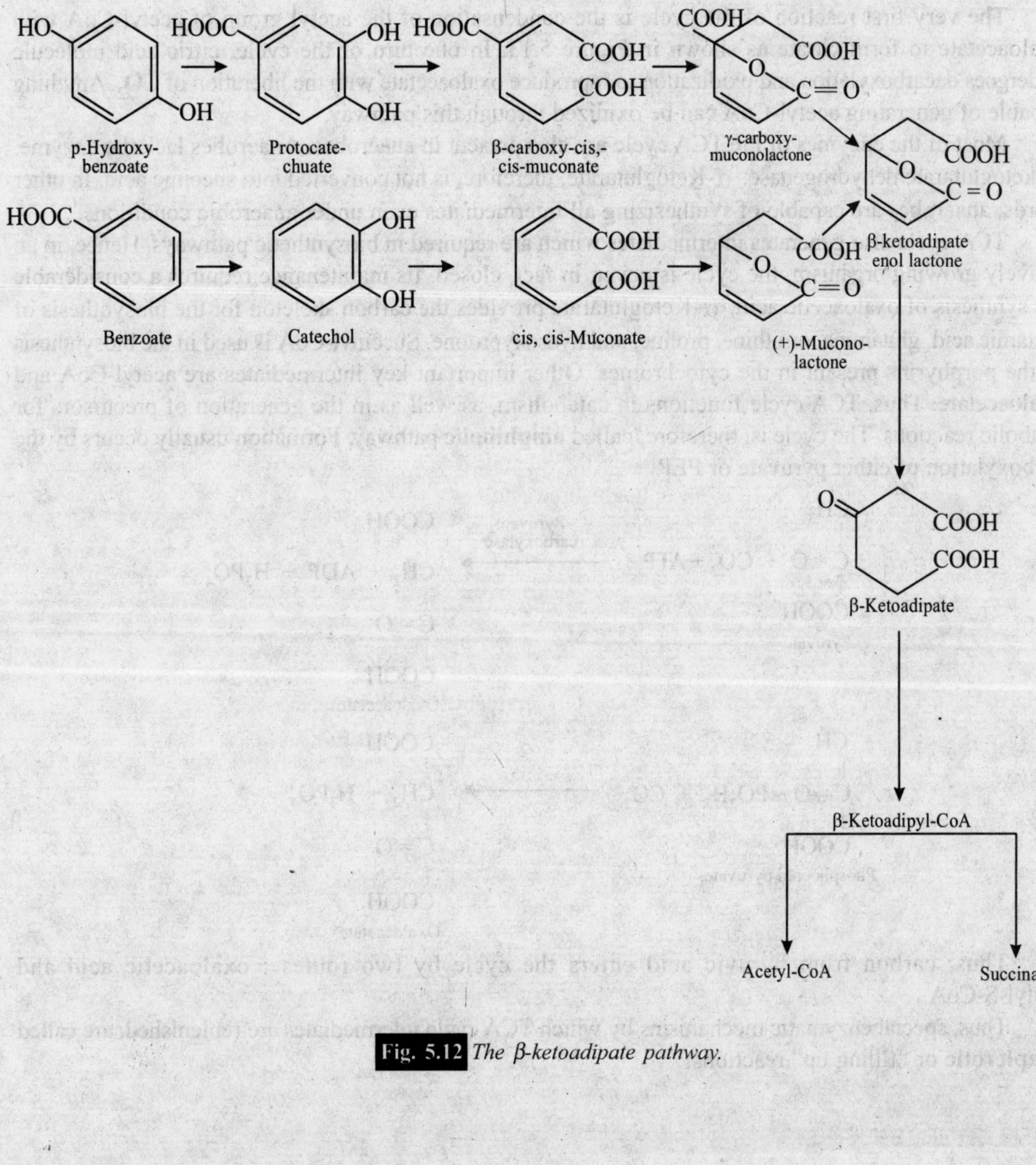

Fig. 5.12 *The β-ketoadipate pathway.*

Fate of pyruvate under aerobic conditions

Pyruvate may enter in TCA cycle where it is oxidized into CO_2 and H_2O.

The citric acid cycle (Krebs cycle)

The citric acid or tricarboxylic acid cycle is the aerobic process, in which active acetate (acetyl-CoA) undergoes complete oxidation to CO_2 and H_2O. This cycle is the major route of ATP generation. The electrons removed from the substrates are transferred from reduced pyridine nucleotide to the final electron acceptor. The final electron acceptor is the molecular oxygen. Thus, large amount of energy is generated through oxidative phosphorylation. On complete oxidation of one molecule of acetic acid to CO_2, three molecules of reduced pyridine nucleotides, one molecule of GTP, and one molecule of $FADH_2$ are produced.

The very first reaction of the cycle is the condensation of the acetyl group of acetyl-CoA with oxaloacetate to form citrate as shown in Figure 5.13. In one turn of the cycle, citric acid molecule undergoes decarboxylation and oxidization to reproduce oxaloacetate with the liberation of CO_2. Anything capable of generating acetyl-CoA can be oxidized through this pathway.

Most of the enzymes of the TCA cycle are also present in anaerobes. Anaerobes lack the enzyme, α-ketoglutarate dehydrogenase. α-Ketoglutarate, therefore, is not converted into succinic acid. In other words, anaerobes are capable of synthesizing all intermediates even under anaerobic conditions.

TCA cycle also generates intermediates which are required in biosynthetic pathways. Hence, in an actively growing organism, the cycle is never, in fact, closed. Its maintenance requires a considerable net synthesis of oxaloacetic acid. α-Ketoglutarate provides the carbon skeleton for the biosynthesis of glutamic acid, glutamine, ornithine, proline, and hydroxyproline. Succinyl-CoA is used in the biosynthesis of the porphyrins present in the cytochromes. Other important key intermediates are acetyl-CoA and oxaloacetate. Thus, TCA cycle functions in catabolism, as well as in the generation of precursors for anabolic reactions. The cycle is, therefore, called **amphibolic** pathway. Formation usually occurs by the carboxylation of either pyruvate or PEP.

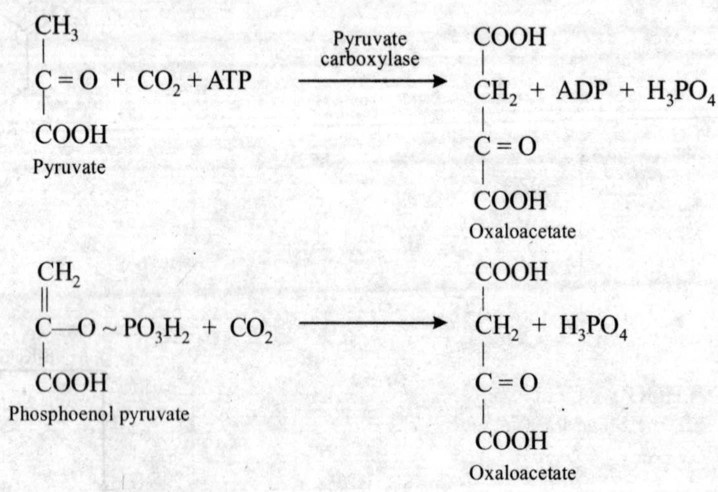

Thus, carbon from pyruvic acid enters the cycle by two routes : oxaloacetic acid and acetyl-S-CoA.

Thus, special enzymatic mechanisms by which TCA cycle intermediates are replenished, are called **anaplerotic** or "filling up" reactions.

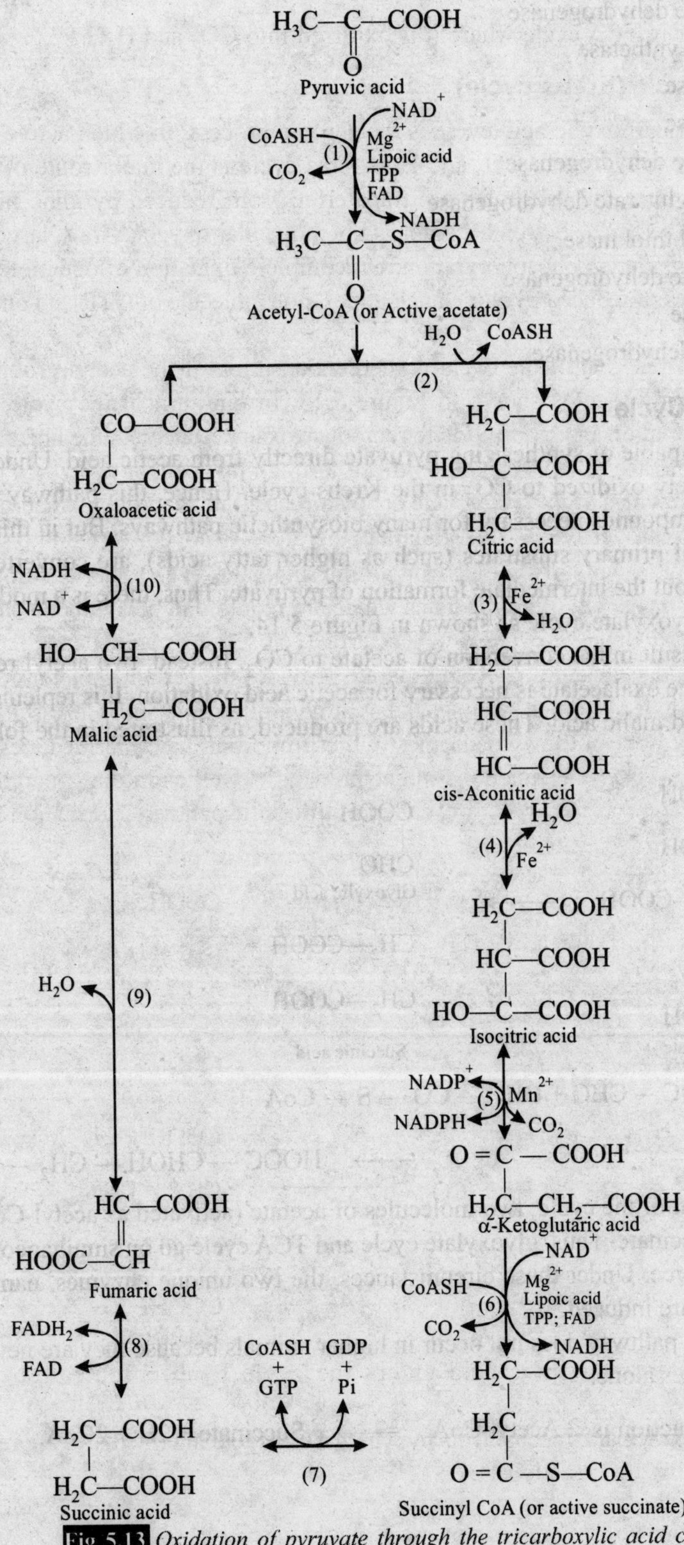

Fig 5.13 *Oxidation of pyruvate through the tricarboxylic acid cycle.*

1. Pyruvate dehydrogenase
2. Citrate synthetase
3. Aconitase
4. Aconitase
5. Isocitrate dehydrogenase
6. α-Ketoglutarate dehydrogenase
7. Succinyl thiokinase
8. Succinate dehydrogenase
9. Fumarase
10. Malate dehydrogenase

The Glyoxylate Cycle

Aerobes are not capable of synthesizing pyruvate directly from acetic acid. Under aerobic conditions, acetate is completely oxidized to CO_2 in the Krebs cycle. Hence, this pathway cannot provide a net synthesis of C_4 compounds necessary for many biosynthetic pathways. But in this cycle, the oxidation of acetic acid or of primary substrates (such as higher fatty acids), are converted into active acetate (acetyl–CoA) without the intermediate formation of pyruvate. Thus, there is a modification of the Krebs cycle, known as glyoxylate cycle as shown in Figure 5.14.

It does not result in the conversion of acetate to CO_2. Instead, two acetyl residues are converted into oxalacetate. The oxalacetate is necessary for acetic acid oxidation. It is replenished by the oxidation of succinic acid and malic acid. These acids are produced, as illustrated in the following two different reactions:

(i) COOH–CHOH–CH(COOH)–CH_2–COOH (Isocitrate) ⟶ COOH–CHO (Glyoxylic acid) + CH_2–COOH | CH_2–COOH (Succinic acid)

(ii) $HOOC—CHO + CH_3—CO—S—CoA \longrightarrow HOOC—CHOH—CH_2—COOH + CoASH$

With each turn of the cycle, two molecules of acetate (activated as acetyl-CoA) are converted to one molecule of succinate. Thus, glyoxylate cycle and TCA cycle go on simultaneously, when acetate is the sole carbon source. Under these circumstances, the two unique enzymes, namely isocitratase and malate synthetase, are induced.

Moreover, the pathway does not occur in higher animals because they are never forced to feed on two carbon molecules alone.

The overall reaction is 2 Acetyl-CoA $\longrightarrow$ Succinate + 2H + 2CoA.

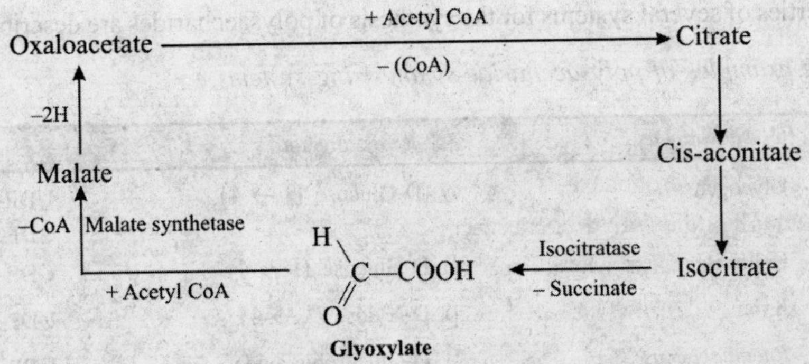

Fig. 5.14 *The glyoxylate cycle.*

6 ANABOLISM

(1) Biosynthesis of some Polysaccharides

A primer is required as an acceptor of additional monomer units in polysaccharide synthesis. In this way, this resembles to DNA synthesis where a small DNA molecule is necessary as an acceptor of deoxynucleoside triphosphates. Glycogen synthesis (Figure 5.15) will explain about the nature of polysaccharide biosynthesis. The following two enzymes are required in the glycogen biosynthesis :

(i) *Glycogen synthetase* and

(ii) *Amylotransglycosylase.*

Glycogen synthetase catalyzes the transfer of glucose moieties from ADP-glucose to the non-reducing end of an α–1, 4-glucan to form a new α–1, 4-glycosyl-glucan. Glycogen synthetase requires a primer molecule to accept the glucose units. It has been found that the primer must contain more than four sugar units to function effectively. Second enzyme is also required to form the α–1, 6 linkage in glycogen. It catalyzes the shift of an oligosaccharide unit of six or seven residues in length to another point in the amylose chain to make the α–1, 6 branch point.

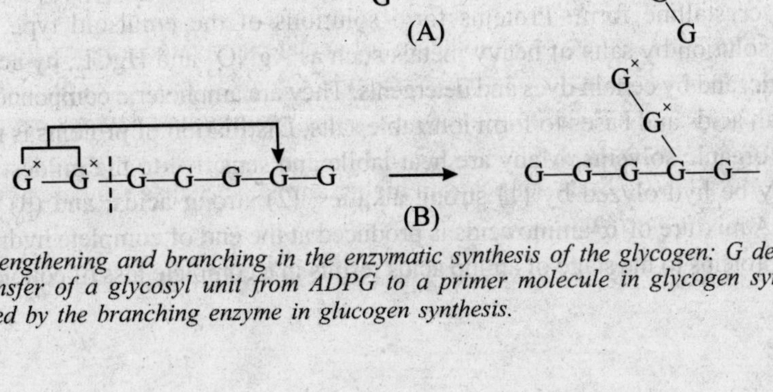

Fig. 5.15 *Chain lengthening and branching in the enzymatic synthesis of the glycogen: G denotes glucosyl units. (A) transfer of a glycosyl unit from ADPG to a primer molecule in glycogen synthesis. (B) reaction catalyzed by the branching enzyme in glucogen synthesis.*

The properties of several systems for the synthesis of polysaccharides are described in Table 5.3.

Table 5.3 *Some examples of polysaccharide-synthesizing systems.*

Sr. No.	*Polysaccharide*	*Repeating unit*	*Precursor*
1.	Glycogen	α-D-Glucose $(1 \rightarrow 4)$	UDP-glucose (animals) ADP-glucose (bacteria)
2.	Cellulose	β-D-Glucose $(1 \rightarrow 4)$	GDP-glucose
3.	Xylan	β-D-Xylose $(1 \rightarrow 4)$	UDP-Xylose
4.	*Pneumococcus* type III capsular Polysaccharide	β-D-Glucuronic acid $(1 \rightarrow 4)$ β-D-Glucose $(1 \rightarrow 3)$	UDP-glucuronic acid UDP-glucose

(2) The Synthesis of Carbohydrates from CO_2

A special mechanism is found in organisms that use CO_2 as the sole source of carbon for the production of intermediates. This occurs in photosynthetic organisms and in bacteria which obtain their energy by the oxidation of inorganic substances. (For further details, see Chapter-4).

(3) Biosynthesis of a cell-wall Peptidoglycan

The rigid portion of a bacterial cell-wall is a polymeric structure known as a *murein, peptidoglycan,* or *mucopeptide*. Peptidoglycans vary in their chemical composition and structure from species to species, but their basic structure is the same. Peptidoglycans are very large polymers. They are composed of three kinds of building blocks :

(i) **acetylglucosamine**

(ii) **acetylmuramic acid**

(iii) a **peptide** consisting of four or five amino acids of limited variety

(B) METABOLISM OF PROTEINS

The name protein is derived from the Greek and means 'of first importance'. Proteins are extremely complex combinations of amino acids. They are essential components of all living cells. They are composed of very large molecules (macro-molecules).

Almost all soluble proteins form colloidal solutions. Most proteins form opalescent solutions. This characteristic suggests that they are macromolecules. Most of proteins are amorphous and some have been obtained in crystalline form. Proteins form solutions of the emulsoid type. Proteins may be precipitated from solution by salts of heavy metals such as $AgNO_3$ and $HgCl_2$, by acids such as tannic and phosphotungstic, and by certain dyes and detergents. They are amphoteric compounds. They, therefore, may react with both acids and bases to form ionizable salts. Distillation of proteins is not possible. They are not soluble in organic solvents. Many are heat-labile and sensitive to the mildest reagents.

Proteins may be hydrolyzed by (1) strong alkalies, (2) strong acids, and (3) certain groups of specific enzymes. A mixture of α-amino acids is produced at the end of complete hydrolysis of proteins. The hydrolysis of proteins to the stage of amino acids results in a complete loss of colloidal characteristics.

There are about 20 different amino acids recognized as constituents of proteins. All amino acids are optically active except glycine (glycocoll). They all are soluble in water except tyrosine and cystine. Tyrosine is sparingly soluble in cold water but more soluble in hot water. Cystine is sparingly soluble in both hot and cold water. They all are soluble in both dilute acids and alkalies except cystine. Cystine dissolves with difficulty in dilute ammonia water. All the amino acids have at least one free amino ($-NH_2$) group and one free carboxyl (–COOH) group except proline and oxyproline. These two amino acids possess an imino (–NH) group in place of an amino ($-NH_2$) group. Most of the amino acids possess one free amino group and one free carboxyl group and are neutral in reaction. Some of them possess two amino groups to one carboxyl group and are alkaline in reaction. Still some of them have one amino to two carboxyl groups and are acidic in reaction.

7 STRUCTURE OF PROTEIN MOLECULE

Amino acids are linked together principally through their α-amino and carboxyl groups, producing peptide linkage (R.CO–NH.R'). For example, in a tripeptide containing glycine, alanine, and serine, the acids are linked end to end. Here the –COOH of one acid combines with the $-NH_2$ of the next with the elimination of a water molecule as under :

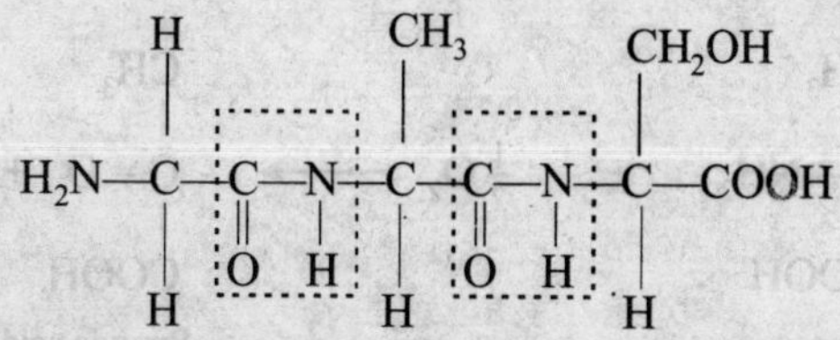

Glycyl–alanyl–serine

Under the action of acids, alkalies, or certain enzymes, peptide chains or residues break apart with the addition of water, resulting in the hydrolysis of the peptide linkage.

Protein molecules are complex in structure. For example, insulin contains 777 atoms. Some proteins may be 50 times larger. All enzymes are also proteins. The enzyme ribonuclease has two-dimensional formula according to Stein and Moore (1961). It contains :

(i) 587 atoms of carbon (C)
(ii) 909 atoms of hydrogen (H)
(iii) 197 atoms of oxygen (O)
(iv) 171 atoms of nitrogen (N)
(v) 12 atoms of sulphur (S)

Thus, it contains a total of 1,876 atoms.

8 CATABOLISM

Proteolytic enzymes may bring about hydrolysis of proteins. It results in the formation of the following fractions in the order named:

Proteins (>10,000) ⟶ Proteoses (~5000) ⟶ Peptones (~2000) ⟶

Peptides (~1000 to 500) ⟶ dipeptides (~200) ⟶ α-amino acids (~100)

Enzymes which hydrolyze peptide bonds are grouped under the hydrolases. Many peptidyl peptide hydrolases are known which may break the bond –CONH– by the addition of water. This results in the release of free –COOH and $-NH_2$ groups. Bacterial peptidyl peptide hydrolases are probably exoenzymes. Their function is to bring about conversion of indiffusible proteins into diffusible polypeptides and dipeptides. Thus, these simpler molecules can easily enter the bacterial cell.

The polypeptide and dipeptide hydrolases may be both intracellular and extracellular. They hydrolyze both polypeptides and dipeptides and reduce them to mixtures of smaller units including the individual amino acids.

9 ACTION ON AMINO ACIDS

Bacteria have power to attack amino acids in variety of ways. They may remove the amino group and use the resulting ammonia as a source of nitrogen. They may destroy the carboxyl group with the formation of amines. These are only two of the many types of reactions produced by bacteria acting on amino acids. Other types of reactions are the following:

(A) Oxidative deaminization

Such reaction leads to the formation of an α-Keto acid :

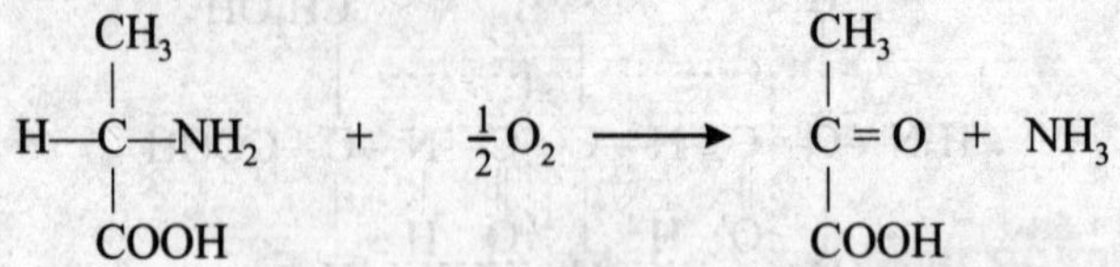

Alanine → Pyruvic acid

Some organisms have ability to produce an oxidative deaminization of the dicarboxylic aspartic and glutamic acids. This results in the formation of the corresponding α-Keto acids. Others are capable of metabolizing the compounds to acetic acid, ammonia, and carbon dioxide as shown under :

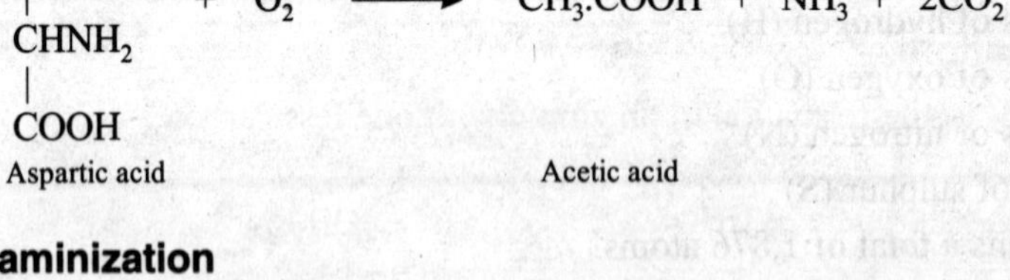

Aspartic acid → Acetic acid

(B) Reductive deaminization

This type of reaction leads to the formation of a saturated acid. This reaction is generally carried out by the strict and facultative anaerobic organisms grown under anaerobic conditions :

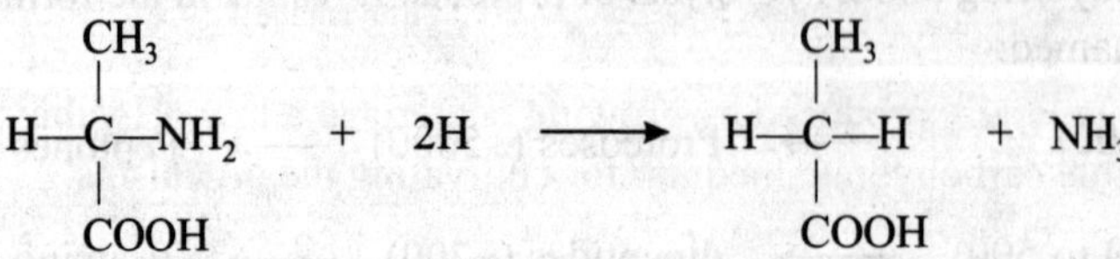

Alanine → Propionic acid

(C) Hydrolytic deaminization

This reaction leads to the formation of a hydroxy acid:

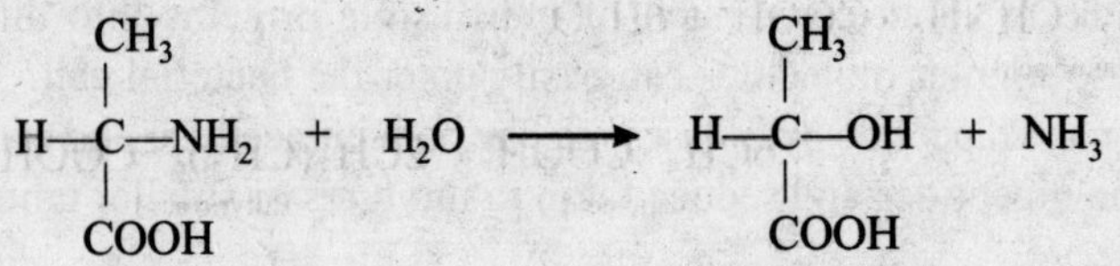

Alanine Lactic acid

Deaminization usually takes place in the bacteria grown in a neutral or slightly alkaline medium.

(D) Deamination and desaturation at the α, β-linkage

This reaction leads to the formation of an unsaturated acid :

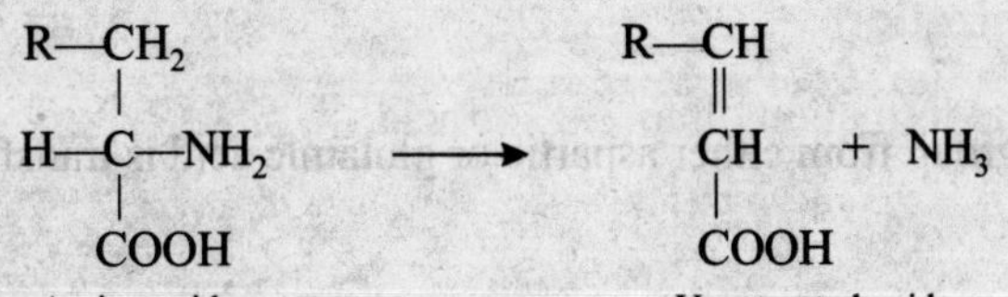

Amino acid Unsaturated acid

(E) Hydrolytic deaminization

This reaction leads to the formation of a primary alcohol with one less carbon atom :

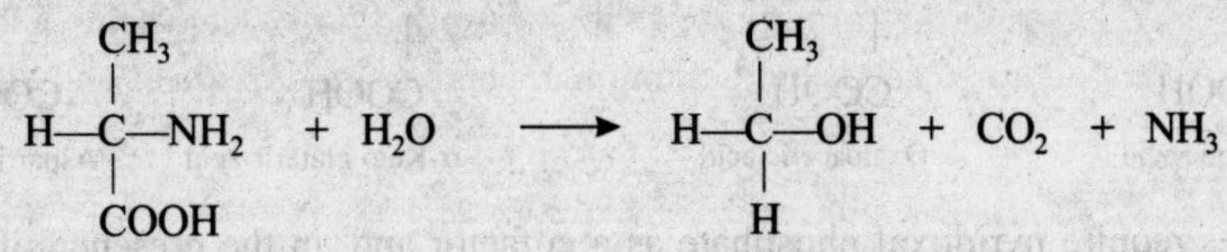

Alanine Ethanol

(F) Decarboxylation

This reaction leads to the formation of an amine with one less carbon atom :

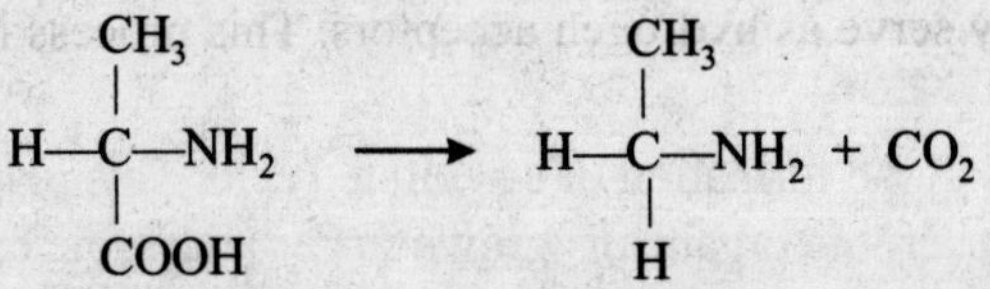

Alanine Ethylamine

Decarboxylases are generally produced by growing bacteria at a low pH value. This can be achieved by employing fermentable carbohydrate medium for cultivating the organisms.

(G) Anaerobic decomposition

There occurs liberation of hydrogen:

$$\underset{\text{Glutamic acid}}{5COOH \cdot (CH_2)_2 \cdot CHNH_2 \cdot COOH} + 6H_2O \longrightarrow$$

$$\underset{\text{Acetic acid}}{6CH_3 \cdot COOH} + \underset{\text{Butyric acid}}{2CH_3(CH_2)_2 \cdot COOH} + 5CO_2 + 5NH_3 + H_2$$

(H) Decomposition

This results in the oxidation of one molecule of an amino acid and the reduction of another :

$$\underset{\text{Alanine}}{3CH_3 \cdot CHNH_2 \cdot COOH} + 2H_2O \longrightarrow$$

$$\underset{\text{Propionic acid}}{2CH_3 \cdot CH_2 \cdot COOH} + \underset{\text{Acetic acid}}{CH_3 \cdot COOH} + 3NH_3 + CO_2$$

(I) Transamination

In this reaction, an amino group from either aspartic or glutamic acid is transferred to the α – position in an α-keto acid:

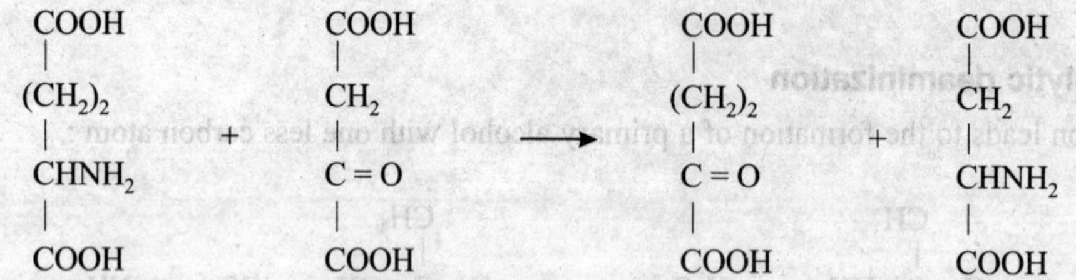

Glutamic acid　Oxaloacetic acid　α-Keto glutaric acid　Aspartic acid

The transaminases require pyridoxal phosphate as a cofactor and, in the presence of the enzyme, the coenzyme forms a Schiff's base with the amino acid.

(J) Stickland reaction

Some amino acids (alanine, valine, leucine, phenylalanine) have been shown to be oxidizable. They serve as hydrogen donors while some other amino acids (glycine, proline, hydroxyproline, ornithine, arginine) are reducible. They serve as hydrogen acceptors. This process has already been discussed in Chapter 4.

10 ANABOLISM

Proteins are made of amino acids. Amino acids are called building units. Their arrangements are specific in specific proteins. Twenty amino acids are required for the synthesis of proteins. All amino acids are derived from branched pathways except histidine. Histidine has a completely isolated biosynthetic origin. All twenty amino acids are synthesized from a relatively small number of central intermediary metabolites. Five families may be formed on the basis of their biosynthetic origin as shown in Table 5.4 given on next page.

Table 5.4 *Biosynthetic derivation of amino acids.*

Sr. No.	*Precursors*	*Amino acids*	*Family*
(1)	α-Ketoglutarate	→ Glutamate → Arginine, Proline; Glutamate ↔ Glutamine	'Glutamate' Figure 5.16
(2)	Oxaloacetate	→ Aspartate → Methionine, Threonine, Isoleucine, Lysine[x]; Aspartate → Asparagine; Threonine → Isoleucine	'Aspartate' Figure 5.17
(3)	Phosphoenol pyruvate + Erythrose-4-phosphate	→ Tryptophan, Phenylalanine, Tyrosine	'Aromatic' Figure 5.18
(4)	3-phosphoglycerate	→ serine → Glycine, Cysteine	'Serine' Figure 5.19
(5)	Pyruvate	→ Alanine, Valine, Leucine	'Pyruvate' Figure 5.20
(6)	Phosphoribosyl pyrophosphate + ATP	→ Histidine	Figure 5.21

[x] In certain algae and fungi, lysine is synthesized from α-ketoglutarate.

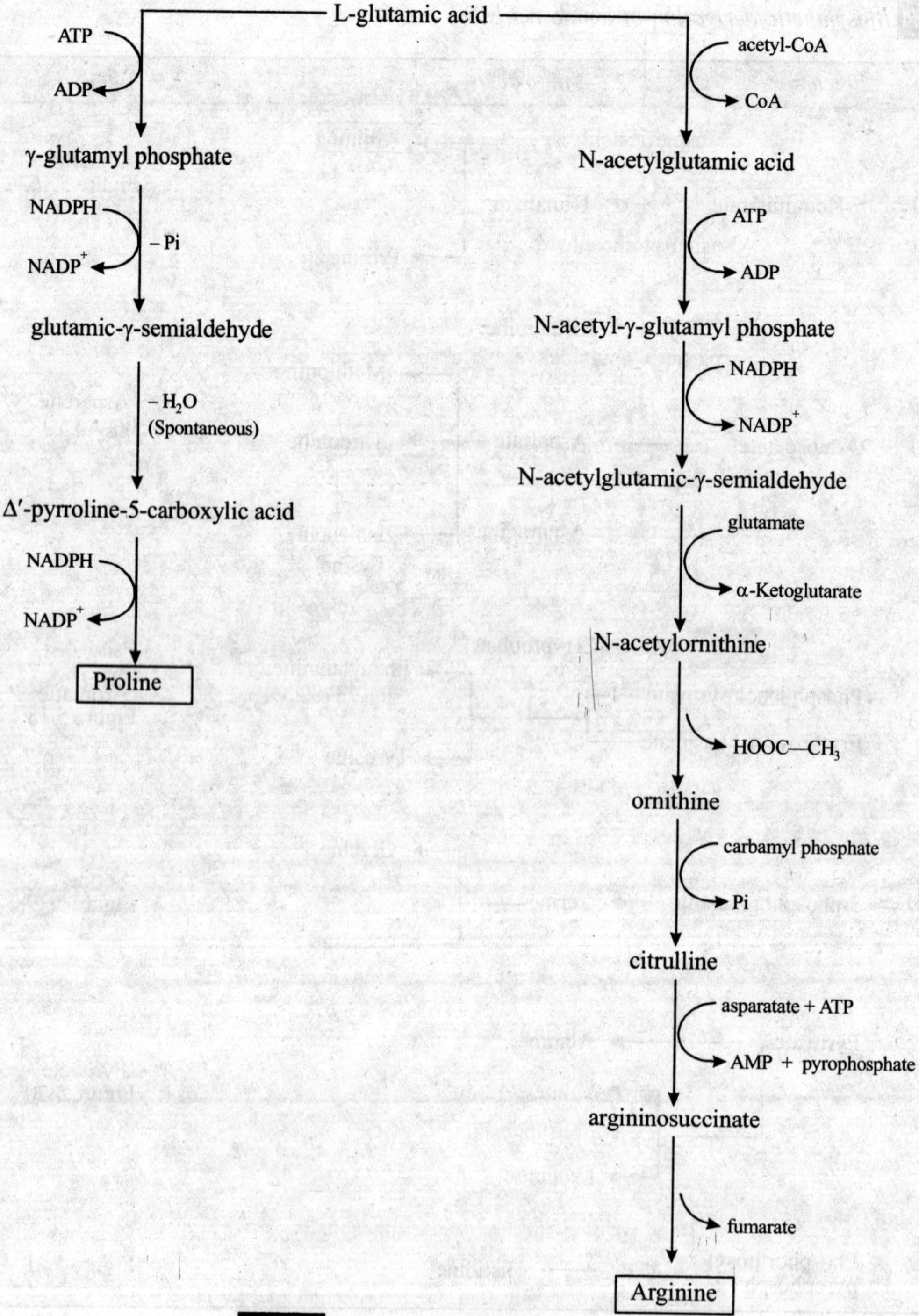

Fig. 5.16 *Biosynthesis of proline and arginine.*

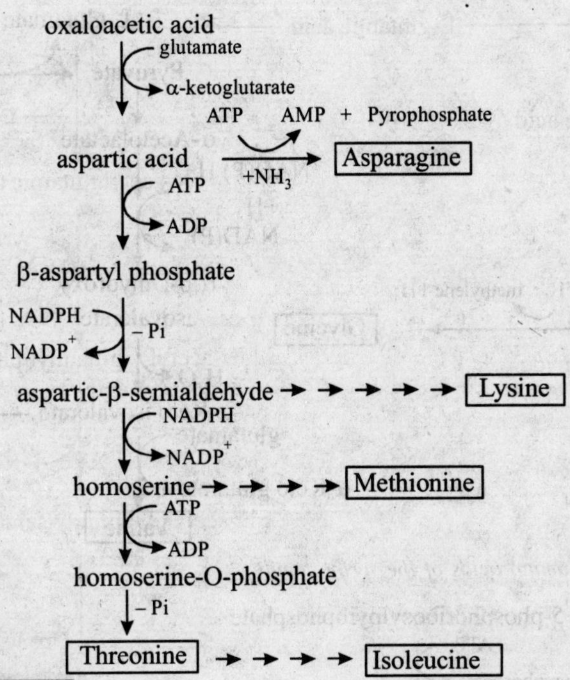

Fig. 5.17 *Biosynthesis of amino acids of the aspartate family.*

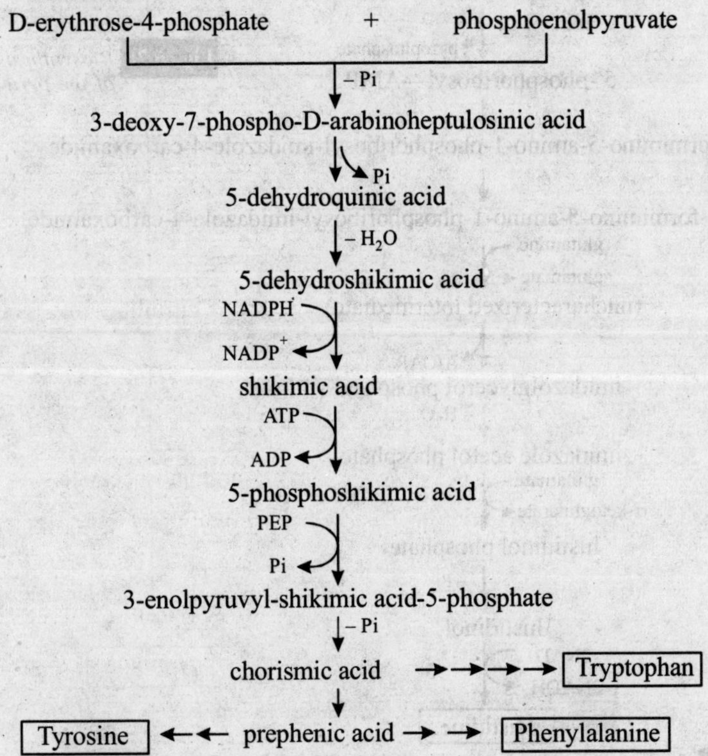

Fig. 5.18 *Biosynthesis of amino acids of the aromatic family.*

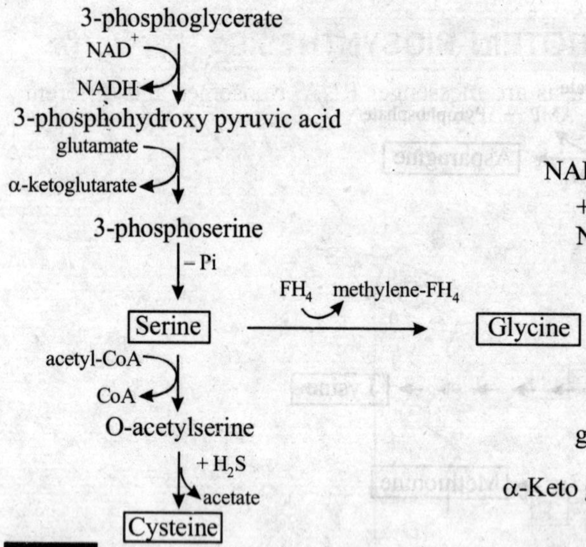

Fig. 5.19 *Biosynthesis of the amino acids of the serine family.*

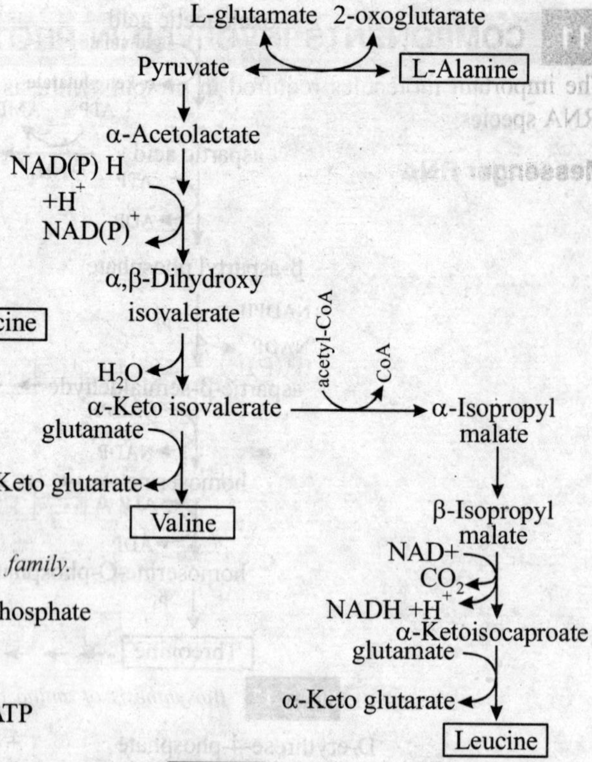

Fig. 5.20 *Biosynthesis of the amino acids of the pyruvate family.*

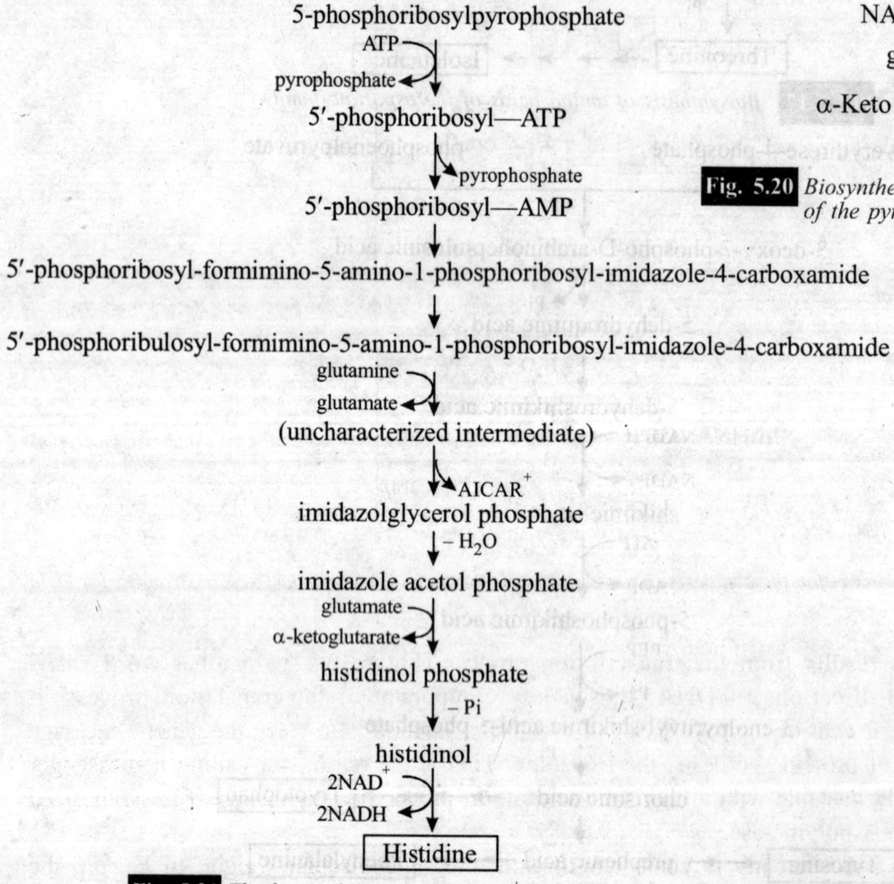

Fig. 5.21 *The biosynthesis of histidine.* $^{+}$AICAR = aminoimidazole carboxamide ribotide

11 COMPONENTS INVOLVED IN PROTEIN BIOSYNTHESIS

The important molecules required in protein synthesis are messenger RNA, ribosomes and different tRNA species.

Messenger RNA

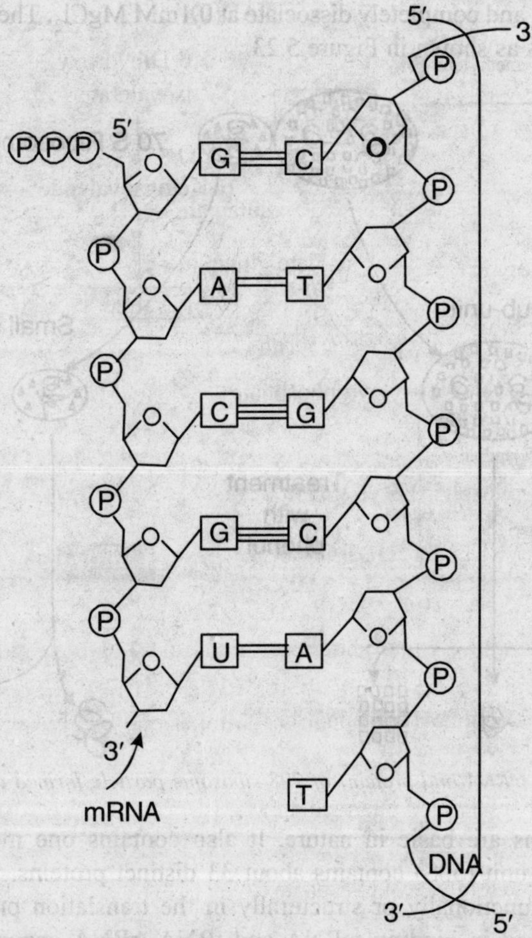

Fig. 5.22 *Transcription process for the synthesis of m-RNA complementary to one strand ('sense' strand) of a DNA template.*

Messenger RNA results from the transcription process (Figure 5.22). In other words, it is synthesized under the directions of DNA. It is a key component of the translation process. It comprises only a few per cent of the total RNA of a cell. Its function is to carry the genetic message from DNA to the site of protein synthesis, the ribosome. This is the reason for calling it messenger RNA. It is metabolically unstable with a high turnover rate in procaryotic organisms. It is synthesized by DNA-dependent RNA polymerase.

Messenger RNA varies greatly in chain length, and thus, in molecular weight. In *E. coli*, the average size of mRNA is 900-1500 nucleotide units, and codes for more than one polypeptide chain.

Ribosomes

Ribosomes are large ribonucleoprotein particles. They are considered to be the site for protein synthesis. The actual translation process occurs on these molecules. In procaryotic organisms, they occur in the free form as monosomes, or are associated with mRNA as polysomes. An average bacterial cell contains about 10^4 ribosomes.

All ribosomes consist of a larger sub-unit and a smaller sub-unit. They associate themselves at 10mM $MgCl_2$ concentration and completely dissociate at 0.1mM $MgCl_2$. The 30S sub-units of ribosome contains 21 distinct proteins as shown in Figure 5.23.

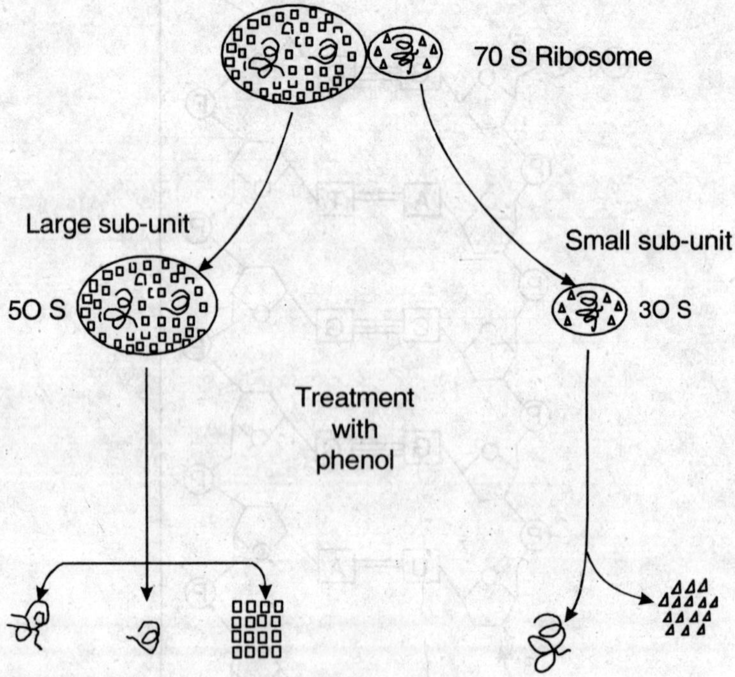

Fig. 5.23 *The building units (rRNA and protein) of 70S ribosome particle formed after treatment with phenol.*

Most of these proteins are basic in nature. It also contains one molecule of RNA, the 16S component. The larger sub-unit (50S) contains about 33 distinct proteins. Evidence suggests that all the proteins are involved functionally or structurally in the translation process. Specific ribosomal proteins are directly involved in binding mRNA and tRNA. rRNA apparently does not participate directly in these binding sites. rRNA evidently serves as a structural polymer, holding the multiprotein particle in a compact configuration.

The two ribosomal subunits have different binding properties. Thus, the *E. coli* 30S subunit binds specific tRNA's. The 50S sub-unit does not associate with mRNA in the absence of the 30S sub-unit, but will nonspecifically bind tRNA. Each 70S ribosome contains two different binding sites for tRNA molecules. Site B (amino acyl site) is involved in the positioning of the specific incoming aminoacyl tRNA with its matching codon on the mRNA. Site A binds the growing polypeptidyl tRNA. The peptidyl transferase which is responsible for the formation of peptide bond, is located on the 50S ribosomal particle. The 50S sub-unit also has a site that hydrolyzes GTP to GDP during the translocation process.

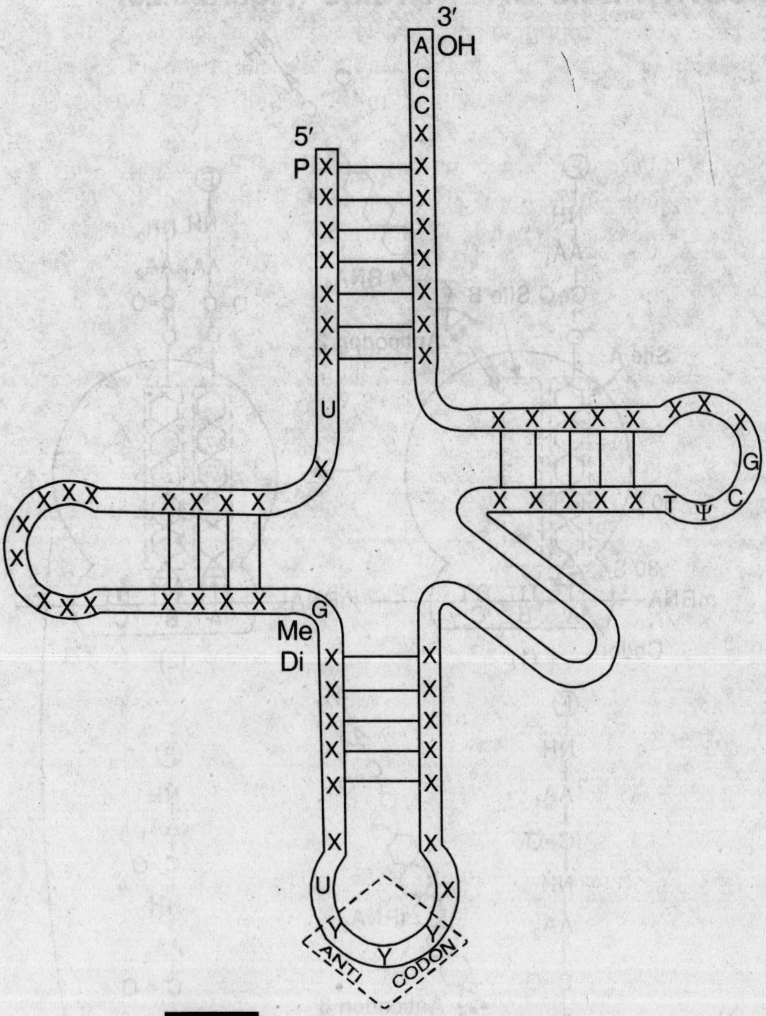

Fig. 5.24 *Clover-leaf conformation of tRNA.*

Transfer RNA

Each bacterial cell contains about 60 different tRNA species. Of these, over 20 tRNAs have had their base sequences determined. Several of them have been crystallized. Chain lengths of known tRNA species vary between 73 and 88 nucleoside residues. The molecular weights range between 25,000 and 30,000. The nomenclature $tRNA^{ala}_{y_{east}}$ signifies a transfer RNA specific for alanine and obtained from yeast (Figure 5:24).

Over 85 per cent of the tRNA's have their 5′ terminus as the base guanine while the remainder have cytosine as the terminal base. All tRNAs that are capable of being charged with amino acids have their 3′ terminal sequence as cytosine-cytosine-adenine.

Transfer-RNA is a single-stranded polymer molecule and measures about 250Å in length. There is believed to be a different t-RNA for each amino acid. The tRNA and amino acid are joined together by the action of a specific enzyme. The tRNA then deposits the given amino acid at a position in the polypeptide chain specified by m-RNA.

12 THE BIOSYNTHESIS OF PROTEINS (Figure 5.25)

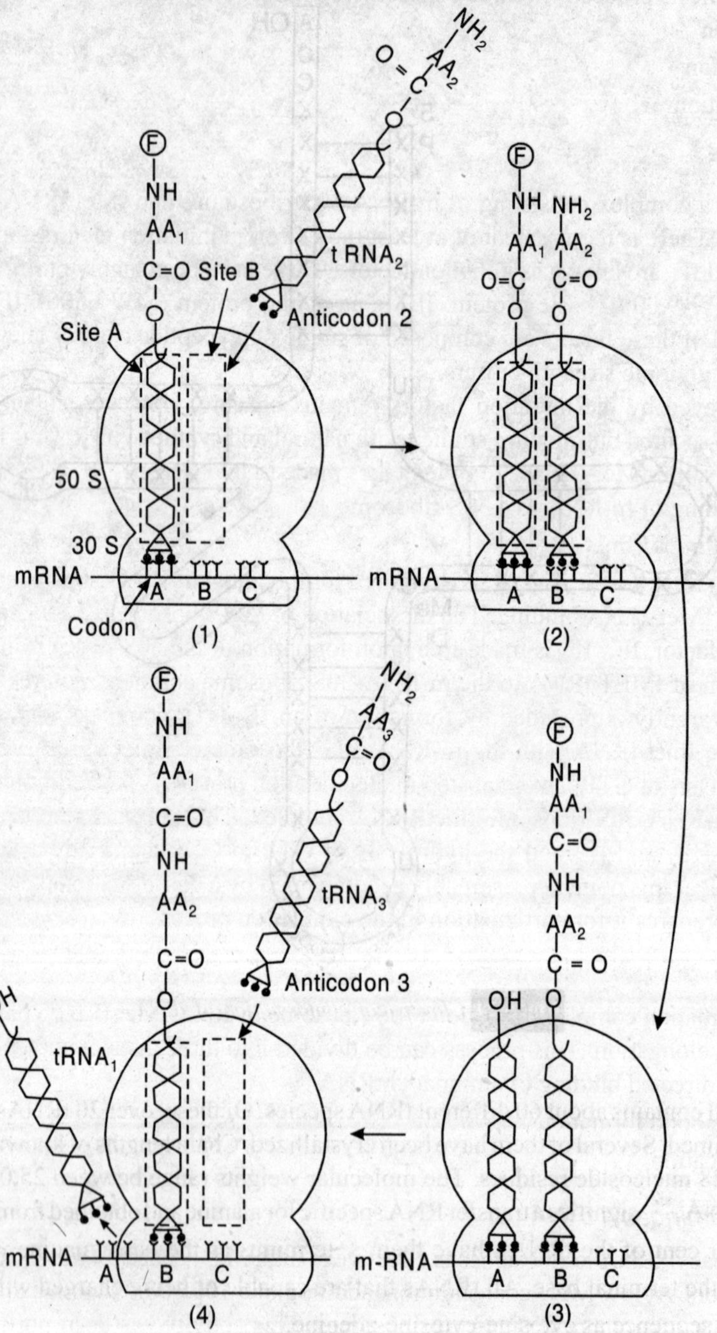

Fig. 5.25 *Four steps in the synthesis of a polypeptide chain: (1) tRNA$_1$ molecule carrying AA$_1$ binds to site. A. The anticodon present on tRNA$_1$ is complementary to codon $_A$. (2) tRNA$_2$ molecule carrying AA$_2$ binds to site B. It has anticodon complementary to codon $_B$. (3) An enzyme complex catalyzes the transfer of AA$_1$ to the amino group of AA$_2$. (4) The ribosome moves to the right, so that site A and B are now opposite codons $_B$ and $_C$. Thus, tRNA$_1$ is discharged and tRNA$_2$s moves to site A. Site B may be occupied by tRNA$_3$ carrying AA$_3$.*

Protein synthesis proceeds in three consecutive phases. They are:

(i) initiation

(ii) elongation

(iii) termination

(i) Initiation

The formation of a complex consisting of m-RNA 30S ribosomes and fMet tRNA is the beginning of protein synthesis. There is requirement of at least three protein initiation factors and the codon AUG or GUG. IF_1, IF_2, and IF_3 are known as initiation factors. These have been highly purified and characterised. IF_1 is a small (MW 9,000), basic protein. IF_2 is an acidic protein (MW 6500). IF_3 is a basic protein (MW 29,000). All of these factors are composed of single poplypeptide chains. They can be eluted from ribosomes with high ionic strength buffers.

The role played by the initiation factors remains obscure. However, some knowledge of their function has been secured during vitro studies with natural and synthetic m-RNAs. The formation of the m-RNA-30S ribosome-fMet--tRNA complex takes place in two steps:

(1) the binding of m-RNA to a 30S ribosome

(2) the binding of fmet-tRNA

It appears as if factors IF_1 and IF_3 are required for the initial m-RNA-ribosome interaction, whereas factor IF_2 directs fMet-tRNA binding. The dissociation of 70S ribosomes to 50S and 30S ribosomes is promoted by the factor, IF_3. IF_3 is made free upon formation of the 30S-m-RNA initiation complex.

The addition of fMET-tRNA to the m-RNA-30S ribosome complex requires factor IF_2 and GTP. This reaction apparently is preceded by formation of an IF_2-GTP complex. This complex directs the association of the fmet-tRNA with the m-RNA-30S ribosome complex. This later reaction does not require the hydrolysis of GTP. The final step in the initiation process is the addition of the 50S ribosome sub-unit to the m-RNA-30S-ribosome-fmet-tRNA complex. This terminal step is accompanied by the release of factors IF_1 and IF_2 and the hydrolysis of GTP to GDP and Pi by a GTPase activity. The hydrolysis of GTP at this point is presumably accompanied by a conformational change in the initiation complex which prepares it for participation in the elongation process.

(ii) Elongation

Completion of initiation complex formation (70S ribosome.m-RNA.fMet-tRNA) prepares the ribosomes for peptide chain elongation. This process can be divided into three distinct and consecutive steps:

(i) codon-directed binding of aminoacyl tRNA

(ii) peptide bond formation

(iii) translocation

(i) Binding of aminoacyl tRNA

Binding of aminoacyl tRNA to ribosomes is effective at high Mg^{2+} concentrations (20mM) in the absence of additional factors. At low Mg^{2+} concentration, addition of a 100,000 g supernatant and GTP stimulates the rate of binding of aminoacyl tRNA to ribosomes. This activity has been purified from *E.coli* and *Bacillus stearothermophilus*. In *E. coli*, it accounts for 2-3 per cent of the soluble protein during logarithmic growth and has been called EF-T. The T designation derives from the role these factors play in transferring aminoacyl-tRNA from the cytoplasm to the m-RNA-ribosome complex.

Upon passage through a DEAC-cellulose column, *E.coli* EF-T is separated into two distinct components: EF-T_S (stable, MW 28,500-35,000), eluted at 0.1 M KCl; and EF-T_U (unstable, MW 42,000) eluted at 0.2 M KCl. More recent determinations of the molecular weights of the E. *coli* components by sedimentation equilibrium yielded 47,000 for the EF-T_U. GDP complex, 34,000 for EF-Ts, and 66,000-68,000 for EF – T_U + T_S, is either native or reconstituted. In *B. stearothermophilus* both proteins are stable and have been called S_1 (MW 37,000) and S_3 (MW 51,000), respectively. EF-T_U has a very high affinity for EF-T_s (K=3×10^{-9}M). The EF-T complex has been crystallized. EF-T_U binds GDP (K = 4 × 10^{-9}M) and GTP with lower affinity (K = 3 × 10^{-7}M). It has been crystallised in the presence of GDP. Addition of GTP to the EF-T complex produces dissociation into its components, EF-T_U and EF-T_S, determined by sucrose gradient centrifugation.

Aminoacyl tRNA binds to the GTP.EF-T_u complex, forming a ternary complex containing GTP.EF-T_U.aa-tRNA. Since aminoacyl tRNA is unable to bind to EF-T_U in the absence of GTP, this experiment illustrates the existence of an EF-T_U.GTP complex. One GTP is bound per aminoacyl tRNA. The EF-T from *E. coli* forms ternary complexes with GTP and all aminoacyl $tRNA_S$ of *E.coli* except the initiator, fMet-tRNA.

What is the function of EF-T_S in this process? It may be somewhat similar, although not identical, to the role of IF_1 in intiation. EF-T_S is bound neither to ribosomes nor to the GTP.EF-T_U.aa-tRNA complex. This complex transfers aminoacyl tRNA to the ribosomes at the same rate and to the same extent in the presence or absence of antibody prepared against EF-T_S. EF-T_S increases the rate of polyphenylalanine synthesis at low EF-T_U concentration or in the presence of GDP.

In the absence of EF-T_S the action of EF-T_U is stoichiometric; in its presence crtalytie. Thus, EF-T_S participates in the recycling of EF-T_U, promoting the formation of EF-T_U. GTP from EF-T_U.GDP in the presence of free GTP, and the release of GDP from EF-T_U.GDP in the absence of GDP, so that EF-T_U can be used for the attachment of another aminoacyl tRNA to the ribosome.

The role of GTP in this process is obscure. However, it is known that 5′-guanylyl-methylenediphosphonate can replace GTP and promote transfer of aminoacyl-tRNA to the protein-synthesizing complex. In this latter case, the analogue and T_U remain bound to the complex and subsequent peptide bond formation is prevented.

(ii) Peptide Bond Formation

Upon binding of the aminoacyl tRNA to the initiation complex, at the B site of the ribosome, peptide bond formation occurs. The activity, called peptidyltransferase, has been localized exclusively on the 50S ribosome. No soluble factors are needed; mRNA and the small sub-unit stimulate transferase activity. The peptide bond formation takes place between the esterified carboxyl group of the formylmethionine or the growing peptide chain of the tRNA at the donor site and the amino group of the amino acid bound by the tRNA at the acceptor site. This process results in occupancy of the donor site (A site) by a discharged tRNA and peptidyl-tRNA at the acceptor site (B site). It should be emphasized that peptide bond formation, at this stage, does not require the expenditure of energy, since the esterified carboxyl group at the donor site is at a sufficiently high energy level to take part in group transfer. This is verified by the observation that isolated 50S ribosomes in the presence of 33 per cent ethanol will catalyze peptide bond formation between aminoacyl-tRNAs in the absence of mRNA, 30S ribosomes, and GTP.

(iii) Translocation

For protein synthesis to continue, translocation has to occur. This involves three processes :

(a) removal of the deacylated tRNA from the A site

(b) movement of the peptidyl tRNA from the B to A site so that a new aminoacyl tRNA can bind

(c) movement of message by one codon

Elongation factor G, EF-G (for GTP ase), and GTP hydrolysis are required for three processes. EF-G (MW 72,000-84,000) from *E. coli* has been crystallised. The molecular weight of EF-G in guanidine.HCl or in SDS is that of the native enzyme. EF-G comprises 2 per cent of the soluble *E.coli* proteins during logarithimic growth.

Highly purified preparations of EF-G do not bind GTP and do not by themselves manifest GTPase activity. For this to occur, ribosomes (70S or 50S) are absolutely required. No activity is detected with 30S sub-units alone, although they stimulate the activity of 50S. Aminoacyl tRNA and m-RNA are not required. Under these conditions, the activity has been called 'uncoupled GTPase'.

Free GTP is necessary for translocation; GTP from the GTP.EF-Tu.aa tRNA complex cannot be used in the reaction. During elongation, two molecules of GTP are hydrolyzed for each peptide bond formed.

Termination

The termination step consists of two processes :

(i) the recognition of a termination signal in the m-RNA

(ii) the hydrolysis of the final peptidyl tRNA ester linkage to make free the nascent protein

The termination condons are UAA, UAG and UGA. This step requires three protein factors. They are R_1, R_2, and R_3. The R_1 factor is required for the recognition of the condons UAA and UAG. The R_2 is required for the recognition of UAA and UGA. The R_3 factor does not take part in the release activity. It seems that the R_3 factor is helpful in terminator codon recognition. The emerging picture indicates that the termination reaction can be divided into two events:

(a) a terminator codon-dependent R_1 or R_2 factor binding reaction

(b) a hydrolytic reaction, in which, either R_1 or R_2 converts the peptidyl transferase activity at site A into a hydrolytic reaction. This hydrolytic reaction brings about the transfer of the peptidyl tRNA to water, rather than to another amino acyl tRNA. A final factor, TR, may be required to discharge the residual tRNA from site A. The 70S ribosome undergoes dissociation from the m-RNA into a 30S and a 50S sub-unit

The nascent protein, presumably, has a formyl methionyl NH_2–terminus. It must be removed before the protein completes its folding sequence in the final stage, two enzymes may take part in the following two reactions.

(i) a specific deformylase :

Formyl methionyl peptide $\longrightarrow$ Formic acid + Methionyl peptide

(ii) a specific aminopeptidase :

Methionyl peptide $\longrightarrow$ Methionine + Peptide

(C) METABOLISM OF NUCLEIC ACIDS

There are two types of nucleic acids, namely, DNA and RNA. These molecules are very important for any living system.

Catabolism

Nucleic acids may be hydrolysed by DNAase and RNAase. Degradation leads to the formation of free nucleoside monophosphates.

Anabolism

Purine and Pyrimidine nucleoside triphosphates are the building blocks of nucleic acids. They have the same general structure as shown in Figure 5.26.

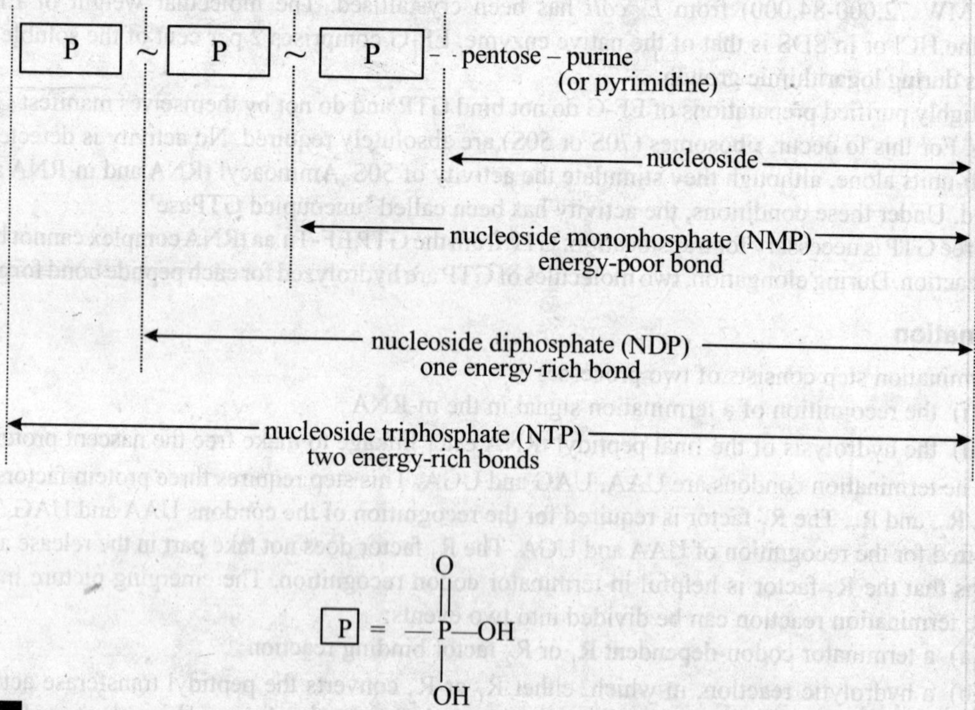

Fig. 5.26 *The general structure of nucleoside triphosphates. A curve bond (~) indicates high-energy phosphate bonds. A straight bond indicates low-energy phosphate bonds.*

A Purine or Pyrimidine base is attached through nitrogen atoms to a pentose. Linking of phosphate groups occurs to the 5′ position of the nucleoside.

Specific nucleosides are shown in Figure 5.27 with their names and structures. Nucleotides are symbolized by letters, A, G, U, C, or T, to indicate the purine or pyrimidine base, present in them. MP, DP, or TP indicates whether they are mono-, di- or triphosphates. Letter d indicates deoxynucleotides (e.g. dGTP= 2′ deoxyguanosine triphosphate). In a DNA molecule, there are two purine (dATP and dGTP) and two pyrimidine (dCTP and dTTP) nucleoside triphophates as its building blocks. Similarly, in RNA molecules, the two purine (ATP and GTP) and two pyrimidine (CTP and UTP) nucleoside triphosphates are present as its precursor molecules. The formation of deoxyribonucleotides occurs by the reduction of the corresponding ribonucleotides.

Synthesis of ribonucleotides

The first step in the synthesis of ribonucleotides is the formation of 5-phosphoribosyl-1-pyrophosphate (PRPP). PRPP is synthesized from ribose-5-phosphate and ATP as under:

$$\text{Ribose-5-phosphate + ATP} \xrightarrow[\text{synthetase}]{\text{PRPP}} \text{PRPP + AMP}$$

BASE		RIBONUCLEOSIDES		2′ DEOXYRIBONUCLEOSIDES	
Name	*Base structure*	*Pentose structure*	*Name*	*Pentose structure*	*Name*
Purines adenine	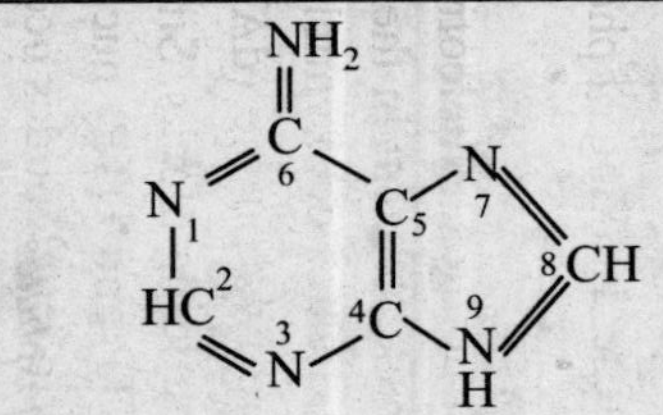		adenosine		2′-deoxyadenosine
Purines guanine	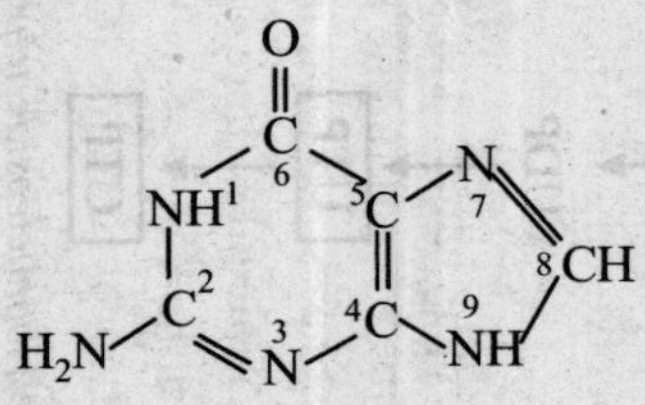	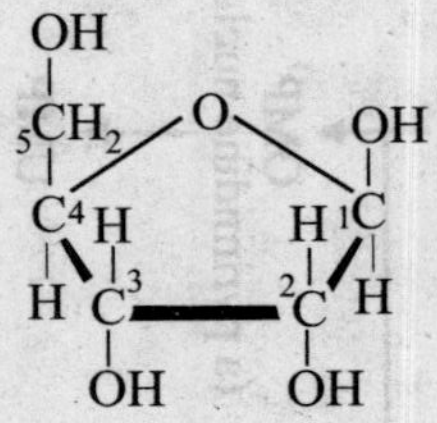 ribose	guanosine	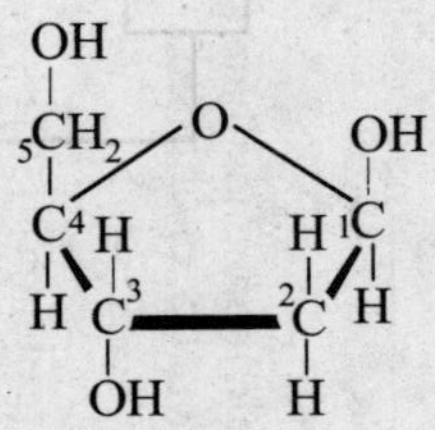 2′-deoxyribose	2′-deoxyadenosine
Pyrimidines uracil	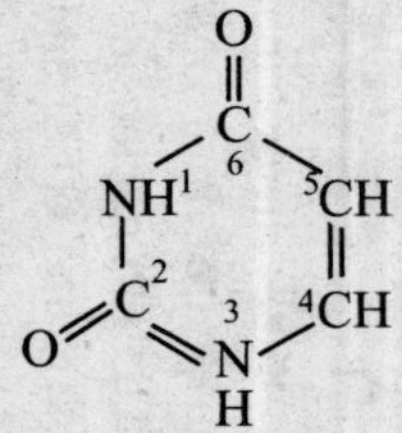		uridine		———
Pyrimidines cytosine	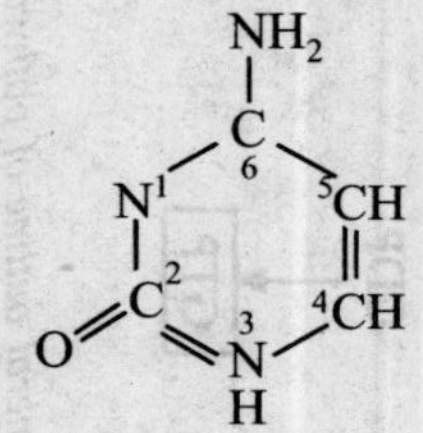		cytidine		2′-deoxycytidine
Pyrimidines thymine	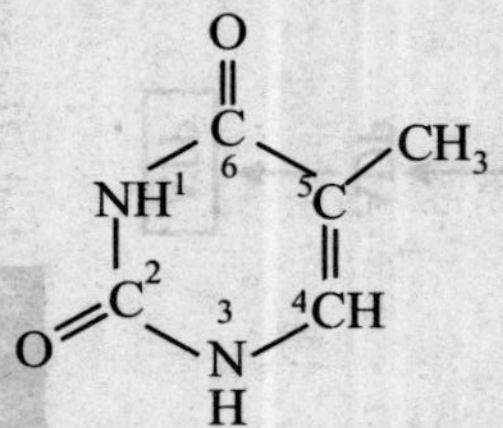		———		2′-deoxythymidine

Fig. 5.27 *Names and composition of nucleoside triphosphates. Purines at the 9 position, and pyrimidines at the 3 position. They are attached to the 1 position of pentoses to form nucleosides.*

The source of ribose-5-phosphate is the hexose monophosphate shunt. PRPP is the starting point of the pathway of bio-synthesis of the purine ribonucleotides. There occurs gradual incorporation of amino groups and small carbon-containing groups for the biosynthesis of the nine-membered purine ring as shown in Figure 5.28 (left). All intermediates of the pathway are ribonucleotides.

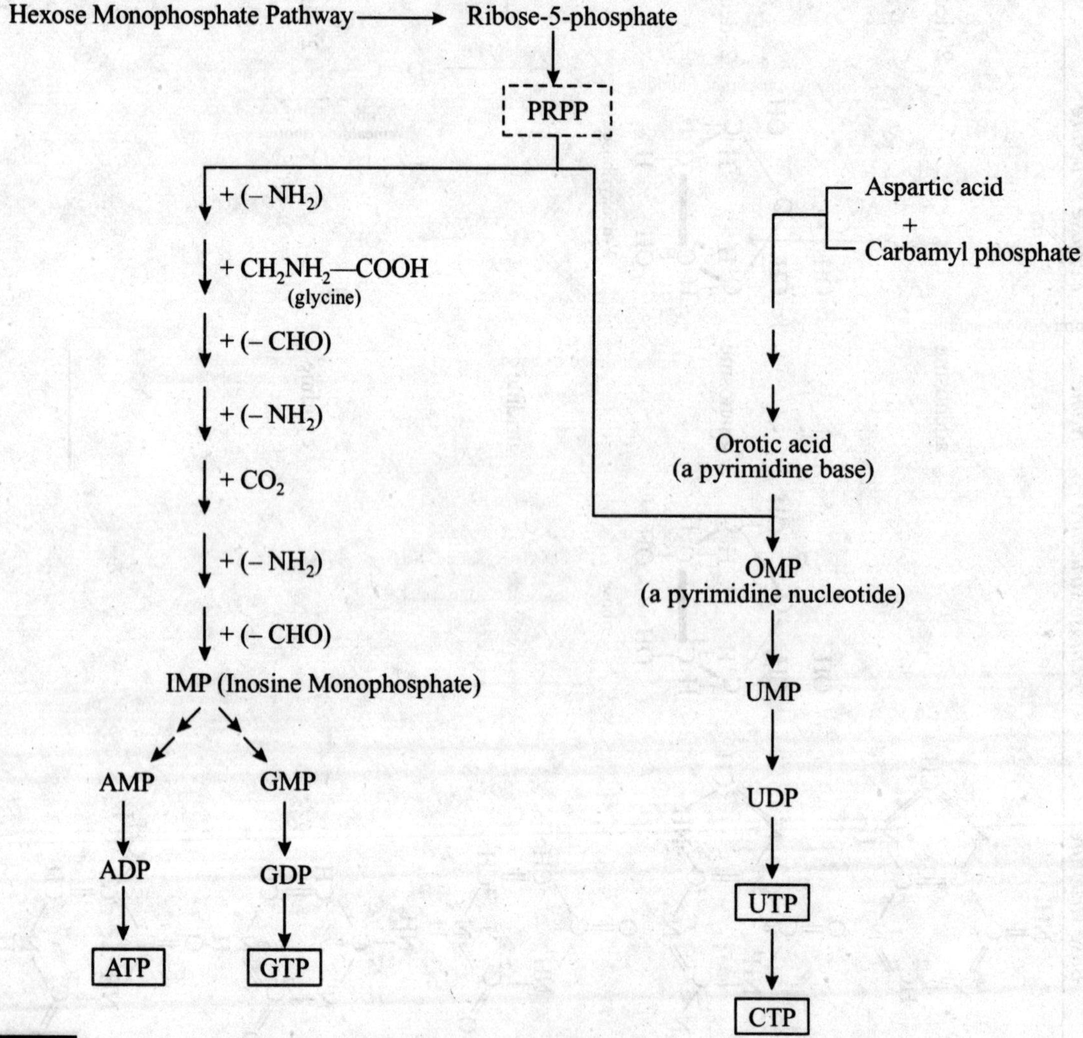

Fig. 5.28 *Left: The general outline of pathway of synthesis of purine ribonucleoside triphosphates. Right: The general outline of the pathway of synthesis of pyrimidine ribonucleoside triphosphates.*

In constrast to this, incorporation of the ribose-phosphate moiety of the pyrimidine ribonucleotides occurs only after the six-membered ring has been completely formed. The condensation between aspartic acid and carbamyl phosphate leads to the formation of the six-membered ring, as shown in Figure 5.28 (right).

Biosynthesis of all nucleoside triphosphates occurs from the corresponding nucleoside monophosphates except CTP.

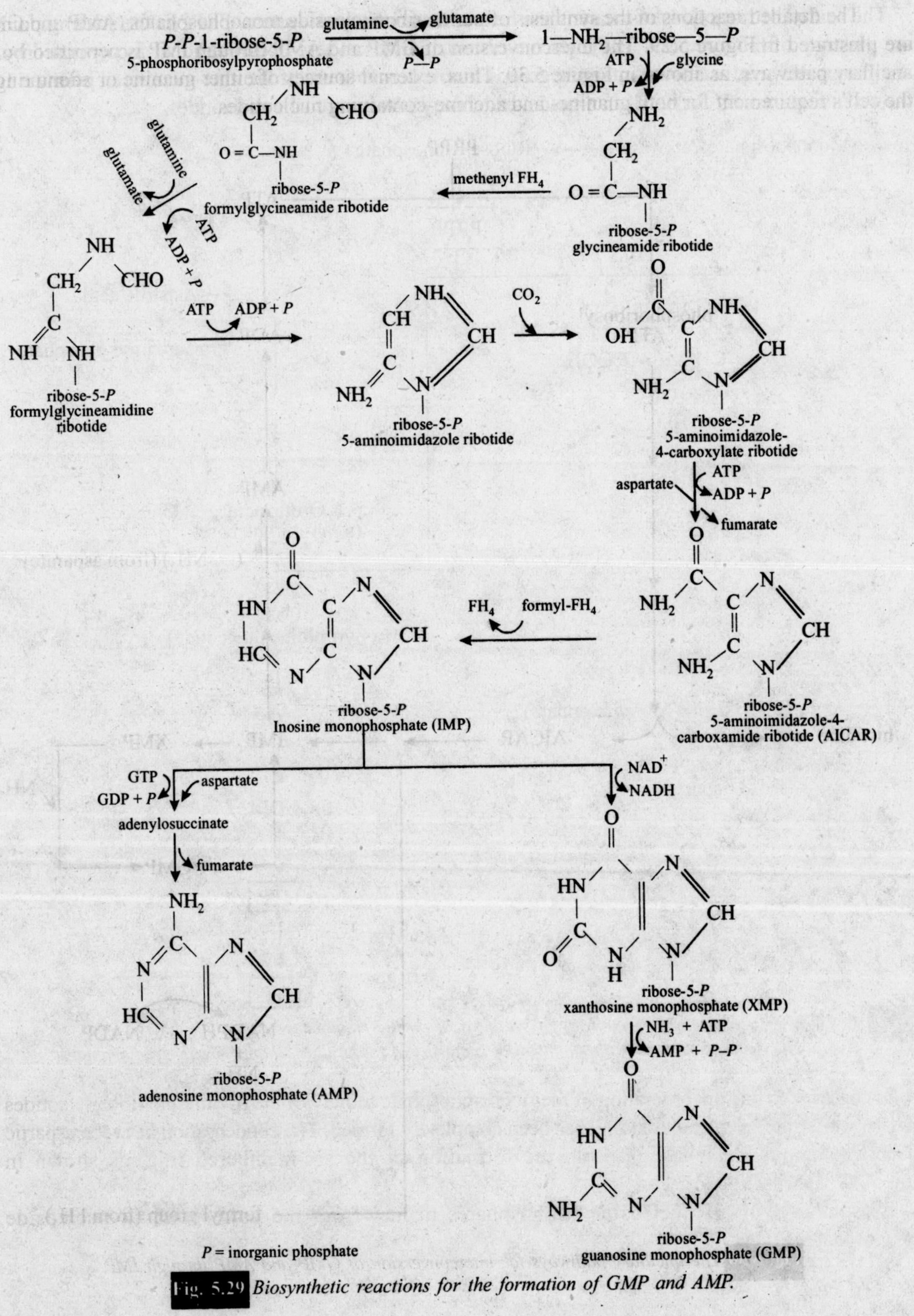

Fig. 5.29 *Biosynthetic reactions for the formation of GMP and AMP.*

The detailed reactions of the synthesis of purine ribonucleoside monophosphates (AMP and GMP) are illustrated in Figure 5.29. The interconversion of GMP and AMP through IMP is permitted by the ancillary pathways, as shown in Figure 5.30. Thus, external sources of either guanine or adenine fulfil the cell's requirement for both guanine- and adenine-containing nucleotides.

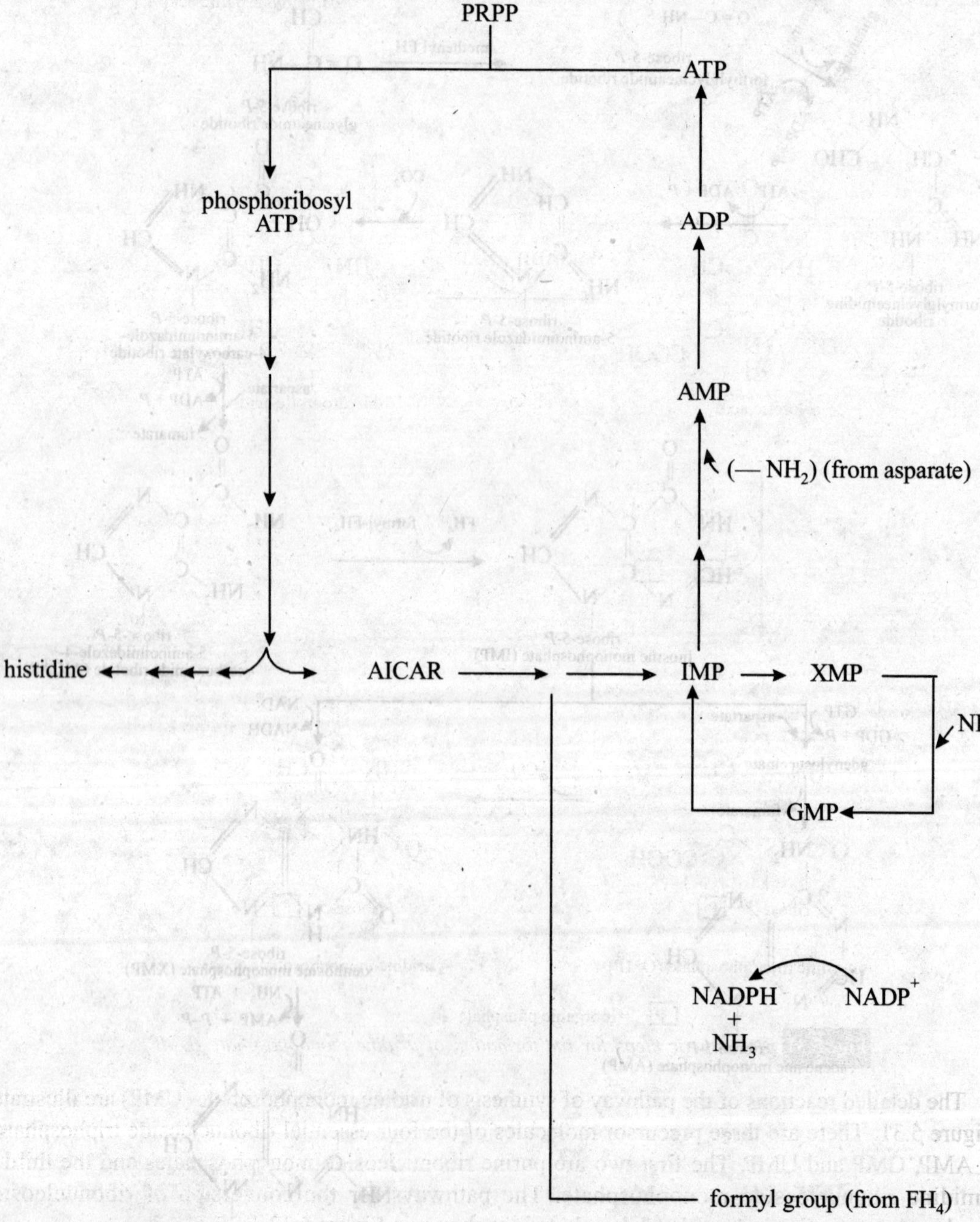

Fig. 5.30 *The ancillary pathways for interconversion of GMP and AMP through IMP.*

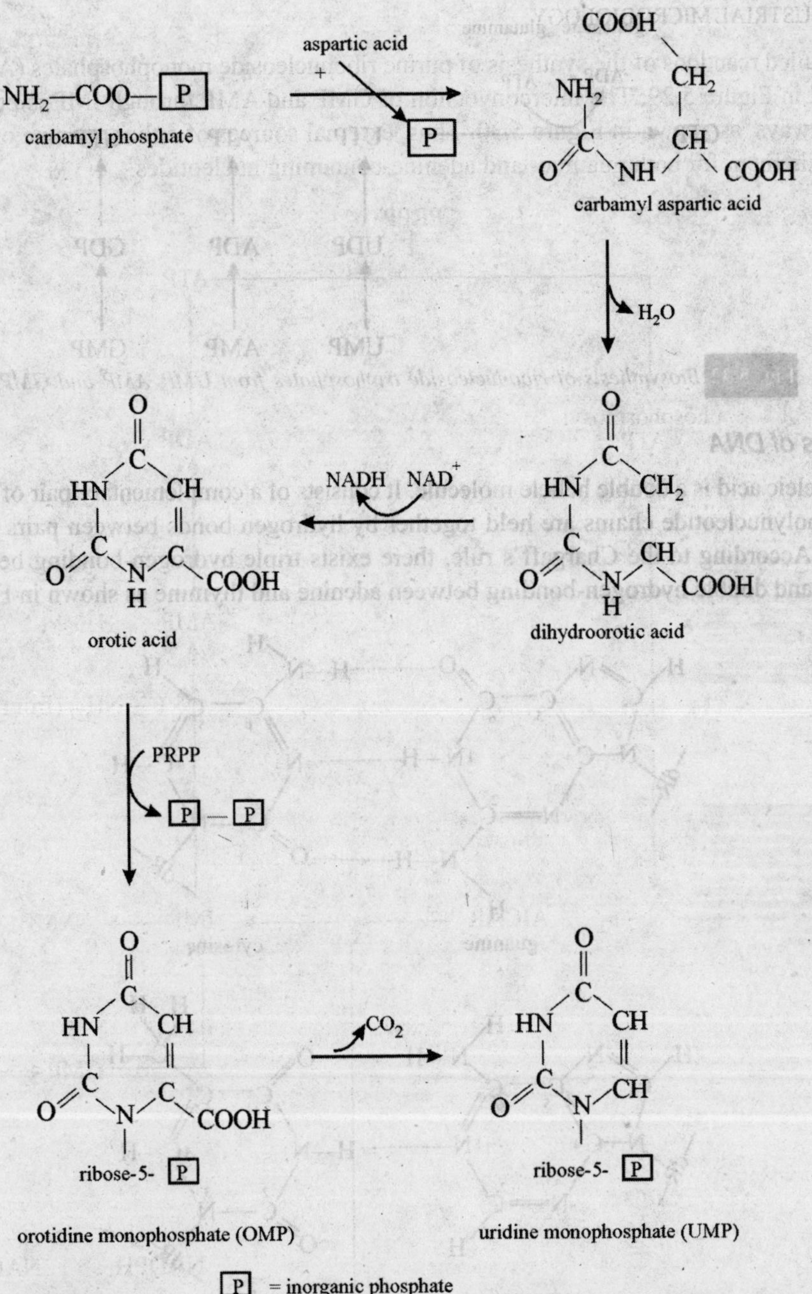

Fig. 5.31 *Biosynthetic steps for the formation of uridine monophosphate (UMP).*

The detailed reactions of the pathway of synthesis of uridine monophosphate (UMP) are illustrated in Figure 5.31. There are three precursor molecules of the four essential ribonucleoside triphosphates, viz., AMP, GMP and UMP. The first two are purine ribonucleoside monophosphates and the third is pyrimidine ribonucleoside monophosphate. The pathways for the conversion of ribonucleoside monophosphates to ribonucleoside triphosphates are shown in Figure 5.32.

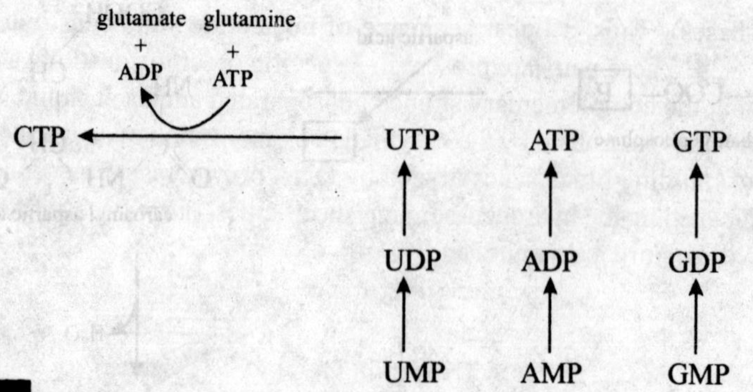

Fig. 5.32 *Biosynthesis of ribonucleoside triphosphates from UMP, AMP and GMP.*

Biosynthesis of DNA

Deoxyribonucleic acid is a double helicle molecule. It consists of a complementary pair of polynucleotide chains. The polynucleotide chains are held together by hydrogen bonds between pairs of purines and pyrimidines. According to the Chargaff's rule, there exists triple hydrogen-bonding between guanine and cytosine and double hydrogen-bonding between adenine and thymine as shown in Figure 5.33.

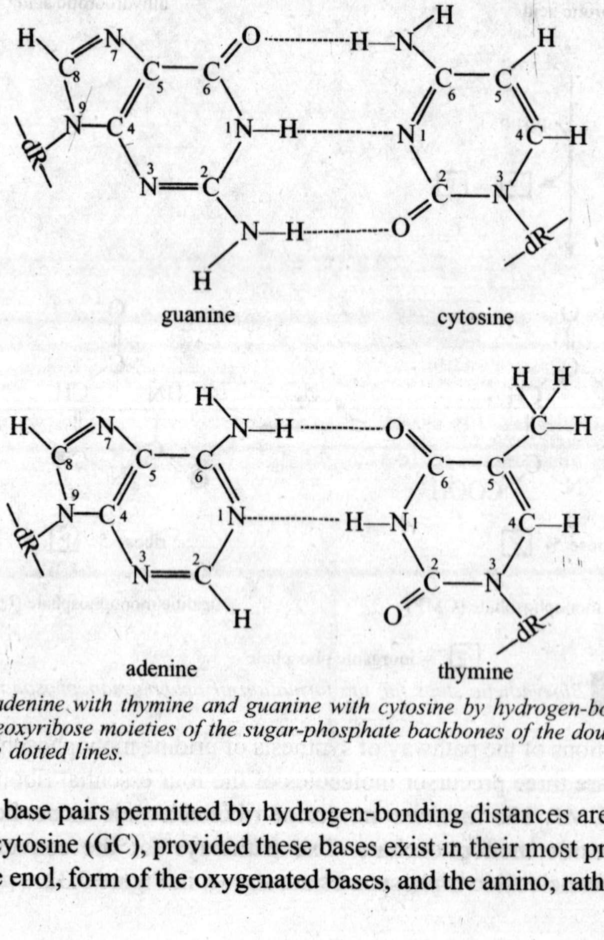

Fig. 5.33 *The pairing of adenine with thymine and guanine with cytosine by hydrogen-bonding. The symbol—dR— represents the deoxyribose moieties of the sugar-phosphate backbones of the double helix. Hydrogen bonds are indicated by dotted lines.*

The only possible base pairs permitted by hydrogen-bonding distances are adenine with thymine (AT) and guanine with cytosine (GC), provided these bases exist in their most probable energetic forms (the keto, rather than the enol, form of the oxygenated bases, and the amino, rather than the imino, form of the aminated bases). Thus, a linear sequence of nucleotide pairs constitutes the DNA molecule as shown in Figure 5.34. These pairs are present in a specific order to constitute a genetic message. It is an established fact that the complementary strands get separated and each strand acts as a template. A new complementary strand is assembled by the enzymatic polymerization of deoxyribonucleoside triphosphates. Polymerization of building blocks is catalyzed by DNA polymerase. The sequence of bases in the new strand is rigidly dictated by the hydrogen-bonding specificities, (i.e. whenever the template carries adenine the new strand will acquire a thymine, and so on).

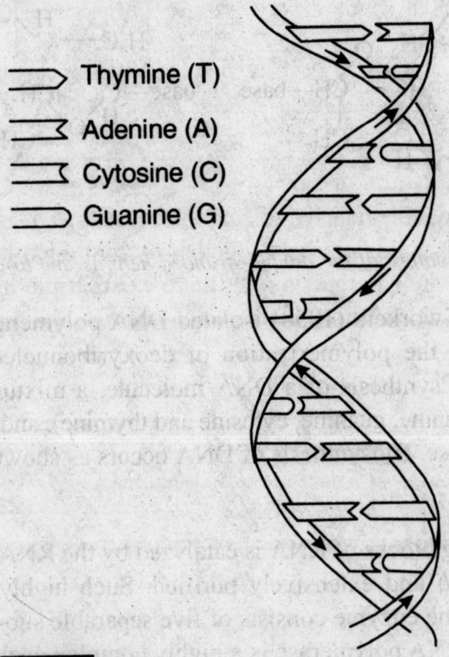

Fig. 5.34 *The double-stranded helicle structure of DNA.*

Each strand of the double helix is a polarized structure. This polarity arises due to the sequential linkage of polarized subunits, the deoxyribonucleotides. Each nucleotide has a 5′–phosphate end and a 3′–hydroxyl end as illustrated in Figure 5.35. When a series of nucleotides are linked through phosphodiester

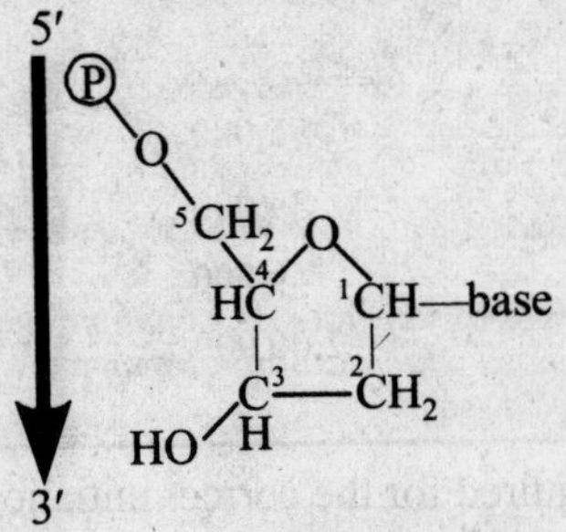

Fig. 5.35 *A deoxyribonucleotide. The molecule is polarized, having a 3′-hydroxyl group at one end and a 5′-phosphate group at the other.*

bonds, the resulting chain is polarized with a 5′-phosphate end and a 3′-hydroxyl end. The nature of the double helix is antiparallel. In other words, the two strands run in opposite directions as shown in Figure 5.36, (i.e. the left-hand base runs from 5′ to 3′ reading from top to bottom, while the right-hand base runs from 3′ to 5′).

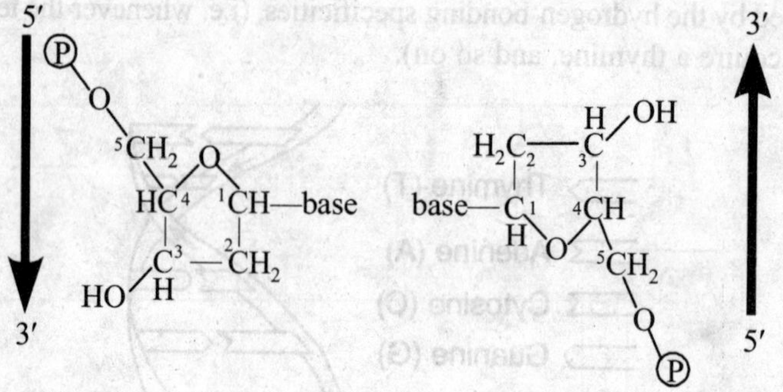

Fig. 5.36 *The antiparallel nature of the double-stranded helix is due to opposite directions of bases.*

A. Kornberg and his co-workers (1958) isolated DNA polymerase from the extracts of *E. coli* cells. This enzyme catalyzes the polymerization of deoxyribonucleoside triphosphates to form a polynucleotide chain. For the synthesis of a DNA molecule, a mixture of the four triphosphates as building blocks (containing adenine, guanine, cytosine and thymine), and some primer DNA are required along with the DNA polymerase. Biosynthesis of DNA occurs as shown in Figure 5.37.

Biosynthesis of RNA

The polymerization of building blocks of RNA is catalyzed by the RNA polymerase. RNA polymerase has been isolated from *E. coli* and extensively purified. Such highly purified preparation of RNA polymerase has revealed that the enzyme consists of five separable sub-units in the porportions shown in Table 5.5. In other words, RNA polymerase is a highly complex enzyme.

Table 5.5 *Sub-units of E. coli RNA polymerase.*

Sr. No.	*Sub-unit*	*Molecular weight*	*Number*	*Function*
1.	β′	1,65,000	1	DNA binding
2.	β	1,55,000	1	initiation and site catalytic
3.	σ	95,000	1	initiation
4.	α	39,000	2	not known
5.	ω	9,000	1	not known
6.		2,00,000		termination factor

The sigma factor (σ) is required for the correct initiation of RNA synthesis at a specific site of the DNA template. Without the σ protein, the enzyme is called the core polymerase; with the σ factor, it is called the holoenzyme.

A polynucleotide strand is synthesized, which is complementary to one of the two strands of DNA (Figure 5.22). This process is called **transcription**. The building blocks are linked together by 5′ ⟶ 3′ phosphodiester linkages. Thus, linking of ribonucleoside triphosphates is analogous to DNA.

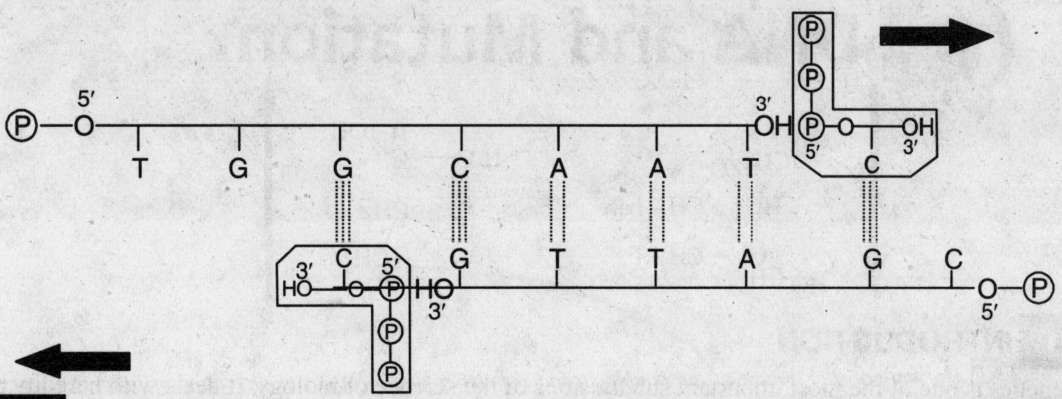

Fig. 5.37 *Diagram showing a short fragment of double-stranded DNA, with single-stranded regions at each end. The polynucleotides are in the process of being lengthened by the action DNA polymerase, which catalyzes the addition of deoxyribonucleotide triphosphates to the 3′-hydroxyl ends of the chains. The triphosphate that is added is governed by the base in the opposite chain, which thus serves as template. The arrows show the direction of sequential addition of deoxyribonucleotides; inorganic pyrophosphate (P—P) is split off in the process.*

The synthesis of RNA is also possible outside the cell, (i.e. in vitro). This requires the presence of the ribonucleoside triphosphates of adenine, guanine, cytosine and uracil, as well as a DNA template, along with the RNA polymerase. If the fragments of the double-stranded DNA are present as template, the synthesis of RNA molecules complementary to both strands of the DNA molecule occurs. When circular double-stranded DNA, (e.g. replicative form of the single-stranded DNA virus, ϕX174) is present as template, only one of the two strands is synthesized. The copying of one strand of DNA is called *asymmetrical transcription* whereas copying of both template strands of DNA is termed as *symmetrical transcription*. The core enzyme will transcribe a DNA template symmetrically (both strands of DNA can serve as templates). However, the transcription process is slow and non-specific. With the σ factor, the holoenzyme transcribes DNA asymmetrically, initiating the RNA chains at specific promotor sites. Thus, it seems that in fact DNA possesses specific sites to begin the RNA synthesis. These sites orientate the polymerase and produce copies of one specific strand. It seems that artificial initiation sites are created through the fragmentation of DNA molecule. Thus, both strands of DNA are copied in the presence of DNA fragments.

There are three types of RNAs namely; ribosomal RNA (rRNA), transfer RNA (tRNA) and messenger RNA (m-RNA). All these RNAs are produced by the process of **transcription**. They are, directly or indirectly, important in protein synthesis. The amino acid sequence in a polypeptide chain is decided by the m-RNA. This, in turn, is decided by the sequence of bases in m-RNA molecule. This process is called **translation**. Amino acids scattered in the cytoplasm are transferred to the protein-syntesizing site, (i.e. ribosomes) by soluble RNA molecules (also known as tRNA or adaptor RNA). Ribosomal RNA and ribosomal protein constitute the ribosomal particle.

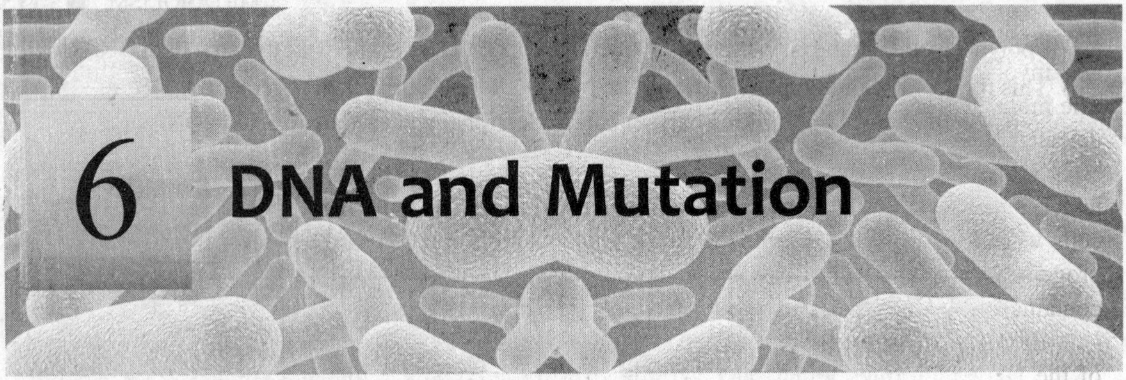

6 DNA and Mutation

1 INTRODUCTION

Genetics is one of the most important subdivisions of the science of biology. It deals with heredity and variation among related organisms. The term 'genetics' was proposed by Bateson in 1906. The word 'genetics' has been derived from the Greek root, 'genesis', meaning origin, and when literally applied it refers to the origin of the various kinds of living forms that populate the globe. Genetics deals not only with the way in which characteristics are transmitted from one generation to the next, but also with the actions of the units of heredity.

The basic principles of heredity were discovered by an Austrian monk, Gregor Mendel in 1865. He performed his experiments on garden peas (*Pisum sativum*). The results of his experiments received no recognition from contemporary biologists. It was at the drawn of this century that three independent investigators—de Vries in Holland, Correns in Germany and Tschermak in Austria rediscovered Mendel's laws of heredity which form the foundation of modern genetics. Prior to the time of Mendel, the characteristics of parents were believed to blend in the offsprings. Mendel was the first to visualize and demonstrate that the heredity was transmitted from parents to offsprings as individual units, which maintained their integrity from generation to generation. Mendel called these hypothetical units of heredity as "factors". They have later been called as "genes" by the Danish botanist, named Wilhelm Johannsen.

The term "gene", has been used to designate the submicroscopic, intracellular determiners of the inherited characteristics of an organism. Experiments have revealed that the genes are arranged in linear sequence along the length of the chromosomes. The sequence in which genes are arranged tends to be the same for all the individuals of a particular species. It has been estimated that there are several thousand genes within one chromosome. This makes genes very small and it is not possible to see them even under the highest powers of the microscope. It was not possible to isolate and examine them physically, chemically or in any other way. However, since 1969-70 it has been made possible to isolate and synthesize the gene.

Genes are the entities responsible for the development of the adult traits which characterize the existing organisms. A gene does occupy a definite position within a chromosome. This is referred to as its "locus". The family of elements or units which are always found at the same locus are known as alleles or allelomorphs or allelic. There are many thousands of these different loci.

DNA Replication

The mechanism of DNA replication was proposed by J.D. Watson and F.H.C. Crick in 1953. It has been supported by a large number of experiments with bacteria, algae and other organisms. Such experiments also involved the use of isotopes. The commonly used isotope was ^{15}N (a heavier species of nitrogen). The molecules of DNA containing ^{15}N are denser and heavier and this difference in density could be detected by centrifugation.

Experiments using ^{15}N labelled DNA were carried out in 1958 by M. Meselson and F.W. Stahl, undergraduates at California Institute of Technology. They used *E. coli* cells in their experiments. The cells were bred for several generations on a medium in which the nitrogen was present as ^{15}N. The progeny cells had DNA fully labelled with ^{15}N. Such cells were then transferred to a medium that contained only ^{15}N as a nitrogen source.

If DNA replication involved strand separation, definite predictions could be made about the density of the DNA molecules found after various growth intervals in a medium containing ^{14}N. After one generation of growth, all the DNA molecules should contain one heavy (^{15}N) and one light (^{14}N) strand of intermediate hybrid density. Experiments demonstrated exactly the same compositions. Similarly, after two generations of growth, half the DNA molecules are light and half hybrid, as strand separation predicts (Figure 6.1). The conclusion has, therefore, been drawn that the DNA replication is *semi-conservative*.

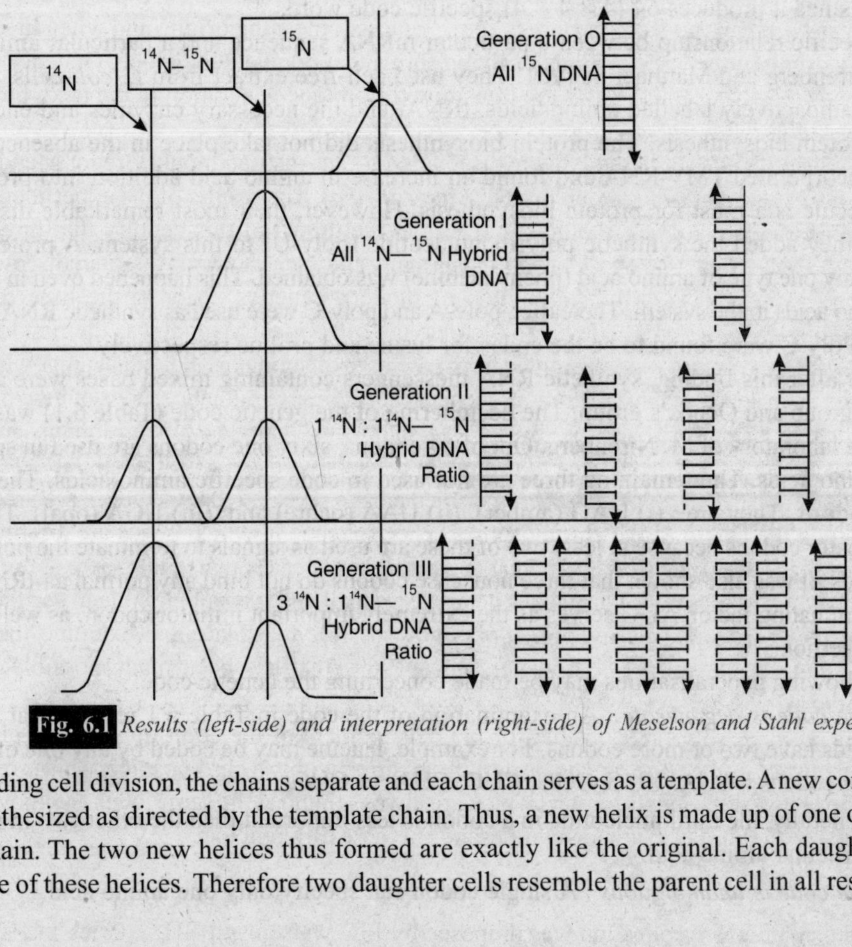

Fig. 6.1 *Results (left-side) and interpretation (right-side) of Meselson and Stahl experiment.*

Preceding cell division, the chains separate and each chain serves as a template. A new complementary chain is synthesized as directed by the template chain. Thus, a new helix is made up of one old chain and one new chain. The two new helices thus formed are exactly like the original. Each daughter cell then receives one of these helices. Therefore two daughter cells resemble the parent cell in all respects.

Cairns in 1963 presented an important evidence on this subject. With the help of radioactive labelling experiment, he has shown that the *E. coli* chromosome is double stranded and circular (> 1000 μ in circumference) and consists of only a single piece of DNA. It has also been shown that the two circular component strands split during replication, with each strand duplicating individually. The fact that this replication produces a Y type of joint, indicates that unwinding of the two complementary DNA strands is not completed before replication begins. But strand-separation proceeds simultaneously with the replication.

2 THE GENETIC CODE

The specification of amino acids during protein biosynthesis is made by the code words. These code words present in the mRNA molecule are called the *genetic codes*.

The genetic code is a dictionary. There are 20 different amino acids which may take part in constituting a protein molecule. Therefore, there must be at least 20 code words which comprise the genetic code. Since there are only four different nucleotides in mRNA, each code word must consist of more than a single purine or pyrimidine nucleotide. Code words consisting of 2 nucleotides each, could provide for only 16 (4 × 4) specific code words. Code words consisting of 3 nucleotides each seem more likely, since it produces 64 (4 × 4 × 4) specific code words.

The specific relationship between a particular mRNA sequence and a particular amino acid was shown by Nirenberg and Matthaei in 1961. They used cell-free extract from *E. coli* cells. It contained ribosomes, radioactively labelled amino acids, tRNA, and the necessary enzymes and energy sources (ATP) for protein biosynthesis. The protein biosynthesis did not take place in the absence of mRNA. They then incorporated TMV-RNA and found an increase in amino acid addition into proteins. Thus, mRNA molecule is a must for protein biosynthesis. However, their most remarkable discovery took place when they added the synthetic polyribonucleotide (poly-U) to this system. A protein molecule containing only one type of amino acid (phenylalanine) was obtained. This happened even in the presence of other amino acids in the system. Thereafter, poly-A and poly-C were used as synthetic RNA messengers. Poly-A and Poly-C were found to be the codes for lysine and proline respectively.

Shortly after this finding, synthetic RNA messengers containing mixed bases were also used by Nirenberg's group and Ochoa's group. The deciphering of the genetic code (Table 6.1) was carried out largely in the laboratory of M. Nirenberg. Out of 64 codons, sixty one codons are used in specifying 20 standard amino acids. The remaining three are not used to code specific amino acids. They are called **nonsense codons**. They are : (i) UAG (amber), (ii) UAA (ochre) and, (iii) UGA (opal). They are also called terminator codons because at least two of these are used as signals to terminate the polymerisation of amino acids. It was also shown that three nonsense codons do not bind any normal aa-tRNA, but they do bind a terminating factor. AUG serves as the extremely important initiator codon, as well as, internal methionyl insertion.

The following generalisations may be made concerning the genetic code:

(1) *The code is degenerate* : An examination of the code in Table 6.1 reveals that most amino acids have two or more codons. For example, leucine may be coded by any one of the following codons : UUA, UUG, CUU, CUC, CUA, or CUG.

Generally, the third nucleotide in a codon is less important than the other two in specifying a particular amino acid.

(2) *The code is unambiguous* : A single codon can specify only one amino acid.

Table 6.1 *The genetic code.*

THE CODE
SECOND BASE OF TRIPLET CODE

FIRST BASE OF TRIPLET CODE	URACIL		CYTOSINE		ADENINE		GUANINE		THIRD BASE OF TRIPLET CODE
URACIL	Phenylalanine UUU	Phe	Serine UCU	Ser	Tyrosine UAU	Tyr	Cysteine UGU	Cys	URACIL
	Phenylalanine UUC		Serine UCC		Tyrosine UAC		Cysteine UGC		CYTOSINE
	Leucine UUA	Leu	Serine UCA		UAA →	Ochre	UGA →	Opal	ADENINE
	Leucine UUG		Serine UCG		UAG →	Amber	Tryptophan UGG	Tyr	GUANINE
CYTOSINE	Leucine CUU	Leu	Proline CCU	Pro	Histidine CAU	His	Arginine CGU	Arg	URACIL
	Leucine CUC		Proline CCC		Histidine CAC		Arginine CGC		CYTOSINE
	Leucine CUA		Proline CCA		Glutamine CAA	Glu-N	Arginine CGA		ADENINE
	Leucine CUG		Proline CCG		Glutamine CAG		Arginine CGG		GUANINE
ADENINE	Isoleucine AUU	Ileu	Threonine ACU	Thr	Asparagine AAU	Asp-N	Serine AGU	Ser	URACIL
	Isoleucine AUC		Threonine ACC		Asparagine AAC		Serine AGC		CYTOSINE
	Isoleucine AUA		Threonine ACA		Lysine AAA	Lys	Arginine AGA	Arg	ADENINE
	Methionine AUG	Met	Threonine ACG		Lysine AAG		Arginine AGG		GUANINE
GUANINE	Valine GUU	Val	Alanine GCU	Ala	Aspartic Acid GAU	Asp	Glycine GGU	Gly	URACIL
	Valine GUC		Alanine GCC		Aspartic Acid GAC		Glycine GGC		CYTOSINE
	Valine GUA		Alanine GCA		Glutamic Acid GAA	Glu	Glycine GGA		ADENINE
	Valine GUG		Alanine GCG		Glutamic Acid GAG		Glycine GGG		GUANINE

The unambiguous but degenerate code can be characterised in molecular terms. The recognition of specific codons in the mRNA by the tRNA molecules is dependent upon their anticodon region and the base pairing rules. Each tRNA has a specific sequence and it is complementary to a codon. For a given codon in the mRNA, only a single species of tRNA molecule possesses the proper anticodon. As each tRNA molecule can be charged with only one specific amino acid, each codon, therefore, specifies only one amino acid. However, some tRNA molecules can use the anticodon to recognise more than one codon.

(3) *The code is non-overlapping*: The reading of the genetic code does not involve any overlapping of codons. The following Figure 6.2 shows overlapping and non-overlapping of the codon.

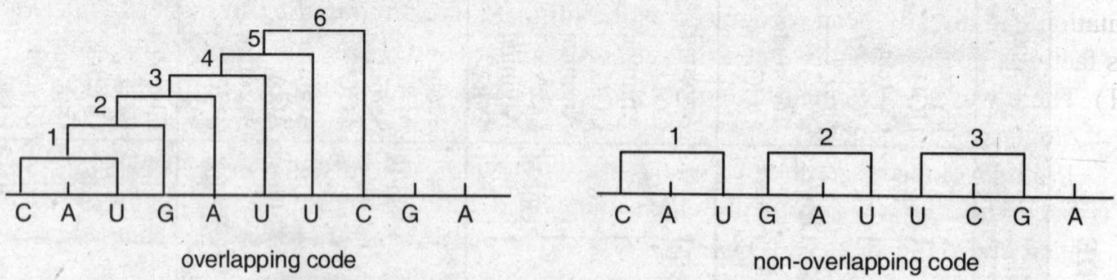

Fig. 6.2 *Overlapping code and non-overlapping code.*

(4) *The code is commaless* : Once the reading is started at a specific codon, there is no punctuation between codons. The message is read in a continued sequence of nucleotide triplets until a terminator codon is reached.

(5) *The code is universal* : All procaryotes and eucaryotes use the same codons to specify each amino acid.

(6) *The code is colinear* : Code words are arranged linearly in a single-stranded RNA molecule.

3 EARLY CONCEPTS OF BACTERIAL VARIATION

In the early days of bacteriology, bacteria were assumed to have capability for unlimited morphological and physiological alterations. This concept was called the doctrine of pleomorphism. Some proponents of this concept considered that very few bacteria were present in nature. The diverse morphological forms represented different developmental stages of a few bacteria. The concept of pleomorphism was discarded after the invention of pure culture technique by Robert Koch. He demonstrated that each type of bacterium breeds true when grown in pure culture. Thus, a new concept of monomorphism became predominant. According to this concept, each bacterium possesses constant morphological and physiological properties.

In the early period of twentieth century, it was known that a certain amount of variation always occurred in the pure bacterial cultures grown under controlled conditions. There were two types of variations to be considered here :

(A) Adaptation

The characteristics of a bacterial culture are greatly affected by the environmental conditions. In other words, they adapt themselves to the new environment of increased fitness. For example, bacteria grown for many generations in the presence of drugs may develop drug-resistance.

(B) Dissociation

This type of variation was first noted with virulent pathogens. Upon primary isolation from infected animal host, they produce typical colonies. On the very first plate, colonies are smooth and glistening with even margins. These colonial characteristics are attributable to the presence of specific soluble substance present on the surface layers of the bacterial cells. Thus, environment of the animal body is fit for the capsule formation. Upon further serial transfer on laboratory media, first a mixture of smooth and rough colonies, and finally only rough colonies appear. The colonial features of rough colonies are wrinkled and dull with uneven margins. On the laboratory medium, cells lose the ability to synthesize the capsular material. In other words, environment of the laboratory medium does not favour the formation of capsule. This type of variation is referred to as dissociation.

Now-a-days it is recognized that both types of variations (adaptations and dissociations) occur from the selection of rare mutant types. The rare mutant types arise by chance in the bacterial culture. The mutation had already been recognised as the ultimate basis of variation in higher living systems. But this fact was not accepted by bacteriologists for a long time for the following two reasons:

(1) There was a widespread opinion that bacteria lacked nuclei like the higher forms. Some biologists had frequently reported the presence of nuclear bodies in bacteria. C. Robinow in 1942 developed a staining technique for the nucleus and thus universal presence of nuclei in bacterial cells was accepted. In the same period, thin sections of bacterial bodies were examined under the electron microscope. Electron microscopic study revealed that bacterial 'nuclei' are different from higher organisms. Bacterial 'nuclei' are not surrounded by the nuclear membrane as in the case of higher forms. Also, they do not contain chromosomes like those of typical plants and animals. The nuclear body consists of simply deoxyribonucleic acid (DNA). It looks like a single and very long DNA molecule. Thus, bacterial nucleus is of a primitive type.

(2) Frequency of occurrence of bacterial variation is extremely rare (at the rate of about 1 for every 10 million cell divisions). The main difficulty lies with the identification and isolation of the mutant cell. This difficulty is experienced owing to the smallness and simplicity in the structure of bacteria. Bacterial cells do not possess many morphological features like those of flowers, feather and so on. The recognition of bacterial mutation becomes possible only when mutant type becomes a major part of the bacterial population.

Thus, bacterial variation had its basis in the spontaneous gene mutation. The first evidence for this opinion was provided by introducing the fluctuation test in 1943 (Luria and Delbrück).

4 EVIDENCES FOR OCCURRENCE OF SPONTANEOUS MUTATION

There are many experiments to prove the spontaneous mutation in bacteria. Some of them have been discussed as under:

The Fluctuation Test (Figure 6.3):

The fluctuation test was developed by S. Luria and M. Delbrück in 1943. This test is administered to prove the spontaneous nature of bacterial mutation.

Procedure:

(1) A young suspension of a bacterium (500 *E. coli* cells/ml.) is divided into two large cultures, A and B, of 10 ml. each.

(2) Large culture A is further subdivided into 50 small cultures (Each small culture contains 0.2 ml.).

(3) All the 50 small cultures as well as one B large culture are incubated until a suitable population density is reached.

(4) The contents of each of the 50 small cultures are plated separately on phage-coated agar plates. Also, fifty equal samples (0.2 ml.) from B large culture are likewise plated separately.
(5) All the plates are incubated until the development of phage-resistant colonies.
(6) Number of phage-resistant colonies on each plate is recorded.

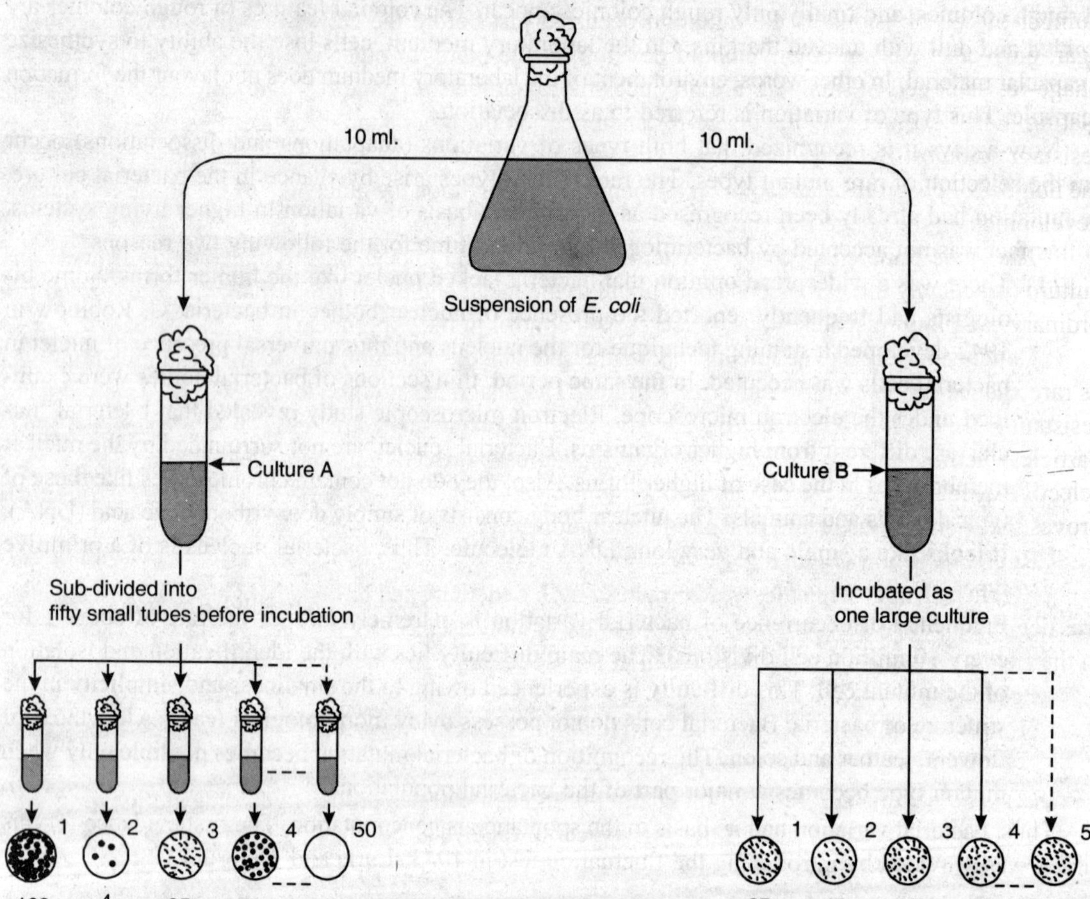

Fig. 6.3 *The fluctuation test. It is an indirect evidence for the occurrence of mutation.*

Result : A large fluctuation is noticed between the numbers of phage-resistant colonies developed from each of the 50 small cultures. But, roughly the same number of phage-resistant colonies is found on each of the 50 plates prepared from B large culture.

Interpretation : Luria and Delbrück reasoned that if bacterial resistance is developed by direct adaptation to the phage, there should be no significant differences in the numbers of phage-resistant colonies on any of the plates made from both large cultures. By this theory, it is made evident that it does not matter whether the cultures were incubated in a series of separate tubes, or were all grown together in one large tube, since all the bacterial cells are presumed to be sensitive until brought into contact with phage.

If, on the other hand, phage resistance resulted from spontaneous mutations during growth in phage-free medium, there should be a large fluctuation between the numbers of phage-resistant colonies which have developed from each of the 50 small cultures. This fluctuation in the numbers of resistant colonies is a consequence of the randomness in the time of the occurrence of spontaneous mutations. If the mutation to resistance occurred early during incubation period, there would be many resistant bacteria present at the time of plating. This happens because the resistant mutant cell would pass through a large number of cell divisions. If the mutation occurred just before the cells were plated on phage-coated agar, only one resistant colony should develop on the plate. In other words, early development of a phage-resistant cell in small culture produces more number of phage-resistant colonies on subsequent plating. Late development of a phage-resistant cell, on the other hand, produces less number of phage-resistant colonies. Between these two extremes, a phage-resistant mutant cell develops randomly and the number of phage-resistant colonies is found depending upon the quickness of the phage-resistant cell development in the phage-free medium.

The same number of phage-resistant colonies should appear on the plates made from B large culture (practically the variance is about equal to the mean and is not larger than will be expected from ordinary sampling mistakes. Thus, it serves as a control in the experiment).

Conclusion : This experiment supports the spontaneous mutation theory. Bacterial mutations occur as rare chance events in the phage-free cultures. The phage has nothing to do with the development of resistant cells, but it simply acts as a selective agent in phage-coated plates. The selective agent (phage particles) permits the development of only phage-resistant cells on phage-coated agar plates. Other selective agents, like streptomycin, may be employed in the test. Thus, the fluctuation test indirectly proves the spontaneous nature of bacterial mutation.

Replica Plating (Figure 6.4):

'Replica plating' technique was introduced by J. Lederberg and E. Lederberg in 1952. This technique was devised to prove the spontaneous origin of bacterial resistance. Now-a-days, it has many applications in the field of molecular engineering, (e.g. in the isolation of auxotrophic mutants).

Procedure :

(1) About 10^8 cells of a phage-sensitive strain of *E. coli* are spread on a non-phage nutrient agar plate.

(2) The plate is subjected to incubation until each cell has produced a colony containing a few hundred progeny cells. This will appear to the eye as confluent growth on the surface of the agar.

(3) The petri-dish with lawn of bacterial cells is inverted on the velvet surface as shown in the Figure 6.4. Then the plate is gently pressed against the velvet to sample every colony of cells in the lawn.

(4) The petri-dish is removed, and two or more fresh phage-coated agar plates are pressed against the velvet in turn. The original 'master plate' is preserved, and copied phage-coated plates are incubated for the development of phage-resistant colonies.

(5) A colony appearing at the identical position on every replica plate is presumed to have arisen from an inoculum of phage-resistant cells. The phage-resistant cells from a phage-resistant clone on the master plate are transferred via the velvet. Therefore, the investigator marks this position on the preserved master plate.

(6) Some material from that marked colony on the master plate is picked with care. The suspension is prepared in non-phage broth by suspending the picked material. Let us assume that this inoculating needle removes 10^5 cells, of which two or three are phage resistant.

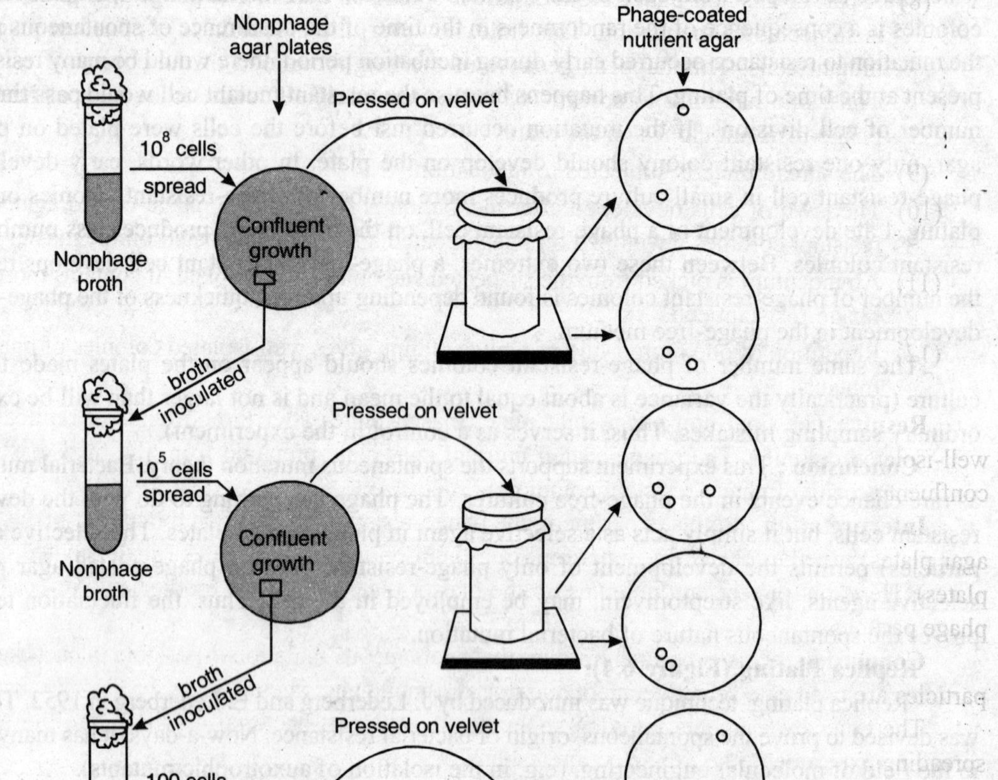

Fig. 6.4 *The Replica plating technique. It is direct evidence for the occurrence of spontaneous mutation in bacteria.*

(7) The non-phage broth culture is incubated for several hours to increase the total cell number.

(8) A part of this non-phage broth culture is spread on a fresh non-phage agar plate, and the plate is subjected to incubation. This time, however, the inoculum is enriched in phage-resistant mutant cells. This happens because of the neighbourhood of the colony of resistant cells. Thus, one has only to spread 10^5 cells, instead of 10^8 cells, to be sure of having several phage-resistant clones on the plate.

(9) The spread plate is subjected to incubation.

(10) The lawn of cells developed after incubation is replicated to phage-coated agar. The phage-resistant colonies developed on phage-coated agar may be located on the master plate.

(11) An inoculum is picked from the suspected site on the master-plate. It contains more number of phage-resistant cells.

(12) Repetition of one more cycle of replica plating gives well-isolated colonies of phage-resistant mutants.

Result : The final non-phage agar plate (master-plate) after three rounds of replica plating has well-isolated colonies. The phage-coated nutrient agar plate prepared from the final master plate has confluent growth of phage-resistant colonies.

Interpretation : In this experiment, the phage-resistant colonies on the master plates (non-phage agar plates) may be located by superimposing them on the replica plates (phage-coated nutrient agar plates). Here, phage-resistant colonies on the master plates are never brought into direct contact with phage particles.

Conclusion : This experiment proves that spontaneous mutation is possible in bacteria. The phage particles are used as a selective agent and not as a mutagenic agent.

The other experiments, proving spontaneous nature of bacterial mutation, are the Newcombe's spreading experiment, the Sib-selection test and others.

5 THE CHEMICAL BASIS OF MUTATION

Mutagenesis

Escherichia coli and many phages for which *E. coli* is a normal host are exploited in the study of molecular biology as research tools. They have been utilized to understand the mechanisms of mutation. They have been employed for the following reasons:

(1) They require less space so that experiments can be performed in the test-tube. Many cultures may be handled at a time.

(2) They require less nutrients for their growth and multiplication. Therefore experiments are less expensive.

(3) The generation time is very short. For example, the generation time of *E. coli* is about 20 minutes in the nutrient broth. Hence one does not have to wait for a long time to study the changes in the subsequent generations.

(4) Like eucaryotic organisms, sexual instinct exists in *E. coli. E. coli* cells harbouring fertility factor are called male cells. Those cells which lack this factor are called female cells.

Thus, the general principles of genetics derived from these micro-organisms are applicable to higher organisms.

Various mutagenic agents have been used to study the mechanisms of mutation. They may be grouped into three major classes. They are:

(*A*) *Chemical agents*

1. pyrimidine analogues (see Figure 6.5a)
2. purine analogues (see Figure 6.5b)
3. acridine dyes (see Figure 6.5c)
4. alkylating agents
5. peroxides and epoxides
6. nitrous acid
7. hydroxylamine
8. formaldehyde
9. diethyl sulfate
10. β-propiolactone

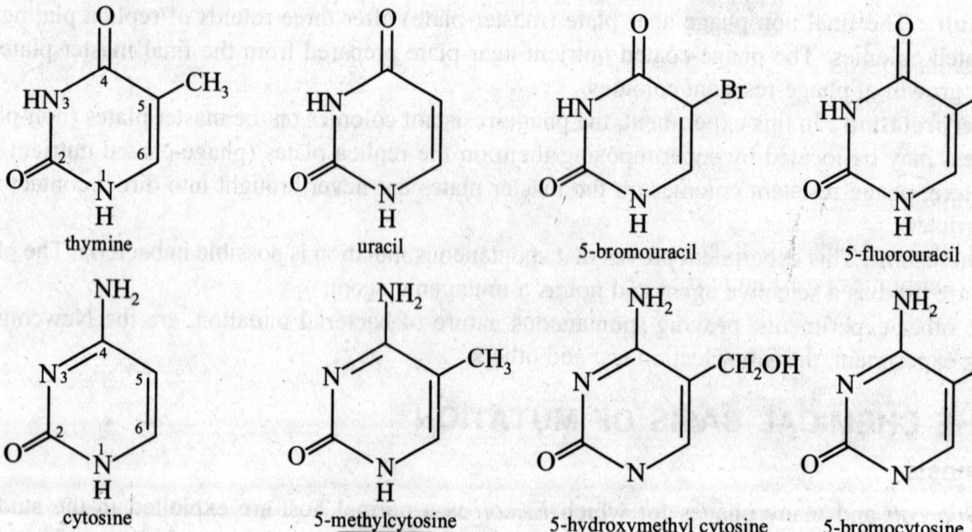

Fig. 6.5a *Pyrimidines and their analogues.*

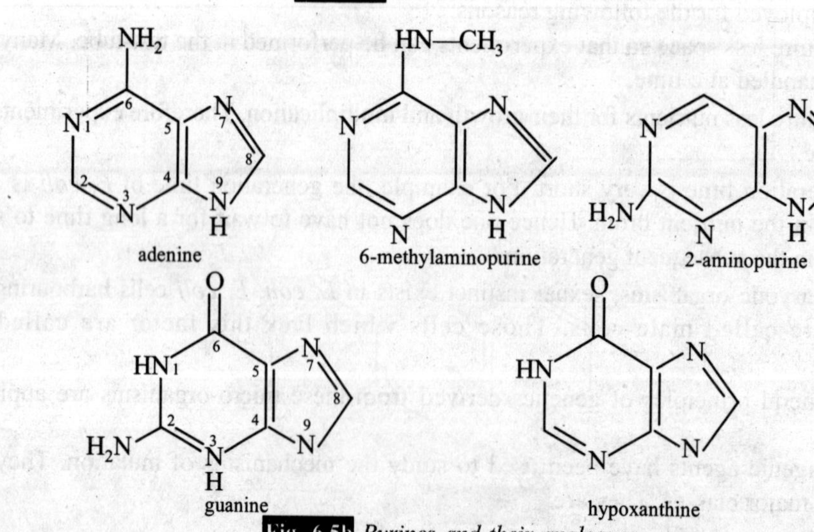

Fig. 6.5b *Purines and their analogues.*

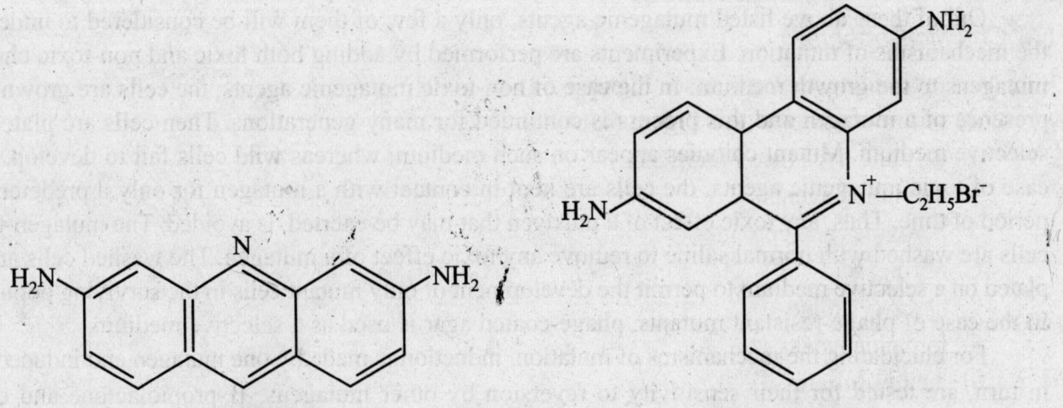

(i) Proflavin (ii) Ethidium bromide

Fig. 6.5c *Acridine dyes : (i) Proflavin and (ii) Ethidium bromide.*

(*B*) *Physical agents*

1. ultraviolet light
2. X-rays

(C) *Biological agents*

1. bacteriophages, (e.g. *mu*)
2. antibiotics (see Figure 6.5d)

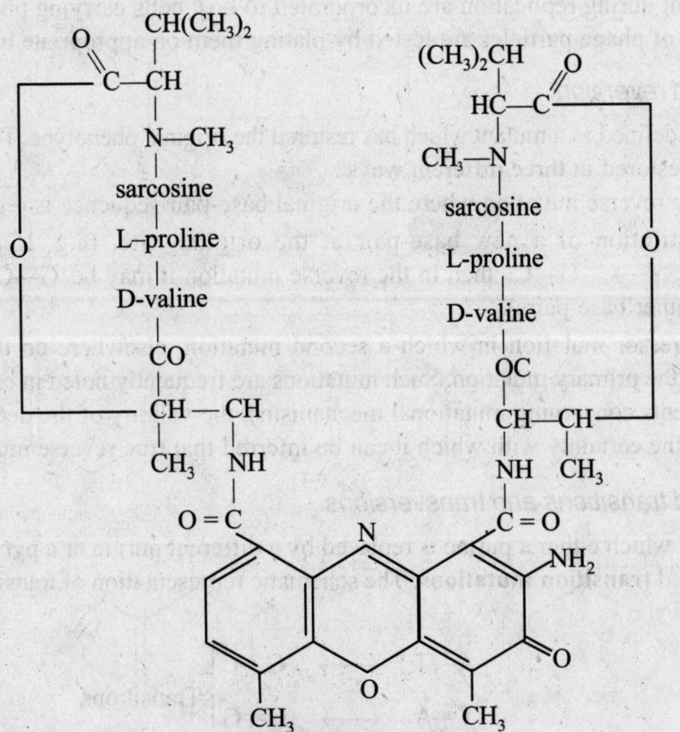

Fig. 6.5d *Actinomycin-D.*

Out of these above listed mutagenic agents, only a few of them will be considered to understand the mechanisms of mutation. Experiments are performed by adding both toxic and non-toxic chemical mutagens to the growth medium. In the case of non-toxic mutagenic agents, the cells are grown in the presence of a mutagen and this process is continued for many generations. Then cells are plated on a selective medium. Mutant colonies appear on such medium, whereas wild cells fail to develop. In the case of toxic mutagenic agents, the cells are kept in contact with a mutagen for only a predetermined period of time. Thus, any toxic effect of a mutagen that may be exerted, is avoided. The mutagen-treated cells are washed with normal saline to remove any toxic effect of a mutagen. The washed cells are then plated on a selective medium to permit the development of only mutant cells in the surviving population. In the case of phage-resistant mutants, phage-coated agar is used as a selective medium.

For elucidating the mechanisms of mutation, induction is made by one mutagen and induced cells, in turn, are tested for their sensitivity to reversion by other mutagens. β-propiolactone and diethyl sulfate may be used for testing the sensitivity to reversion. Such a reversion test is generally performed by spreading a washed suspension of a mutant strain (primary mutation) on agar which is selective for the revertant type. For example, auxotrophs may be reverted by spreading them on agar selective for prototrophic revertants. This can be accomplished by keeping a drop of a solution of a mutagenic agent on the agar medium. If the mutant type is sensitive to reversion by that mutagen, revertant colonies in the form of a ring appear surrounding the drop of a mutagen. Similar experiments are also performed with phages to study the mode of action of mutagens on DNA to induce mutation.

The chemical mutagens are added directly to suspensions of phage particles and allowed to react for a specified period of time. The phage particles are then plated on lawns of appropriate bacterial strains to detect mutant particles if present, (e.g. altered host range or plaque-type mutants). Mutagens that must be present during replication are incorporated to host cells carrying phage particles in them. The liberated crop of phage particles are tested by plating them on appropriate bacterial strains.

Mutagen-induced reversion

A revertant may be defined as a mutant which has restored the original phenotype. The original phenotypic character may be restored in three different ways :

(1) By a true reverse mutation where the original base-pair sequence is regained.

(2) By substitution of a new base-pair at the original site, (e.g. If primary mutation is A—T ⟶ G—C, then in the reverse mutation it may be G—C ⟶ T—A or similar other base-pairs).

(3) By suppressor mutation in which a second mutation, elsewhere on the DNA, cancels the effect of the primary mutation. Such mutations are frequently noted in bacteria. In most of the experiments concerning mutational mechanisms, the validity of the decisions that are drawn rests on the certainty with which it can be inferred that true reverse mutation has occurred.

Mutagen-induced transitions and transversions

Those mutations in which either a purine is replaced by a different purine or a pyrimidine by a different pyrimidine are called **transition mutations**. The schematic representation of transition mutation may be made as under :

$$\left.\begin{array}{l} \text{A—T} \longleftrightarrow \text{G—C} \\ \text{T—A} \longleftrightarrow \text{C—G} \end{array}\right\} \text{Transitions}$$

Transition mutation may be understood by considering the effect of 5-*bromouracil* (a thymine analogue) as shown in Figure 6.6(i) and (ii). If this chemical mutagen is added in its keto form, A—T base pair is replaced by G—C as shown in Figure 6.6(i).

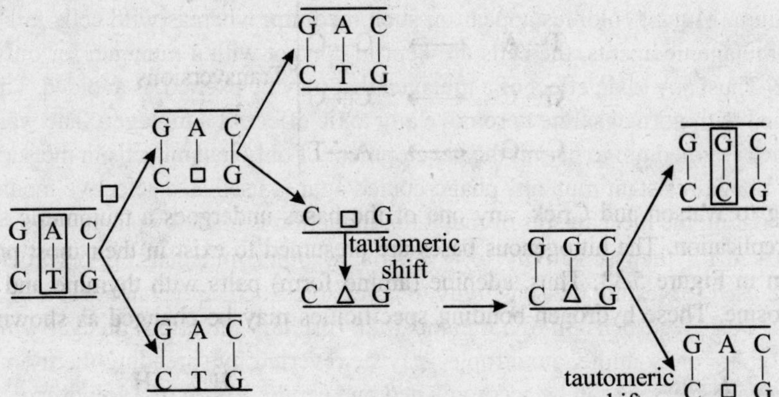

(i)

Fig. 6.6 (i) *In the illustration, squares are used to indicate keto form of 5-bromouracil and triangles indicate enolic form of 5-bromouracil. When 5-bromouracil is added in its keto form, A — T ⟶ G — C transition takes place. Also note that tautomerization takes place during replication, and thus, hydrogen-bonding specificity is changed.*

On the other hand, G—C base pair may be replaced by A—T by the addition of the enolic form of the same agent as shown in Figure 6.6 (ii) Thus, transition mutation in either direction is possible by 5-bromouracil.

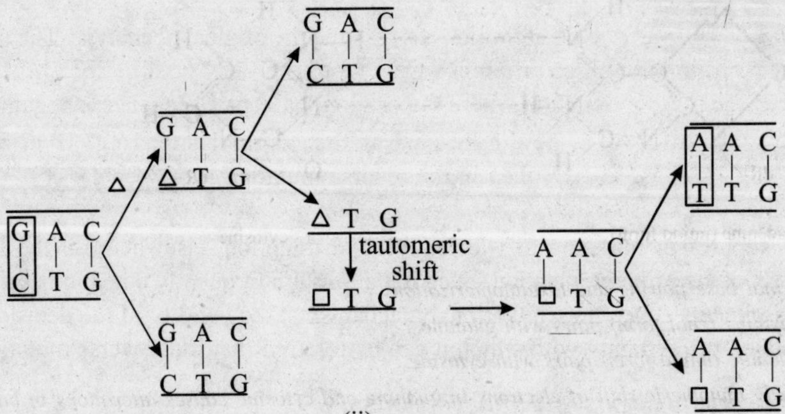

(ii)

Fig. 6.6 (ii) *In the illustration, triangles indicate the enolic form of 5-bromouracil and squares indicate keto form of 5-bromouracil. If 5-bromouracil is added in its enolic form, G — C ⟶ A — T transition takes place. Here, cytosine is first replaced by enolic 5-bromouracil (Δ). During replication, Δ undergoes tautomerization, resulting in the formation of keto form of 5-bromouracil (□). □, in turn, forms a bond with adenine, which subsequently leads to the formation of A—T pair.*

Those mutations in which a purine is substituted for a pyrimidine, and vice versa are referred to as transversion mutations. It may be illustrated as under :

A—T	⟷	T—A	Transversions
T—A	⟷	G—C	
G—C	⟷	C—G	
C—G	⟷	A—T	

According to Watson and Crick, any one of the bases undergoes a tautomeric shift of electrons at the time of replication. The nitrogenous bases are presumed to exist in their most probable energetic forms as shown in Figure 5.33. Thus, adenine (amino form) pairs with thymine and guanine (amino form) with cytosine. These hydrogen-bonding specificities may be changed as shown in Figure 6.7.

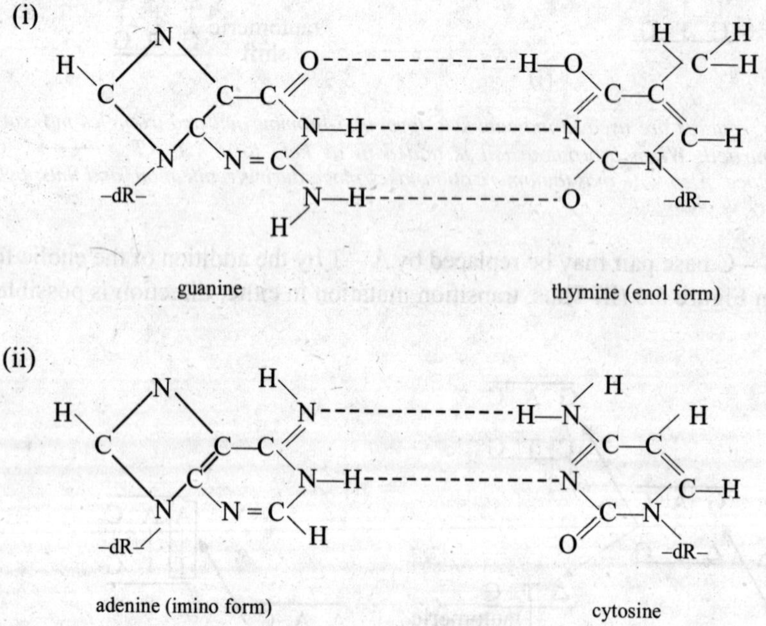

Fig. 6.7 *Abnormal base pairing due to tautomerization.*
(i) Thymine (enol form) pairs with guanine.
(ii) Adenine (imino form) pairs with cytosine.
Similarly tautomeric shift of electrons in guanine and cytosine causes alterations in base pairing.

It is clear that thymine in the enolic form pairs with guanine, instead of adenine. Adenine in the imino form pairs with cytosine, instead of thymine. Similarly, guanine in the imino form pairs with thymine, instead of cytosine. Cytosine in the enolic form pairs with adenine, instead of guanine.

Nitrous acid is also an important mutagenic agent. It can bring about transition mutations in either direction as shown in Figure 6.8, (i.e. A—T ⟶ G—C or G—C ⟶ A—T). If deamination of the adenine of an A—T pair occurs, an A—T → G—C transition occurs. Deamination of the adenine leads to the formation of hypoxanthine. At further replication cycle, hypoxanthine will pair with cytosine. Finally, cytosine pairs with guanine in further replication cycle. Thus, A—T → G—C transition results. On the other hand, deamination of cytosine to uracil brings about G—C → A—T transition.

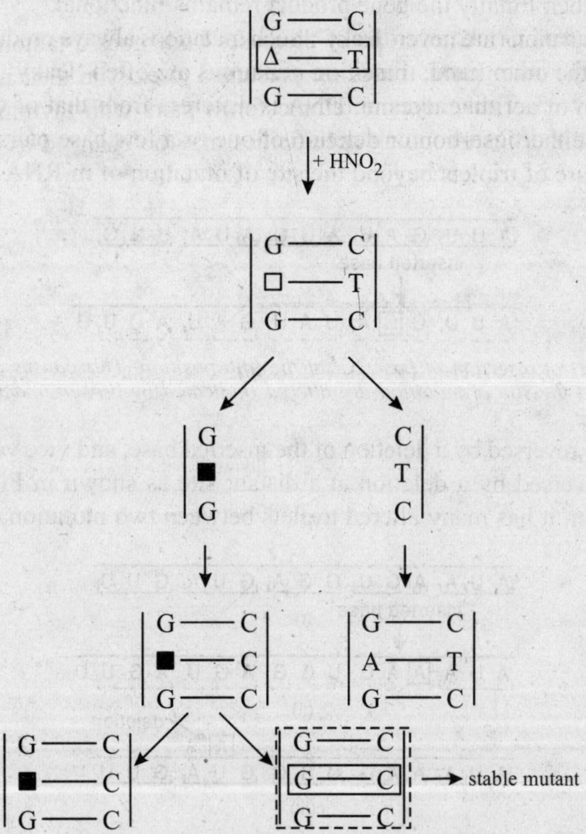

Fig. 6.8 *A—T ⟶ G—C transition by the action of nitrous acid. First of all adenine (Δ) is deaminated with the formation of enol form of hypoxanthine (□). Enol form of hypoxanthine, in turn, undergoes tautomerization forming keto form of hypoxanthine (■) and it pairs only with cytosine. Thus G—C pair is formed after one replication cycle.*

Hydroxylamine brings about G—C → A—T transition only. It reacts only with cytosine. The amino group of cytosine is replaced by a hydroxylamino group. The hydroxylamino derivative of cytosine seems to undergo frequent tautomerization. This results exclusively into G—C → A—T transitions.

Mutagen-induced insertions and deletions

Acridine dyes have been used to induce a specific type of mutation. Proflavin, ethidium bromide and propidium di-iodide are some examples of acridine dyes. Mutations induced by acridine dyes revert spontaneously. They may be induced to revert at a higher rate by acridine dyes. They are never reverted by other mutagens, namely, base analogues, HNO_2 or hydroxylamine. A second-site mutation (or suppressor mutation) within the same gene takes place during reversion. The suppressor mutation is similar to the primary mutation. The suppressor mutation inactivates the gene product and may be reverted spontaneously or by acridine dyes. Thus, two acridine—induced mutations within the same locus may cancel each other. Finally the gene product remains functional.

Acridine-induced mutations are never 'leaky'. Such mutations always produce a total loss of function of the gene product. On the other hand, transition mutations are often 'leaky'.

The mode of action of acridine dyes on DNA is different from that of the fore-going mutagens. The acridine dyes cause either insertion or deletion of one or a few base pairs in DNA. This results in the alterations of the nature of triplets beyond the site of mutation of m-RNA as shown in Figure 6.9.

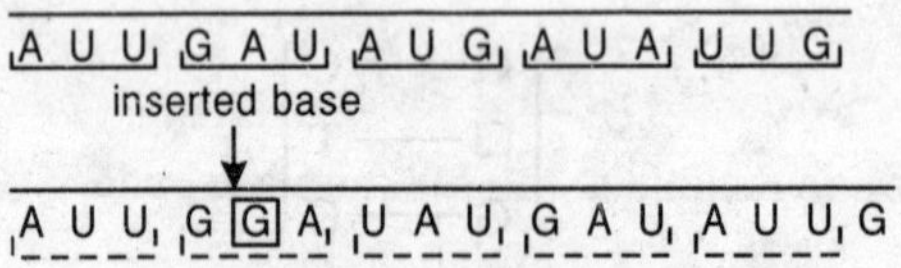

Fig. 6.9 *It depicts the effect of insertion of base 'G' at the fifth position. This causes a shift in the reading frame. All triplets beyond the site of mutation are altered (indicated by broken underlining).*

An insertion can be reversed by a deletion of the inserted base, and vice versa. It has been observed that insertion may be reversed by a deletion at a distant site as shown in Figure 6.10. The revertant strain is normal. However, it has many altered triplets between two mutational sites.

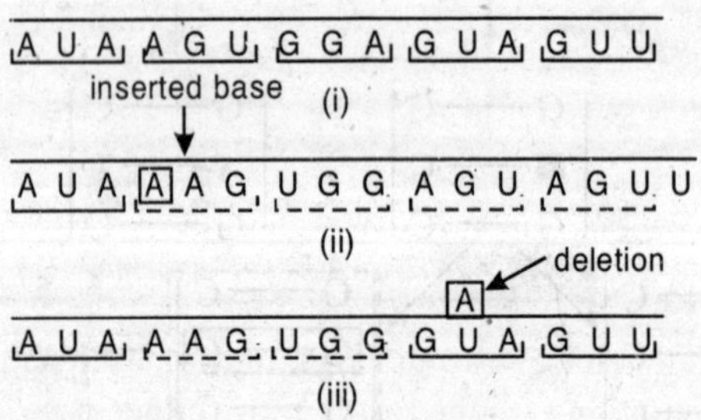

Fig. 6.10 *(i) A segment of m-RNA consisting of five triplets.*
(ii) Insertion of base 'A' at fourth position.
(iii) Deletion of base 'A' at a distant site. The resultant segment of m-RNA has some triplets in between insertion and deletion sites.

The induction of insertions and deletions is due to the ability of acridine dyes to intercalate between the stacked bases of DNA.

6 ULTRAVIOLET LIGHT—A PHYSICAL MUTAGEN

Albert Kelner (1948, U.S.A.) explained the mode of U.V. action and the subsequent repair of damage in the following way :

There are mainly two possible types of damages to DNA upon exposure to U.V. light:

(1) Formation of two unwanted chemical bonds between pyrimidine bases adjacent to each other on one strand of the DNA molecule. Of the three possible types of pyrimidine dimers, dimer formation between adjacent thymine bases is most common as shown in Figure 6.11.

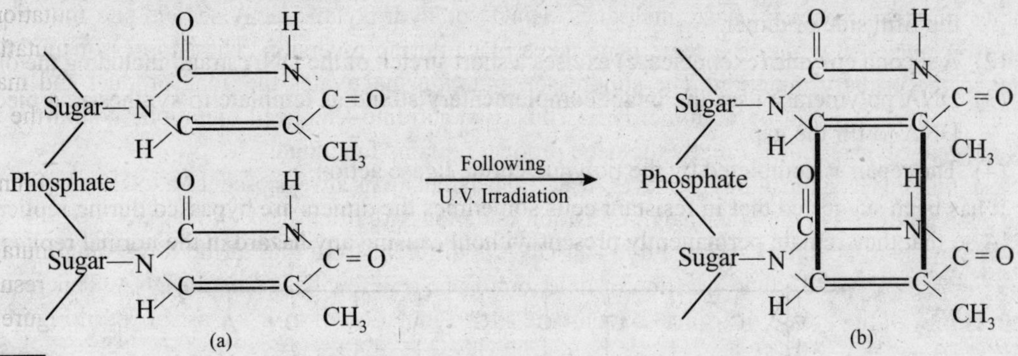

Fig. 6.11 *(a) Two thymine bases present adjacent to each other.*
(b) Dimer formation between adjacent thymine bases after irradiation with U.V. light.

(2) Hydration of pyrimidine residue at the 5, 6, double bond. For example, cytosine is hydrated by U.V. light by the insertion of a water molecule into the C=C double bond as shown in Figure 6.12.

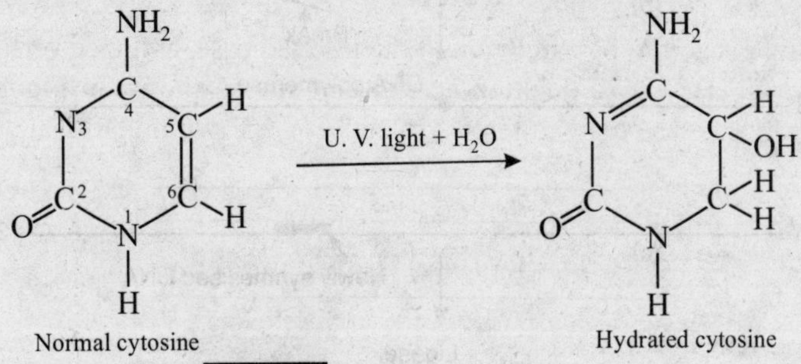

Fig. 6.12 *Hydrolysis of cytosine.*

Thus, it can safely be said that lethal effect of U.V. light is attributable to the formation of pyrimidine dimers, (e.g. T—T dimer). Mutations might result from either type of chemical alteration in DNA molecule. Transitions might occur as a result of the hydration of pyrimidine bases. Many U.V. light induced mutations in phage can be reverted by base analogues. The mechanism of mutations caused by pyrimidine dimers has not been understood. Richard B. Setlow (U.S.A) and his co-workers showed that the dimers block normal replication of DNA and bacteria with even a few such defects fail to divide. Claud S. Rupert (U.S.A.) and his co-workers explained that exposure to visible light (300 to 400 nm) results in the activation of an enzyme. The activated enzyme selectively binds to DNA irradiated with U.V. and facilitates the cleavage of pyrimidine dimers to restore the original DNA configuration. Thus, mutation frequency as well as lethality are greatly decreased. This phenomenon is called *photoreactivation*.

Yet another repair system which facilitates repair of U.V. light-damaged DNA is *dark reactivation*. It is so called because light has no role to play in the repair process. E. Witkin (1946) isolated a strain of the bacterium, *E. coli* (B/r strain), which is particularly resistant to radiation. There are, however, certain strains which are sensitive to radiation (Bs-1 strain). It has been suggested that the radiation-sensitive bacteria lack one or more enzymes needed for repair of U.V. light-damaged DNA. It is probable that dark reactivation takes place in the following way (see Figure 6.13):

(1) In the first step, an endonuclease recognizes the local distortion so as to introduce a nick on the fifth side of dimer.

(2) A second enzyme (exonuclease) excises a short stretch of the DNA strand including the dimer.

(3) DNA polymerase uses the intact complementary strand as template to synthesize a piece of DNA to fill the gap.

(4) The repair is completed by the polynucleotide ligase action.

It has been suggested that in resistant cells sometimes the dimers are bypassed during replication of DNA so that they remain permanently present without causing any hazard in the normal replication.

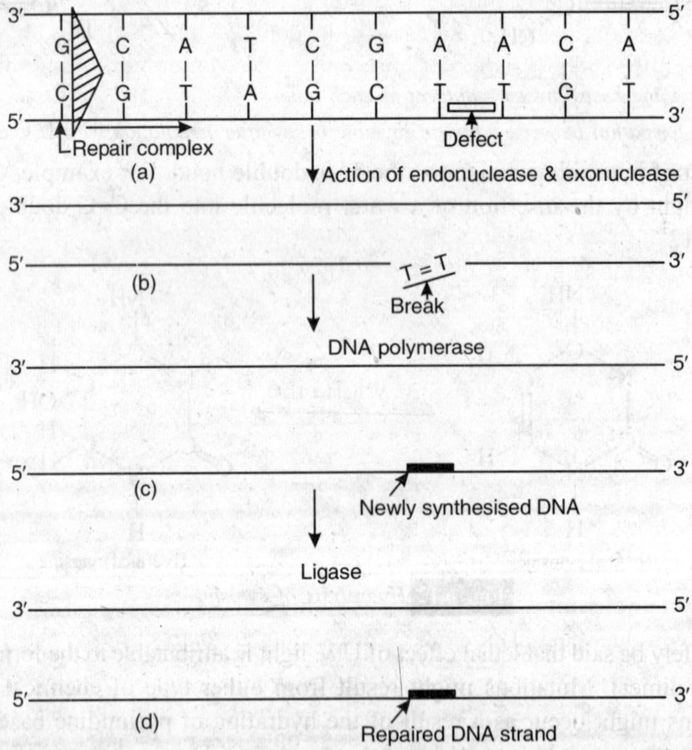

Fig. 6.13 *The four possible stages of dark reactivation of U.V. light-damaged DNA. (a) and (b) Excission of dimer by sequential action of the repair complex, (c) Insertion of new complementary bases in the presence of DNA polymerase and (d) Final joining effected by polynucleotide ligase.*

The occurrence of a dimer-induced mutation depends on the outcome of a race between the repair enzyme system and the replicating enzyme system. If the replicating enzymes reach the dimer first, a mutation may result. If the repair enzymes reach the dimer first, the mutation does not occur.

7 MUTATION RATE

The mutation rate may be defined as the probability of a mutation of any one cell in a specified period of time. There are many methods to measure the mutation rate. The first estimation of the mutation rate was made by Luria and Delbrück (1943). The undermentioned data are essential in the estimation of mutation rate:

(A) The number of mutations that occur.

(B) The number of generations that have passed.

(C) The mean number of bacteria present over a specified time interval.

The mutation rate is equal to the average number of mutations per cell generation. Therefore, the mutation rate may be calculated by the following equation:

$$a = \frac{m}{\text{cell generations}} \qquad \ldots\text{(i)}$$

where a = mutation rate and

m = average number of mutations during a rise in the number of cells from n_0 to n.

Now, a cell generation means an increase in the cell size and finally, to undergo cell division to form two daughter cells. There is a rise of only one cell in the culture during each cell–generation. Therefore, the number of cell generations equals the net increase in cells during the incubation period. It is represented by the following expression:

$$n - n_0 \qquad \ldots\text{(ii)}$$

where n is the final number of cells at the end of cultivation and n_0 is the number of cells at zero time. The cells are in varying stages of completion of their next division cycle at the end of incubation. Therefore, small correction is to be made in the (ii) expression to accomplish the true number of cell generations by the culture.

Now, $$\frac{n - n_0}{\text{In } 2} = \frac{n - n_0}{0.69} \qquad \ldots\text{(iii)}$$

where In 2 is the logarithm of 2 to the base e. Substituting (iii) in equation (i)

We get, $$a = \frac{m}{(n - n_0)/0.69}$$

$$= (0.69)\frac{m}{n - n_0}$$

It is very easy to determine m. The cells are allowed to undergo mutations on a solid medium. Here, each mutation produces a mutant clone. Since a mutant clone is fixed on a solid medium, it can be detected as a single colony.

Practically a population of cells must be allowed to undergo a limited number of cell divisions on a solid medium. Thereafter, environmental conditions must be changed, where only the mutant clones can continue their growth. For example, a limited number of cell divisions of an auxotrophic strain occur on a minimal medium containing the required growth factor (the growth is controlled by the amount of the required growth factor). Only prototrophic mutants are able to continue their growth. Thus, visible colonies of prototrophic mutants appear. Substraction of number of such mutants in the inoculum must be made. This number is determined by plating one set of plates without the required growth factor in the agar.

In the above example, the number of colonies per plate (correction is made by substracting the number of mutants present in the inoculum) equals the number of mutations per plate. The number of cell generations is determined by washing the cells off several plates in the known volume of a liquid. A viable count is then performed from the liquid.

Now, the number of cell generations $= \dfrac{n - n_0}{0.69}$

where n = the average number of cells per plate at the time when the parental population is inhibited or killed and n_0 = the average number of cells per plate in the original inoculum.

Also Luria and Delbrück suggested a statistical method to estimate the number of mutations taking place in a liquid culture.

The following three conditions should be fulfilled for a satisfactory estimation of a mutation rate:

1. A large population of cells must be used in the method. It increases the reliability of the estimate, since it allows a better chance for observing true rates.
2. An accurate estimation of back mutation must be made.
3. There should be equal coexistence of the mutant and the parental types in the environment where mutation takes place. Moreover the investigator should invariably be able to detect each mutant cell.

8 GENOTYPE AND PHENOTYPE

Genes control all the characteristics of an organism. Each gene remains relatively stable and such a gene is called *wild allele*. The changes in the environment may bring about the modifications in the wild allele. Such a modified allele is called *mutant allele*. Thus, there are various forms of a given gene. The existence of mutant alleles is a result of mutational changes in the nucleotide sequence of a given wild-type allele. These sudden changes in the genetic constitution are called mutations. **Genotype**, therefore, may be defined as the complete array of the allelic forms of all the genes possessed by an organism. Genotypic changes are inheritable. Genotypic changes occur only in the limited number of cells in the population. Genotype of an organism determines its phenotype. **Phenotype** may be defined as, the array of all characteristics exhibited by an organism at a given time. Thus, a remarkable change in the cell structure and function may be brought about by altering the structure of a single protein. Phenotypic changes are not inheritable. Phenotypic changes, simultaneously, affect most of the cells in the population. There are three major types of phenotypic changes. They may be summarized as under :

(a) Morphological modifications

Morphology of an organism is influenced by the growth conditions, like nature of medium, temperature of incubation and so on. For example, milk is the best medium for the production of very large capsules, whereas capsule formation does not occur in the nutrient broth by a given bacterial species.

(b) Cultural modifications

This may be well explained in connection with the pigment production in *Serratia marcescens*. The culture produces a distinct brick-red pigment at room temperature. The pigment production is inhibited at body temperature of 37°C. All the cells in a culture will promptly resume pigment production when the incubation temperature is lowered.

(c) Physiological modifications

This may be understood by considering the case of an inducible enzyme in *Escherichia coli*. β-galactosidase is an inducible enzyme present in *E. coli*. It hydrolyzes lactose to its constituent monosaccharides, glucose and galactose. β-Galactosidase can be detected within the genotypically lac$^+$

cells grown in a medium containing lactose or chemically related compounds. But, β-galactosidase cannot be detected within genotypically lac$^+$ cells grown in a medium devoid of lactose or chemically related compounds. Such *E. coli* cells are phenotypically lac$^-$. In this case, only the potential for the utilization of lactose is inherited.

9 TYPES OF BACTERIAL MUTANTS

Bacterial properties are not stable. They are changed as a result of gene mutation. The type of alteration taking place in the bacterial properties depends on the type of gene undergoing mutation. Cells with altered properties from that of wild-type parental cells are called mutants. There are many types of bacterial mutants. Many of them have been isolated and subjected to elaborate study.

The list of major types of bacterial mutants is as under:

(1) auxotrophic or nutritional mutants
(2) drug-resistant mutants
(3) fermentation mutants
(4) bacteriophage-resistant mutants
(5) pleiotropic mutants
(6) antigenic mutants
(7) temperature-sensitive mutants
(8) mutants that exhibit changes in colonial form or ability to produce pigments
(9) mutants that exhibit some change in morphological properties, e.g. loss of ability to synthesize components of the cell surface

Auxotrophic and drug-resistant mutants are of practical interest and will be discussed with their isolation methods.

9.1. Auxotrophic mutants

Mutants of micro-organisms which are nutritionally deficient, i.e. which cannot synthesize one or more growth factors (amino acids, purines and pyrimidines, vitamins, etc.) in contrast to the wild type organism are called auxotrophic mutants or simply auxotrophs. This state of growth-factor dependence is termed *auxotrophy*. On the other side, the state of growth-factor independence, a characteristic of the wild type organism, is termed *prototrophy*. This means that the wild type organism can grow in a medium containing the following ingredients:

(i) a carbon source, e.g. sugars, starch, etc.
(ii) an inorganic nitrogen source, e.g. ammonium chloride, ammonium sulphate, potassium nitrate, etc.
(iii) some inorganic salts like those of magnesium, calcium, manganese, iron, copper, etc.

Such a growth medium of the wild type organism is called minimal medium. An auxotroph derived from the same wild type organism cannot grow in that minimal medium unless supplied with one or more specific growth factors. An auxotrophic mutant requiring the amino acid valine may be called valine auxotroph (or simply, amino acid auxotroph), the auxotroph requiring the vitamin nicotinic acid as nicotinic acid auxotroph (or vitamin auxotroph) and so on. Therefore, the minimal medium is to be supplemented with the required growth factor/growth factors by an auxotroph. This supplemented minimal medium is called a complete medium. Auxotrophic mutants can be visually identified by their growth only in the complete medium and failure to grow in the minimal medium.

The cause of nutritional deficiency or auxotrophy is the formation of a block, due to mutation, in the biosynthetic pathway of the metabolite. The block is due to the failure of the organism to synthesize an enzyme operative at a particular step in the biosynthetic pathway. This results in an interruption in the biosynthetic sequence, and consequently, failure of the particular metabolite formation. So the auxotrophic mutants of an organism, requiring the same metabolite for growth may be biochemically different, on account of being blocked at different steps of the pathway. For example, the loss of any enzymatic reactions leading to chorismic acid formation in the aromatic pathway as shown in Figure 6.14, causes the cell to require phenylalanine, tyrosine, tryptophan, *para*-hydroxybenzoate, *para*-aminobenzoate, and 2,3-dihydroxybenzoate for growth. This pattern of auxotrophy is that of multiple requirements. All the six aromatic end products may be replaced by *shikimic acid*, provided the biosynthetic pathway is interrupted prior to the formation of this compound.

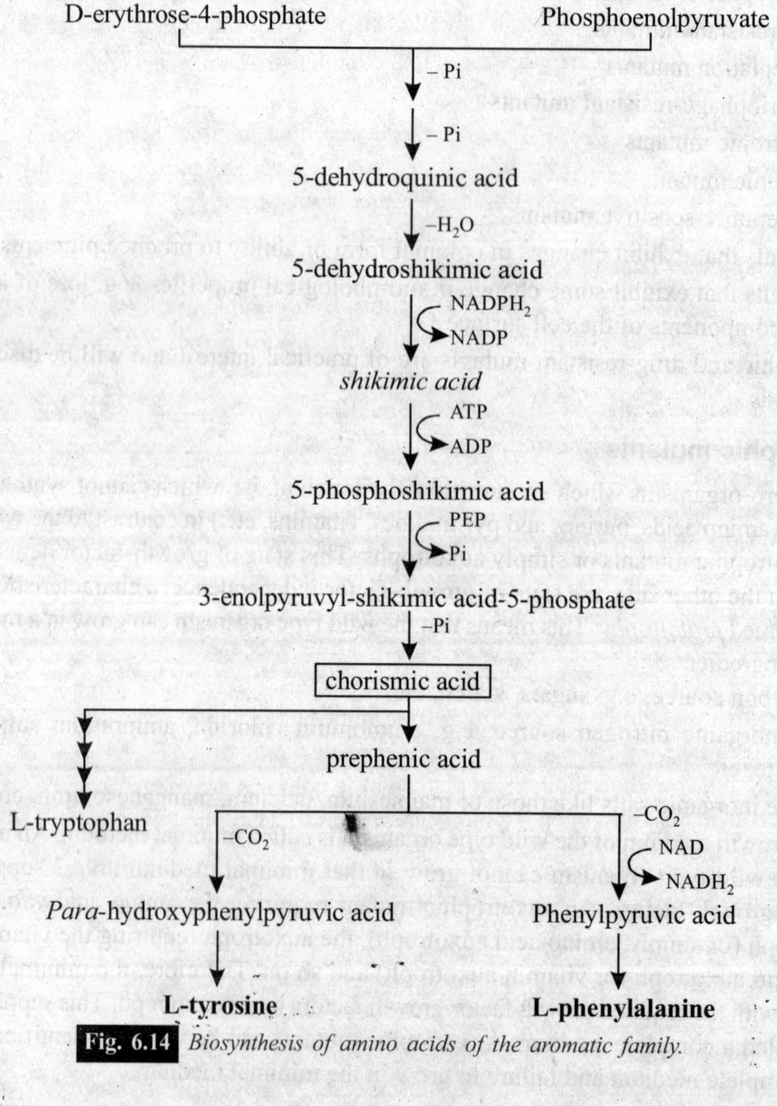

Fig. 6.14 *Biosynthesis of amino acids of the aromatic family.*

Like other mutants, auxotrophic mutants may arise spontaneously in nature. But they can also be induced artificially by the use of some physical or chemical agents called the mutagens, e.g. X-rays, ultraviolet light, nitrosoguanidine, nitrous acid (HNO_2), etc. The rate of induced mutation is much higher than that of spontaneous mutation. Auxotrophic mutants can back mutate, i.e. can revert again to the normal type which are then called the prototrophs. This back mutation may occur spontaneously and can also be induced artificially by the use of mutagens.

In isolation of auxotrophic mutants, **"penicillin enrichment"** technique is generally used. This method was developed independently by Lederberg and Zinder (1948) and by Davis (1948). This method makes use of the selective action of penicillin which kills only actively growing cells, and practically has no effect on non-growing ones. This can be illustrated by the following experimental procedure for the isolation of auxotrophs of *E. coli*:

(1) The protorophic strain of *E. coli* is treated with ultraviolet light. This increases the number of mutants in a normal population of cells. Also most of the cells are killed during the process.

(2) U.V. light-treated cells are transferred to a complete medium. This medium supports the growth of survivors in the irradiated cells.

(3) The culture from step 2 is transferred to a minimal medium containing penicillin. Minimal medium does not support the growth of auxotrophs and, therefore, penicillin does not exert its bactericidal effect against them. On the other hand, prototrophs can grow on this medium, and hence, are killed by the penicillin. Thus, this manipulation removes many organisms which are of no importance in this experiment.

(4) The penicillin is removed after the prototrophs are killed (as in step 3). The surviving auxotrophs are recovered by plating them onto a nutritionally complete medium. Living cells will form visible colonies.

(5) Now each colony formed on a complete medium (step 4) is checked to determine whether the population of cells in each colony in fact, represents, an auxotrophic strain. Replica-plating technique is employed to accomplish this result (see Figure 6.4). The petri dish with the lawn of bacterial cells is inverted and kept on the metal block over which velveteen cloth is stretched. The surface of a petri dish is lightly pressed against the velveteen cloth. Thus, each colony is sampled at the same time. The original master plate is removed and preserved. Now, the petri dish containing minimal medium is gently pressed against velveteen cloth, and then subjected to incubation. In this manner many plates can be inoculated in a one-step manipulation. All these plates are called replica plates.

(6) The replica plate after incubation is superimposed on the preserved master plate. While superimposing the replica plate on the master plate, some of the bacterial colonies of the master plate are not found on the replica plate. These colonies are of auxotrophic mutants. The suspension of cells from such a colony is prepared. They are subjected to elaborate study to determine their exact nutritional deficiency.

9.2. Drug-resistant mutants

Microbes may develop resistance against chemicals as well as physical antimicrobial agents. Antimicrobial agents used in chemotherapy are commonly called "drugs". There are various mechanisms for the development of drug-resistance :

(1) An enzyme produced by the microbes may either degrade or inactivate the drug. For example, **penicillinase** produced by some bacteria, (e.g. *Staphylococcus aureus*) confers penicillin-

resistance on them. Here, penicillin is converted into an inactive penicilloic acid by the action of penicillinase.

(2) The drug may fail to enter the cell. This happens due to some change in the cell-membrane, making it impermeable to the drug. Thus, the drug cannot reach to the target site in the cell.

(3) The primary target of drug action may acquire a reduced affinity for the drug. This is due to alteration of ribosomal protein structure.

(4) There may be the development of an alternate metabolic pathway. This alternate pathway bypasses some reaction that would normally be inhibited by the drug.

(5) The drug-resistance may develop due to competitive inhibition between an essential metabolite and an antimetabolite (drug). For example, the sulphanilamide inhibits the enzymatic incorporation of *para*-aminobenzoic acid into folic acid. The addition of sufficient *para*-aminobenzoic acid will compensate for the sulphanilamide inhibition. Thus, bacteria can grow even in the presence of inadequate amount of sulphanilamide as compared to that of *para*-aminobenzoic acid.

Drug-resistant mutants may develop spontaneously. They can also be induced with the help of physical or chemical mutagens, (e.g. X-rays or ultraviolet light).

Drug-resistant mutants may be isolated by the "**gradient-plate**" technique. For example, the steps for the isolation of an oxytetracycline-resistant mutant are as under:

(1) Tilt a sterile petri dish by placing a glass rod or wood stick (approximately 1/16 inch in diameter) under it.

(2) Pour melted trypticase-soy agar into the petri dish.

(3) Allow the agar to solidify by slanting the petri dish just enough to cover the entire bottom surface. The plate prepared in this way is called a gradient agar plate (see Figure 6.15).

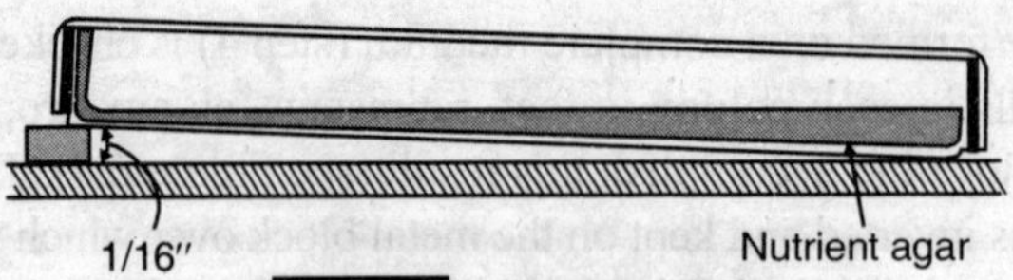

Fig. 6.15 *The gradient agar plate.*

(4) After the agar has hardened, place the plate in the normal horizontal position.

(5) Add 0.1 ml. of oxytetracycline into the tube of melted trypticase-soy agar. Mix the contents well by rotating the tube between the two palms of hands.

(6) Pour the melted agar (prepared as in step 5) on the surface of the gradient agar plate. Allow this added agar to solidify. Thus, the oxytetracycline concentration in the agar varies throughout the plate. The oxytetracycline concentration is in proportion to the thickness of the layer of oxytetracycline as under :

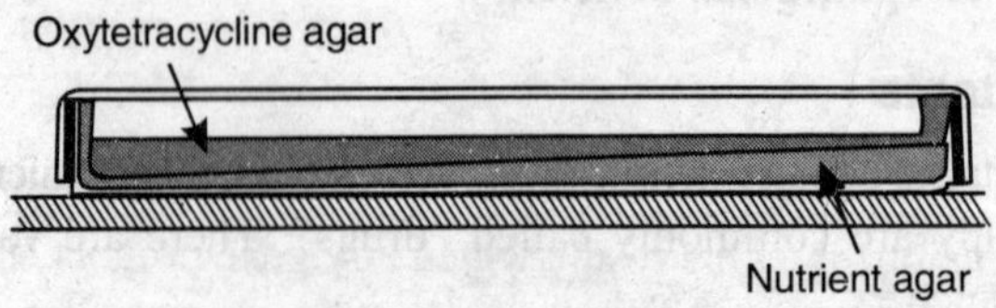

(7) Pipette approximately 0.3 ml. of *S. aureus* onto the agar surface. Streak the culture over the agar surface with the help of a sterile glass spreader. Incubate at 37°C until the next laboratory period.

(8) Observe keenly to notice the development of resistant colonies in the areas of higher oxytetracycline concentration. Choose two or three colonies growing at the highest concentration of oxytetracycline and streak these colonies with a sterile loop in the direction of higher oxytetracycline concentration. Incubate the plate at 37°C until the next laboratory period.

(9) Make a suspension in broth from the colonies growing in the areas of highest oxytetracycline concentration. This broth suspension contains an oxytetracycline-resistant mutant cells.

"Replica-plate" technique can also be used for the isolation of drug-resistant mutants.

9.3. Fermentation mutants

Those mutants which are unable to ferment a given sugar are called fermentation mutants. For example, cells of *E. coli* may undergo mutation and, therefore, the genotype of the lac^+ cells is changed. In this case, the hereditary potential for β-galactosidase production is lost. These fermentation mutants and their progeny are no longer capable of fermenting lactose, regardless of its presence or absence in the environment.

Fermentation mutants may be isolated by **"penicillin enrichment"** technique. The given culture is incubated in the medium containing the sugar in question as only fermentable carbon source. The normal fermenting cells will be killed by the penicillin, since they are actively growing cells in the medium. Fermentation mutants fail to grow and, therefore, are not affected by the penicillin. Now, penicillin is removed and surviving fermentation mutants can be detected by plating them onto a medium containing sugar in question and a chemical indicator. For example, the colourless compound **tetrazolium** can be used as an indicator. It is reduced intracellularly to the brilliantly red, insoluble substance (formazan) only within a narrow pH range. Therefore, in a sugar-supplemented medium, cells capable to ferment that sugar lower the pH to the point where the indicator is not reduced, and form colourless colonies. On the other hand, mutant cells unable to ferment the added sugar, however, reduce the tetrazolium intracellularly to formazan (brilliantly red substance) and produce bright red colonies. By this technique, it is possible to detect a single fermentation-deficient mutant colony among 10^5 wild type colonies on a plate.

9.4. Bacteriophage-resistant mutants

Those mutant cells which are not affected by their bacterial viruses are called bacteriophage-resistant mutants. The mechanism in the development of phage-resistance is due to some alteration in the receptor sites. The bacterial viruses, therefore, cannot be adsorbed on the cell-surface, and hence, are not affected by the existing phage-particles. The receptor sites are located on different cell structures. For example, in some Gram-negative bacteria, receptors are present on the lipoprotein outer layer of the cell-wall. For some phages, the receptor sites are present on cell appendages (i.e. flagella or pili).

Bacteriophage-resistant mutants can be isolated by plating on agar previously spread with a suspension of phage. Thus, only the desired type will form colonies in the presence of a selective agent (i.e. bacteriophage).

9.5. Pleiotropic mutants

Mutants with altered two or more different phenotypic properties are called pleiotropic mutants and this phenomenon is termed as pleiotropy. This results when any gene product, either primary or secondary, has more functions than one. For example, the loss of capsule formation due to a mutation leads to the two secondary phenotypic alterations :

(1) rough colony formation

(2) the loss of virulence

The remaining types of bacterial mutants (in the list) are of little significance, and hence, they do not require any detailed discussion in this text.

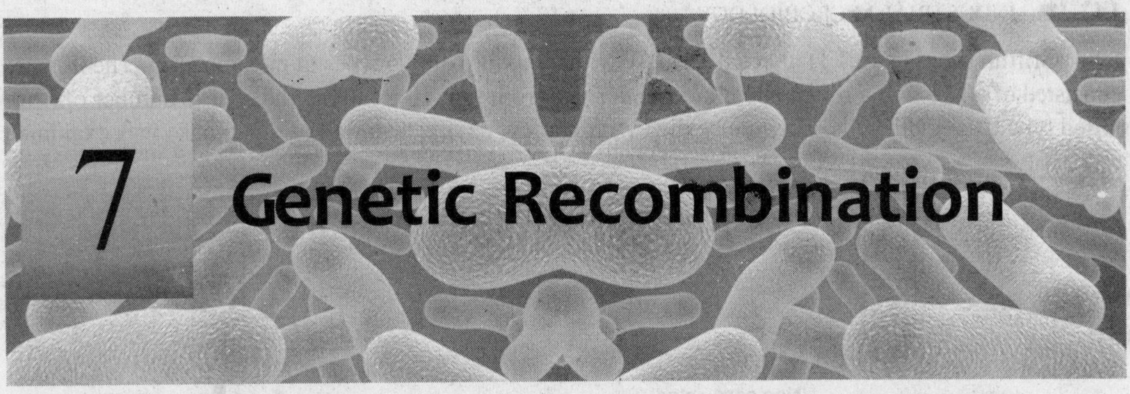

7 Genetic Recombination

Genetic recombination may be defined as the occurrence of offspring whose combinations of genes differ from those that are present in the parents. In higher plants and animals, recombination follows the coming together within a single cell of the haploid nuclei from two gametic cells, and subsequent formation of a diploid nucleus. There is involvement of complete genomes from both the gametic cells. Genetic recombination can occur from the independent assortment of chromosomes or by the process of crossing-over between homologous chromosomes. On the other hand, bacteria are haploid organisms. They are not differentiated into gametic and somatic cells. Each cell can behave as a gamete. In those bacteria that have displayed sexuality, the cells have been differentiated into two functional types called *donors* (males) and *recipients* (females). As a general rule, a part of the genetic material of a donor cell is transferred to a recipient cell. Here, the cytoplasm of a recipient cell also takes part in the process. Ultimately the recipient cell becomes incompletely partially diploid. Such an incomplete diploid is called a *merozygote* (*meros*-partial; *zygote*-diploid). Thus, the formation of partial diploids is a unique property of bacteria. Recombination takes place when a new recombinant chromosome is formed from DNA contributed by two different organisms. The transfer of DNA from donor cells to recipient cells may give rise to recombinant forms. There are three distinct mechanisms known for the transfer of DNA. These mechanisms may be listed as under in the order of their discovery:

(1) bacterial transformation
(2) bacterial conjugation
(3) transduction by bacteriophage

1 BACTERIAL TRANSFORMATION

1.1. Discovery

Bacterial transformation was discovered by Fred Griffith (1928) in England while working with two strains of pneumococci (*Diplococcus pneumoniae*). Virulent pneumococci are always surrounded by large capsules. The capsules are constituted of polysaccharide. On the basis of chemical differences between their capsular polysaccharides, pneumococci can be divided into many types. Thus, almost 100 different pneumococcal types have been reported. The different types are designated as type I, type II, type III, and so on. Pneumococci isolated from the sputum or the tissues of a victim of pneumonia are encapsulated and they form smooth (S), glistening colonies on artificial medium. After repeated subculture on laboratory medium, some of the smooth organisms undergo mutation, and subsequently become noncapsulated. These noncapsulated cells form rough (R), wrinkled colonies and are avirulent.

Griffith (see Figure 7.1) injected a mouse with a mixture of two types of pneumococci. The mixture consisted of a few living rough pneumococci derived from type II-S culture and a large number of heat-killed type III smooth cells. The mouse died within a few days. The blood of this mouse was examined

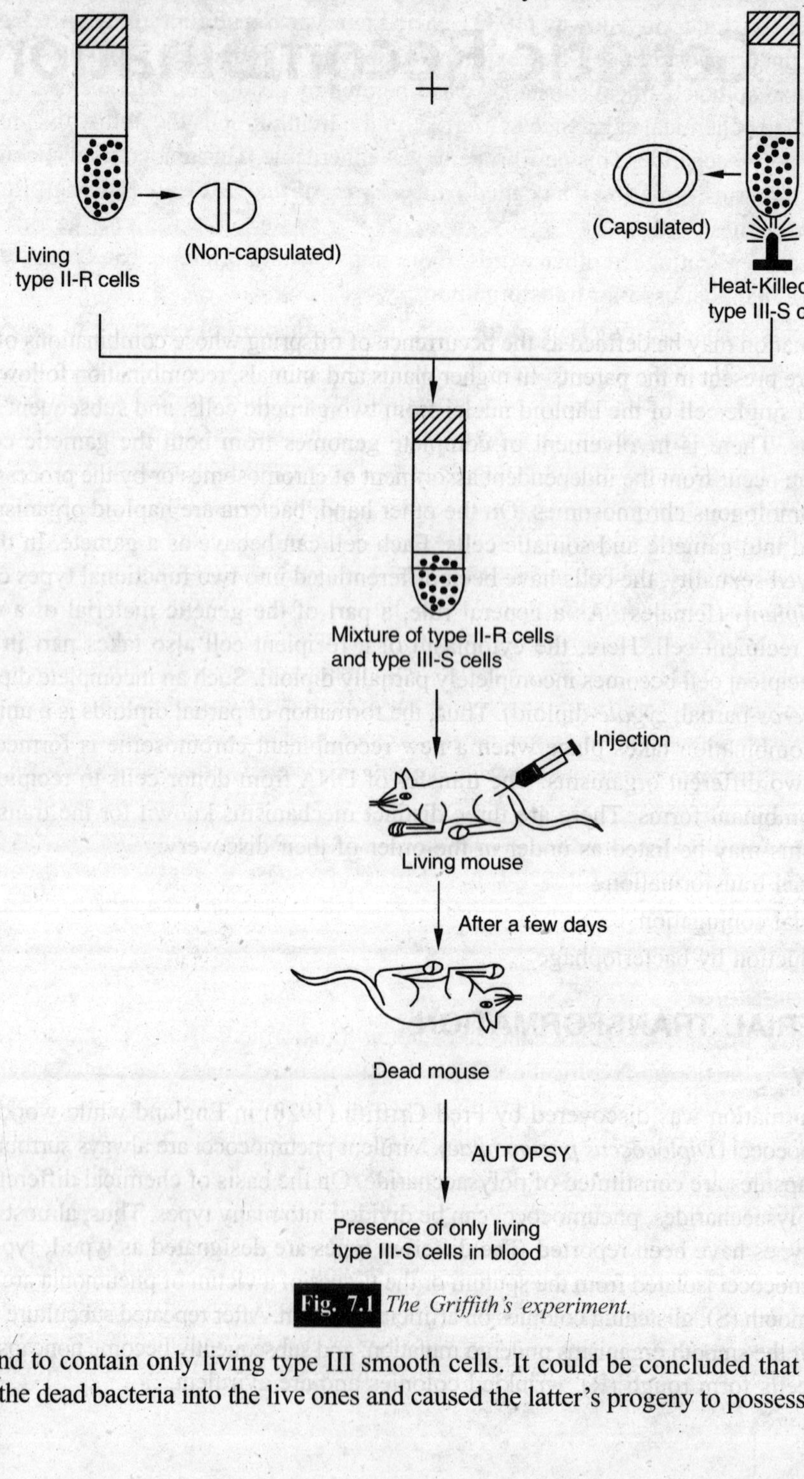

Fig. 7.1 *The Griffith's experiment.*

and was found to contain only living type III smooth cells. It could be concluded that some substance passed from the dead bacteria into the live ones and caused the latter's progeny to possess new characters;

namely, to possess capsules and become pathogenic to mice. Griffith, however, failed to interpret this phenomenon. Such *type* transformations were successfully carried out by other biologists. Dawson and Sia (1931) performed type transformation in vitro by mixing living type II-R pneumococci and heat-killed type III-S cells. Later on Alloway (1933) carried out type transformation in vitro by exposing the living type II-R pneumococci to cell-free extracts of the type III-S cultures. From these experiments, it was concluded that some chemical substance was liberated by dead smooth (S) cells and it entered the rough (R) cells. This chemical substance conferred on the living R cells the ability to synthesize a new type of capsular polysaccharide. This new property was inheritable. This unidentified chemical substance responsible for the transformation was called *transforming principle*. Transforming principle has the following two properties:

(1) It is self-replicating. In other words, more transforming principle can be extracted from the culture than that used for transformation.
(2) It directs a specific function of the cell, (e.g. production of one type of polysaccharide in pneumococci).

O.T. Avery, C.M. MacLeod, and M. McCarty (1944) successfully purified the transforming principle from pneumococcal strain. In their simple experiment, they found that transformation was not accomplished with the DNAase-treated transforming principle. Thus, they identified the nature of the transforming principle to be DNA. Prior to their report, it was generally believed that the protein portion of nucleoprotein determined genetic specificity.

Several characteristics, other than type specificity in pneumococci, have since been shown to be transformable, (e.g. transformation of fermentation and antibiotic resistance markers). The occurrence of transformation has been reported in many bacterial genera. The notable genera undergoing transformation are *Haemophilus*, *Neisseria*, *Streptococcus*, *Rhizobium*, and *Bacillus*. *Bacillus subtilis* has been studied extensively for transformation process because (i) it can be cultivated and transformed into simple chemically defined media and (ii) it has been possible to easily obtain and transform many nutritional markers. Moreover, genetic studies on sporulation have become possible by using *B. subtilis* and it has also been possible to easily infect recipient cultures of this species with the pure form of the DNA extracted from bacterial virus. Oishi and his colleague showed that *E. coli* K-12 strains undergo transformation in the presence of high concentrations of calcium ion in the medium. The ability to undergo transformation is greatly increased by the use of mutant strains lacking certain DNAases.

1.2. Genetic markers

All mutant loci of a recipient cell are capable of being transformed. The DNA is usually sheared into fragments during extraction and purification. These fragments have rarely a molecular weight that exceeds 1×10^7. This represents about 0.3 per cent of the bacterial genome. It corresponds to about 15 genes. In other words, the transforming DNA is usually very small and rarely carries more than one genetic marker. "Linked transformations" have occasionally been marked. It is possible to obtain the linked transformation of many markers by the use of crude preparations of DNA. Thus, it is possible to transfer as much as a third of a chromosome.

Recombination by transformation usually occurs with a relatively low frequency. The frequency varies from species to species. For this reason selective markers, (e.g. drug resistance and nutritional independence) must be used for successful transformation. For example, transformation of even a few cells of the recipient culture can be detected by plating the DNA-treated cells on streptomycin agar.

1.3. Uptake and integration of transforming DNA

Uptake of transforming DNA (exogenote) takes place only during the late log phase. This physiological state of the recipient cells is called *competence* and the cells are referred to as the competent cells. There are two hypotheses put forward to explain the phenomenon of competence. According to one hypothesis, the permeability of the cell-wall is changed, allowing large molecules like transforming DNA. The other hypothesis states that synthesis of specific receptor sites takes place on the bacterial cell surface. These receptor sites are used for adsorbing the transforming DNA. It appears that a protein is synthesized during the development of competence, since competence does not develop in the presence of agents, (e.g. chloramphenicol) which block protein synthesis. The uptake of transforming DNA is inhibited in the presence of the agents which interfere with energy metabolism. This suggests that the uptake process is an energy-requiring mechanism.

The uptake process has been extensively studied in both Gram-positive, (i.e. *Diplococcus pneumoniae*) and Gram-negative, (i.e. *Haemophilus influenzae*) organisms. There are, generally three stages in the process as shown below:

(1) Binding of double-stranded transforming, DNA to receptor sites of the recipient culture. In the case of *Diplococcus pneumoniae*, continuous addition of choline (a component of membrane phospholipids) is necessary for the DNA binding.

(2) Cleavage of externally bound DNA at random sites with the help of enzymes. Thus, fragments with a median molecular weight of 4 to 5×10^6 are formed.

(3) Penetration of transforming DNA into the cell. It should be remembered that fragments with mol. weights less than about 5×10^5 are not taken up.

Two major differences between *D. pneumoniae* and *H. influenzae* have been observed:

(1) *D. pneumoniae* can take up calf thymus DNA as efficiently as homologous DNA. But recombinants are not formed. On the other hand, *H. influenzae* can take up only homologous DNA. In other words, *D. pneumoniae* is not specific with respect to the type of DNA that they take up, but *H. influenzae* is specific in that character.

(2) In the case of *D. pneumoniae*, a double-stranded DNA is converted into a single-stranded one during its penetration. But it is not so in the case of *H. influenzae.* The conversion of a double-stranded DNA into a single-stranded one occurs during integration.

The entered DNA integrates with endogenote in the region of homology. Exogenote is integrated by a series of enzymatic steps. Very little is known about the integration of exogenote. During integration, replacement of an existing part of the recipient's DNA takes place with the corresponding part of the previously added exogenote.

Transformation process plays an important role in nature in the evolution of bacterial species. It occurs at low frequency.

2 BACTERIAL CONJUGATION

2.1. Discovery

The main method of reproduction among living beings is sexual reproduction. In the higher organisms, it is the only method. In lower forms of life, sexual reproduction is rare. Bacteria are such organisms in which the main method of reproduction is by cell division. However, recent work has shown that sex does exist in bacteria, though there are no visible parts to perform this function. Sexual reproduction in bacteria occurs through a process called **conjugation**. This process was first demonstrated by J. Lederberg and E.L. Tatum in 1946 while working in the Department of Botany at Yale University, U.S.A. They used polyauxotrophic mutants of *Escherichia coli* K-12 in their experiment.

E. coli K-12 requires no growth factors. The normal prototroph of it can grow easily on a glucose-salts medium. In other words, it can synthesize all its necessary constituents from a simple minimal medium. By mutagenesis, these biologists derived two polyauxotrophic strains of *E. coli* K-12. One strain required the amino acids namely biotin and methionine, whereas the other required the amino acids namely threonine and leucine. Thus, these polyauxotrophic mutants were deemed essential for this experiment because it is almost impossible for them to completely mutate to the normal prototrophic state. This is because back mutation or spontaneous reversion to the wild type occurs at the low frequency (about once per 10^{6-7} cell divisions).

The genotypes of polyauxotrophic mutants used by Lederberg and Tatum can be partially represented as follows:

Polyauxotrophic strain I :	Bio^-	Met^-	Thr^+	Leu^+
Polyauxotrophic strain II :	Bio^+	Met^+	Thr^-	Leu^-

The symbols (Bio, Met, Thr, and Leu) stand for biotin, methionine, threonine, and leucine respectively. The plus superscript indicates the ability of the strain to synthesize the amino acid. The minus superscript indicates its inability to synthesize the amino acid.

Both cultures were first grown separately on complete media, (i.e. In the first case biotin and methionine were present in the medium, whereas threonine and leucine had been added in the second case). Both cultures were then washed and about 10^8 cells of each type were mixed together and plated on a minimal medium lacking the four growth factors. A few hundred colonies of prototrophs, (i.e. having the genotype Bio^+ Met^+ Thr^+ Leu^+) appeared on this medium. This is because auxotrophs cannot grow on a minimal medium unless and until supplemented with the required growth factors. Several experiments were devised to test whether or not transformation reaction occurred to bring about the observed genetic changes, (e.g. U-tube experiment of Davis in 1950). But all attempts to find a diffusible chemical substance responsible for transformation were uniformly negative. It was finally established that recombination in *E. coli* K-12 occurs by sexual conjugation, (i.e. cell-to-cell contact). Electron microscopic studies have revealed that a tube or bridge is formed between the mating cells through which the genetic material is passed from one cell to the other.

2.2. Fertility (F) factor

Fertility (F) factor is an extrachromosomal particle present only in the donor cells. It is a small circular piece of DNA. It replicates independently of the F^+ cell's chromosomal DNA. Generally only one copy of sex factor DNA per bacterial genome is present. Fertility factor is about 1/100 as large as the bacterial genome. It contains only sufficient DNA to specify about 40 genes. These genes control fertility factor replication and synthesis of sex pili. Fertility factor is also called sex factor. Thus, sex differentiation in *E. coli* has become possible. Those bacteria harbouring fertility factor are called male, donor (F^+) bacteria. On the other hand, bacteria lacking this factor are designated as female, recipient (F^-) bacteria.

Crosses between two cells lacking F-factor (F^-) do not occur. But cells having F-factor (F^+) can be made to mate easily with F^- cells. F^+ cells, however, also mate with F^+ cells but the frequency with which they do so is very poor. The frequency of formation of recombinants by such crosses is of the order of 10^{-5} to 10^{-6} per cell pair. F^+ cells also have the ability to transfer F-factor into recipient (F^-) cells, thereby making the F^- recipients into F^+ donors. This can only be done by cell-to-cell contact. Extracts of F^+ cells are unable to accomplish this feat. The segregant population after the conjugation process and the transfer of genetical material in part or whole turns into F^+ population. F^+ donors on treatment with acridine orange and other inhibitory agents lose this factor and thereby are converted back to F^- recipients. The donor state could also get lost spontaneously or through irradiation. This process is called 'curing'. The donor condition once lost cannot be recovered spontaneously. The only way in which it can be brought back to the previous state is by conjugation with a donor. The F-factor, thus, behaves as the genetic determinant for the donor state, and is also capable of self-replication. Even though the F-factor is transferred very efficiently in $F^+ \times F^-$ crosses, transfer of chromosomal genes from donor to recipient strains is a very rate event.

2.3. Hfr (high frequency of recombination) cells (see Figure 7.2)

Another donor type can arise from F^+ cells. This is capable of transferring chromosomal markers at a very high frequency (about 20,000 times higher than their parental F^+ strains) and is called Hfr. Hfr can appear in F^+ population due to integration of F-factor to the bacterial chromosome. Hfr behaves as a donor or male cell in the mating process. The rate of undergoing recombination is 10^3 times greater than $F^+ \times F^-$ cells. Hfr can revert to F^+ form showing that F^+ is retained in Hfr in some tightly bound or concealed form. The recombinants formed after mating between F^+ and F^- cells are all invariably F^+ (males), whereas in mating between Hfr and F^- cells the recombinants are F^-, only very minute portions are of Hfr type. Hfr donors unlike F^+ donors do not transfer randomly the chromosome of the donor strain but they transfer particular segment from the donor chromosome. Moreover, the donor ability of Hfr is not eliminated by growing them in the presence of acridine orange.

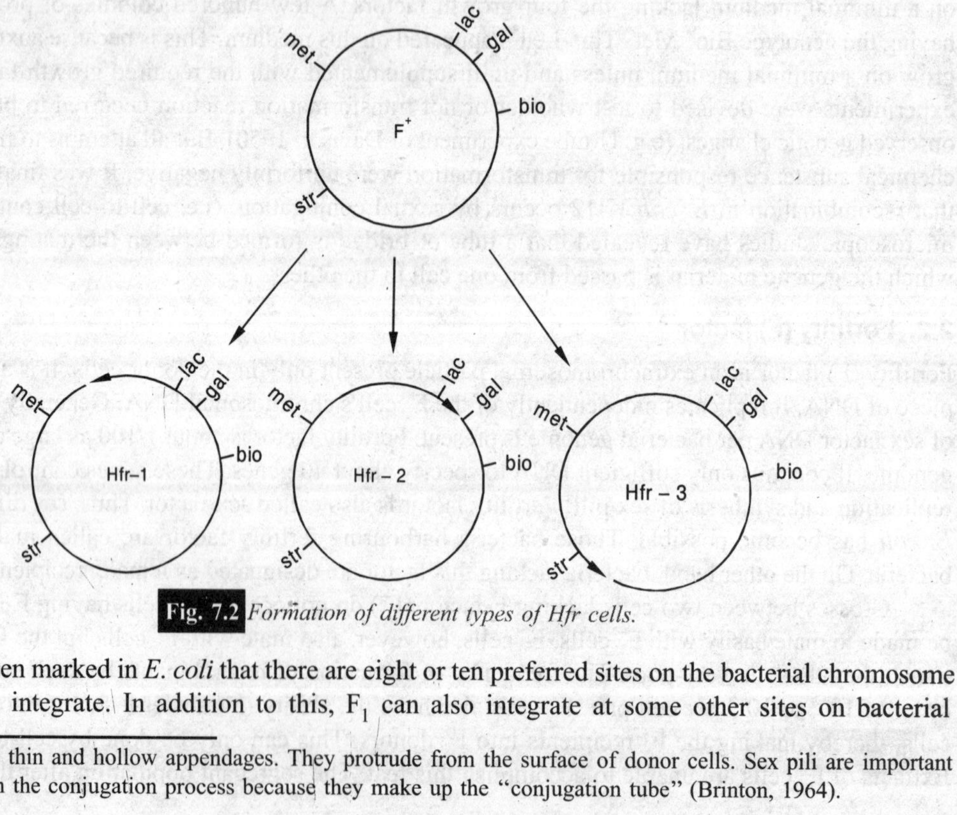

Fig. 7.2 *Formation of different types of Hfr cells.*

It has been marked in *E. coli* that there are eight or ten preferred sites on the bacterial chromosome where F_1 can integrate. In addition to this, F_1 can also integrate at some other sites on bacterial

+ Sex pili are thin and hollow appendages. They protrude from the surface of donor cells. Sex pili are important structures in the conjugation process because they make up the "conjugation tube" (Brinton, 1964).

chromosome with extreme rarity. It takes place by crossing over at a rate of about once per 10^5 cells at each generation. The integration of F–factor with bacterial chromosome is a reversible process. In other words, a few F^+ cells are present in a population of Hfr cells. The rate being the same as that of Hfr formation from F^+ population.

E. Wollman and F. Jacob (1956) performed ingenious experiments using Hfr mutants. They prepared a mating mixture of Hfr and F^- strains. Samples from such mating mixture were withdrawn at specific intervals and then subjected to violent agitation in a blender to interrupt conjugation. Each agitated suspension was plated on different selective media to determine the numbers of different types of recombinants. If the mating pairs are permitted to mate for a longer time, the more genetic material is transferred from the Hfr cells to the F^- cells. In the case of *E. coli*, the time required for injecting the entire genome is 90 minutes at 37°C. From these experiments, it was reported that chromosomal genes from Hfr cells were transferred in a definite order and at a fixed rate. Also, they indicated that the chromosome of *E. coli* is really a circle. Good maps have been established for several different bacteria.

2.4. F′ cells

The integration of fertility factor and chromosome is a reversible process. The fertility factor can detach in three possible ways :

(A) When recombination occurs within the original region of pairing between fertility factor and chromosome, normal F^+ cells are regenerated (see Figure 7.3a).

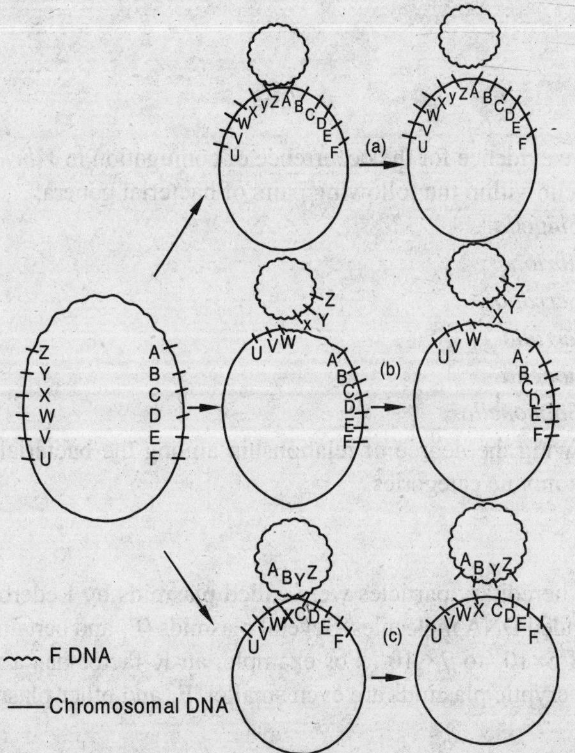

Fig. 7.3 (a, b, c) *It illustrates the detachment of fertility (F) factor from chromosome in different ways forming F^+ cell (a) and primary F′ cell (b and c).*

(B) When pairing in an exceptional region takes place, primary F′ cells are formed (see Figure 7.3b). F–genote carries the chromosomal markers, i.e. X, Y and Z.

(C) When exceptional pairing takes place in a different region, it also leads to the formation of primary F′ cells (see Figure 7.3c). In this case, F-genote contains the chromosomal markers, i.e. A, B, Y and Z.

Cells which harbour F-genotes (F-genote means fertility factor carrying a segment of chromosome) are called primary F′ (*prime*) cells. When primary F′ cells are mated with F^- recipients, fertility factor is transmitted very efficiently together with the added chromosomal segment. The cells thus formed after receiving such F-genotes from primary F′ cells are called secondary F′ cells. F-genote entered in secondary F′ cell reestablishes its circular form and becomes an autonomous replicon. Thus, the secondary F′ cell is a partial diploid, since it carries chromosomal genes in F-genote, in addition to the host chromosome. The secondary F′ cells will express these genes which may be detected by normal selective techniques. This is an important mechanism of transmitting bacterial genes by sex factor from the donor to the recipient cell and this phenomenon is called **sexduction** by Jacob and Wollman (1961). The secondary F′ cell is a moderately efficient donor of chromosomal genes. This is because of the ready occurrence of pairing and recombination in the region of diploid.

2.5. Occurrence

The conjugation process has been marked in the following listed bacterial genera:

(i) *Escherichia*
(ii) *Salmonella*
(iii) *Pseudomonas*
(iv) *Serratia*

There is suggestive evidence for the occurrence of conjugation in *Vibrio*. Intergeneric conjugation has been reported to occur within the following pairs of bacterial genera:

(i) *Escherichia-Shigella*
(ii) *Salmonella-Vibrio*
(iii) *Escherichia-Serratia*
(iv) *Salmonella-Serratia*
(v) *Shigella-Salmonella*
(vi) *Escherichia-Salmonella*

This helps in knowing the degree of relationship among the bacterial genera. Now-a-days it is used for establishing taxonomic categories.

3 PLASMIDS

The extrachromosomal hereditary particles were called plasmids by Lederberg in 1952. All plasmids are circular, double-stranded DNA molecules. Several plasmids (F_1 and certain R factors) have molecular weights in the range of 5×10^7 to 7×10^7. For example, an R factor has a molecular weight of only 1×10^7, and some of the cryptic plasmids are even smaller. F_1 and other plasmids may contain as many as 100 genes.

Plasmids can be characterized in terms of mean base composition. For example, the DNA of F_1 has been found to contain two regions of different base composition: one comprising 90 per cent of the molecule, contains DNA of 50 per cent G+C. The remaining 10 per cent contains 44 per cent G+C. This property facilitates to separate the plasmids from one another or from chromosomal DNA. This can be accomplished by ultracentrifugation in cesium chloride density gradients.

The general features of plasmid replication are assumed to be similar to those of chromosomal replication. In both cases, the circular DNA is assumed to be attached to the bacterial membrane at the replicating fork.

A number of different plasmids are able to integrate with the host chromosome. A plasmid having this ability is called an **episome**. In most cases, it is not yet known whether such integrations are mediated by integrase-like enzymes or by recombinases. The host chromosome must possess an appropriate site if plasmid integration is to occur. Thus, the ability of a plasmid to behave as an episome is host-specific. For example, F_1 behaves as an episome in *E. coli*, but not in *Proteus mirabilis*. It is known that F_1 DNA has some homology with the *E. coli* chromosome, but hasn't any with the *P. mirabilis* chromosome.

Once F has integrated with its host chromosome, a second adjoining crossover event will bring about its detachment. During a second crossover event, exchange of the DNA segments takes place.

Some Col factors and R factors are also able to integrate with the chromosomes of certain hosts. They are also able to detach themselves from host-chromosome by a second crossover.

A large number of bacterial host properties have been determined by plasmid-carried genes. Plasmids have been classified into different classes as under :

(1) Fertility factors (for "fertility")
(2) R factors (for "resistance")
(3) Col factors (for "colicinogeny")
(4) Penicillinase plasmids
(5) "Degradative" plasmids
(6) Cryptic plasmids

Those plasmids which confer on their hosts the ability to transfer chromosomal markers are called F-factors. This group includes the original F factor discovered in *E. coli* K-12. It has been designated as F_1. However, many plasmids including some R factors and Col factors, can mobilize the chromosome for transfer. Hence, some authors have used the term **sex factor** to describe any plasmid possessing this potential.

R factors mediate the drug resistance property of the host. The number of inhibitory substances for which resistance may be mediated by R factors has grown to eight or more antibiotics, several heavy metals (mercury, cadmium, nickel, cobalt), and sulphonamides. Different R factors have different combinations of resistance genes, ranging from one to as many as seven or eight numbers.

Many bacterial strains liberate toxins. They are called **bacteriocins**. Toxins of this type which are liberated by strains of *E. coli* are called colicins. Most of them are simple proteins. *E. coli* strains are generally immune to the action of colicins which they produce themselves. The genetic determinants for a number of different colicins have been shown by P. Fredericq to be found on plasmids. These plasmids are called **colicinogeny** factors. The principal ones that have been studied are Col B, Col E_1, Col E_2, Col I, and Col V.

All the three genetic elements (chromosome, plasmid, and phage) are capable of recombining with one another as indicated in the following diagram:

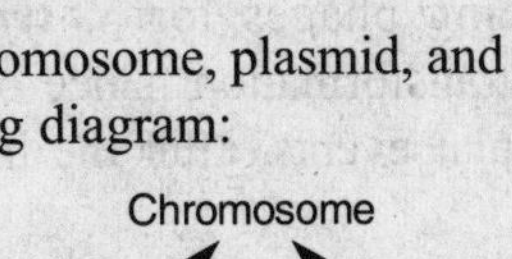

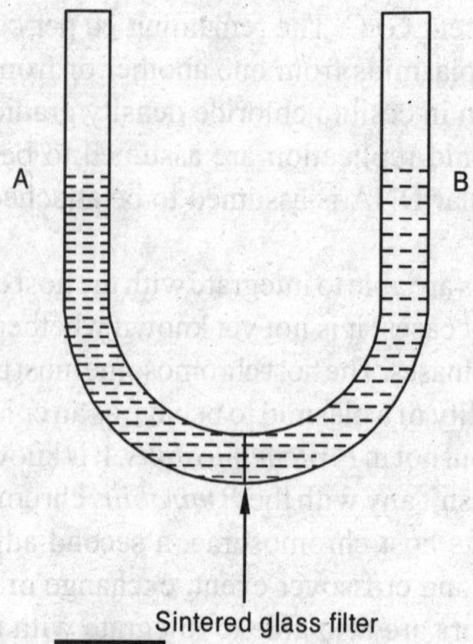

Fig. 7.4 *U-tube experiment.*

4 TRANSDUCTION BY BACTERIOPHAGE

4.1. Discovery

Transduction was discovered accidently by N. Zinder and J. Lederberg (1952) while searching for sexual conjugation among *Salmonella* species (Figure 7.4). They prepared a mixture by taking auxotrophic mutants and plated them on selective nutritional media for detecting recombinants. They found the development of prototrophic recombinant colonies at low frequency. They further carried out the U-tube experiment by using a parental auxotrophic strain in each arm (i.e. A and B). Both the arms of the U-tube were separated by a sintered glass filter. Finally, prototrophs were recovered in one arm of the tube. The bacteriophage was released from a lysogenic culture. The phage penetrated through the filter and infected the other donor strain, lysing it. The phage took up parts of the bacterial genome with it while replicating. Again, it passed through the filter, together with the added gene(s) and imparted to the recipient cell. Thus, bacteriophage serves as a vector in transferring a part of DNA from one bacterium to another. Usually a temperate phage can carry only one gene at a time during the process. If the genes are present very close together on the donor genome, both of them can be transferred by a temperate bacteriophage.

4.2. Generalized transduction

Generalized transduction is the type of transduction in which any gene on a bacterial genome can be transferred by a temperate phage. The probability for any gene of being transferred is roughly the same. Generalized transduction is found to occur in *Salmonella* and *Escherichia coli.* The phages, P_{22} and P_1, are generalized transducing phages for *S. typhimurium* and *E. coli*, respectively. Generalized transduction is similar to transformation; since only small fragments of bacterial DNA are transferred. Moreover, the intracellular events, after the injection of foreign DNA, are much the same as in transformation.

The generalized transducing phage particles contain only small amounts of phage DNA. Their formation follows a lytic cycle of phage growth.

4.3. Specialized transduction

Specialized transduction is the type of transduction in which only those bacterial genes lying immediately adjacent to the prophage attachment site on a genome are transferred by a temperate phage. It is also called restricted transduction. λ (lamda) phage of *E. coli* is a well-known specialized transducing phage particle (Morse et al. 1956). This phage particle attaches itself between the *bio* and *gal* loci on the chromosome (see Figure 7.5a). Therefore, λ was found to transduce either the *bio* loci or *gal* loci. λ cannot transduce loci lying away from the prophage attachment site.

Specialized transducing phage particles are produced following the induction of a lysogenic culture. Usually, the detachment of prophage occurs by a recombinational event within the normal area of pairing between a genome and the prophage. Pairing rarely occurs between two different sites, followed by a recombinational event. Ultimately, a circle of DNA composed of a large part of λ, together with a segment of chromosomal DNA is formed (see Figure 7.5b). If *gal* genes are present,

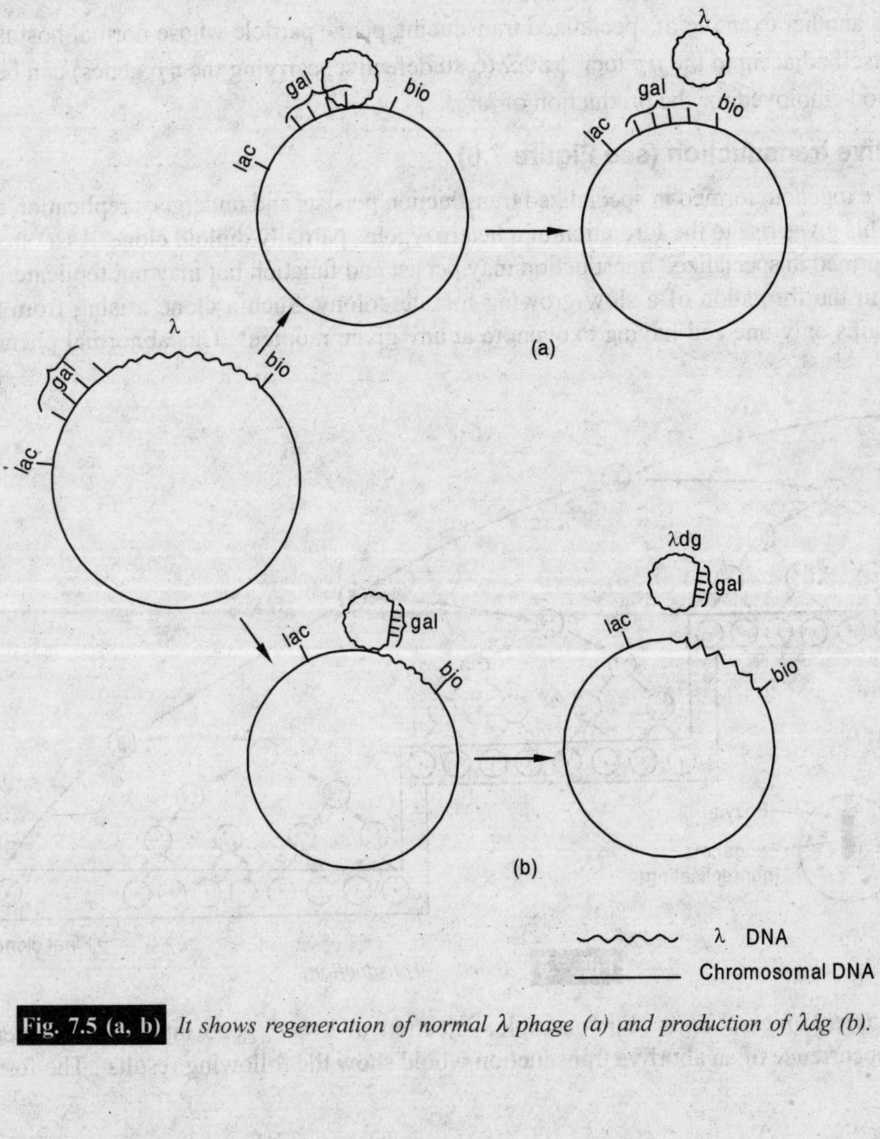

Fig. 7.5 (a, b) *It shows regeneration of normal λ phage (a) and production of λdg (b).*

in the excised DNA, the resulting phage particle is called λ*dg*. λ*dg* stands for "λ defective, carrying the *gal* loci." Similarly, λ*dbio* is produced in which *bio* genes are included in the excised DNA. A phage particle can include a limited amount of DNA. Therefore, if *gal* or *bio* genes are included in the phage, equivalent amount of λ DNA must be missing. This helps to understand why the transducing particle lacks some λ genes. Moreover, it also explains why *gal* and *bio* genes are not present in the same transducing phage particle. The lysate produced after inducing a λ-lysogenic gal^+ culture contains about one λ*dg* for every 10^5 normal phage particles. If the gal^- culture is infected with λ*dg* particles, as well as normal λ particles, the cell receiving both types of phage particles becomes a double lysogen. When such a double lysogenic culture is induced, λ*dg* can mature due to provision of missing phage functions by normal λ. Hence the lysate contains equal numbers of both λ and λ*dg* phages. Such a lysate contains roughly half the phage particles capable of transducing the gal^+ loci. This phenomenon is called **high-frequency transduction**.

ϕ 80 is another example of specialized transducing phage particle whose normal host is *E. coli*. It can attach itself adjacent to the *trp* loci. ϕ 80*dt* (ϕ 80 defective, carrying the *trp* genes) can be produced by the method employed for the production of λ*dg*.

4.4. Abortive transduction (see Figure 7.6)

Usually, the exogenote formed in specialized transduction persists and undergoes replication as part of a prophage. This gives rise to the formation of a heterozygous, partially diploid clone. As a rare event, the exogenote formed in specialized transduction may persist and function but may not replicate. Therefore, this results in the formation of a slow-growing minute colony. Such a clone arising from the partial zygote contains only one cell having exogenote at any given moment. This abnormal phenomenon is

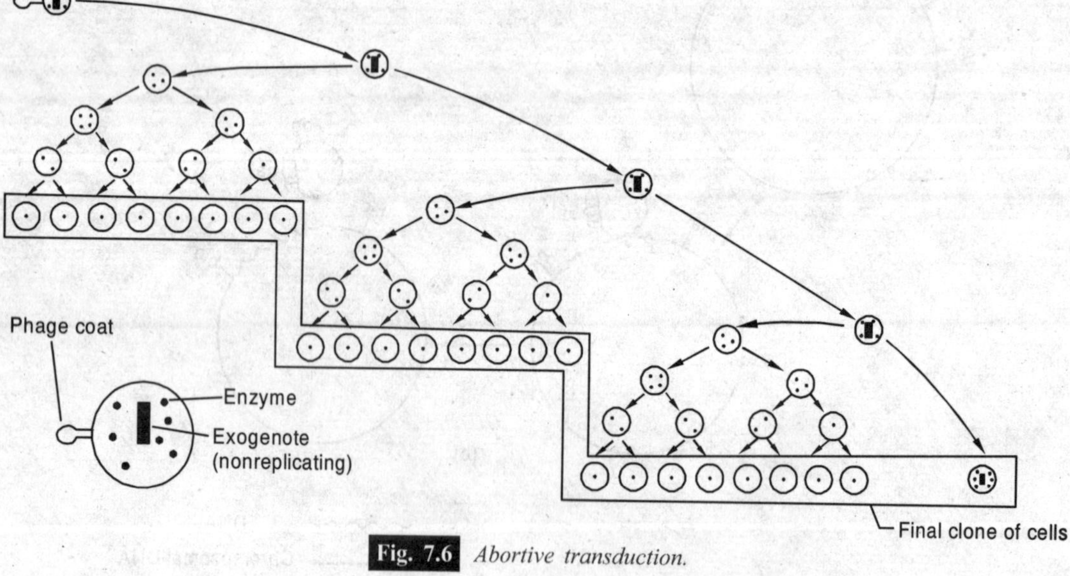

Fig. 7.6 *Abortive transduction.*

termed as *abortive transduction*. For example, injection of a his^+ gene into a his^- bacterium and subsequent occurrence of an abortive transduction would show the following results : The formation of enzyme necessary for the synthesis of histidine is continued, since the *his*$^+$ gene is functional. The partial zygote can grow and divide itself on agar lacking histidine. But the exogenote in such a partial zygote does not undergo replication. Ultimately, one of the two daughter cells does not receive the *his*$^+$ gene and, therefore, it can divide till the initial supply of enzyme is diluted by growth. The remaining daughter cell is a heterogenote, and behaves like its parent cell. Such a cell can continue to form enzyme as it is provided with the exogenote. This process is repeated during the growth cycle of a bacterium and, finally, one can find only one cell containing the exogenote in the final clone.

The abortive transduction results due to the failure of recombination event between the exogenote and the endogenote. This defect of abortive transduction can be corrected by exposing the transduced cells to the ultraviolet light. Here, U.V. light stimulates recombination in many diploid systems and thus, abortive transduction is converted into normal transduction.

4.5. Phage conversion

In some bacteria, certain properties are controlled by the injected phage genes. For example, toxin is produced by the phage-infected *Corynebacterium diphtheriae* (Groman, 1955). It should be remembered that a particular strain of phage should infect the culture for the production of the toxigenic strain of *C. diphtheriae*. Similarly, production of certain antigenic components of the cell wall of *Salmonella* depends on the presence of a particular phage. Thus, the bacterium acquires a new property soley due to the infection by a particular phage, and such a phenomenon is called *phage conversion*. Thus, phage conversion differs from specialized transduction.

8 Genetic Aspects of Enzyme Regulation

1 INTRODUCTION

Genes, which control all life processes in living organisms, act by directing the synthesis of various kinds of proteins. We also know that the cells of a single living organism retain the same set of genes during the entire life cycle, unless a change is brought about by way of mutation. The different genes are meant for different functions and, therefore, are not supposed to take part in protein synthesis at all stages of the life cycle. Therefore, it was argued that some sort of controls must be present within the cell which would direct the latter to synthesize specific proteins needed at a particular time. Considerable work in this direction has been done using microbial systems and some information is also available now for higher organisms. One of the earliest models proposed to explain the control of the gene expression is popularly known as the operon model. The operon model proposed more than twenty years ago has now been further modified.

2 'LAC' OPERON

The operon model was proposed by F. Jacob and J. Monod of Pasteur Institute, Paris, in the year 1961. In its simplest form, an operon consists of (1) one operator gene and (2) one or more structural genes. The structural genes are meant for the synthesis of specific enzymatic gene products. Several structural genes belonging to the same operon have related functions. An operator gene which always lies adjacent to the structural genes controls the expression of the latter. The operator gene and the structural genes together constitute an operon.

How does the operator gene control the expression of structural genes? The operator gene is the site of binding of another class of molecules called repressors. The repressors are synthesized under the control of a set of genes known as regulator genes. When the repressor attaches to the operator gene, the structural genes are unable to synthesize m-RNA (messenger RNA), an intermediate product in protein synthesis. Figure 8.1 shows a simple plan of an operator gene, structural genes and a regulator gene. It was proposed on the basis of Jacob and Monod's work on 'lac' operon in the bacterium *Escherichia coli*. 'Lac' operon is responsible for the synthesis of enzymes needed in utilizing lactose. It has three structural genes responsible for the synthesis of three enzymes, namely, (1) β-galactosidase gene 'z', (2) galactoside-permease gene 'y' and (3) galactoside acetylase gene 'a'.

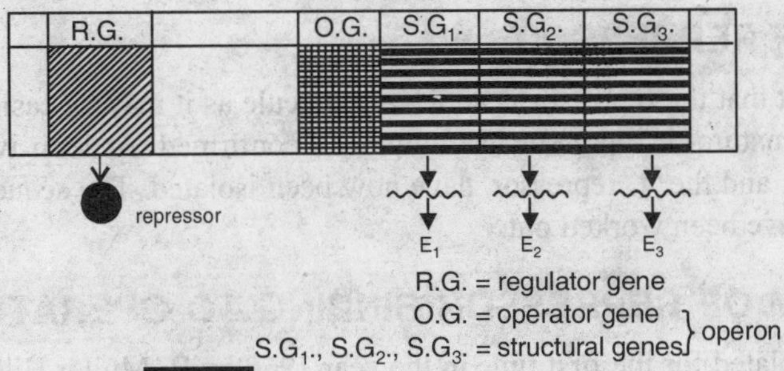

Fig. 8.1 *An operon and regulator gene.*

Although in some cases repressors may control the synthesis of only one enzyme, often a single repressor affects the synthesis of several enzymes. For instance, galactosidase repressor controls at least three enzymes listed above for the 'lac' operon. Moreover, although initially it was thought that the repressors were specific for single operons, cases are now known where the same repressor may control more than one operon, or else, more than one repressor may control the same operon. For instance, the genes responsible for arginine biosynthesis in *E. coli* are known to be distributed among three unlinked operons and one single repressor can control all the three operons in this case.

3 NEGATIVE CONTROL *vs* POSITIVE CONTROL

The control of protein synthesis through the activity of repressors is an example of negative control, because in such a control a mechanism for checking the synthesis of enzyme has been provided. In the absence of this mechanism, the synthesis will continue and no control is required to switch on the machinery. There are other systems known now where a positive control is required to switch on the machinery. In such positive control systems, the synthesis of enzyme will not take place unless a particular gene product is available. Arabinose operon of *E. coli* is a classic example of positive control. In this operon, on adding arabinose, enzyme synthesis increases but not in the absence of a particular gene product. It is presumed that the gene product along with arabinose makes a complex which works as an activator for the synthesis of m-RNA. A diagrammatic representation of arabinose operon is given in Figure 8.2. Protein synthesized through C gene (P1) works as a repressor in the absence of arabinose and, therefore, in that case hardly any enzyme synthesis takes place. The most important point which distinguishes positive control from negative control is that the regulator gene product is necessary for activation in the positive control, while it is not required in the negative control. The biochemical basis of positive control is still unknown.

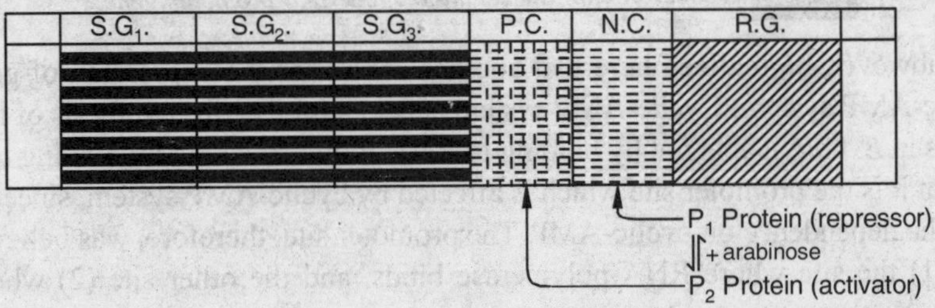

S.G1., S.G2. and S.G3. = structural genes
P.C. = site for positive control
N.C. = site for negative control
R.G. = regulator gene

Fig. 8.2 *Arabinose operon in E. coli showing positive and negative controls.*

4 NATURE OF REPRESSOR

Initially, it was thought that the repressor is an RNA molecule as it may be easily attached to operator DNA, but the protein nature of repressor has now been confirmed. At least two repressors, *E. coli.* galactosidase repressor and the λ repressor, have now been isolated. The sequence of amino acids in these two repressors have been worked out.

5 MECHANISM OF REPRESSOR BINDING TO OPERATOR GENE

'Lac' repressor was isolated for the first time in the year 1966 by B. Muller Hill and his colleagues of the Institute of Genetics, University of Koln, Germany. This repressor has now been characterized as a tetramer with four subunits, each having a molecular weight of 40,000 so that the repressor has a molecular weight of 160,000. In June 1972, Muller Hill and his colleagues tried to show how 'lac' repressor might bind to lac operator DNA. They suggested that there should be protrusions in the subunits of the repressor which should fit into the grooves of operator DNA. In making these protrusions, the first fifty amino acids on the 5' terminal region may take part. It was also suggested that the sites on the repressor for binding to the operator gene are different.

6 PROMOTER GENE AND ITS COMPONENTS

In the year 1965, another site was discovered which was called 'promoter' site. This is believed to be a site where RNA-polymerase, an enzyme responsible for the synthesis of RNA on DNA template, would bind. Mutations at this site result in reduced efficiency of the operon. The map given in Figure 8.3 shows a revised model where a promoter gene has been proposed to be present adjacent to the operator. In such a situation it was suggested that RNA-polymerase attaches to the promoter and the repressor attaches to the operator gene.

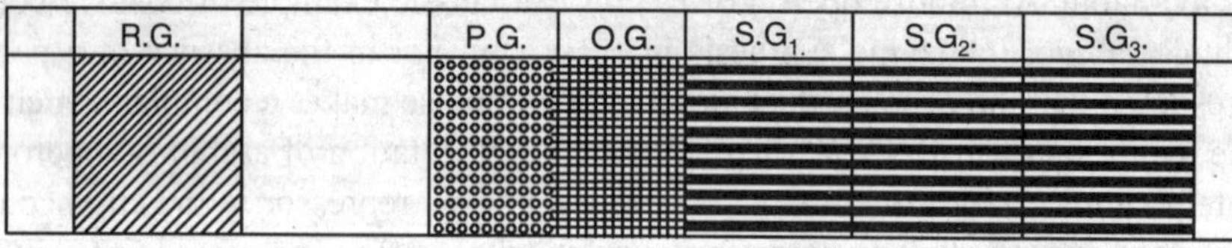

R.G. = regulator gene
P.G. = promoter gene
O.G. = operator gene
S.G_1., S.G_2., S.G_3. = structural genes

Fig. 8.3 *An operon with the regulator gene and promoter gene.*

It should, however, be realized here that another control of the expression of gene is exercised through the cyclic-AMP system. Cyclic-AMP system acts to stimulate the synthesis of m-RNA from all inducible operons in *E. coli*; it acts not by binding to RNA-polymerase but by binding to DNA. It could also be shown that it is the promoter site which is affected by cyclic-AMP system, since mutation in this site will change the dependence on cyclic-AMP. The promoter site, therefore, was believed to consist of two elements : (1) the site where RNA-polymerase binds, and the other site (2) where cyclic-AMP system acts.

In June 1972, F.R. Blattner and his coworkers from the University of Wisconsin, U.S.A also tried to show that RNA-polymerase must bind to the promoter but starts transcription at a point some 200 nucleotides away from the promoter, i.e. from the site of RNA-polymerase binding. Therefore, for RNA-polymerase, there should be a binding site and another site which will be the starting point for transcription. The question then would be how the gap between the binding site and the starting point is covered. Blattner *et al.* (1972) suggested two models. The first envisages that DNA takes up some teritiary conformation so that the two points are brought together. This may be helped by some protein which would then act as a positive regulator protein. The role of this protein would be to enable RNA-polymerase to move from binding site to the starting point. The second model presumes that RNA-polymerase drifts from the binding site to the starting point without synthesizing RNA. Therefore, if the operator lies between the binding point and the starting point it could act without itself being transcribed into RNA. It is not known where the operator site lies in relation to the binding site and the starting point. It is possible that the promoter may contain the operator site also and thus may have the characteristics of the operator site, i.e. to bind RNA-polymerase, initiate transcription, and control the expression of the structural genes. Figure 8.4 shows a map of an operon alongwith the regulator gene and the detailed structure of the promoter. The position of the operator gene with respect to RNA-polymerase binding site and starting point is not known at present, but if the operator is the site for the repressor binding and

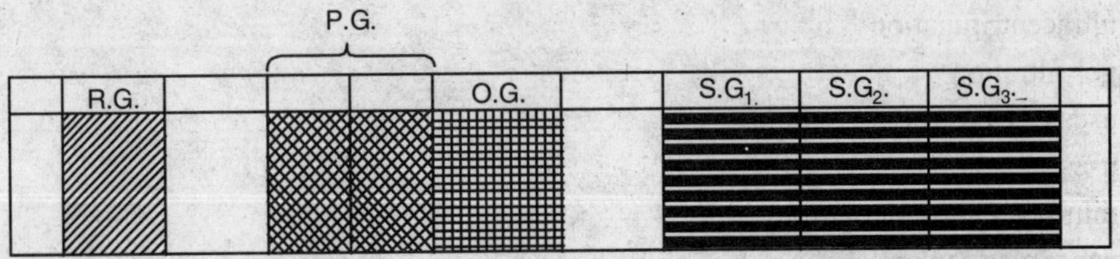

R.G. = regulator gene
P.G. = promoter gene
O.G. = operator gene
$S.G_1$., $S.G_2$., $S.G_3$. = structural genes

Fig. 8.4 *An operon with regulator and structure of the promoter gene.*

if the repressor has to act by blocking the passage of RNA-polymerase before it reaches the starting point, the operator will have to be located somewhere in between the binding site and the starting point.

9 Some Important Techniques

There are many techniques which are exploited in research work. They have helped in knowing many facts, (e.g. path of CO_2 in photosynthesis). Some of these methods have been discussed in this chapter. They may be listed as under:

(1) ultracentrifugation
(2) gel filtration
(3) uses of auxotrophic mutants
(4) uses of isotopes
(5) thin-layer chromatography
(6) paper electrophoresis

1 ULTRACENTRIFUGATION

With the help of this technique, certain properties of a molecule can be determined. Thus, molecular weight, shape, size and density can be determined. Moreover, the number of components in a protein solution can be known.

A small volume (less than 1 ml.) of protein solution is taken in a quartz cell. The cell is subjected to ultracentrifugation. The movement of the macromolecules in the centrifugal field is recorded by means of optical and photographic systems. The operation of ultracentrifuge machine is done at the speed of about 55,000 r.p.m. Initially, the solute molecules are uniformly distributed throughout the solution. During ultracentrifugation, the solute molecules are pushed towards the bottom of the cell by the centrifugal field. As a result of this, only solvent molecules are found to be present at the top of the cell. The movement also leaves a region in the cell where the solute concentration is uniform. A boundary is set up in the cell between solvent and solution in which concentration varies with the distance from the axis of rotation. The measurement of the movement of the boundary is the basis of the analytical method. By the data obtained, namely, the sedimentation rate, the Svedberg unit (S) can be calculated. A Svedberg unit has been named in honour of T. Svedberg, the Swedish pioneer in the field. It is defined as the velocity of sedimenting molecule per unit of gravitional field or 1×10^{-13} cm./sec./dyne/g. The typical S values are 4.4 for bovine serum albumin, 1.83 for cytochrome C. Molecular weights can be readily calculated, provided diffusion coefficient is known. The relationship between S and molecular weight is as indicated in the following equation:

$$\text{Mol. Wt.} = \frac{RTS}{D(1-V_D)}$$

where, R = the gas constant
T = the absolute temperature
S = the Svedberg unit
D = the diffusion constant
V = the partial specific volume
D = the density of the solution

A simple centrifugation is required in determining the number of components in a solution. In this case, the number of boundaries based on concentration gradient peaks is determined. Diffusion coefficient measurements need not be made.

2 GEL FILTRATION

The technique of separating molecules of different sizes by passing through a gel column is called **gel filtration**. Dextran is carefully cross-linked to give small beads of a hydrophilic, insoluble nature. These beads are placed in water to swell. Thus, an insoluble gel is formed. The commercial name for the gel is Sephadex. The Sephadex excludes solutes of large molecular size. For diffusion, it is accessible to molecules of small molecular size.

It is possible to fractionate proteins on gel filtration. The selection of a particular type of Sephadex beads is important in separating different fractions. The column is filled with the selected Sephadex beads. For example, Sephadex G-25 excludes molecules of the molecular weight of 3,500–4,500, Sephadex G-50, 8,000–10,000, Sephadex G-75, 40,000–50,000 and so on. Sephadex G-100 and G-200 gels may be used for higher molecular weight proteins. The solution is applied from the top of the column and thus, elution is made.

The column-gel-filtration technique may also be used in determining the molecular weights. A suitable gel is selected on the basis of the molecular weight of protein. Sephadex G-100 or G-200 is selected. The column is filled with the selected Sephadex beads. Elution volumes of pure proteins with the known molecular weights and stabilities are determined to prepare a standard curve. The protein whose molecular weight is to be determined is applied from the top of the column, and its elution volume is determined. This is performed under conditions similar to those used in standard proteins. The results are plotted and thus, the molecular weight of test protein can be determined.

3 USES OF AUXOTROPHIC MUTANTS

(a) As indicator organism in bioassay tests

The presence of any particular metabolite in a given sample can be detected by employing an auxotroph requiring that particular metabolite for growth. For example, presence of riboflavin in a sample can be detected by adding the sample in a suitable proportion to a minimal medium, and then allowing a riboflavin auxotroph to grow in this medium. Growth of the auxotroph confirms that the test sample contains riboflavin. In fermentation microbiology this bioassay test is of much importance. Organisms producing a desired matabolite, e.g. an amino acid or vitamin, can be detected by employing this test using the corresponding auxotroph.

(b) In the study of bacterial genetics

Auxotrophs are also used to study the recombination process among bacteria, i.e. transfer of genetic material from one bacterium to another forming a recombinant cell. The negative character of auxotroph, i.e. the deficiency of a nutritional factor is used to test whether the corresponding positive character from a related bacterium is transmitted to the auxotroph making it nutritionally independent. For example, suppose two auxotrophs A and B, the former requiring methionine and thiamine and the latter requiring threonine and biotin (but A synthesizing threonine and biotin and B synthesizing methionine and thiamine normally), are mixed together. And if from the resulting mixture any nutritionally independent type which has no requirement for any of the four growth factors can be isolated, it should be taken as a proof of the transfer of genetic material among the two auxotrophs.

(c) An elucidation of biosynthetic pathways

By isolating many nutritional mutants requiring the same metabolite for growth, it is possible to obtain some auxotrophs which are blocked at different points in the biosynthetic pathway of the particular metabolite. In many cases auxotrophs accumulate the compound which is the substrate of the blocked reaction. By thorough examination of the auxotrophs and their accumulated compounds it becomes possible to explain the biosynthetic pathway. Let us assume a biosynthetic pathway and some auxotrophic mutants, blocked at different steps as in Figure 9.1. A is the starting material or substrate, E is the end product metabolite while B, C, D are the intermediate compounds.

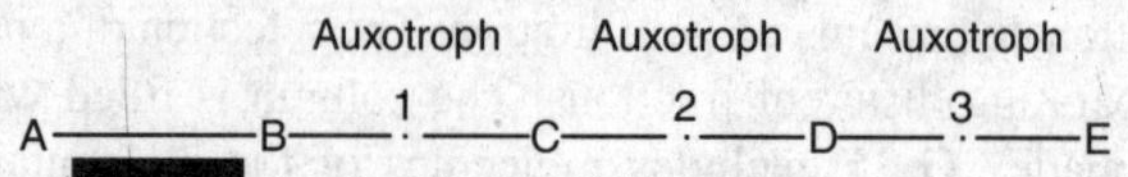

Fig. 9.1 *An elucidation of biosynthetic pathways.*

The auxotroph 3 can grow only when supplied with the product E, auxotroph 2 can grow when supplied with either D or E and auxotroph 1 can grow if supplied with either C or D or E. These three auxotrophs 1, 2 and 3 can accumulate the compounds B, C and D respectively. By identifying the accumulated compounds and determining the growth response of the auxotrophs to different intermediates, the sequence of the pathway becomes clear. It should be noted that an auxotroph blocked at a step is deficient in the enzyme of that particular step only while other enzymes of the pathway are synthesized normally.

(d) As producer organisms of fermentative products

As mentioned above, many auxotrophs accumulate the immediate precursor of the blocked reaction. This phenomenon can be industrially exploited for large scale production of some metabolites. Again, in a branched biosynthetic pathway (that is, a pathway in which after first few steps of starting from a substrate an intermediate compound is diverted through different routes resulting in different end products) if any branch is blocked by induction of auxotrophy, the synthesis of end product of the other branch may be increased, provided the nutritional factor required by the auxotroph is supplied to the organism to maintain normal growth. An example may be found in fermentative production of amino acids where high yields of lysine are obtained by using homoserine auxotroph or methionine+threonine auxotroph. Similarly, threonine can be produced by auxotrophs requiring either lysine, or methionine or both. Moreover, in fermentation microbiology other types of auxotrophs which have no direct relationship with the biosynthetic pathway of the fermentation product are also being employed, for example, in the production of glutamic acid by glycerol auxotroph of *Corynebacterium alkanolyticum*. Such auxotrophs are said to increase the accumulation of the fermentation product by enhancing the cellular permeability.

4 USES OF ISOTOPES

The rapid progress in the last two decades in biology, especially at the molecular level, has provided new and more fundamental insights into living systems. Many important discoveries which have redefined and broadened our concepts of biological phenomena are attributable to the development of new techniques and instrumentation. One of the most important of these has been the wide use of isotopes as labels or analytical tools in biological research, e.g. in tracing and elaborating the major steps in the metabolic systems.

4.1. What is an isotope?

A 'radioactive isotope' or briefly 'radioisotope' is the radioactive counterpart of a stable element. In other words, they are atoms exactly identical in all respects to normal elements except in one property-radioactivity. Isotope tracer atoms, which may also be non-radioactive, have a 'tag' or label on them which makes them instantly recognizable amongst a large number of normal atoms. This tag may either be the radioactivity or the different mass of the atom. It enables one to observe the course the atoms follow in chemical or bio-chemical reactions. For all practical purposes, chemical or bio-chemical behaviour of an isotope remains identical to that of the original element. This is the most important condition for the use of isotopes as tracers. The radioactive molecule under investigation expresses its distinctiveness only at the moment of disintegration of the isotopic atom within it. An element can be made radioactive by irradiation with high energy particles from nuclear reactors.

The rate of reactions is to some extent influenced as the isotopes differ in mass from the normal ones. This sometimes causes difficulty with complex biological systems where reactions do not usually reach completion. But this has again proved beneficial to researchers in elucidating the mechanisms of enzymatic reactions.

Another problem with radioisotopes is that the emission of radiation causes ionisation which, in turn, causes cell damage. Beta emitters are very dangerous to tissues, because this radiation is readily absorbed by an aqueous medium and may cause genetic damage. So, to decide the proper dose for a particular experimental animal or a homogenate, the investigator has to take into account two main factors—the radioactivity and the quantity.

Apart from these, some ill effects upon health are also noticed which can be overcome by proper shield protection, control and careful handling of waste disposals and by properly designing the laboratory.

4.2. Selection of isotopes

The isotopes used as tracer elements are usually stable, i.e. they do not easily degenerate. Some of the widely used stable isotopes are those of carbon, hydrogen, nitrogen, oxygen, sulphur, etc.; their choice depending upon the nature of the investigation and the extent to which the labelled isotopes are lost in the process. One of the most important factors for selection of a radioisotopes is its life period. Very short-lived isotopes provide only a limited time for observation and hence are not of much use. On the other hand, long-lived isotopes emit only negligible amounts of radiation which are difficult to detect. Further, some isotopes emit large quantities of radiation and may render irrepairable damage to the tissue under investigation or the observer. Therefore, most useful radioisotopes are those which have long life periods and emit radiation continuously. Such isotopes usually have life periods from a few hours to several thousand years.

Detection of radioisotopes is comparatively simpler than finding out their concentration because they emit one or the other type of radiation. Usually the method employed to locate the radiating substance is to expose some radiation-sensitive paper or film to the tissue containing the radioisotopes. The process is called autoradiography. To find out concentration of radioactive assimilate, the quantum of radiation is measured—a method which is also dependent upon the type of radiation emitted. The common measuring devices are: electroscope, ionization chamber, scintillation counter, proportional counter, (e.g. Geiger Muller counter), etc.

4.3. Applications

(a) Nitrogen fixation

The insectivorous plants receive nitrogen from proteinaceous matter through their leaves. The other plants, however, have their nitrogen supply in the form of nitrates which are absorbed from the soil. The production of nitrate in the soil takes place through the agency of bacteria which are symbiotic to some plants. For example, *Bacterium radicicola* is capable of fixing atmospheric nitrogen in legume plants. There are also some free-living soil bacteria, e.g. *Azotobacter* which are able to utilize gaseous nitrogen.

The metabolic pathways of nitrogen fixation in *Azotobacter* have been studied by using ^{15}N as a tracer. It is found that if ^{15}N is used, it is rapidly incorporated into $^{15}NH_3$ which can be then expelled into the environment. Secondly, there is high accumulation of ^{15}N in glutamic acid, suggesting that this amino acid is situated on the metabolic pathway of nitrogen fixation. Thirdly, the external supply of $^{15}NH_3$ results in similar condition of amino acid production, (i.e. of glutamic acid) in bacteria and formation of nitrites or nitrates in the soil. Thus, with tracer technique it has been possible to understand the nitrogen-fixation cycle.

(b) Carbon pathway in photosynthesis

The first isotope used to replace non-radioactive carbon (^{12}C) in carbon pathway studies was ^{11}C. Its application met with limited success as the life time of the latter was too short for easy detection (half-life of ^{11}C is 20.5 seconds). Just before World War II, Martin Kamen and Sam Ruben at Berkeley, California, discovered another isotope ^{14}C (half-life 5,720 years). Its application has given spectacular results.

Conversion of CO_2 into carbohydrates with the help of reducing agents is the most thoroughly studied work in the field of photosynthesis. The various steps have been elucidated with the help of radioactive tracer carbon. This was done at California by Melvin Calvin, A. A. Benson and J. Bassham in 1961 which won them the Nobel prize. They experimented with a unicellular alga, *Chlorella*, and exposed it to $^{14}CO_2$ for different time periods. The plants were then killed and analysed. It was observed that phosphoglyceric acid (PGA), a 3-carbon compound, was the first photosynthetic product. Further studies revealed that the carbon at position 1 only was replaced by radioactive carbon. When $^{14}CO_2$ was added in the dark, the labelled PGA accumulated as it did not have the chance to be converted into sugars. During this process, however, the radioactivity of one of the compounds decreased due to the increase in PGA concentration. This compound was later identified as ribulose diphosphate (RDP). It was only by using $^{14}CO_2$ that the progress through the intermediate stages of the cycle could be followed.

(c) The Krebs cycle

This cycle is a chain of reactions in which chemical energy is made available to various metabolic processes and CO_2 released. It is actually a series of syntheses but without any net synthesis. Here, a 2-carbon compound enters the cycle of reactions and two CO_2 molecules leave; no weight is gained and product formed. However, energy is produced as adenosine triphosphate (ATP) which accompanies CO_2 liberation.

Though this cycle was discovered without the help of isotopes, details of the processes particularly that of participation of acetic acid could be studied only with labelled compounds. Either carboxyl-labelled (CH_3C^*OOH) or methyl-labelled (C^*H_3COOH) acetic acid was usually incorporated in such experiments. Apart from this, some side chain reactions of the metabolic pool, e.g. synthesis of glutamic acid from α-ketoglutaric acid (a chief product of the Krebs cycle), aspartic acid from oxaloacetic acid, and 'haeme' group of haemoglobin from succinic acid were also elaborated. The distribution of radioactive carbon amongst different products also indicated the priority of a reaction over others, as well as the rates at which the equilibria of different reactions were reached (Figure 9.2). In some micro-organisms it has been further confirmed that valine, an amino acid, is derived from pyruvate and that all carbon atoms of valine are donated by the latter (Figure 9.3). This was possible only by isotope incorporation and analysis of isotope distribution in the product from a single precursor-pyruvate labelled in three different positions.

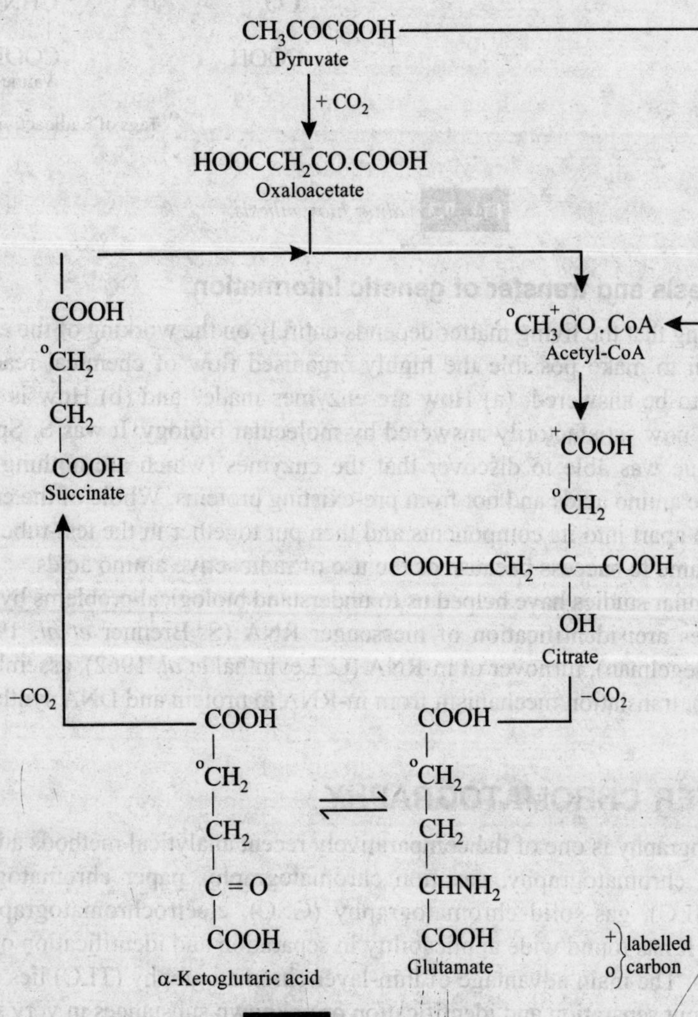

Fig. 9.2 *The Krebs cycle.*

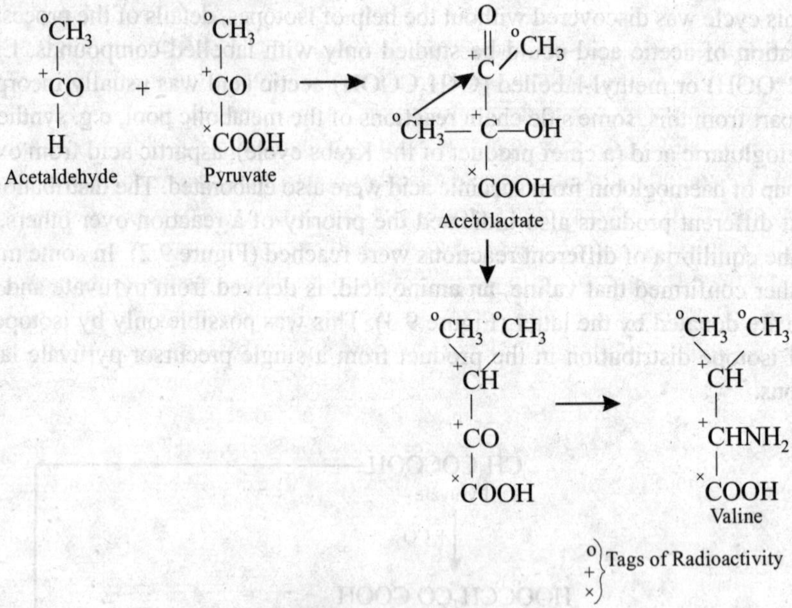

Fig. 9.3 *Valine biosynthesis.*

(d) Protein synthesis and transfer of genetic information

With the understanding that the living matter depends entirely on the working of the enzymes which act in close coordination to make possible the highly organised flow of chemical reactions, two major questions remained to be answered: (a) How are enzymes made? and (b) How is enzyme synthesis controlled? Both are now satisfactorily answered by molecular biology. It was S. Spiegelman who by radioisotope technique was able to discover that the enzymes (which are nothing but proteins) are synthesized from free amino acids and not from pre-existing proteins. Whole of the enzymatic machine was examined, taken apart into its components and then put together in the test-tube to make proteins. The entire venture came to success because of the use of radioactive amino acids.

Many more similar studies have helped us to understand biological problems by the use of isotope techniques. Examples are: identification of messenger RNA (S. Brenner *et al.* 1961), nucleic acid hybridization (S. Spiegelman), turnover of m-RNA (C. Levinthal *et al.* 1962), assembly of amino acids (M.H. Dintzis, 1961), translation mechanism from m-RNA to protein and DNA synthesis (A. Kornberg 1961), etc.

5 THIN–LAYER CHROMATOGRAPHY

Thin–layer chromatography is one of the comparatively recent analytical methods added to the list that includes adsorption chromatography, partition chromatography, paper chromatography, gas-liquid chromatography (GLC), gas-solid chromatography (GSC), electrochromatography, ion-exchange chromatogrphy, etc. It has found wide applicability in separation and identification of both organic and inorganic substances. The main advantage of thin-layer chromatography (TLC) lies in the fact that the technique can carry out separation and identification of unknown substances in very small amounts and in a very short time. None of the other methods can match it in this regard. Even paper chromatography which has earned the reputation of carrying out separation in very small amounts has been exceeded by this method. For example, TLC can carry out separation or identification in still smaller amounts and in much less time than is taken by paper chromatography.

Like paper chromatography, thin-layer chromatography is a micro-chromatographic technique in which glass plates are coated with thin layers of adsorbents like alumina, silica gel, cellulose powder, etc. These adsorbents adhere to the glass plates firmly due to the presence of binding materials added to the adsorbent slurry. The adsorption plates are sometimes called chromato-strips or chromatoplates and are used in a similar manner as paper strips in paper chromatography but they are much easier to handle.

The credit of discovering this new technique goes to two Russian workers, N.A. Izmailov and M.S. Schraiber, who in the year 1938 attempted the separation of plant extracts on alumina-coated glass plates and got a satisfactory separation. This discovery did not attract the attention of other workers till 1949. In that year, J.E. Meinhard and N.F. Hall showed some satisfactory separation with the new method. They were also the first to describe an apparatus for coating glass plates with adsorbents. But the technique could not gain popularity because of the difficulties involved in the preparation of suitable adsorbents and applying them in uniform layer on glass plates.

The method which was practically given up was rejuvenated by the untiring efforts of Egon Stahl, a German worker, who not only found a practical equipment for coating glass plates with adsorbents but also demonstrated the usefulness of this technique on a wide range of materials. Since 1958, the use of TLC has been so widespread that it has entered almost every phase of analysis. This is because the technique has been developed to such an extent that good and reproducible results are obtained readily and in a comparatively less time. Since thin-layer chromatography is based fundamentally on adsorption and separation chromatography, the same theoretical principles are involved in it.

TLC has been used in separation and identification of a large variety of organic compounds– alkaloids, terpenes, vitamins, hormones, dyes, drugs, sugars, alcohols, aldehydes, acids, and many others. TLC has also found useful application in the analysis and identification of inorganic compounds.

5.1. Technique and materials

Adsorbents and applicators are the two essential items of the TLC technique, although other accessories like glass plates, aligning tray, activating oven, developing chamber, etc., play their individual roles. The efficiency of the former two plays a very important part in the success of the technique.

Applicator : There are various types of applicators for spreading the adsorbent in slurry form over the glass plates which are used in the chromatography. Adjustable applicators are available for spreading adsorbent layers of various thicknesses ranging from 250μ to 2 m.m. over glass plates.

Adsorbents : The character of the adsorbent is of the greatest importance in this technique. Silica gel and alumina are mostly used and have proved to be quite suitable. They are used alongwith some binding materials like plaster of Paris to fix the thin and porous layer on the glass plates. Besides, kiesel gel has also been used for making thin layers. Cellulose powders with and without binding materials are also in use. Various types of cellulose powder including ion-exchange cellulose powder are in use.

5.2. Preparation of thin layers

The adsorbents are mixed thoroughly with requisite quantities of water or mixtures of water and alcohol with the help of a mechanical stirrer for about two minutes to form a homogeneous slurry. Then immediately it is spread over well-cleansed glass plates of standard size (20 × 20 cm. or 20 × 5 cm.) arranged in a row on an aligning tray with the help of the applicator. The applicator is first adjusted for the thickness of the layer, after that the slurry is put inside the applicator and the latter is drawn by hand over the glass plates arranged on the aligning tray. The coated glass plates are then allowed to dry at room temperature for 10 minutes to 2 hours during which the binder present in the adsorbent sets the adsorbent firmly over the glass plates.

After air drying, the coated glass plates are activated by drying in an oven at a suitable temperature depending on the adsorbent. The heating period similarly varies. After activation the plates are stored in a vacuum desiccator until they are used. If the plates are not stored in a desiccator and exposed to air or kept for a long time without being used, the efficiency of the plates decreases.

The samples to be analysed are applied in the same way as in paper chromatography. Usually a small amount, e.g. 4 microlitres of the solution of the substance or mixtures in a micro-pipette is applied to the starting point near one of the edges (narrow edge) of the chromatoplates in a manner such that the drop becomes as small as practicable, and then dries up. The smaller the diameter of the spot the better is the resolution. 5-10 microgram is sufficient for one operation. Substances are usually applied about 2 c.m. from one side of the plate.

The plates are then put in the chromatography chamber, usually much smaller than paper chromatography chamber. The chamber should be saturated with the solvent vapour sufficiently ahead of the actual operation. Its walls should be lined up with filter paper soaked with the solvent to maintain steady saturation inside the chamber. The solvent is put inside the chamber to a thickness of 1 cm. The spotted chromatoplates are developed by the ascending technique. The plates are held vertically inside the chamber. They can also be developed by the descending technique with certain modifications of the chamber. The development takes much less time than in case of paper chromatography. When the development is complete, the plates are taken out and allowed to dry in air inside a clean, dust-free chamber at room temperature.

Detection of the movement of spots and determination of Rf values are similar to those in paper chromatography. Detection of spots can be made by spraying the chromatogram with visualizing reagents, (e.g. ninhydrin) or in some cases it may be done by seeing the chromatogram in U.V. light.

Partition and ion-exchange TLC can be performed in the same way. Two-dimensional thin-layer chromatography has also been achieved with success by developing the single plate in two directions at right angles to each other utilizing two different solvent systems. Thin-layer electrophoresis has shown good success in separation of mixtures of various types of compounds.

6 PAPER ELECTROPHORESIS

Recent advances in analytical techniques have effectively been used in obtaining and identifying the substances in the high state of purity. Physical methods like fractional precipitation, distillation, and crystallisation have been used in the separation and purification of chemical compounds. These methods worked quite successfully in many cases but difficulty arose where the individual components of the compounds had similar physical and chemical properties. For example, fractional distillation method could be safely used in the case of mixture of liquids which have a good range of boiling point differences. However, this method could be used in the case of liquid air as well as rare gases where the boiling points of individual components are very close to each other. Likewise, after treating the biological materials by usual classical methods, one is usually left with a number of compounds such as mixture of amino acids which are similar in properties to each other. In such cases the extremes of temperature, pH, organic solvents and the use of oxidising and reducing agents are avoided as these may irreversibly change the structure of the molecule and destroy their biological activity.

Such complicated separations were achieved successfully by the techniques of chromatography and paper electrophoresis. These methods resolve the individual components under relatively mild conditions and utilise differences in the basic physical properties of the individual molecules such as their mass, size, shape, charge and adsorption properties. Paper electrophoresis method has an edge over paper chromatographic method as in the latter substances with low distribution coefficients are not properly separated. Such difficulties are not faced with the technique of paper electrophoresis which involves migration of charged substances under the influence of electric current. Paper electrophoresis is an incomplete form of electrolysis which is widely used in the separation of biological materials and many other rare and costly substances. In the compounds or mixture of compounds where some ionise and others do not, the degree of separation can be effectively made by the use of paper electrophoresis.

This method owes its development to the pains-taking research of Ar ne Wilhelm Kaurin Tiselius of Sweden who was awarded the Nobel Prize in 1948.

6.1. Technique

The paper to be used for electrophoresis is first wet with an electrolytic solution which is normally a buffer solution. The wet paper is placed in a vessel made up of anodic and cathodic compartments in such a manner that the two ends dip inside the solution. Test solutions are put on the middle portion of the paper, which is dried by pressing in between filter papers and the test solutions are put at the middle portion of the paper. A current of the order of (2-10) V/cm. is then passed through the solution. Buffer solutions serve the purpose of conductors and the wet paper as conducting bridge between the two compartments. The substances move either to the cathode or anode depending on the charge which they possess.

The movement of the substances is expressed by electrophoretic mobility (μ) which is defined as :

$$\mu = \frac{\text{distance moved/unit time}}{\text{electrical field strength}}$$

$$\text{whose unit is} = \frac{\text{cm./sec.}}{\text{volt/cm.}}$$

$$= \text{cm.}^2 \text{ volts sec.}^{-1}$$

The movement of the substances depends on many factors which are :

Size and charge: The bigger the charge the faster will be the movement, and bigger the size slower the movement.

Environmental factors: The concentration of electrolyte, ionic strength, dielectric properties, chemical properties, temperature, viscosity and the presence of non-polar molecules have their own individual and independent role.

pH effect: The direction of the movement of the substances will depend on the pH of the solution especially with those substances which have both positive and negative charges like amino acids. If we represent amino acid as:

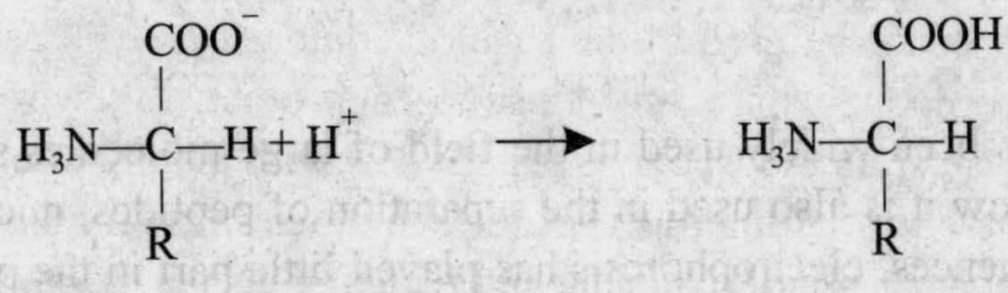

We see that the addition of HCl will cause the following reaction :

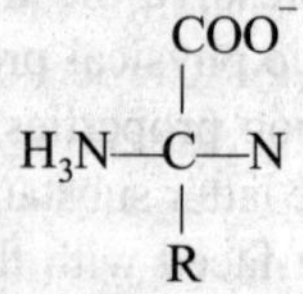

Whereas addition of NaOH results in

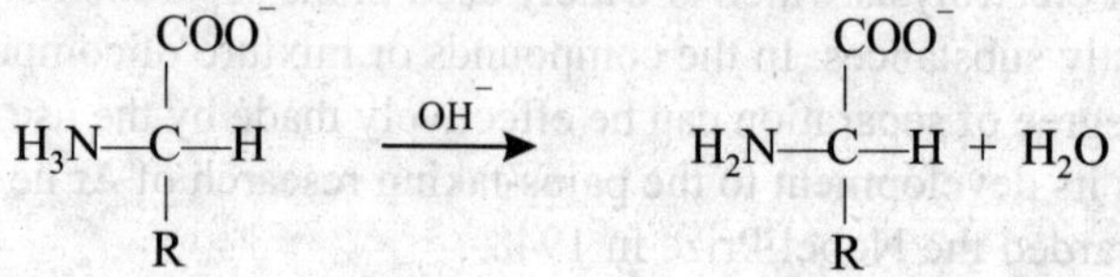

Thus, when the pH is low enough amino acid will be positively charged and will migrate towards the cathode, and at sufficiently higher pHs it will be negatively charged and will migrate towards the anode. At some intermediate pH value, termed the isoelectric point (I.E.P), no migration in an electrical field will occur in either directions. Different amino acids have different R groups. Their net charges at a given pH value will differ offering the possibility of separation by electrophoresis.

Heating effect

Paper moistened with buffer solution has resistance which depends on the nature and amount of buffer on the paper. The resistance is constant in the beginning and by Ohm's law $I = V/R$. The power consumed in watts is $V \times I$ or $I.I.R = I^2.R$.

Heat generated for 2 seconds = $I^2.Rt4.18$ Coulombs

Thus, it is evident that the passage of current produces or generates heat on the paper. The heat so produced causes evaporation of the water of buffer solution. As the solvent evaporates, salt is left behind on the paper. The increase of voltage increases the amount of salt on the paper which affects the electrophoretic mobility of substances. It has been found that the usual maximum amount of heat that can be tolerated is 0.15 watts/cm^3.

Diffusion

When the samples are put on the paper they tend to diffuse along the paper. The diffusion depends on the molecular size. It has been observed that the diffusion occurs even when no potential is applied to the system.

Electro–osmosis

When buffer solution is placed in the compartments it moves towards the cathode which means that the water is positively charged. In the test substance, the positive and negative charges move towards cathode and anode respectively. Thus the speed of the anions and cations would be affected. However, in general, this effect is ignored in electrophoresis experiments.

6.2. Applications

Paper electrophoresis has been widely used in the field of large molecules such as proteins, enzymes and nucleic acids, etc. Now it is also used in the separation of peptides, nucleotides and amino acids. Outside the biological sciences, electrophoresis has played little part in the problems of ion separation possibly because quantitation is more important and because good ion exchange procedures are available. Much work was done on the best methods of separation of serum proteins. The method is considered to be one of the most standard hospital biochemical procedures. Various diseases are shown to exhibit specific patterns following electrophoresis. The diseases affecting the liver and kidneys, malignancies, etc., show a change in quantity of different bands and occasionally a band in a different position which is of diagnostic significance. The proteins of the other component of blood, e.g. the red cell could be separated by similar procedures.

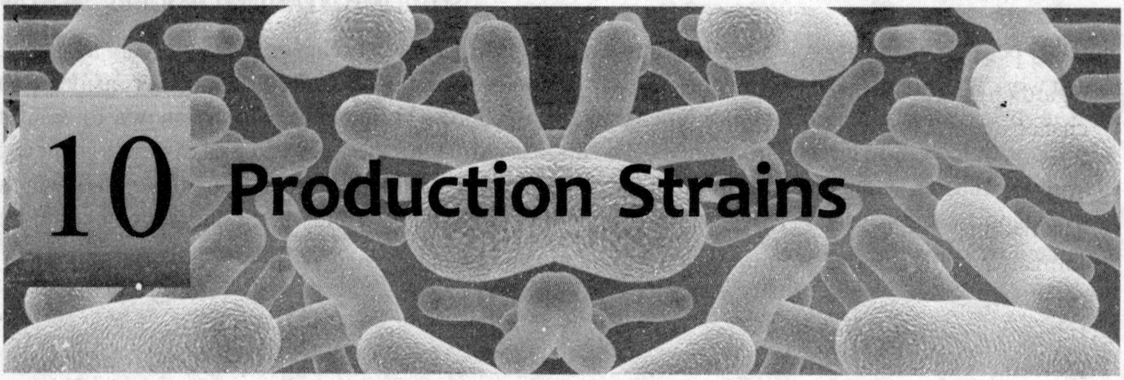

10 Production Strains

1 INTRODUCTION

The most important factor for the success of any fermentation industry is of a production strain. It is highly desirable to use a production strain possessing the following four characteristics:

(i) It should be a high-yielding strain.

(ii) It should have stable biochemical characteristics.

(iii) It should not produce undesirable substances.

(iv) It should be easily cultivated on a large-scale.

2 SCREENING TECHNIQUES

Both detection and isolation of high-yielding species from the natural source material, such as soil, containing a heterogeneous microbial population is called screening. There are many screening techniques. Usually screening programmes include primary screening and secondary screening.

2.1. Primary Screening

This consists of some elementary tests required to detect and to isolate new microbial species exhibiting the desired property. With antibiotic producers, primary screening programmes serve to remove worthless micro-organisms on the basis of relatively simple, fundamental criteria. The important selection criteria are the activity of antibiotics *in vitro*, and possibly *in vivo*, against a small number of the most important test organisms. Primary screening is also needed in the case of other useful microbial species, (e.g. micro-organisms capable of producing organic acids, amines, vitamins, etc.). The evaluation of the primary screening in industrial research programmes may be made by citing some specific examples of screening procedures as under:

The crowded plate technique: The crowded plate technique is the simplest screening technique employed in detecting and isolating antibiotic producers. It consists of preparing a series of dilutions of the soil or other source material for the antibiotic producing micro-organisms, followed by spreading of dilutions on the nutrient agar plates. The agar plates having 300 to 400 or more colonies per plate are considered since they are helpful in locating the colonies producing antibiotic activity. The ability of a colony to exhibit antibiotic activity is indicated by the presence of a zone of growth inhibition surrounding the colony. Such a colony is subcultured to a similar medium, and purified by streaking, before making stock cultures. It is necessary to carry on further testing to confirm the antibiotic activity associated with a micro-organism, since the zone of inhibition surrounding the colony may, sometimes, be due to other causes.

The crowded plate technique has a limited application, since it merely provides information regarding the inhibitory activity of a colony against the unwanted microbes that may be present by chance on the plate. Therefore, the technique has been improved upon by introducing the use of a 'test organism'. In this modified technique, agar plates which give well isolated colonies (roughly **100 to 300** colonies per plate) after incubation are flooded with a suspension of the test organism. Then the plates are subjected to further incubation to allow the growth of the test organism. The formation of inhibitory zones around certain colonies indicates their antibiotic activity. The diameter of the zones of inhibition are measured in millimeters, to obtain a rough approximation of the relative amounts of antibiotic(s) produced by various colonies. The colonies of the antibiotic producers must be isolated and purified before further testing.

Auxanography: This technique is largely employed for detecting micro-organisms able to produce growth factors, (e.g. amino acids and vitamins) extracellularly. The two major steps of the technique are as under:

(A) Preparation of First Plate

1. A filter paper strip (1.5 × 12 cm) is put across the bottom of a petri dish in such a way that the two ends pass over the edge of the dish.
2. A filter paper disc of petri dish size is placed over paper strip on the bottom of the dish.
3. The nutrient agar (45°C) is poured on the paper disc in the dish and allowed to solidify.
4. Microbial source material such as soil, is subjected to dilution such that aliquots on plating will produce well isolated colonies.
5. Plating of aliquots of properly diluted soil sample is done.

(B) Preparation of Second Plate

1. A minimal medium lacking the growth factor under consideration is seeded with the test organism.
2. The seeded medium, is poured on the surface of a fresh petri dish.
3. The plate is allowed to set.

The agar in the first plate, as prepared in step (A), is carefully and aseptically lifted out with the help of tweezers and a spatula and placed, without inverting, on the surface of the second plate as prepared in major steps (B).

The growth factor(s) produced by colonies present on the surface of the first layer of agar can diffuse into the lower layer of agar containing the test organism. The zones of stimulated growth, of the test organism around the colonies is an indication that they produce growth factor(s) extracellularly. Productive colonies are sub-cultured and are further tested.

Enrichment culture technique: This technique was designed by a soil microbiologist, Beijerinck, to isolate the desired micro-organisms from a heterogeneous microbial population present in soil. Either medium or incubation conditions are adjusted so as to favour the growth of the desired micro-organisms. On the other hand, unwanted micro-organisms are eliminated, or develop poorly since they do not find suitable growth conditions in the newly created environment. Today, this technique has become a valuable tool in many screening programmes meant for isolating industrially important strains.

Generally, it consists of the following steps:

1. Nutrient broth containing a unusual substrate, (e.g. Cellulose powder) is inoculated with the microbial source material, (e.g. soil) and incubated.
2. A small portion of inoculum from step (1) is plated onto a solid medium having the same composition. Well isolated colonies appear after incubation.
3. Suspected colonies from plate of step (2) are sub-cultured on fresh media and they are also subjected to further testing.

An example of screening of enzyme producing micro-organisms may be cited. Micro-organisms excreting alkaline proteases may be detected from the soil as under:

(i) Soil is subjected to serial dilution.

(ii) All soil dilutions are heated at 80°C for 10 minutes. This treatment kills vegetative cells but spores remain unaffected.

(iii) The plating of heat treated samples is done by spreading the samples (usually 0.1 ml.) from dilutions onto the surface of nutrient agar containing casein at pH 10-12.

(iv) The colonies surrounded by a clear zone are sub-cultured.

Use of an indicator dye: The pH indicating dyes may be employed in some screening methods for detecting micro-organisms capable of producing organic acids or amines, since a pH indicating dye undergoes colour changes according to its pH. Such dyes, (e.g. neutral red, bromthymol blue, etc.) are added to the poorly buffered nutrient agar media. The change in the colour of a dye in the vicinity of the colony suggests the capability of colonial cells to produce either organic acid(s) or amine(s), depending upon the nature of a reaction. Such colonies are subcultured to make stock cultures. However, further testing is needed, since inorganic acids or bases are also potential metabolic products of microbial growth. In other words, these methods are not foolproof.

This differential technique may also be employed in finding out whether micro-organisms are capable of certain microbial transformations or not, e.g. Δ —7 Oestrogen formation by bioconversion.

2.2. Secondary Screening

Secondary screening is strictly essential in any systematic screening programme intended to isolate industrially useful micro-organisms, since primary screening merely allows the detection and isolation of micro-organisms that possess potentially interesting industrial applications. Moreover, primary screening does not provide much information needed in setting up a new fermentation process. Secondary screening helps in detecting really useful micro-organisms in fermentation processes. This can be realized by a careful understanding of the following points associated with secondary screening:

1. It is very useful in sorting out micro-organisms that have real commercial value from many isolates obtained during primary screening. At the same time, micro-organisms that have poor applicability in a fermentation process are discarded. It is advisable to discard poor cultures as soon as possible, since such studies involve much labour and high expense.
2. It provides information whether the product produced by a micro-organism is a new one or not. This may be accomplished by paper, thin layer, or other chromatographic techniques. The compound under consideration is compared with previously known compounds.
3. It gives an idea about the economic position of the fermentation process involving the use of a newly discovered culture. Thus, one may have a comparative study of this process with processes that are already known, so far as the economic status picture is concerned.

4. It helps in providing information regarding the product yield potentials of different isolates. Thus, this is useful in selecting efficient cultures for the fermentation processes.
5. It determines the optimum conditions for growth or accumulation of a product associated with a particular culture.
6. It provides information pertaining to the effect of different components of a medium. This is valuable in designing the medium that may be attractive so far as economic consideration is concerned.
7. It detects gross genetic instability in microbial cultures. This type of information is very important, since micro-organisms tending to undergo mutation or alteration in some way may lose their capability for maximum accumulation of the fermentation products.
8. It gives information about the number of products produced in a single fermentation. Additional major or minor products are of distinct value, since their recovery and sale as by-products can markedly improve the economic status of the prime fermentation.
9. Information about the solubility of the product in various organic solvents is made available. This knowledge is useful in the recovery and the subsequent purification of the product.
10. Chemical, physical and biological properties of a product are also determined during secondary screening. Moreover, it reveals whether a product produced in the culture broth occurs in more than one chemical form.
11. It reveals whether the culture is homofermentative or heterofermentative.
12. Determination of the structure of the product is done. The product may have a simple, complex, or even a macromolecular structure.
13. With certain types of products, (e.g. antibiotics) determinations of the toxicity for animal, plant or man are made, if they are to be used for therapeutic purposes.
14. It reveals whether micro-organisms are capable of a chemical change or of even destroying their own fermentation products. For instance, micro-organisms that produce the adaptive enzyme, decarboxylase, can remove carbon dioxide from the amino acid, leaving behind an organic amine.
15. It tells us something about the chemical stability of the fermentation product.

Thus, secondary screening gives answers to many questions that arise during the final sorting out of industrially useful micro-organisms. This is accomplished by performing experiments on agar plates, in flasks or small bioreactors containing liquid media, or by a combination of these approaches.

A specific example of antibiotic producing *Streptomyces* species may be taken for an understanding of the sequence of events during a screening programme.

Those streptomycetes able to produce antibiotics are detected and isolated in a primary screening programme. These streptomycetes exhibiting antimicrobial activity are subjected to an initial secondary screening where their inhibition spectra are determined. A simple, 'giant-colony', technique is used to do this. Each of the streptomycal isolates is streaked in a narrow band across the centres of the nutritious agar plates. Then, these plates are incubated until growth of a streptomycete occurs. Now, the test organisms are streaked from the edges of the plates up to but not touching the streptomycete growth. Again, the plates are incubated. At the end of incubation, growth inhibitory zones for each test organism are measured in millimeters. Thus, the 'microbial inhibition spectrum' study extensively helps in discarding poor cultures. Ultimately, streptomycete isolates that have exhibited interesting microbial inhibition spectra need further testing. With streptomycetes suspected to produce antibiotics with poor solubility in water, the initial secondary screening is done in some different way which is out of the scope of this book.

Further screening is carried out employing liquid media in flasks, since such studies give more information than that which can be obtained on agar media. At the same time, it is advisable to use accurate assay techniques, (e.g. paper disc-agar diffusion assay) to exactly determine the amounts of an antibiotic present in samples of culture fluids. Thus, each of the streptomycete isolates is studied by using several different liquid media in Erlenmeyer flasks provided with baffles. These streptomycete cultures are inoculated into sterilized liquid media. Then, such seeded flasks are incubated at a constant temperature. Usually, such cultures are incubated at near room temperature. Moreover, such flasks are aerated by keeping them on a mechanical shaker, since the growth of streptomycetes and production of antibiotics occur better in aerated flasks than in stationary ones. Samples are withdrawn at regular intervals under aseptic conditions and are tested in a quality control laboratory. Important tests to be carried out include:

(i) checking for contamination
(ii) checking of pH
(iii) estimation of critical nutrients
(iv) assaying of the antibiotic
(v) other determinations, if necessary

The result of the above tests points out which medium is the best for antibiotic formation, and at which stage the antibiotic yields are greatest during the growth of the culture on the various media.

After performing all necessary routine tests in the screening of an actually useful streptomycete for a fermentation process, other additional determinations are made. They are:

1. Screening of fermentation media through the exploitation of which the highest antibiotic yields may be obtained.
2. Determination of whether the antibiotic is new.
3. Determination of the number of antibiotics accumulated in the culture broth is made.
4. Effect of different bioparameters on the growth of streptomycete culture, fermentation process and accumulation of an antibiotic.
5. Solubility picture of an antibiotic in various organic solvents. Also, it is to be determined whether an antibiotic is adsorbed by adsorbent materials, (e.g. ion-exchange resins or activated carbon). This knowledge is essential in the recovery and purification of an antibiotic from the fermented broth.
6. Toxicity tests are conducted on mice or other laboratory animals. An antibiotic is also tested for the adverse effects if any, on man, animal or plant.
7. The streptomycete culture is characterized and is classified up to species.
8. Further studies are made on a selected individual streptomycete culture. For example, mutation and other genetic studies for strain improvement are carried out.

In conclusion, tests are designed and conducted in such a way that production streptomycete strains may be obtained with least expenses.

Similar screening and analytical techniques could be employed for the isolation of microbial isolates important in the production of other industrial chemical substances.

3 STRAIN DEVELOPMENT

It is highly desirable that the industrial fermentation process should be made more and more economical. This largely depends upon the efficiency of the production strain involved in the fermentation process. Therefore, a person interested in starting a fermentation industry or in competing with other industries must procure an efficient strain. Thus it is clear that, the use of a high-yielding strain in any fermentation process is the most critical factor. Usually, newly isolated strains obtained by screening techniques are not so efficient as could be used in industrial fermentation processes. Therefore, such strains require improvement, as far as the yield of a particularly desired compound is concerned. This is accomplished by producing the mutant fermentation strains with the help of physical or chemical methods. These mutants may be grouped into two major categories:

(i) auxotrophic mutants

(ii) mutants resistant to analogues

Micro-organisms, usually, have regulatory mechanisms that control the amount of metabolites synthesized. Therefore, micro-organisms cannot synthesize excess of the metabolites overlimiting the cells' requirements. Obviously, suppression of these regulatory mechanisms is necessary to develop the strains for higher yields of the desired metabolites. Microbial cultures which have multivalent mechanisms, concerted repression or feed-back inhibition may be used for strain improvement. Subsequently, a search is made for mutants which have lost the ability to synthesize one of the end products capable of feed-back inhibition or repression. This may be explained by considering a situation where three end products (E.P.$_1$, E.P.$_2$, and E.P.$_3$) are synthesized via a branched biosynthetic pathway from an intermediary metabolite (A) as shown in Figure 10.1.

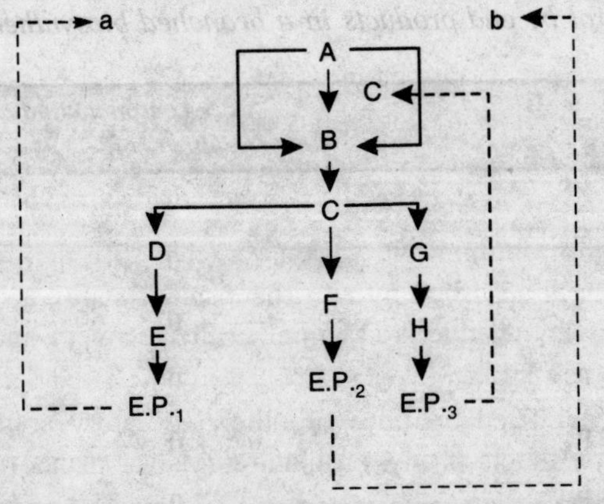

Fig. 10.1 *Regulatory system of iso-enzymes involved in a branched biosynthetic pathway.*

There are two main regulatory mechanisms that differ from each other. There may be three distinct iso-enzymes (a, b and c) capable of effecting the first reaction in the pathway (A to B). And each may be inhibited or repressed by one of the three end products.

With the multivalent or concerted regulatory mechanism, repression is only apparent if all the three end products are present together as illustrated in Figure 10.2.

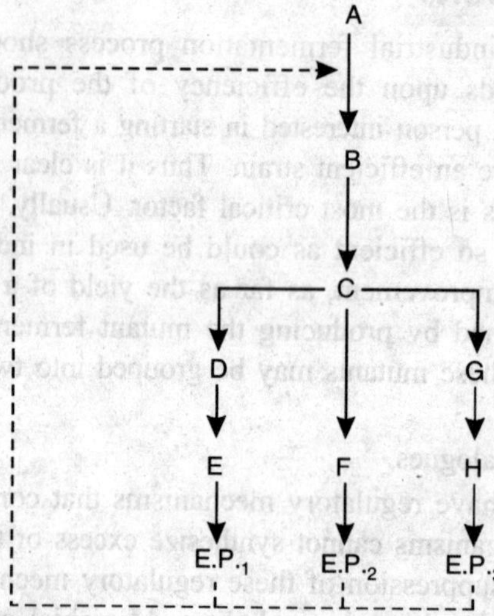

Fig. 10.2 *Multivalent (or concerted) regulatory mechanism of a branched bio-synthetic pathway.*

Here, it is to be noted that, there is only a single enzyme for the reaction A to B. It should also be noted that there is sometimes a cumulative action of the three end products as shown schematically in Table 10.1.

Table 10.1 *Feed-back control by end products in a branched biosynthetic pathway.*

End Product (in slight excess)	*Per cent inhibition or repression (multivalent)*	*(cumulative)*
$E.P._1$	0	50
$E.P._2$	0	50
$E.P._3$	0	50
$E.P._1 + E.P._2$	0	75
$E.P._2 + E.P._3$	0	75
$E.P._1 + E.P._3$	0	75
$E.P._1 + E.P._2 + E.P._3$	100	87.5

Different types of industrially important mutants have been summarized as under:

(a) A mutant strain of *Corynebacterium glutamicum* can excrete about 60 g. of lysine per litre in a medium based on glucose and minerals. This mutant strain needs homoserine. On the other hand, wild strain of this bacterium does not need homoserine and fails to excrete lysine. This discrepancy can be well explained by the schematic illustration as shown in Figure 10.3.

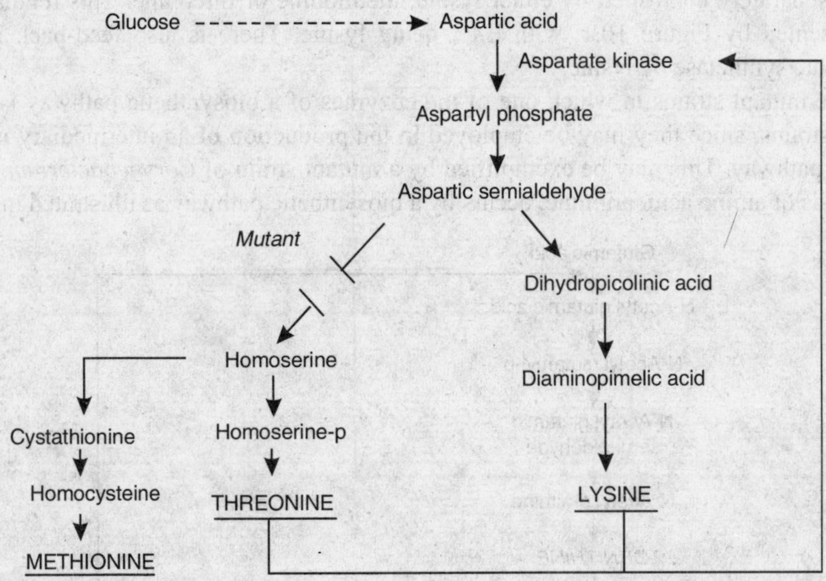

Fig. 10.3 *A branched biosynthetic pathway and control of lysine biosynthesis in Corynebacterium glutamicum.*

In the case of a wild strain, there is a common biosynthetic pathway to the biosynthesis of lysine and threonine, for the first few reaction steps. This pathway is subject to feed-back inhibition by a mixture of lysine and threonine controlling the activity of aspartate kinase. But a mutant strain requiring homoserine can no longer synthesize threonine. Moreover, feed-back inhibition no longer occurs, and lysine gets accumulated in the medium. Optimum production of lysine takes place in a medium containing 400 μg of homoserine per ml., and a high concentration of biotin (20μg/l). It may be shown that inhibition due to threonine is increased by methionine. Methionine reacts competitively with a regulatory site on aspartate kinase. The ratio of threonine to methionine also plays an important role as shown in Table 10.2.

Table 10.2 *Specific inhibition of lysine production by a mutant of Corynebacterium glutamicum.*

Amino acid added	*Concentration (in M)*	*Lysine produced (μM)*	*Control (Per cent)*
Lysine	10^{-2}	12.3	64
Methionine	10^{-2}	15.9	83
Threonine	10^{-2}	0	0
Homoserine	10^{-2}	0	0
Lysine	10^{-3}	28	145
Methionine	10^{-3}	21.8	113
Threonine	10^{-3}	0	0
Homoserine	10^{-3}	0	0
Threonine	10^{-4}	14.2	74
Methionine	10^{-4}	10.6	55
Homoserine	10^{-4}	2.4	12

(b) It should be noted that in the wild strains of *Escherichia coli,* the biosynthetic pathway to lysine is the same, but its regulatory mechanism is different. In this case, there are three aspartate kinases, each separately controlled by either lysine, methionine or threonine. This regulatory system may be represented by Figure 10.1, with $E.P._3$ being lysine. There is also feed-back inhibition of dihydropicolinate synthetase by lysine.

(c) Those mutant strains in which one of the enzymes of a biosynthetic pathway is missing are also valuable strains, since they may be employed in the production of an intermediary metabolite of that particular pathway. This may be exemplified by a mutant strain of *Corynebacterium glutamicum.* The biosynthesis of amino acid, arginine, occurs by a biosynthetic pathway as illustrated in Figure 10.4.

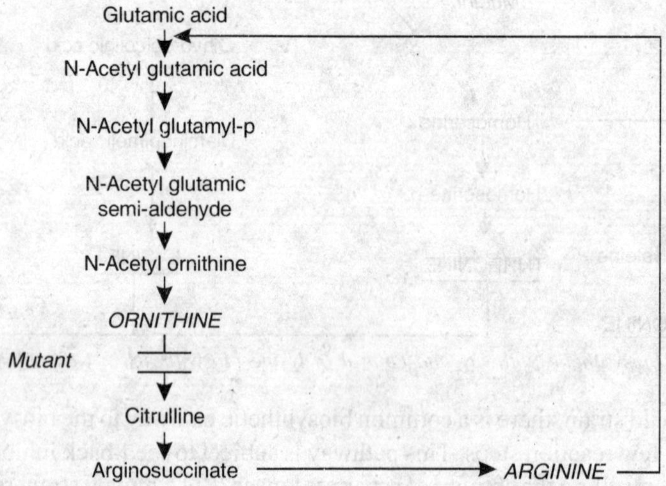

Fig. 10.4 *The biosynthetic pathway for arginine and its regulatory mechanism in Corynebacterium glutamicum.*

Now, a mutant strain which has lost the enzyme acting on ornithine will excrete that amino acid so long as just sufficient arginine is provided for growth, without enough being present to cause feed-back inhibition. Optimum production of ornithine occurs in a medium containing 200μg of arginine/ml., and 5μg of biotin per litre. Moreover, this medium should be rich in glucose and ammonium salts.

(d) There are some mutant strains with enzymes that offer resistance to feed-back control. Looking to the regulatory mechanism of feed-back inhibition, interaction between the end product and the regulatory site of an enzyme changes the enzyme configuration. Subsequently, the enzyme becomes non-functional. A mutant strain may be produced having the enzyme with an altered regulatory site. Such an altered regulatory site fails to interact with the inhibitor. Therefore, feed-back inhibition does not take place.

(e) It is also possible to use an analogue in the selection of industrially important strains. An analogue can interact with the regulatory site associated with feed-back inhibition. Such an analogue often exerts toxic effect. And, this toxicity eliminates all sensitive mutant cells in a population. For example, α-amino β-hydroxyvaleric acid is the analogue of threonine. By the use of this antimetabolite, the selection of a mutant strain may be done at the following two stages:

(1) The analogue of an amino acid, threonine, is added during the preparation of a nutrient agar medium that is poured into a sterile petri-dish. Then, the medium is allowed to solidify at an angle as shown under:

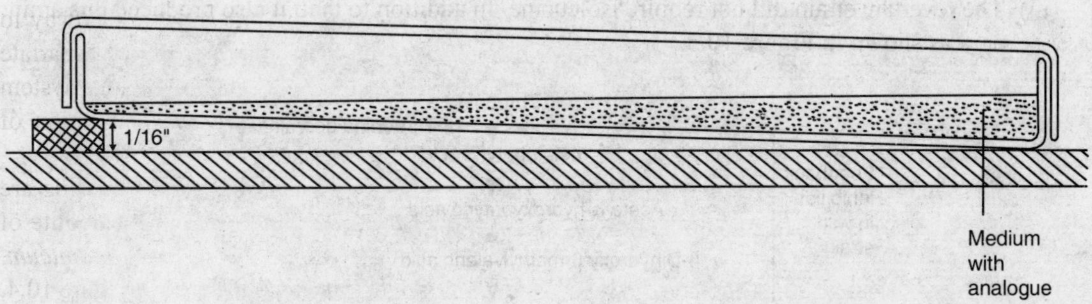

When the wedge is set, a second layer of the same medium, without analogue is poured onto it and allowed to set with the plate level as under: After some time, diffusion of an analogue into the upper

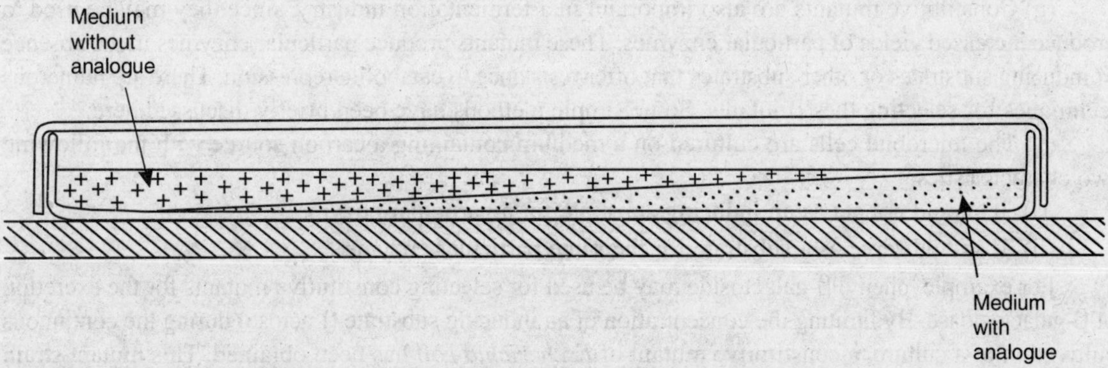

layer of the medium takes place. As a result of this, there is development of a concentration gradient at the surface. Now, a culture, previously treated with a mutagen, is spread on the surface of this medium. Then, selection of any mutants offering resistance to high concentrations is done.

(2) Lastly, a search is to be made to find out resistant mutants capable of producing threonine. This may be accomplished by inoculating the mutants, as point cultures, onto an agar medium seeded with a threonine dependent culture. The growth of seeded culture, (i.e. threonine requiring culture) around each colony of threonine excreting mutant strain may occur. The diameter of the zone of seeded culture growth depends upon the quantity of threonine produced by the mutants. Thus, analogue-resistant mutant strains excreting higher yields of threonine may be obtained.

Using the above technique, a mutant strain of *Brevibacterium flavum* capable of excreting threonine up to 12.6 g. per litre is obtained.

(f) Mutant strains may, sometimes, undergo reversion since mutations are not always stable. As a result of this, revertants may develop in the microbial population of a mutant strain. The revertant strain possesses an enzyme different from that which has been lost due to the previous mutation. Also, the enzyme is not sensitive to feed-back inhibition. This may be exemplified by the threonine deaminase of a revertant strain of the bacterial genus, *Hydrogenomonas*. The situation is represented as under:

(i) The wild strain did not require the amino acid, isoleucine. Moreover, it did not produce this amino acid.

(ii) The auxotrophic mutant required isoleucine for growth and multiplication.

(iii) The revertant strain did not require isoleucine. In addition to that, it also produced this amino acid as shown in Figure 10.5.

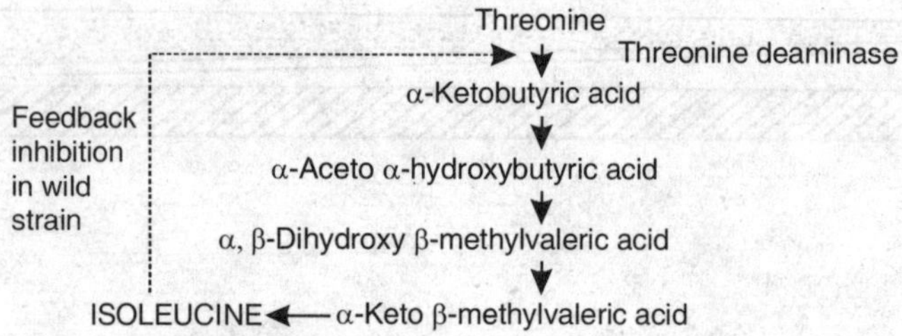

Fig. 10.5 *Biosynthesis of isoleucine from threonine.*

(g) Constitutive mutants are also important in a fermentation industry, since they may be used to produce increased yields of particular enzymes. These mutants produce particular enzymes in the absence of inducing substrates or other substrates that offer resistance to catabolite repression. There are numerous techniques for selecting these mutants. Some simple methods have been briefly discussed here:

(1) The microbial cells are cultured on a medium containing a carbon source with the following two characteristics:

(i) It should not act as an inducing substrate for a particular enzyme.

(ii) It should serve as a substrate for the same particular enzyme.

For example, phenyl β-galactoside may be used for selecting constitutive mutants for the excretion of β-galactosidase. By limiting the concentration of an inducing substrate (Lactose) during the continuous cultivation of a culture, a constitutive mutant of *Escherichia coli* has been obtained. This mutant strain produces the enzyme, β-galactosidase, as 25 per cent of its total protein.

(2) The microbial culture is cultivated in a cyclic manner, alternatively with and without an inducing substrate. For example, the microbial culture may be grown in the presence of glucose and then of lactose in a cyclic manner. After a certain number of cycles, the bacterial population will contain an increased proportion of constitutive mutants.

(3) Use of inhibitors of the inducer in a medium may also be made in the selection of constitutive mutants. For example, 2-nitrophenyl β-fucoside is an inhibitor of lactose and it may be used in selecting mutants constitutive for the production of β-galactosidase. Thus, mutants of *Escherichia coli* produce β-galactosidase without induction, provided the medium contains lactose as the sole source of carbon.

(4) It is possible to isolate constitutive mutants which offer resistance against toxic substances. The enzyme, for which the mutants are constitutive destroys the toxic substances present in the environment. For example, when a culture of the photosynthetic bacterium, *Rhodopseudomonas spheroides,* is repeatedly exposed to 0.1 M H_2O_2, 25 per cent of the surviving mutants are constitutive mutants for the production of catalase.

(5) Toxic antimetabolites may be used for selecting constitutive mutants having ability to produce an increased yield of enzymes involved in the bio-synthesis of the metabolite concerned. For example, mutants of *Lactobacillus casei* resistant to dichloroamethopterin excrete 80 times more thymidylate synthetase than those of the parent culture.

Apart from different methods for inducing high-yielding mutant strains, there is also another method for obtaining high-yielding strains. In this case, the genetic constitution of the microbial cells is changed.

This is accomplished by transferring all, or part of the DNA to the recipient culture from the donor culture. Again, there are many techniques for the transfer of genetic material. They have been listed as under:

(i) transformation
(ii) transduction
(iii) lysogeny
(iv) conjugation
(v) parasexuality

4 PRESERVATION OF MICRO-ORGANISMS

All practising microbiologists have felt the need to preserve the viability of micro-organisms with which they work. In addition, all the cultural characteristics of a culture, as they were at the time of preservation, must be conserved. The nature of work being done will determine whether the preservation requirement is only very short-term, (e.g. a few days) or for an unlimited time period, (e.g. many years). Long-term preservation of a culture is required if a culture is to be deposited in one of the service culture collections with a view to preserving something of scientific value "for perpetuity". Many methods of preservation for micro-organisms have been developed. Here, it is to be noted that there exists different types of micro-organisms (bacteria, viruses, algae, protozoa, yeasts and moulds). Therefore, there are two criteria for selecting a method of preservation for a given culture. They are:

(i) the period of preservation desired
(ii) the nature of a culture to be preserved

With the increasing importance of micro-organisms to industry, (e.g. in biochemical and antibiotic production, bio-assay, as spoilage microbes, and the like), human, animal and plant pathologists, geneticists, taxonomists and teachers have felt the need for culture collections. There are several large public service collections. These serve as repositories for cultures and as sources of their distribution.

The best known of these are the Central Bureau voor Schimmelcultures (C.B.S.), founded in 1906, the American Type Culture Collection (A.T.C.C.), founded in 1925 and the collection of the Commonwealth Mycological Institute (C.M.I.), founded in 1947. Several other countries are developing their own national collections, and there are large collections belonging to industrial concerns as well as specialized government departments.

However, any biologist dealing with living material must at least temporarily maintain his own cultures during the course of his studies and preserve them until they are ready for depositing them in one of these major collections. This depositing of important strains is most desirable as in the past, many organisms which have been the subject of intensive investigation, have been discarded at the end of the work or on the death of the biologist. Thus, much valuable material has been lost. There are three basic aims in maintaining and preserving the micro-organisms. They are:

(i) to keep cultures alive
(ii) uncontaminated
(iii) as healthy as possible, both physically and physiologically, preserving their original properties until they are deposited in any major collections, (i.e. unchanged in their properties). For very long-term preservation, involving stocks of the strains (as opposed to single specimens of each strain) and where withdrawals from stocks are regularly made, a fourth aim is to have adequate stocks and appropriate systems for replenishing stocks when necessary. This fourth

aim is very much the concern of service culture collections, of course, but the other three are the concern of any maintenance and preservation programme.

The running of the collection and the methods of maintenance used are designed to minimize the following hazards to which cultures are exposed:

(i) Repeated transfer selection can occur, either of a mutant strain or of a purely vegetative non-sporulating form. The transfer should, therefore, be done as far as possible by an expert with an eye for the wild strain. However, the fewer transfers made, the less is the risk.

(ii) Some strains, sometimes, tend to become attenuated under the artificial conditions of culture. Others deteriorate to wet slimy disintegrated mycelium or spores. Simmons (1963) suggests that this may be due to virus infections and there is considerable evidence to support his theory.

(iii) The maintenance processes to which the micro-organism, (e.g. the fungus) is subjected are selective, and only adaptive strains survive. These may have somewhat a typical characteristic.

(iv) Cultures are subject to contamination, infection with mites and adverse conditions of temperature, light, humidity, etc., responsible for their contamination. The latter may arise through breakdown of apparatus, or by incomplete understanding of the organism.

(v) Adequate documentation of the strains must be made. In a culture collection of long standing the strains may well survive several generations of microbiologists, so as to assist in maintaining them in their original condition, a clear description of the cultural characteristics supported by dried cultures should be provided at the time of depositing them.

4.1. Serial Subculture

This is the simplest and most common method of maintaining microbial cultures. Microbes are grown on agar slants and are transferred to fresh media before they exhaust all the nutrients or dry out. An exception to this is aerobic *Streptomyces* spp. where drying-up of the medium has been found successful, provided the initial growth showed the production of aerial hyphae. The drying-up of the medium appeared to encourage good sporulation and the preserved specimen became simply a dried out strand of agar coated with spores which remained viable for a few years at room temperature. For some microbial cultures, no other methods have been found satisfactory, but for the majority of species other methods are available.

There are several factors to be borne in mind while choosing a suitable medium. Solid media should be chosen in preference to liquid media, so that growth of a contaminant can be more readily observed. However, bacteriophages are often successfully maintained as suspensions in liquid media. Also, anaerobes, especially *Clostridium* spp., are frequently maintained in a liquid medium, (e.g. Robertson's cooked meat). Some technicians prefer stab cultures for maintenance. But there do not appear to be any published data to show these to be any better than slope cultures. Obviously, if the micro-organism is oxygen sensitive, a stab culture would be suggested as an extra safeguard while handling on the bench. While a rich medium may give the best initial growth for heterotrophs, it may also run the risk of accumulating toxic end-products of metabolism. Therefore, the best medium for growth may not necessarily be the best for maintenance and preservation of micro-organisms. Besides a suitable substratum, other factors affect the growth of cultures for storage. They are: light intensity, temperature, humidity, standard growth conditions, method of transfer, culture vessel and storage. For further details of these factors, the reader is requested to refer to *Methods In Microbiology,* Vol. 4 (1971), edited by C. Booth.

The time period appropriate for subculture may range from a week to even a few years (Table 10.3). Under normal conditions, cultures have to be re-grown at fairly frequent intervals, (e.g. every four, six or eight months). With a large collection, this requires much labour. Moreover, there is a risk of occurring hazards as discussed previously, everytime a culture is handled. To cut down the frequency of handling of the cultures, it is desirable to prolong the intervals between subculturing. There are various means to accomplish this, (e.g. cold storage and mineral oil storage).

Table 10.3 *Procedure for the preservation of some bacteria by serial subculture.*

Bacteria	*Medium*	*Transfer time*	*Incubation temperature °C*	*Storage temperature °C*
Bacillus spp.	Nutrient agar	12 months or longer	28	10
Pseudomonas spp.	Nutrient agar	3 months	28	10
Clostridium spp.	Robertson's cooked meat medium	6 months or longer	28	Room temperature

4.2. Preservation by overlaying cultures with mineral oil

This method of preservation is a modification of serial subculture technique. It was first extensively used by Buell and Weston (1947). Of 2000 fungus strains maintained under oil for 10 years at the C.M.I., only 45 were lost.

This method is cheap and easy, since it does not require special skills or apparatus such as a centrifuge, dessiccator, or vacuum pump. The steps involved in this method are:

(i) First of all, inoculation of the agar slant contained in a screw-cap tube with a given culture is practised.

(ii) Inoculated agar slant is subjected to incubation until good growth appears.

(iii) Using sterial technique, a healthy agar slant culture (from above step) is covered with sterile mineral oil to a depth of about 1 cm. above the top of the agar slant. If a short slant of agar is used, less oil is required.

(iv) Finally, oiled culture from step (iii) can be stored at room temperature. But better viability is obtained when stored at lower temperatures, (e.g. 15°C).

The oil used should be of good quality. British pharmacopoeia medicinal paraffin oil of specific gravity 0.865 to 0.890 is quite satisfactory. Sterilization of oil at the C.M.I. is done in McCartney bottles for 15 minutes at 15lb/in^2. However, Fennel (1960) insists that the oil be autoclaved at 15lb/in^2 for 2 hours and then dried in the oven at 170°C for 1 to 2 hours. Simmons (1963) also stresses the need for high quality oil, initial sterility and dryness.

The covering of the culture with oil prevents drying out. The oil allows slow diffusion of gases so growth continues at a reduced rate. This may induce change due to adaption to growth in oil. Some fungus isolates appear stable and survivals of over 20 years have been obtained at the C.M.I. Others change rapidly, producing a typical culture in a few months, (e.g. *Fusarium* species).

The depth of oil of 1 cm, is fairly critical (Fennel, 1960), as the oxygen transmission by layers of mineral oil in excess of 1 cm, becomes less favourable. If less oil is used, strands of *mycelium* may be exposed which allows the cultures to dry out (Dade, 1960; Fennel, 1960). If the McCartney bottles are used the rubber liners should be removed from the metal caps as the oil tends to dissolve the rubber and this can be toxic to the cultures.

This method has the following advantages:

(i) Practically all bacterial species or strains tested live longer under oil than in the control tubes without oil. Some bacterial species have been preserved satisfactorily for 15 to 20 years.

(ii) Transplants may be prepared when desired without affecting the preservation of the stock cultures.

(iii) The method is especially advantageous when working with unstable variants where occasional transfers to fresh media or growth in mass cultures results in changes in the developmental stages of the strains.

This method also appears to be an ideal method of storage for a busy laboratory with limited funds and a relatively small collection. Many workers have reported their experiences with the oiled fungus culture maintenance method (Sherf, 1943; Norris, 1944; Wernham, 1946; Buell and Weston, 1947; and others).

4.3. Lyophilization or freeze-drying

Lyophilization is the most satisfactory method of long-term preservation of micro-organisms. It is universally used for the preservation of bacteria, viruses, fungi, sera, toxins, enzymes and other biological materials. While, it offers a convenient technique for preserving a large number of cultures, it is by no means the perfect method for storing yeasts with completely unchanged characteristics.

The process of lyophilization was first applied to microfungi on a large-scale by Raper and Alexander in 1942. They were successful in processing the cultures at the N.R.R.L. (Northern Regional Research Laboratory) at Peoria and presented their work in 1945.

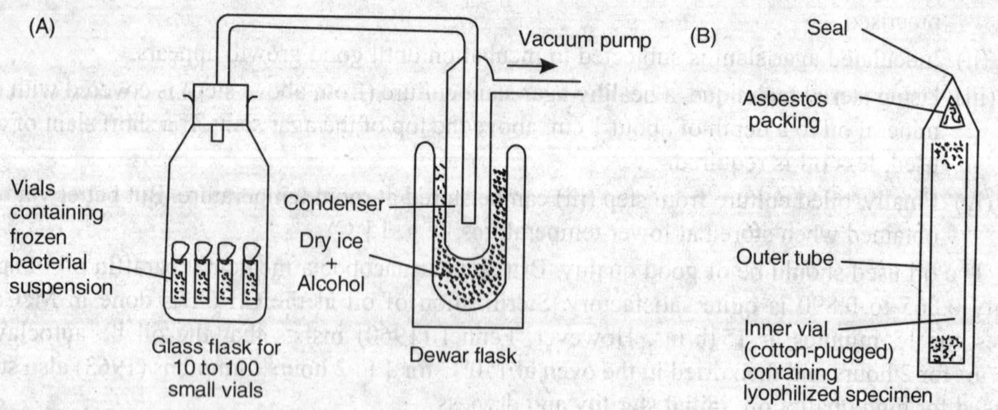

(A) Small cotton-plugged vials containing frozen suspensions of the micro-organisms are placed in the glass-flask, which is attached to a condenser. The condenser is connected with a high-vacuum pump, and this system brings about desiccation of the cultures.

(B) After desiccation of the cultures as in (A), the vials are removed, placed individually in a larger tube, covered with asbestos packing, and sealed under vacuum.

Fig. 10.6 *Lyophilization process for preservation of cultures.*

Lyophilization is perhaps the most popular form of suspended metabolism. It consists of drying cultures or a spore suspension from the frozen state under reduced pressure. There are various kinds of equipment available to do this. This can be accomplished in several ways.

Major steps involved in this technique are (Figure 10.6):

(i) A cell or spore suspension is prepared in a suitable protective medium (at the Commonwealth Mycological Institute 10% skimmed milk and 5% inositol in distilled water is found suitable).

(ii) Using a sterile technique, the suspension from (i) is distributed in small quantities into glass ampoules.

(iii) The ampoules are connected with a high vacuum system usually incorporating a desiccant, (e.g. phosphorus pentoxide, silica gel or a freezing trap), and immersed into a freezing mixture of dry ice and alcohol (–78°C).

(iv) The vacuum pump is turned on and the ampoules are evacuated till drying is complete, after which they may be sealed off.

The details of the methods used vary from one laboratory to another. Relatively simple apparatus can be constructed for processing a few ampoules, though quite complex machines are available for larger scale work. At the C.M.I. a centrifugal two-stage freeze dryer is used. In this the cooling is by evaporation and no freezing mixture is required. After drying, the small vials can be filled with sterile dry nitrogen instead of sealing under vacuum.

Dewald (1966) stressed the importance of the elimination of air and moisture from lyophilized cultures prior to sealing of the ampoules. Similar results were reported by Nei et al. (1966).

The methods of revival vary from one laboratory to the other. In case of fungi, dry pellets may be transferred to a suitable liquid and allowed to dissolve before it is streaked out on agar. At the C.M.I., a volume of sterile water equal to the original volume of the spore suspension is placed in the ampoule at room temperature. Then, the ampoule is left for about 20 to 30 minutes for the water to be absorbed slowly before streaking out. This delay seems to produce more satisfactory cultures. The degree of viability is assessed visually or by spore counts.

Factors affecting the viability of freeze-dried cultures include:

(i) Chemical composition of the protective (suspending) medium;

(ii) Addition of certain compounds to the culture suspension before freeze-drying to give protection to the culture against the toxicity exerted by moisture and oxygen when stored in unsealed ampoules;

(iii) Sealing the ampoules after freeze-drying to stop access of oxygen and moisture;

(iv) Insufficient elimination of oxygen and moisture on the survival rate of freeze-dried culture; and

(v) Storage temperature of freeze-dried and sealed cultures.

This method possesses the following advantages:

(i) As the ampoules are sealed there is no risk of contamination or infection with mites.

(ii) The prepared ampoules are easily stored, they are not readily broken and most species remain viable for many years (more than 20 years in case of many bacterial species).

(iii) There is less opportunity for cultures to undergo changes in characteristics, (i.e. they remain unchanged during storage period).

(iv) Owing to the small size of glass ampoules, hundreds of lyophilized cultures can be stored in a small storage space. In addition to this, the ampoules' small size makes them ideal for postage.

(v) Lyophilization cuts down the number of transfers.

Checking of viability may be done at long intervals. This may be done when cultures are required or by routine sampling. Usually, many replicates are made so that the material from a constant source can be supplied over a considerable period. It makes the 'Seed Stock' system possible, which is used at the American Type Culture Collection (Clark and Loegering, 1967) for conserving living reference microbes over long periods. Clark and Loegering describe this method as follows:

> 'In this system the ampoules of freeze-dried or frozen culture material from original stocks are set aside as seed stock, and stored under optimum temperature conditions. This material is never distributed. Periodically an ampoule of seed stock is opened, and the culture grown from it re-frozen or re-lyophilized in sufficient quantity for distribution. It is also used to prepare new seed stock if the viability of original material declines or the original seed stock becomes nearly depleted. In this way original material can be conserved over a long period.'

4.4. Storage at very low temperatures or Nitrogen Storage

This method is also called *cryogenic storage.* It is like lyophilization, a satisfactory method for the long-term preservation of micro-organisms. It has also been successful with many specimens which cannot be preserved by lyophilization. The maintenance of microbes is done by suspended metabolism.

According to Meryman (1956), life is regarded as 'at a standstill' at –130°C and below, so at the temperature of liquid nitrogen (–196°C), provided the cultures survive the treatment, the period of preservation should be indefinite. Thus, long-term preservation (storage) without any change in the cultural characteristics is now attainable.

Hwang protects her material with 10% glycerol and cools it slowly at the rate of about 1°C per minute to a temperature of –35°C. Thereafter, further cooling of the temperature from –165°C to –196°C was uncontrolled and faster. The procedure and apparatus are fully described (Hwang, 1966). She is able to preserve mycelial forms, *Saprolegniaceae, Pythiaceae, Entomophthoraceae* and other fungi which would not survive the vigorous process of lyophilization as well as the more resistant moulds.

Major steps involved in the performance of this method are:

(i) The culture is suspended as a cell or a spore suspension, as finely broken-up particles of mycelium, or as a piece of fungus mycelium in a suitable suspending medium (10% glycerol is used at the Commonwealth Mycological Institute). As dimethyl sulphoxide is less pleasant to handle, it would appear most satisfactory to use glycerol as a suspending medium as a routine practice and dimethyl sulfoxide should only be used when revivals are poor.

(ii) The suspension as prepared in step (i) is distributed into ampoules (These must be resistant to cold-shock). At the A.T.C.C. 1.2 ml. Wheaton 'Goldband' prescored borosilicate glass ampoules have been found satisfactory. The manufacturers of ampoules warn us that they must not be overfilled (less than 0.5 ml, being recommended).

(iii) Ampoules filled with a culture suspension are frozen and are hermetically sealed. Hwang (1966) recommends precooling to 7°C before freezing begins. Freezing can be attained by plunging the ampoules straight into the liquid nitrogen by suspending them over the liquid nitrogen for a short period and then lowering them into the liquid nitrogen, or by controlled cooling. There are many interacting factors concerning the choice of the method of freezing and it will depend largely on the micro-organisms to be frozen, the degree of revival required and the apparatus available.

(iv) The frozen ampoules prepared as in (iii) are usually clipped on metal (aluminium) canes, one above the other and six to each cane. The canes, in turn, are packed in metal boxes or canisters (aluminium), which hold about 20 canes. These are perforated to allow the free running of the liquid nitrogen. The cultures are revived by removing from the container, rapidly thawing and culturing them in the usual way.

Precautions: Cracked or faulty ampoules are dangerous. The reason is that the liquid nitrogen may penetrate and fill the ampoules. At the time of thawing, such ampoules (filled with the liquid nitrogen) may explode due to the sudden expansion of the nitrogen into gas. The storage nitrogen may also become infected. If the thawing is done in water and if an explosion occurs, the water will tend to hold the fragments of glass. But it is recommended that gloves and eye or face shield be worn when thawing out ampoules. The nitrogen itself should be handled carefully because of the risk of cold burns. Great care should be taken while handling any metal parts in direct contact with the liquid nitrogen. There is a small risk of asphyxiation while handling it in a confined space, since the gas is odourless and colourless. Hence, the presence of nitrogen in the atmosphere may not be noticed. Therefore, the nitrogen containers should be housed in a well-ventilated room. The method has the following advantages:

(i) It is an effective method of preservation.

(ii) No subculturing is required.

(iii) The cultural characteristics remain unchanged. This could be of particular importance while storing highly specialized strains employed in industrial processes.

(iv) The ampoules are not open to contamination or infection by mites, since they are sealed.

(v) The living material of a type which would not normally grow in a culture and would not be preserved in a culture collection can be retained in a viable state.

The method also has some disadvantages. They are:

(i) The method is expensive, since a costly apparatus is required.

(ii) A reliable supply of nitrogen is needed.

(iii) There is a possibility of a minor accident, since the method involves the use of explosive gas (the liquid nitrogen).

A recent short review of current research has been made by Calcott (1978).

4.5. Other Methods of Storage for Fungi

Every laboratory cannot afford lyophilization or liquid nitrogen storage. Many technicians lack the time for the periodic transfer to fresh media and they would find oil storage messy. Indeed, in these days, when the safety of the personnel is the first consideration, the spatter made by oily needles when they are sterilized is undesirable. Therefore, various other methods of preservations have been developed.

Soil cultures: This method is particularly applied for the preservation of sporing microbes, (e.g. *Bacillus, Penicillium, Aspergillus* and *Streptomyces*).

Steps involved in this method are listed as under:

(i) A spore suspension is first prepared which may involve the use of a special medium.

(ii) A mixture of soil (20%), sand (78%) and calcium carbonate (2%) is prepared and distributed into tubes (a few grams per tube). They are sterilized for 8 to 15 hours at 130°C and then cooled.

(iii) A spore suspension is added to the sterilized loam as prepared in step (ii) and allowed to grow for about 10 days.

(iv) The inoculated tubes as in step (iii) are kept in a desiccator under vacuum. The reason behind this is to evaporate the excess water. Then the tubes are sealed.

(v) The culture tubes are stored in a refrigerator at about 5 to 8°C temperature.

Thus, spores may be preserved for many years without germination.

Silica gel cultures: A spore suspension in 5% skimmed milk is added to anhydrous silica gel in screw capped bottles. This is allowed to dry for about 14 days. Then the caps are screwed down. Usually the bottles are stored in a refrigerator, though storage at room temperature can be considered quite satisfactory. The precaution to be taken is to cool everything well before starting the process. The reason behind this is heat liberation when water is added to silica gel. Very good revivals are obtained over several years. This is an adaption of the method of Roberts following Ogata (1962) who used it for the preservation of genetic stocks of *Neurospora*.

Water storage: Surprising success has been reported (Boeswinkel, 1976; Marx and Daniel, 1976) for storage of fungus cultures in water.

Steps involved in this simple method are:

(i) The fungus is grown on a weak medium in a Petri-dish and then cut into small pieces.

(ii) By using sterile technique, the pieces are transferred to sterile distilled water in small screw capped bottles.

(iii) The cap is screwed down.

(iv) The cultures are stored, preferably at a low temperature.

Other methods of preservation are: storage of spores in sand, drying cultures on filter paper and sealing tubes with paraffin wax.

5 PREPARATION OF INOCULUM

The microbial cells or spores required to inoculate large quantities of production medium in a fermentor, (e.g. 30,000 gallon tank) are in high numbers, since a loopful of culture cannot meet the requirement of completing the fermentation run in the economically shortest time. Therefore, microbes in the stock culture are serially transferred to a special inoculum medium with increasing the volume at each step in the sequence. Except the first step of inoculation, (i.e. inoculation from a stock culture), all steps require the transfer of about 0.5 to 5 per cent inoculum by the volume from the previous step in the sequence. Inoculum media and incubation conditions imposed during inoculum build-up are such that, rapid cell or spore growth is favoured without or with accumulation of a negligible quantity of product. For example, special sporulating medium is used in the inoculum production in penicillin fermentation. This sporulating medium favours the rapid production of spores and not the formation of penicillin.

Looking to the chemical composition, inoculum media usually differs from production media. In those fermentations where enzymatic adaption is desired, the presence of an inducing substrate in the inoculum medium is of paramount importance. This precaution eliminates a distinct lag period in the production tank immediately after inoculation while adaption is being accomplished. These media usually contain a considerably lower level of the main nutritive carbon source. In some fermentations, it is desirable to separate the microbial cells or spores from the inoculum medium before they are used for inoculation purposes in a production tank. This may be accomplished as under:

(i) Exhaustion of a carbon source from the inoculum medium just before the harvest of the inoculum can cause certain bacteria to flocculate and settle from the medium. Subsequently, either the cells may be drained from the bottom of the culture vessel or the supernatant liquid may be siphoned off.

OR

(ii) The inoculum cells may be separated from the inoculum medium by centrifugation, sedimentation, or in some other related way.

Unfortunately, any particulate matter of the inoculum medium accompanies with the cells if the second method is adopted. Therefore, it is advisable to use completely soluble components, while designing the inoculum medium. This precaution gives the inoculum cells without adhering to the inoculum medium particles. This practice of using only the inoculum cells is valuable where high-level inoculum additions, (e.g. 20 per cent) are required. And this prevents excessive dilution of the production medium, unwanted pH alterations, or carry over of the undesired matter. Examples pertaining high-level inoculum addition may be given as under: (i) bisulfite-yeast-glycerol fermentation (ii) α, ∈-diaminopimelic acid-*Enterobacter aerogenes*-L-lysine fermentation

There is a possibility for a microbial strain to undergo mutation during the inoculum preparation, since microbial cells have to pass through many generations. It does not raise a serious problem if the micro-organism employed in the fermentation process is not a mutant strain. The reason behind this is the infrequent occurrence of a particular mutation and, if it occurs, it usually is detrimental to the microbial cell or it provides no selective advantage to the microbial cell in its growth competition with wild (non-mutated) cells. On the other hand, it becomes a serious problem if mutated fermentation strain is used in the fermentation process. The reason for this is the high frequency of the occurrence of a back mutation. As a result of high frequency of a back mutation, the final inoculum may contain a great number of back mutants. This may be overcome by employing media and incubation conditions which tend to select the growth of the mutant strain. For example, *Escherichia coli* mutant strain requiring L-lysine for growth and multiplication should not be allowed to exhaust the supply of this amino acid during the inoculum preparation. If the inoculum medium does exhaust the supply of L-lysine, there occurs a back mutation at a high frequency. As a result of this, the number of back mutants become very high in comparison with mutant cells in the inoculum. In other words, the back mutant becomes dominant in the cellular population of the inoculum. Therefore, it is of utmost importance that such mutated fermentation strains should be carefully maintained in stock-cultures, so that the least possible number of back mutants will develop in the inoculum.

The technique of transferring the microbial growth from the stock-cultures to the liquid inoculum media and the initiation of growth in the liquid inoculum media depends upon the nature of micro-organism involved in the fermentation process. For example, bacterial cells or spores are suspended in a sterile diluent, (e.g. tap water) for the transfer in the liquid inoculum medium. Moreover, bacterial spores, (e.g. spores of the genus *Clostridium*) often require 'heat shocking' to induce the germination of spores. This may be exemplified by the highly heat-resistant spores of *Clostridium acetobutylicum.* The spores are germinated by placing them in boiling water for about 90 seconds before transferring them in the liquid medium. In contrast to this, spores of fungi and actinomycetes cannot be subjected to heat treatment for germination purpose since they are heat-sensitive. The alternative method to germinate such spores is to use a special medium which may induce germination. Then these germinated spores are transferred to the normal inoculum medium. The preparation of spore suspensions of fungi and actinomycetes is done by adding a suitable diluent to sporulated agar growth, followed by scraping loose of the spores with a sterile loop. In some cases, it is required to add a small quantity of a non-toxic wetting agent in the diluent to make the distribution of spores uniform in the dilution fluid. On the contrary, certain stationary fungal fermentations, (e.g. *Aspergillus niger*—citric acid fermentation) require the spores to be floated on the surface of the liquid medium. This may be accomplished by allowing the spore suspension to drain down the wall of the vessel so that the spores remain on the surface of the liquid medium. In the case of non-sporulating or poorly sporulating fungi or actinomycetes, fragmented hyphae are used to transfer the same to the liquid inoculum medium. The fragmentation of hyphae is done by vigorous scraping of the agar surface with a sterile loop, or by some other method.

Contamination must be checked at each stage of the inoculum build-up. The contamination may be minor or major. With fermentations using the hyphae of fungi and actinomycetes, heavy bacterial contamination is sometimes easily observed in the form of general turbidity and even foaming. Usually, detection of contamination is done by resorting to microscopic observation and by plating samples of the inoculum on a suitable medium so that it permits the growth of suspected contaminating microbes. Inoculum for anaerobic fermentation should be tested for contamination by subjecting the streaked plates to both anaerobic and aerobic incubation conditions so as to detect the presence of facultative anaerobes. For some fermentations, (e.g. *Clostridium acetobutylicum*—acetone-butanol fermentation), a characteristic fermentation pattern provides a clue for the presence of contamination.

We can come to the conclusion that inoculum production is a critical factor in commercial production as well as research studies. Hence, the quality as well as reproducibility should be maintained.

11 Microbial Transformations

1 INTRODUCTION

Economically important microbial products (antibiotics, alkaloids, vitamins, amino acids, industrial solvents, organic acids, nucleosides, nucleotides, fermented beverages, and fermented foods) are made by the use of micro-organisms, where their multienzyme sequences of fermentation work. But a large number of microbial products, (e.g. L-ascorbic acid, steroid hormones, 6-aminopenicillanic acid, various L-amino acids, L-ephedrine, D-fructose, vinegar and malt) are made by simple and chemically well-defined reactions involving compounds that are not related to these valuable products. Microbial transformations encompass practically every type of chemical reaction possible. They are extremely useful since some of these reactions can be carried out more economically by microbial means than by a strictly chemical manipulation. Moreover, bioconversion reactions can be carried out under mild conditions, at molecular positions that are difficult to attack chemically, and with very high yields as shown in Table 11.1.

Table 11.1 *Yields of some bioconversions.*

Substrate	*Product*	*Micro-organism*	*Wt. Yield (%)*
Sorbitol	Sorbose	*Gluconobacter suboxydans*	98
Mannitol	Fructose	*Gluconobacter suboxydans*	95
Glycerol	Dihydroxyacetone	*Gluconobacter suboxydans*	95
Glucose	5-Ketogluconic acid	*Gluconobacter suboxydans*	90
Glucose	2-Ketogluconic acid	*Pseudomonas mildenbergii*	100
Glucose	Gluconic acid	*Aspergillus niger*	97
L-Tyrosine	L-DOPA	*Aspergillus niger*	100

2 TYPES OF BIOCONVERSION REACTIONS

Micro-organisms catalysing bioconversions act as stereospecific catalysts. Thousands of bioconversions are known. Some types of bioconversions are briefly described in this chapter.

(1) *Hydroxylation*: Microbial hydroxylation of steroid intermediates at C-11 helped to meet the sharply increased demand for cortisone and hydrocortisone. The stereospecific microbial hydroxylations at practically all available carbon atoms of the steroid molecule have been reported. The various types of industrially important hydroxylations are: l lα-hydroxylation, 11β-hydroxylation, 16α-hydroxylation and 21-hydroxylation.

(2) *Hydrolysis*: Microbial hydrolysis of a large number of esters, glycosides, epoxides, lactones, β-lactams and amides has been described. Acetylated steroids have been hydrolyzed with varying degrees of selectivity by numerous micro-organisms; (-)-14-acetoxycodeine has been converted to (-)-14-hydroxycodeine and atropine to tropine; the sugar moiety has been removed from heart glycosides and saponins; and L-amino acids have been produced from their optically inactive forms. Tartaric acid is produced from glucose by *Acetobacter suboxydans* and, more recently, a method has been developed for its preparation from maleic anhydride by hydrolysis of the intermediary *cis*-epoxy-succinic acid:

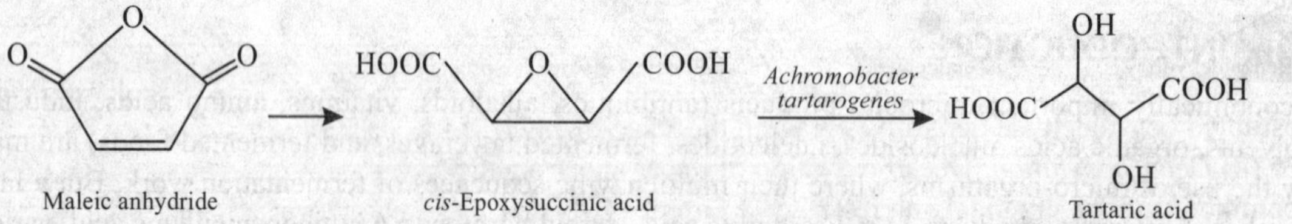

The significance of microbial hydrolysis can be understood in the production of semisynthetic penicillins (Figure 11.1). The acid, 6-aminopenicillanic acid (6-APA), is the principal intermediary in the preparation and manufacture of semisynthetic penicillins. 6-APA is manufactured on a large scale by selective enzymatic hydrolysis of penicillin G or penicillin V. This selective hydrolysis is accomplished by penicillin acylases. Penicillin acylases generally are of two types:

(i) Bacterial acylases which primarily bring about hydrolysis of penicillin G, and

(ii) Fungal acylases which preferentially bring about hydrolysis of penicillin V.

The microbial hydrolysis usually is favoured by high temperatures and an alkaline pH. The reaction can be reversed at an acidic pH, whereby 6-APA is acylated to penicillins in the presence of suitable

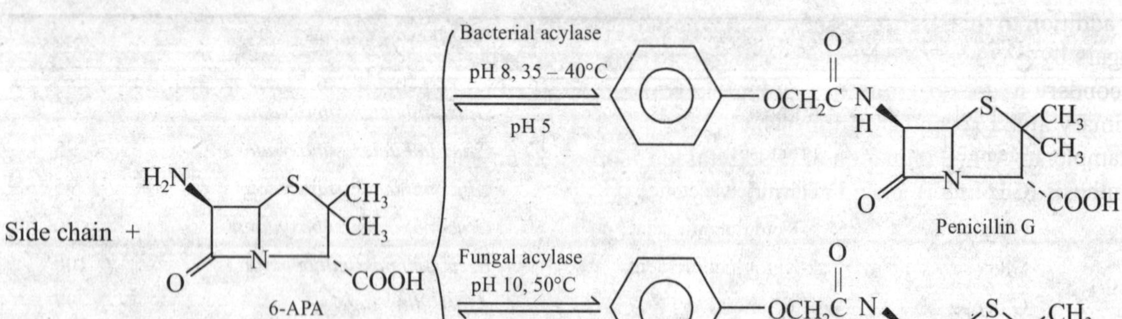

Fig. 11.1 *Production of semisynthetic penicillins by microbial hydrolysis.*

acyl donors. When the enzymatic hydrolysis has been completed, the released side chain, (e.g. phenylacetic acid) is extracted with an organic solvent and 6-APA is isolated by crystallization.

(3) *Condensation:* Asymmetric microbial condensation was discovered in 1921 and was utilized in 1934 in the synthesis of the natural (1R, 2S)-ephedrine. In this process, benzaldehyde is added to the fermenting yeast. It reacts with acetaldehyde (which is produced from glucose by the micro-organisms), and yields (R)-1-phenyl-1-hydroxy-2-propanone. The latter undergoes reductive chemical condensation with methylamine and yields the desired (1R, 2S)-ephedrine:

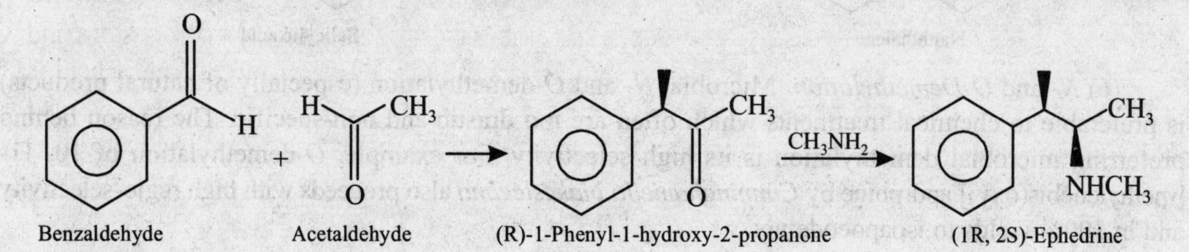

Substituted benzaldehyde derivatives react in the same manner.

(4) *Decarboxylation:* Decarboxylation of linear and aromatic carboxylic acids and of amino acids is common and of practical interest. For example, L-lysine can be synthesized by stereospecific decarboxylation of *meso*-(but not DL)-α, α′-diaminopimelic acid (DAP) to L-lysine. The reaction is catalyzed by *Bacillus sphaericus* and it leads to quantitative yields:

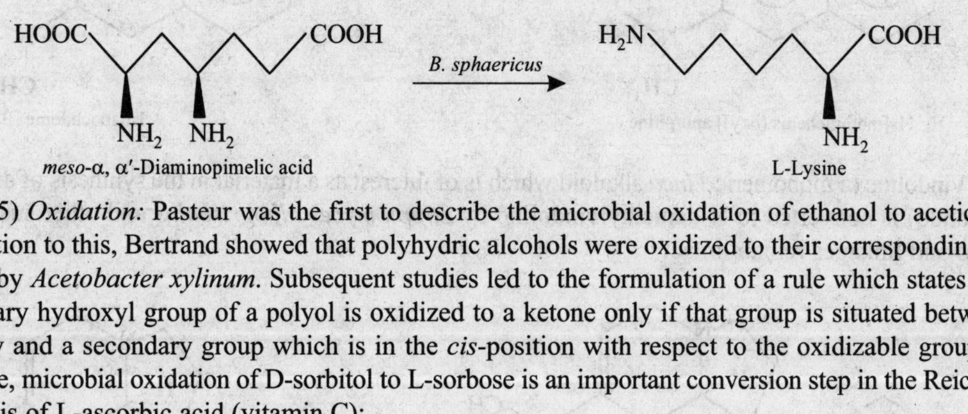

(5) *Oxidation:* Pasteur was the first to describe the microbial oxidation of ethanol to acetic acid. In addition to this, Bertrand showed that polyhydric alcohols were oxidized to their corresponding keto sugars by *Acetobacter xylinum.* Subsequent studies led to the formulation of a rule which states that a secondary hydroxyl group of a polyol is oxidized to a ketone only if that group is situated between a primary and a secondary group which is in the *cis*-position with respect to the oxidizable group. For example, microbial oxidation of D-sorbitol to L-sorbose is an important conversion step in the Reichstein synthesis of L-ascorbic acid (vitamin C):

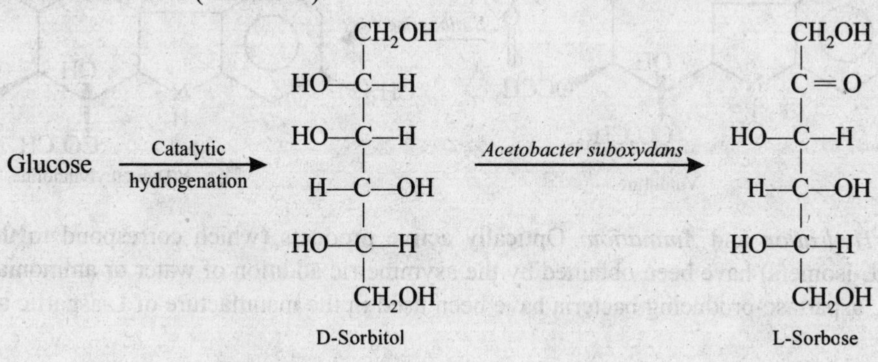

Both mono- and polynuclear aromatic hydrocarbons can be oxidized by different micro-organisms. *p*-Cymene is converted to cumic acid and *p*-xylene to *p*-toluic acid. In addition to this, a high yielding process has been developed in Japan for the production of salicylic acid from naphthalene:

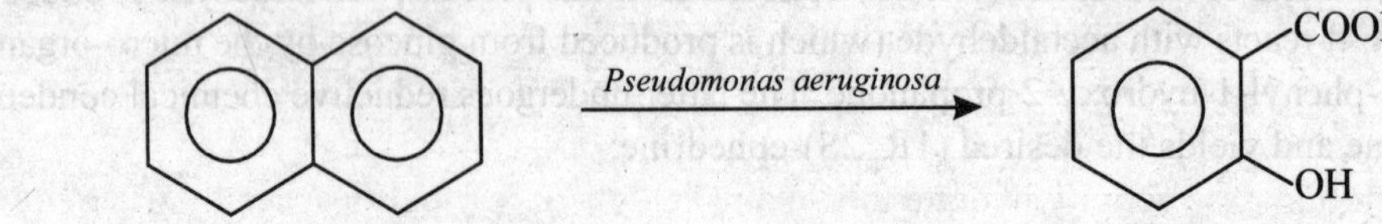

Naphthalene   Salicylic acid

(6) *N*- and *O-Demethylation:* Microbial *N*- and *O*-demethylation (especially of natural products) is preferable to chemical treatments which often are too drastic and non-specific. The reason behind preferring microbial demethylation is its high selectivity. For example, *O*-demethylation of 10, 11-[methylenebis(oxy)] aporphine by *Cunninghamella blakesleeana* also proceeds with high regio-selectivity and in 100% yields to isoapocodeine:

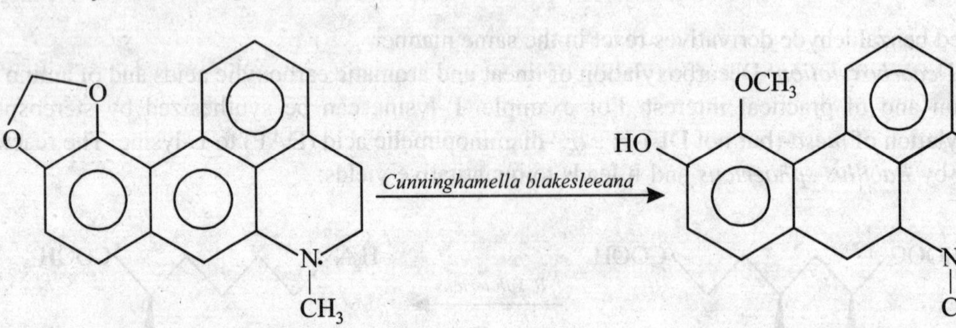

10, 11-[methylenebis (oxy)] aporphine   Isoapocodeine

Vindoline (a monomeric *Vinca* alkaloid which is of interest as a material in the synthesis of antitumor compound) is converted to N-demethylvindoline by *Streptomyces albogriseolus* and other susceptible groups and linkages remain intact :

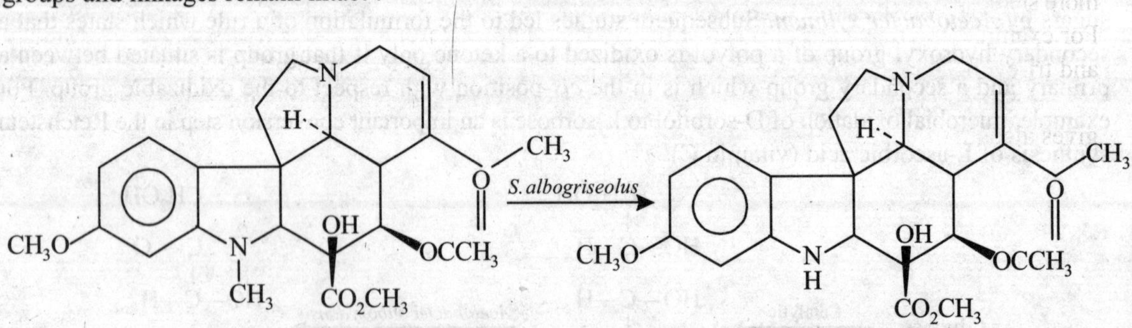

Vindoline   *N*-Demethylvindoline

(7) *Hydration* and *Amination:* Optically active products (which correspond to the naturally occurring L-isomers) have been obtained by the asymmetric addition of water or ammonia to fumaric acid. Thus, aspartase-producing bacteria have been used in the manufacture of L-aspartic acid:

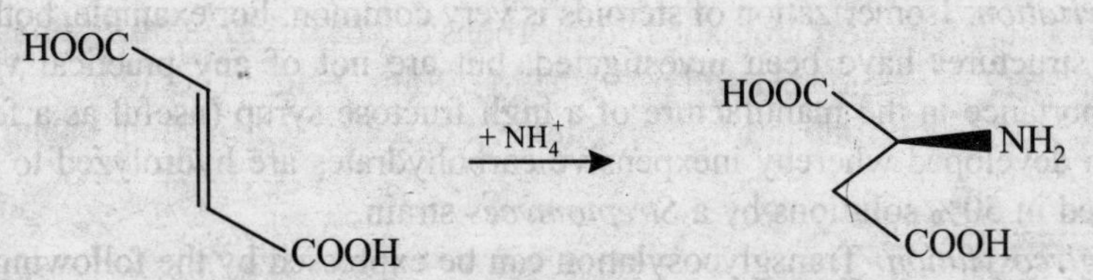

Fumaric acid L-Aspartic acid

Others with high fumarase activity hydrate to L-malic acid in 70 % yields, especially when pre-treated with detergents, (e.g. bile extracts) in order to suppress the formation of succinic acid:

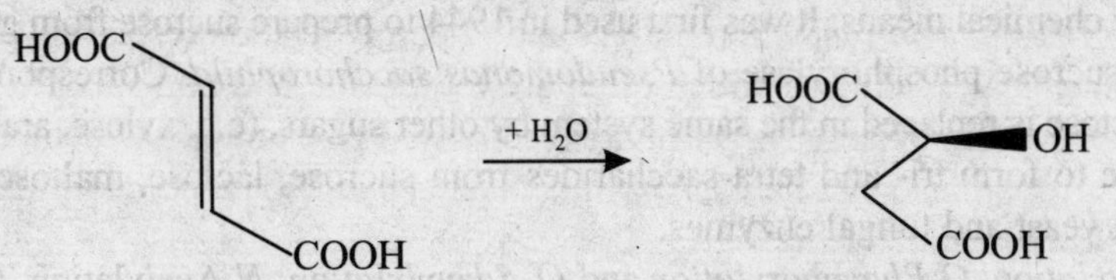

Fumaric acid L-Malic acid

Both reactions are useful in the manufacture of many L-amino acids. Amination is also involved in the manufacture of guanosine-5'-monophosphate (5'-GMP) by means of two mutants of *Brevibacterium ammoniagenes.* The first mutant ferments glucose to xanthosine-5'-monophosphate (5'-XMP) which is converted into 5'-GMP by the second mutant either by sequential operation or by mixed cultures of both mutants.

(8) *Dehydration:* Microbial dehydration of hydroxy fatty acids is quite common. Other compounds, (e.g. elymoclavine, histidine, *cis*-terpine hydrate and L-phenylalanine) also undergo the same reaction.

(9) *Deamination:* Two kinds of microbial deamination that have been observed are:

(i) hydrolytic, e.g. the conversion of L-tyrosine into 4-hydroxyphenylacetic acid in 90% yields, and

(ii) oxidative, e.g. isoguanine $\rightarrow$ xanthine and formycin A $\rightarrow$ formycin B.

(10) *Reduction:* Although the reduction of ketonic substrates usually is only partially asymmetric, more stereospecific microbial reduction depends on the size and nature of the substituents in the substrates. For example, the reduction of racemic decalone and hexahydroindanone derivatives and of related di- and tri-cyclic ketones by *Curvularia falcata* is highly stereospecific.

Stereoselective reduction occurs if the ketone is flanked by a large L and small s groups, and gives alcohol of the S-configuration :

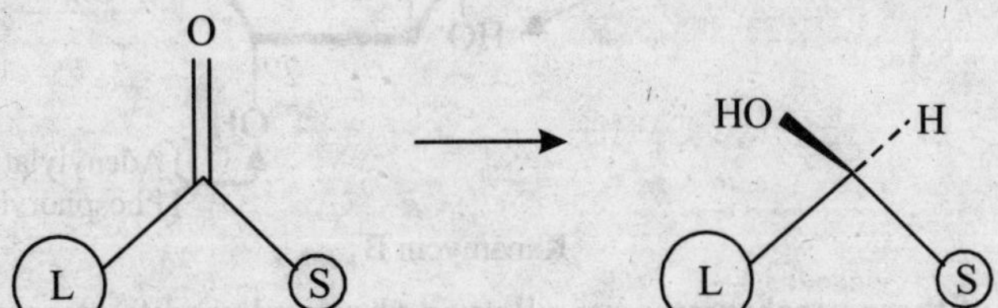

This model may help to predict if the substrate will be reduced to an optically active alcohol, which may be difficult to obtain by purely chemical means. Micro-organisms can also be used to implement selective reductions of β-diketones which are important in steroid syntheses.

(11) *Isomerization:* Isomerization of steroids is very common. For example, both the 5-androstene and 5-pregnene structures have been investigated, but are not of any practical value. Yet, it is of considerable importance in the manufacture of a high fructose syrup (useful as a food sweetener). A process has been developed whereby inexpensive carbohydrates are hydrolyzed to glucose which, in turn, is isomerized in 50% solutions by a *Streptomyces* strain.

(12) *Transglycosylation:* Transglycosylation can be expressed by the following equation:

$$\underset{\text{Donor}}{R—O—R'} + \underset{\text{Acceptor}}{R''OH} \rightleftharpoons \underset{\text{Product}}{R—O—R''} + \underset{\text{By-product}}{R'OH}$$

This reaction permits the enzymatic preparation of various oligosaccharides that cannot be made readily by strictly chemical means. It was first used in 1944 to prepare sucrose from glucose-1-phosphate and fructose by sucrose phosphorylase of *Pseudomonas saccharophila.* Corresponding disaccharides are formed if fructose is replaced in the same system by other sugars, (e.g. xylose, arabinose, or sorbose). It is also possible to form tri- and tetra-saccharides from sucrose, lactose, maltose and cellobiose by various bacterial, yeast and fungal enzymes.

(13) *N-Acetylation, O-Phosphorylation* and *O-Adenylylation: N*-Acetylation, *O*-phosphorylation, and O-adenylylation provide mechanisms by which therapeutically valuable aminocyclitol antibiotics, (e.g. kanamycin, gentamicin, sisomicin, streptomycin, neomycin, and spectinomycin) are rendered either partially or completely inactive. For example, kanamycin B is inactivated at several places by any of the three mechanisms:

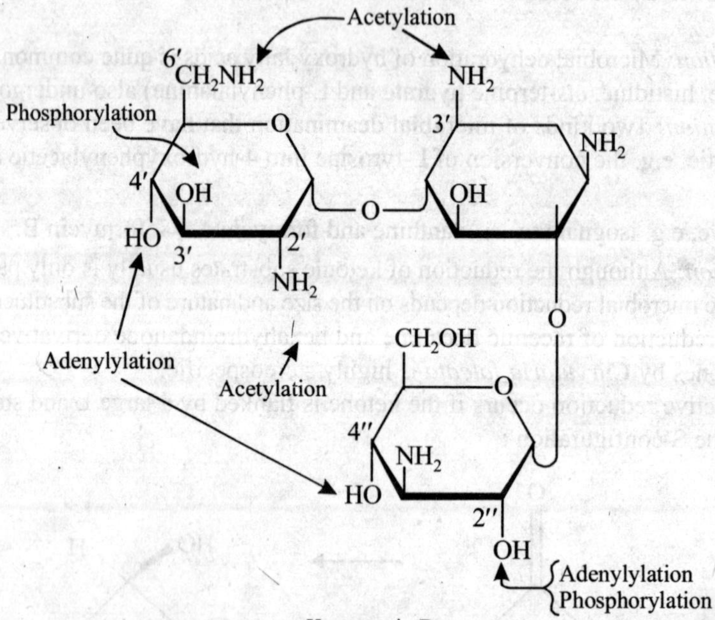

Kanamycin B

The elucidation of these mechanisms has allowed chemical modification of the sites at which the inactivation occurs. Several such bioactive analogues, (e.g. dibekacin and amikacin) have been prepared and are not subject to inactivation. Hence, these inhibit those micro-organisms against which the parent antibiotics are ineffective.

3 CONCLUDING REMARKS

From the above description, it is obvious that there are as many microbial transformation reactions as chemical transformation reactions. Some of them are economically applied in the manufacture of certain valuable products, (e.g. steroidal drugs). In the near future, these reactions will find application in the different fields, and will prove fruitful in many respects, (e.g. economical aspect).

The selection of the micro-organisms that carry out the desired transformation is of paramount importance. However, because only a few systematic examinations have been made of the microbial action on specific classes of organic compounds, there is no assured way to select the micro-organisms that will perform the desired reaction with the substrate of interest. In general, fungi and *Streptomycetes* hydrolyze various esters and carry out specific hydroxylations, yeasts reduce carbonyl groups, and certain bacteria oxidize alcohols and aldehydes.

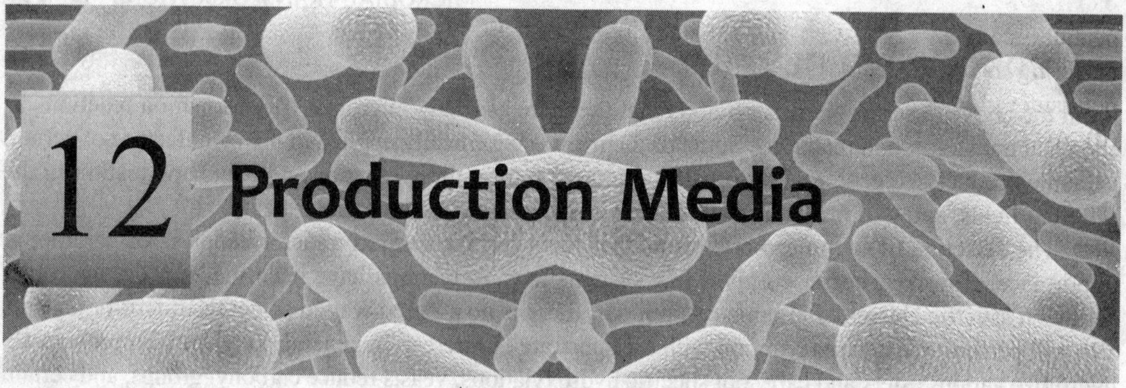

12 Production Media

1 INTRODUCTION

The growth medium meant for the cultivation of a production strain and subsequent production of either microbial cells or a biochemical product is referred to as a production or fermentation medium. One of the problems involved in industrializing a fermentation industry is to design a suitable production medium. This is done by a trial and error method. In a batch fermentation process, the fermentor is filled with the production medium leaving the head space. The production strain is added in a particular way according to the type of fermentation process, (e.g. surface culture or submerged culture method). Usually, production media are used in the liquid state to facilitate the fermentation process.

2 CHARACTERISTICS OF AN IDEAL PRODUCTION MEDIUM

No single production medium is the best or the ideal one for any and all purposes. This is not surprising in view of the variety of conditions under which media may be used, differences in chemical composition, and the many types of products to be manufactured. Had there been an ideal production medium, it would have to possess a formidable array of characteristics. A single production medium possessing these properties may never be found. Nevertheless, the specifications described below can be aimed at in the design of new production media, and they should be considered in the evaluation of the media made available for practical use.

(a) Chemical composition:

The production medium must have a suitable chemical composition. Generally, the medium should contain a source of carbon, a source of nitrogen, growth factors and mineral salts.

(b) Precursors:

In certain fermentations, (e.g. Penicillin fermentation), the medium should supply the required precursor for better yields of a desirable product.

(c) Buffering capacity:

Maintenance of the pH in the optimum range is necessary for making the process successful, since acidic and/or basic compounds depending on the nature of the fermentation process accumulate during the progress of the fermentation. In order to control the pH of the medium, buffers, (e.g. $CaCO_3$) should be added to the medium. Media containing considerable quantities of proteins, peptides, and amino acids possess good buffering capacity in the pH range near neutrality. Additional buffering capacity in this pH range is also provided by phosphates, (e.g. mono- and dihydrogen potassium or sodium phosphates).

(d) Avoidance of foaming:

Foaming is a serious problem in a fermentation industry, since it may aid in contaminating the fermentation medium and possibly cause other problems for the fermentation. Hence, defoamers, (e.g. lard oil mixed with octadecanol for penicillin fermentations) should be used for controlling foam. These defoamers may be added at the stage of media make-up (before sterilization), or they may be incorporated after sterilization or as and when called for during the fermentation.

(e) Toxicity:

Being an ideal production medium, it should be free from any toxic effect on culture or product formation.

(f) Consistency:

In aerobic fermentations, it is necessary to supply sterile air into the medium. Under such circumstances, liquid media allow the diffusion of air throughout the medium under agitation. Thus, this explains why most of the fermentation processes make use of liquid media. In addition to this, fermentation media should not be viscous, since the viscous nature of the medium creates difficulty in the penetration of the air interior of the medium. In other words, air is not easily absorbed by the liquid medium.

(g) Contamination:

Certain conditions of the production medium are helpful to check the contamination. For example, low pH values in citric acid production using *Aspergillus niger* keep the contamination under check. Moreover, media having low pH values may be sterilized at low temperatures.

(h) Recovery:

Recovery of the desired product is an important operation effecting the final cost of the product. Therefore, components of the medium should be such that separation and extraction of the product becomes easy and cheaper.

(i) Availability of raw materials:

The raw materials required for designing of the production medium should be freely available in large quantities at a reasonable price. This explains why non-synthetic media (or crude media) find wide application in large-scale production processes. Synthetic media, on the other hand, are limited for research studies, since they are expensive and low yields are produced. But these media are reproducible, allowing the researcher to study the effect of different components of a medium on the culture and product formation. Again, media may be simple or complex in their composition.

3 RAW MATERIALS

Many different types of raw material are used in different types of industrial fermentation processes. Usually, crude nutritive sources are preferred, since they are economical. Mostly agricultural products are utilized as a source of raw material in fermentation industries. We are interested in agricultural wastes for the following reasons:

(i) Agricultural wastes are produced in large quantities. For example, nearly two billion tons of animal wastes alone are produced annually in the United States.

(ii) Most of these wastes have a high biological oxygen demand (B.O.D.) and, as such, are a potential pollution hazard. It is very expensive to dispose of these wastes within the federal environmental standards.

(iii) These wastes contain a large amount of potentially usable proteins and carbohydrates. In view of the increasing world-wide demand for protein now, and the expected rise in the demand for protein in the next 30 years, it would be desirable to recover the proteins and carbohydrate materials present in the wastes.

(iv) The caloric value of wastes is recoverable as biogas and/or ethanol.

(v) Most of the agricultural wastes are renewable and, hence there is no shortage of fermentable substrate.

(vi) Often, there is a little cost involved in recovering the wastes.

(vii) In nearly all cases, agricultural wastes are not a part of the human food chain, and any process for recycling the wastes will not compete with man's food sources.

As Taiganides pointed out, wastes are really 'resources out of place or out of time' and it is very important that maximum efforts be made to turn our 'effluents into affluence'.

3.1. Saccharine Materials

Sugar cane, sugar beets, molasses, and fruit juices may be included in this category. Molasses and fruit juices will be briefly discussed here.

Molasses: Molasses is a by-product of the cane and beet sugar industry. It is recovered at any one of several stages in the sugar-refining process.

The chemical composition of sugar cane blackstrap molasses is variable. It depends not only on the quality and variety of the cane but also, and to a much greater degree, on the process involved in the manufacture of sugar. The average chemical composition of cane blackstrap molasses is given in Table 12.1.

Table 12.1 *The average chemical composition of cane blackstrap molasses.*

Component	*per cent*
Water	20
Ash	8.0
Total sugar	40-60
Total nitrogen bodies	3.0
Total nitrogen	0.5
Gums	2.0
Free acids	2.0
Combined acids	3.0

About 95 per cent of the total sugar in cane molasses is fermentable. It is particularly rich in biotin, pantothenic acid, thiamine, phosphorus and sulphur. The organic nitrogen content is less than beet molasses, since it does not contain betaine. But this substance is not assimilated by yeasts.

Beet molasses are produced by the same processes employed for cane molasses. The chemical composition of beet molasses are given in Table 12.2 and Table 12.3.

Table 12.2 *Chemical composition of beet molasses (dry weight basis).*

Component	*Per cent*
Sucrose	48-50
Raffinose	1
Glucose + fructose	1
Nitrogen compounds	12-13
Glutamic acid	3.5
Other amino acids (asparagine, aspartic acid, alanine, glycine)	5.5
Betaine	3.25-4.25
Ash (Table 12.3 for approximate composition)	11-12

Table 12.3 *Approximate chemical composition of beet molasses ash.*

Component	*Per cent*
K_2O	45
Na_2O	15
P_2O_5	0.7
CaO	3
MgO	1.8
SO_4^{2-}	6.3
SiO_2	1.1
Cl^-	18
Fe_2O_3	0.2
Al_2O_3	0.8

Vitamins such as biotin, pyridoxine, thiamine, pantothenic acid and inositol are also present. However, beet molasses may be limiting in biotin. Therefore, in fermentations involving yeast culture, a small amount of cane blackstrap molasses or other biotin-supplying material should be incorporated in the production medium. The reason behind this is that yeasts require biotin for their growth.

Cane and beet molasses are widely used in fermentation industries as shown below:

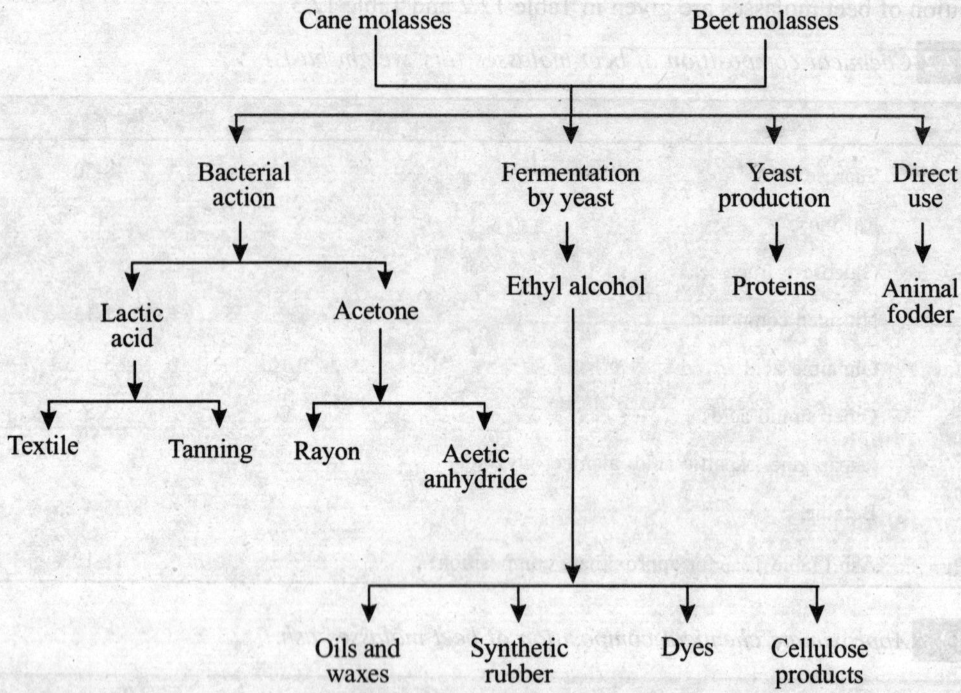

The largest utilization of cane blackstrap molasses in India at present, however, is in the alcohol industry, which utilizes it for the manufacture of spirit (ethyl alcohol), country liquors and other liquors such as rum, brandy, gin and whisky.

Fruit juices: Fruit juices contain soluble sugars. For example, the juice of grapes contains glucose and fructose. Therefore, fruit juices can be used as a source of carbon in fermentation industries. Grapes are used in the production of wine. After the grapes are harvested, they are crushed to give a raw juice (or *must*). The *must* is a highly acidic liquid. It contains 17% w/w sugar, 1% acid (mainly D-tartaric together with a little malic), and 0.3% ash (predominantly containing potassium and phosphorus).

It is desirable to have a low nitrogen content in grapes, since an excess of it may result in undesirable fermentation.

Cheese whey: The straw-coloured liquid produced as a by-product of cheese making is called cheese whey. It is a major waste product for the cheese industry. It cannot be disposed of without proper treatment. Therefore, it is desirable to use it for useful products. Considerable attention has been directed towards discovering uses for it. However, it is often more profitably sold for use as pig feed.

Currently, the cheese industry in the United States produces over 30 billion pounds of liquid whey annually. As recently as 1974, it was estimated that only a little over one-half of the whey produced was utilized. In that year, about 820 million pounds of dried whey was made.

The chemical composition of cheddar whey (Berry) is given in Table 12.4. It has an equally well balanced chemical composition:

Table 12.4 *Chemical composition of cheddar whey (Berry).*

Component	Per cent
Total solids	6.6-7.1
Protein	0.82-0.95
Fat	0.12-0.36
Lactose	4.62-5.01
Ash	0.366-0.649
K	0.135
Ca	0.047
Mg	0.010
P	0.16

Mattick and Cook reported that composition of the whey is influenced by the variety of cheese being produced. In some fermentation industries, whey can serve as an important raw material, (e.g. lactic acid production, SCP production, etc.). The reason is that it contains carbohydrate (lactose), nitrogenous substances including vitamins (e.g. riboflavin) and inorganic salts.

3.2. Starchy Materials

There are two main sources of commercial starches:

(1) cereals, (e.g. wheat, rice, maize, etc.)

(2) roots and tubers, (e.g. potatoes, tapioca, etc.)

The moisture content of the grain is low whereas that of roots and tubers is very high. The approximate chemical composition of cereal grains (dry matter basis) is as shown in Table 12.5.

Table 12.5 *Approximate chemical composition of cereal grains (dry matter basis).*

Cereal	Protein %	Fat %	Soluble carbohydrate %	Crude fibre %	Mineral matter %
Wheat (English)	10.5	2.6	78.6	2.5	1.8
Maize (sweet)	12.1	9.1	74.5	2.2	2.0
Sorghum	12.4	3.6	79.7	2.7	1.7
Millet	13.6	5.4	77.9	1.3	1.8
Rye	13.8	1.4	79.7	2.6	2.2
Barley	11.8	1.8	78.1	5.3	3.1
Rice (brown)	11.0	2.7	83.2	1.2	1.8
Oats (whole grain)	11.6	5.2	69.8	10.4	2.9

The approximate chemical composition of white-potato tuber is shown in Table 12.6.

Table 12.6 *Approximate chemical composition of white-potato tuber.*

Component	Per cent
Starch	15-20
Protein	2-3
Water	75-80
Fat	0.1
Fibre	0.5-1.0
Ash	1.0-1.5

Starches require pretreatment to bring about the conversion to fermentable sugars. This is accomplished either by enzymatic (plant or microbial origin) or chemical agencies. A variety of methods may be used for converting complex carbohydrates to relatively simple materials. Generally, the methods of saccharifying the starchy materials involve the use of enzyme preparations or dilute acids, or a combination of both. The selection of a method of conversion depends upon:

(i) The use intended, and
(ii) The availability of the hydrolytic agents and their relative costs. For example, a continuous process (the Symba process) developed in Sweden for biomass production employed the yeast *(Endomycopsis fibuliger)* for hydrolyzing the starch to fermentable sugars. Subsequently, *Candida utilis* was grown on these sugars.

3.3. Cellulosic Materials

Cellulosic materials are complex carbohydrate materials. The cellulose molecule is made up of the repeating units of β-glucose. The formation of β-cellobiose requires two molecules of β-glucose, which are linked through a 1, 4-linkage. Lastly, 1000 to 10,000 units of cellobiose are required to form a simple linear polymer called cellulose. Here, the units of cellobiose are joined end to end through 1, 4-β-glucosidic linkages as shown under:

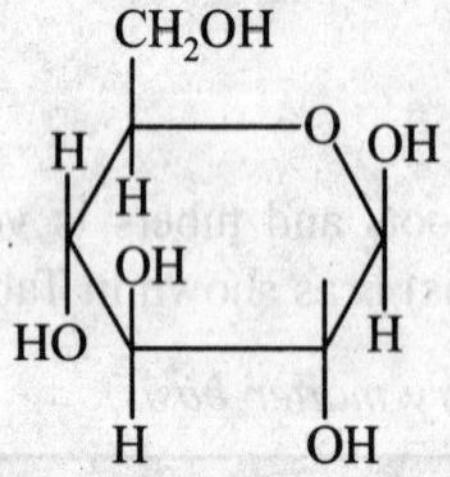

β-Glucose

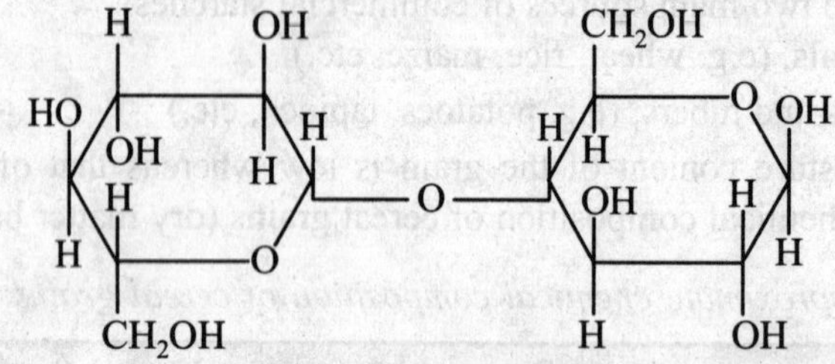

β-Cellobiose

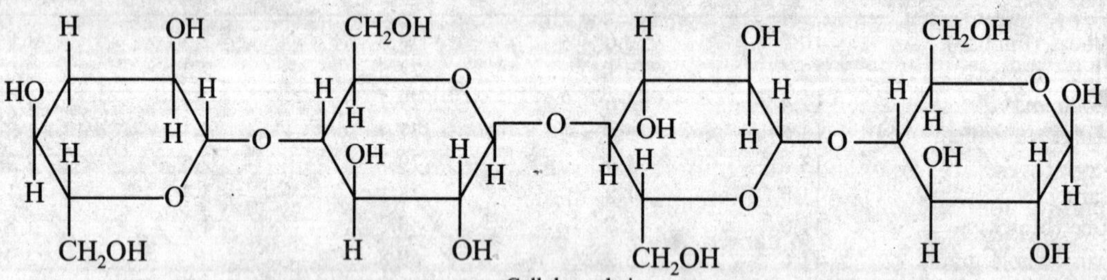

Cellulose chain

For this reason, cellulosic materials require some sort of pretreatment, (i.e. hydrolysis). Sulfite waste liquor, wood molasses and rice straw have been briefly dealt with under cellulosic materials.

Sulphite waste liquor: In the manufacture of paper-pulp, wood is subjected to hydrolysis which is brought about with the help of calcium bisulphite under heat and pressure. This operation is called digestion process. At the end of this process, the spent liquid is left and it is referred to as sulphite waste liquor. It is objectionable to dispose of this liquid without treatment.

Coming to the chemical composition of sulphite waste liquor, we find that it contains 10 to 12 per cent solids, of which sugars make up about 20 per cent. In other words, the sulfite waste liquor is a dilute sugar solution having approximately 2 per cent sugar content. There are mainly two types of monosaccharides:

(i) *Hexoses:* D-glucose, D-galactose and D-mannose, and

(ii) *Pentoses:* D-xylose and L-arabinose.

The relative amount of these monosaccharides present in sulphite waste liquor is variable, since chemical composition of the liquor varies according to the variety of wood being used. For example, angiosperms give sulfite waste liquors containing 3% sugars, of which about 70% are pentoses (principally xylose), whereas gymnosperms produce liquors containing about 2% sugars, about 75% of which are hexoses (principally mannose). Chemical composition of spent spruce sulphite liquor is given in Table 12.7.

Table 12.7 *Chemical composition of spent spruce sulphite liquor [Forss and Passinen (1976)].*

Component	*Content [% (w/v)]*
Lignosulfuric acids	43
Hemilignin compounds	12
Incompletely hydrolyzed hemicellulose and uronic acids	7
Monosaccharides (total)	22
D-Glucose	2.6
D-Xylose	4.6
D-Mannose	11.0
D-Galactose	2.6
L-Arabinose	0.9
Acetic acid	6
Aldonic acids and other substrates	10

Looking to the chemical composition of the liquor, it seems that it may be used as a dilute fermentation medium in some fermentation industries. For example, it is used in the industrial production of ethyl alcohol (using *Saccharomyces cerevisiae*) and in the growth of *Torula utilis* cells for animal feed. One should note that *S. cerevisiae* requires only hexoses, whereas *Torula utilis* uses both hexoses and pentoses. This knowledge is required while using sulfite waste liquor whose composition varies depending upon the variety of wood being used in the manufacture of paper-pulp.

Sulphite waste liquor obtained from the paper-pulp industry cannot be used directly in the fermentation industry. It is necessary to remove the free sulfur dioxide or sulfurous acid present in the waste liquor, since these compounds are toxic to micro-organisms. They may be removed by steam stripping or precipitation with lime.

Wood molasses: Sugars similar to those of sulfite waste liquor may be produced by acid hydrolysis of wood cellulose itself. This may produce 65 to 85% fermentable sugars. Several methods of acid hydrolysis were developed mainly during the Second World War. Usually, sulfuric acid of about 0.5% concentration is used at a temperature range of 150 to 185°C. Using a continuous process, a syrup may be obtained from saw-dust. This syrup may contain 4 to 5% reducing sugars (a mixture of glucose and pentoses) with an overall yield of 45 to 55%. It may be subjected to concentration to give a kind of wood molasses. The chemical composition of wood molasses (% by weight) is given in Table 12.8.

Wood molasses prepared from conifers contains about 85% hexoses and 15% pentoses, and wood molasses produced from broad leaf trees contains about 65% hexoses and 35% pentoses. It should be noted that *Candida utilis* utilizes pentoses readily. Recently, it has been shown that powdered and heated cellulose is susceptible to enzymic hydrolysis. For example, a culture filtrate of *Trichoderma viride* may be employed as the enzyme preparation to do this. In a continuous system (which involves ultrafiltration membranes), it is possible to hydrolyze all the cellulose to glucose which passes through the membrane.

Table 12.8 *Chemical composition of wood molasses (% by weight).*

Component	*Per cent*
Solids	52-60
Reducing sugars (as glucose)	48-50
Other carbohydrates	0.5-1.5
Non-carbohydrate organic compounds	6-8
Ash	2-3
Nitrogen	0.065
Volatile organic acids	1-2

Apart from fermentable sugars obtained from cellulose and their use in fermentation industries, untreated cellulosic wastes may be directly employed to produce microbial protein. This can be accomplished by growing cellulolytic micro-organisms on such untreated cellulosic wastes. In this context, filamentous moulds *(Trichoderma viride)* and bacteria *(Cellulomonas* spp.) are both being studied.

Rice straw and other related agricultural materials can serve as a good source of cellulose. The chemical composition of rice straw is given in Table 12.9.

Table 12.9 *Chemical composition of rice straw.*

Component(s)	*Per cent*
Crude protein	4.5
Ether extract	1.5
Crude fibre	35.0
Lignin	4.5
Cellulose	34.0
Nitrogen free extract	42.0
Total digestible nutrients	43.0
Ash	16.5
Silica	14.0
Ca	0.19
K	1.2
Mg	0.11
P	0.10
S	0.10

Rice straw is a major agricultural by-product in Asia. Its annual production in Asia is 561 million tons which is 90% of the total world production. It is a poor-quality animal feed in its natural state because of its bulkiness, poor palatability, low protein content, and low digestibility. Considerable attention has been directed towards the development of animal feed from straw. Many fermentation methods have been developed to provide micro-organisms or microbial products that improve the nutritional value of animal feed. Numerous organisms are capable of using cellulose for their growth, (e.g. yeasts and mushrooms). Rice straw has been used as a fermentation medium in the production of silage and single cell protein (SCP), mushroom cultivation, etc. At present the work concerning the use of rice straw as animal feed, has been done only till the laboratory or pilot plant stage.

3.4. Hydrocarbons and Vegetable Oils

Hydrocarbons used as fermentation substrates are usually mixtures of various hydrocarbon components. These fermentation raw materials are relatively cheap. However, purified hydrocarbon fractions or hydrocarbon compounds are more expensive.

Hydrocarbon substrates, (e.g. gas oil and *n*-paraffins) are used to produce single-cell protein (SCP) products. In this way, biomass of yeasts, (e.g. *Candida lipolytica, C. kofuensis, C. tropicalis, Candida* spp. ATCC 20, 343, etc.) can be produced on a significant scale under aerobic conditions.

n-Paraffin based processes have been practised on a large pilot plant scale by many companies. *n*-Paraffin used as carbon and carbon sources for yeast SCP processes are prepared by molecular sieve adsorption methods. The resulting *n*-paraffins range in purity from 97.5-99% ($C_{10} - C_{23}$).

Vegetable Oils: Oils obtained by deoiling of vegetable seeds are called vegetable oils. On the basis of their degree of unsaturation, they may be grouped into following three major classes:

(i) Oleic (or 'non-drying' type); These include olive and groundnut oils.
(ii) Linoleic (or 'semi-drying' type); These have a higher content of the double unsaturated fatty acid (linoleic acid) found in maize, sunflower and cotton-seed oils.
(iii) Linolenic (or 'drying' type); These include linseed and soya bean oils containing a fatty acid (linolenic acid) with three double bonds.

These oils may undergo drying if exposed to the atmosphere due to the oxidation of the unsaturated components.

The chemical composition of some typical commercial vegetable oils is presented in Table 12.10.

Table 12.10 *Chemical composition of some typical commercial vegetable oils.*

Oil	*Saponification value*	*Iodine value*	*Percentage Saturation*	*Chief component acids (% w/w)*		
				Oleic acid	*Linoleic acid*	*Linolenic acid*
Olive	189-195	80-85	9-20	65-84	4-9	...
Groundnut	189-196	85-98	18	56-65	17-26	...
Maize	188-193	117-130	12	45-47	40-42	...
Sunflower seed	186-194	127-136	7-10	30-35	55-65	...
Cotton-seed	191-196	103-111	25	25-30	45-50	...
Linseed	189-196	170-185	10-15	15-25	15-20	45-55
Soya bean	190-193	124-133	12-13	25-30	50-55	5-8

Commercial vegetable oils, (e.g. maize oil) may be used in conjunction with surface-active agents as 'anti-foams', or alone as a nutrient source of carbon.

3.5. Nitrogenous Materials

Corn-Steep Liquor (CSL): The used steep water results from the steeping of corn during the manufacture of starch, gluten, and other corn products. This by-product is subjected to concentration to approximately 50 per cent solids, and this concentrate is called corn-steep liquor.

According to Bowden and Peterson (1946), analyses of corn-steep liquor samples from different manufacturers are given in Table 12.11.

Table 12.11 *Analysis of corn-steep liquor samples from different manufacturers.*

Component	*Per cent*
Solids	40-60
Lactic acid	12-27
Total nitrogen	7.4-7.8
Amino nitrogen	2.6-3.3
Reducing sugars, as glucose	1.5-14
Ash	18-20

N.B.: All components except solids were estimated on a dry weight basis.

There is considerable variation in chemical composition for lots from the same manufacturer and lots received from different manufacturers. Therefore, it can lead to poor reproducibility of an industrial fermentation. Thus, it is necessary to estimate certain components, (e.g. a particular amino acid, vitamin, or precursor) of corn-steep liquor to be employed, since they are required at low but critical levels. This is virtually a complete growth medium in itself.

Corn-steep liquor was originally found to be useful for penicillin production specifically. But, it is now recognised as valuable in many fungal antibiotic fermentation media. In addition to this, it is also used in the manufacture of foodstuffs.

Soya bean meal: The material left after deoiling of the soya bean seeds is called soya bean meal. Soya bean meal contains approximately 8% w/w nitrogen. This differs from corn-steep liquor, since soya bean meal is a much more complex nitrogenous source than corn-steep liquor and, therefore, not readily available to microbes. This is used as an ingredient for fermentation media, (e.g. in the production medium for streptomycin).

Pharmamedia: Pharmamedia is nothing but a clean, yellow, finely ground powder prepared from the embryo of cotton seed. It contains 56% w/w protein, 24% carbohydrate, 5% oil and 5% ash. Ash, in turn, contains calcium, iron, chloride, phosphorus and sulfate. It is used as an ingredient for production media, (e.g. in tetracycline production).

Distillers solubles: In the manufacture of alcohol using grain or maize, alcohol is distilled from fermented grain or maize, leaving the residue (containing 6 to 8% w/w total solids). The suspended solids from the residue are eliminated by screening, leaving the effluent. Thereafter, the effluent is subjected to concentration, until the solid content reaches 35% w/w giving 'evaporator syrup'. This syrup is then drum-dried to yield 'distillers solubles'. This may be used as a production medium component, since it supplies nitrogen (in the form of protein), together with many accessory food factors, (e.g. vitamin B complex).

Other natural organic nitrogenous materials are ground-nut meal, fish meal, Bacto peptone, Difco yeast extract and the like.

4 SCREENING FOR PRODUCTION MEDIA

One of the important factors in the success of a fermentation process is the production medium, since costs associated with the production media may comprise a considerable portion of the total cost of an industrial fermentation and for other reasons, (e.g. recovery of fermentation products or microbial cells). It is desired that the production medium should possess most of the features discussed under an ideal production medium. In other words, designing of a new usable production medium requires consideration of the features of an ideal production medium.

The selection of a satisfactory medium may be done after trial and error method. Newly designed production media are subjected to various tests, such as effects of varying the levels of individual components. Such studies enable the investigator to choose one or two good media, and at times, to discard many worthless media. These selected media are further studied from several points of view. For example, quantitative determinations of the residual medium components are made while fermentation is in progress. The reason behind such determinations is to optimize the concentrations of critical carbon and nitrogen nutrients. Moreover, one may have knowledge of the phenomenon of diauxy, if it does occur during the fermentation. Also, a study of the effect of the possible precursor on the formation of the fermentation product is conducted, provided the chemical structure of the product is known.

It is to be kept in mind that most of these studies are carried out without the addition of an anti-foam agent. Defoamer may be added as the inert silicone type, if necessary. A special study of the crude defoamers is made to detect a likely interference or stimulation of product formation by a defoamer. After completing the laboratory study, all parameters, (e.g. aeration and agitation) are rechecked in small fermentors. Finally, the recovery procedure is found out, and its cost is determined. If cost determination of the production medium presents an adverse economic picture, it is necessary to evaluate other media.

Thus, the screening programme for production media is a laborious and expensive programme. Therefore, it remains a trade secret.

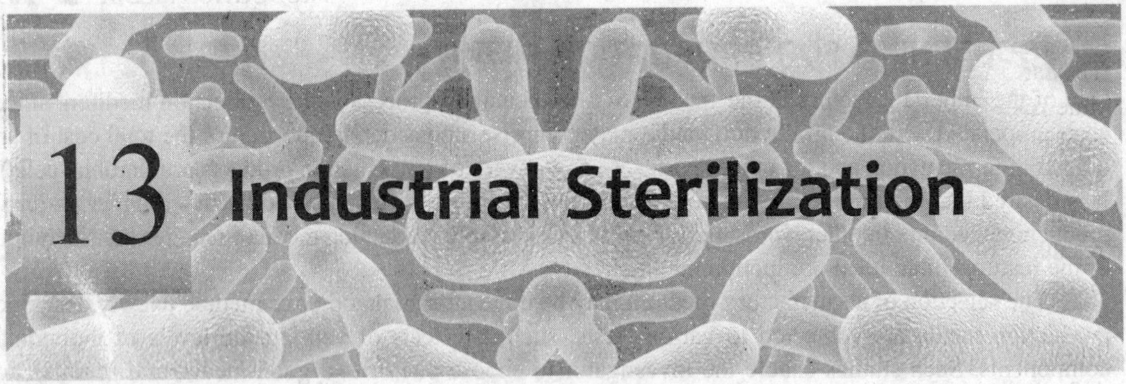

13 Industrial Sterilization

1 INTRODUCTION

Usually, sterilization of equipment, medium, air etc. and the subsequent maintenance of sterility of the whole system undergoing fermentation are critical points in the success of the fermentation process. There are many sterilizing agents, (e.g. steam, U.V. light, chemical agents, etc.) for bringing about sterilization. Steam is preferred to other agents, because it is cheaper for mass sterilization. Moreover, it can penetrate into small awkward sites, as in valves, for efficient sterilization. Sterility of the fermentation process is needed, since contaminants may affect the process adversely and subsequently lower yields of the fermentation product. In certain fermentations, (e.g. antibiotic fermentations), sterility of the fermenting system is strictly to be controlled. Therefore, sterility tests are carried out during the fermentation to detect suspected contaminants.

2 PRINCIPLES OF STERILIZATION

Sterilization removes infecting micro-organisms which would otherwise interfere with the efficient fermentation process. It can also remove pathogenic micro-organisms or spoiling agents which could devalue the products of fermentation. There are two main methods employed for sterilization:

1. destruction of micro-organisms
2. removal of micro-organisms

There are many process operations to achieve each method.

Destructive methods make use of a lethal condition. This may be accomplished either by chemical or physical means. Moist heat is the most common physical agent. It allows for satisfactory industrial sterilization. It is also the least expensive and most reliable to design and operate.

The severity of sterilization conditions that are needed depends on:

(i) the type of fermentation process that is to be protected
(ii) the duration of the production campaign
(iii) the type and extent of contamination of the feed material

Thus, pasteurization, inactivation or destruction of infecting micro-organisms in the feed may be specified.

Infecting micro-organisms may be a combination of yeasts, protozoa, bacteria, spores and phages. They are, more or less, resistant to sterilization agents. With sterlization using heat, there is a temperature/time relationship, determining the degree of sterility. Heat-sterilization also imparts some degree of cooking of the nutrients. This may be beneficial in so far as break-down and solubilization of complex starchy materials are concerned rendering them more available for assimilation by the fermenting micro-organisms. On the other hand, some destruction of heat-sensitive compounds may take place, and this results in a loss in efficiency, impairment of product flavour or scaling in the heat exchanger by charring or precipitation.

The other method of sterilization is the removal of infecting micro-organisms. This is done by filtration. Numerous types of filters are available for this purpose. The statistical probability of eliminating infecting micro-organisms depends critically on (i) the size of the micro-organisms and (ii) the retention efficiency of the filter. Usually, sterilization of gases and biostatic fluids is done by filtration.

3 STERILIZATION OF EQUIPMENT

Steam, a physical agent, is universally employed for the sterilization of fermentation equipment and production medium. There exists a linear relationship between time and temperature. In other words, high temperature application needs less time and vice versa. The seed tank and the fermentor are the central pieces of the equipment. These may be sterilized either in charged (with medium) condition or they may be sterilized empty.

(a) Sterilization of fermentor and medium as a unit:

Steam sterilization of the fermentor and the production medium together is a simple operation. A jacket or coil is fitted to the fermentor and supplied with pressure steam. Water is required to provide a temperature of 120°C for a period of at least 20 minutes and the subsequent cooling back to the fermentation temperature. There is provision of a vent for the air to come out from the fermentor. Also, the head-space above the medium is filled with steam. It is necessary to prevent the formation of a vacuum during the cooling of the system. It is, therefore, necessary to have a supply of sterile air to maintain positive pressure. The presence of a vacuum in a sterile system is always a hazard to sterility. This is due to the danger of drawing non-sterile air or liquid into the sterile system through the seals on the agitator shaft and the spindles of cocks and valves. Apart from sterilization and maintenance of sterile condition of the tanks (fermentor and seed tank), it is equally important to sterilize and to maintain sterility of ancillary equipment attached to the fermentors. Ancillary equipment includes many extra-connections, namely:

(i) Agitator shaft passing through the lid of the vessel. There should be an adequate seal or gland for maintaining pressure in the vessel. Moreover, this prevents the entry of microbes to the vessel from outside (Figure 13.1).

(ii) Sparger, usually beneath the agitator and a pipe for expelling exhausted air from the fermentor.

(iii) A connection between the defoamer tank and the fermentor.

(iv) A connection for introducing the inoculum (Figures 13.2 and 13.3).

(v) A connection for drawing samples during the fermentation (Figure 13.4).

(vi) A connection for unloading the fermentor after the completion of the fermentation.

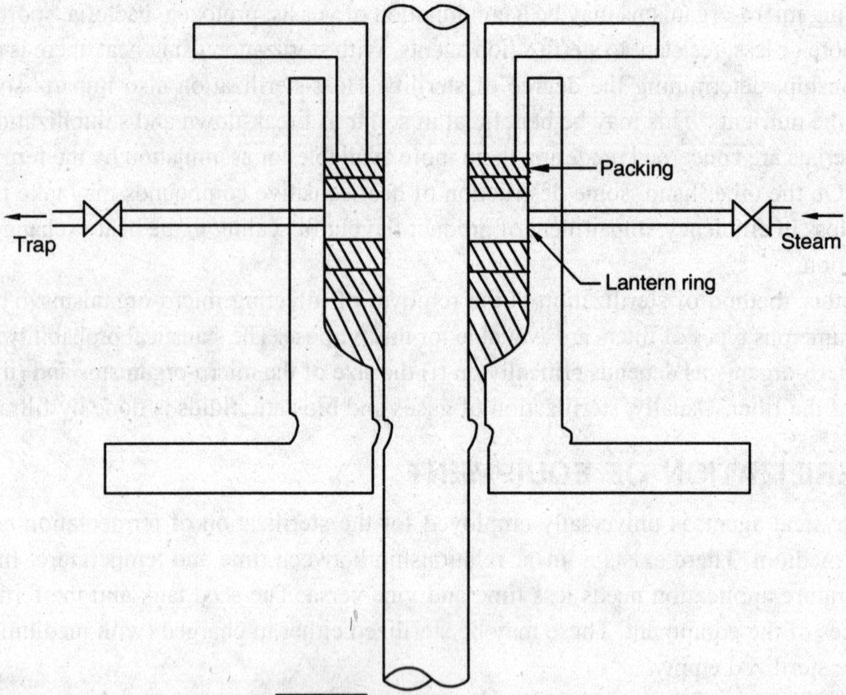

Fig. 13.1 *Steam-sealed gland.*

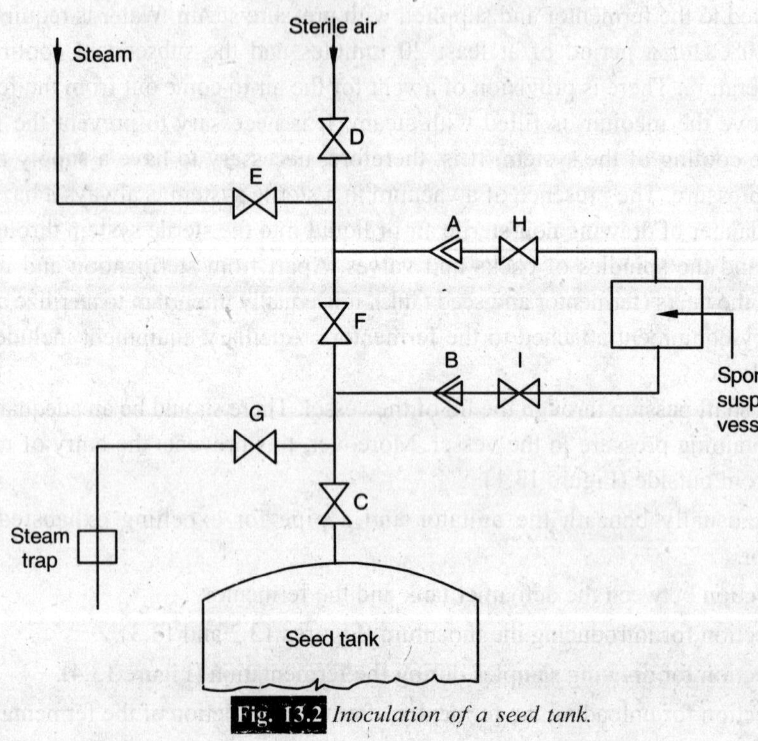

Fig. 13.2 *Inoculation of a seed tank.*

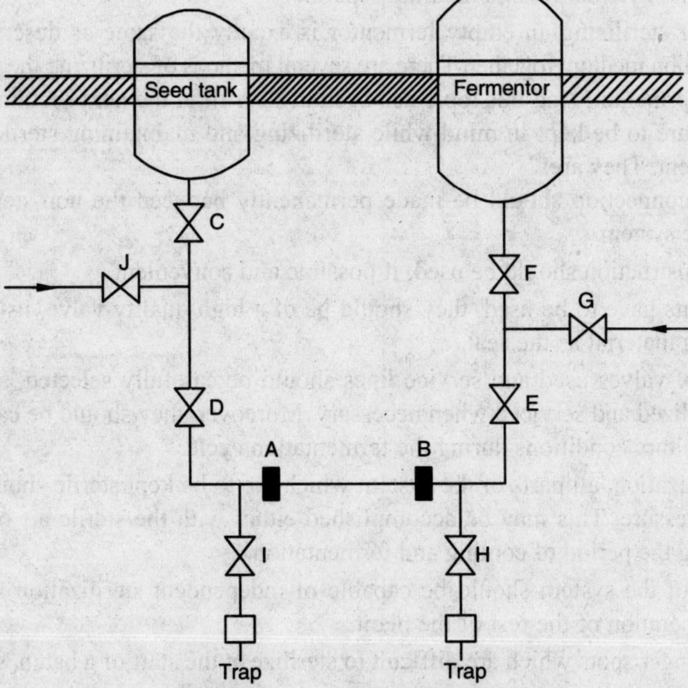

Fig. 13.3 *Inoculation of a fermentor.*

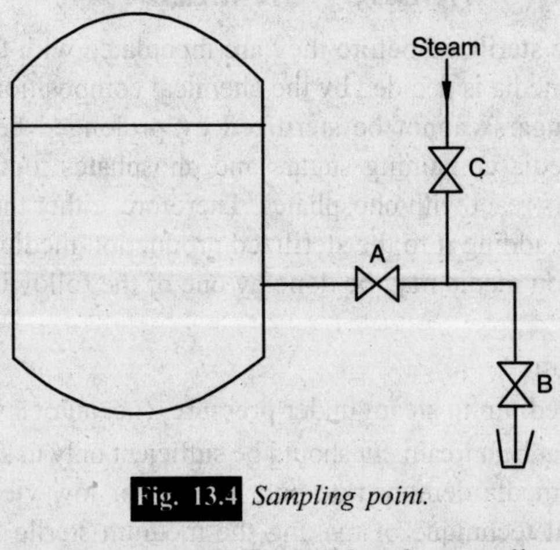

Fig. 13.4 *Sampling point.*

Thus, these above mentioned extra-connections add to the complications of the sterilization of the fermentor (whether empty or with a charge of medium). It is a usual practice to supply each connection with its own presure air-line. This is advantageous, since the ancillary equipment can be steamed out, while the fermentor is held under steam pressure at a temperature of 120°C. Such ancillary equipment is used in aerobic fermentations, requiring much more care for sterility of the system.

These processes involve a considerable loss of production time, for it may require several hours to reach a temperature of 120°C and subsequently cool the medium before inoculation.

(b) Sterilization of fermentor and medium separately:

The method for sterilizing an empty fermentor is exactly the same as described above for the fermentor and production medium together. There are several methods of sterilizing the production medium separately from the fermentor. And this topic has been dealt with in the next section separately.

Certain points are to be kept in mind while sterilizing and maintaining sterile conditions in the fermentation equipment. They are:

(i) No direct connection should be made permanently between the non-sterile and the sterile parts of the system.

(ii) Welded construction should be used, if possible and convenient.

(iii) Where joints have to be used, they should be of a high quality valve, using rubber or other impervious material as the seal.

(iv) The type of valves used and service lines should be carefully selected, so that they can be easily sterilized and serviced when necessary. Moreover, they should be capable of maintaining pure culture conditions during the fermentation cycle.

(v) After sterilization, all parts of the system which are to be kept sterile should be kept under a positive pressure. This may be accomplished either with the sterile air or the sterile liquid used during the period of cooling and fermentation.

(vi) Each part of the system should be capable of independent sterilization without interfering with the operation of the rest of the plant.

(vii) Isolated danger spots which are difficult to sterilize at the start of a batch, should be provided with pressure steam connection for continuous or intermittent use.

4 STERILIZATION OF PRODUCTION MEDIA

Usually, production media are sterilized before they are inoculated with the desired fermentative pure culture. Sterilization of these media is decided by the chemical composition of a particular medium. For instance, media containing sugars cannot be sterilized by prolonged heating, since sugars undergo caramelization. Moreover, media containing sugars and phosphates together cannot be sterilized by prolonged heating, since sugars react with phosphates. Therefore, either the sugar or the phosphate is to be sterilized separately before adding it to the sterilized production medium.

Sterilization of production media may be done by one of the following three methods:

(a) by boiling

(b) by passing live steam

(c) by subjecting the medium to steam under pressure, (i.e. autoclaving)

It is to be kept in mind that heat treatment should be sufficient only to sterilize, avoiding overcooking of the medium. Overcooked media demonstrate poor growth or low yields as compared to properly sterilized media. The classical technique of making the medium sterile by the use of steam may be carried out in two ways:

(i) batchwise in the fermentor

(ii) continuous sterilization

The former method has already been discussed in the previous section. Alternatively, batch cookers may be employed. This is the simplest method of sterilizing production media.. The vessel is equipped with a coil or jacket for heating and cooling. Also, the agitator may be fitted to aid heat-exchange. It is needed to raise the temperature of the whole system, (i.e. both vessel and medium) to 120°C with steam and to maintain this for a period of 20 minutes before cooling the system. There is an interconnecting pipeline between a batch cooker and a fermentor for transferring the sterile medium from the cooker to the previously steam sterilized fermentor. Therefore, the interconnecting pipeline must also be sterilized before making use of it for transfer purpose.

Advantage: The batch cooker method saves the production time, since the fermentor is unoccupied between two fermentation runs.

Limitations: (i) It occupies increased plant space.

(ii) It involves higher cost of the additional equipment required.

(iii) It involves increased steam usage.

The latter method, continuous sterilization, offers more flexibility in the choice of time-temperature conditions to which the medium is exposed. Usually, this method involves passing of production medium through a heat-exchanger, a holding coil and a cooler. The temperature of the medium undergoing sterilization is raised to the desired level in the heat-exchanger. The medium then passes on to a holding coil, where it is maintained at the sterilizing temperature for a predetermined time period. Finally, the medium is rapidly cooled by countercirculating it in the exchanger against the cool input medium, and then against cold water. The possibility of short holding time allows the use of sterilization temperatures higher than 120°C, without any adverse effects on the nutritional qualities of the culture medium (Table 13.1).

Table 13.1 *Parameters involved In continuous sterilization.*

Details	*Steam injection*	*Heat-exchanger*
Holding itme	1-5 minutes	10 minutes
Holding temperature	140°C	120°C
Heating up/Cooling down period	Negligible	1% contribution

A slightly modified system involves the injection of high temperature steam into the medium. Here, the required temperature (< 140°C) is reached in microseconds. The cooling of the medium is carried out in the flash cooler in a vacuum chamber (Figure 13.5).

A continuous retention tube sterilizer is also available. It is widely employed in the antibiotic field.

Advantages: (i) It saves both production time and plant space.

(ii) It gives improved quality of the medium. For instance, an increase of 160% in the yield has been secured in the manufacture of vitamin B_{12}.

(iii) It involves some economy in steam costs, provided the heat-exchange principle is applied.

(iv) It allows the use of lower sterilizing temperature or shorter holding periods, provided the medium has a low pH, or some other characteristic aiding the sterilization.

Synthetic media require a relatively shorter sterilization time period. On the other hand, the cruder media often require a considerably longer sterilization time period because of the greater viscosity of these media, making heat-penetration difficult. In addition, relatively heat-resistant bacterial spores may be present in some of the components of such media. Therefore, it is necessary to determine the conditions, (i.e. temperature and time) for satisfactory sterilization, without over-cooking the medium. Certain medium components, (e.g. vitamins and enzymes) do not allow heat sterilization of the media. These components can be sterilized separately by filtration, (e.g. Seitzfilter). They are then added to the previously sterilized medium.

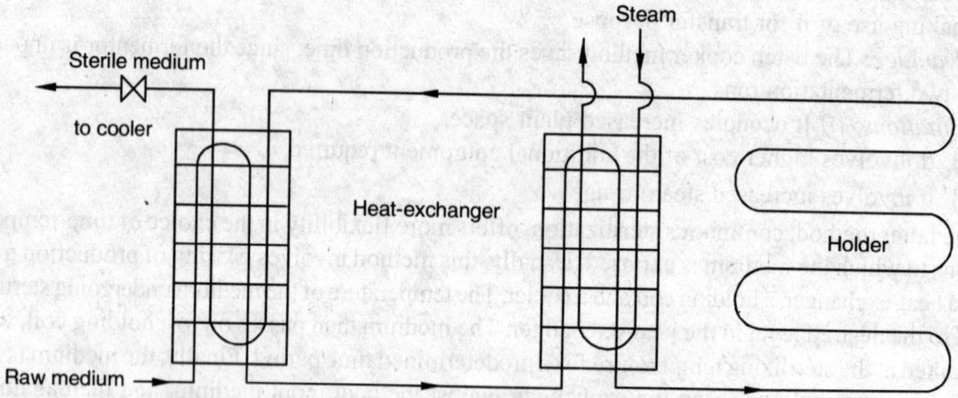

(a) Continuous sterilization: Plate heat-exchanger

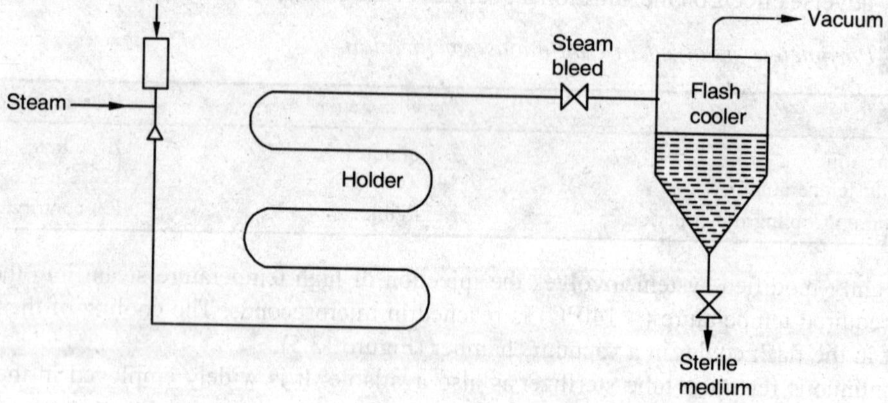

(b) Continuous sterilization: Steam injection, Flash cooling

Fig. 13.5 *Schematic diagram for continuous medium sterilization.*

It is to be brought to one's notice that, some fermentations are carried out without the sterilization of the media, (e.g. certain yeast fermentations that are conducted at relatively low pH values).

5 STERILIZATION OF AIR

With aerobic fermentations, continuous supply of sterile air is vital for successful fermentation. Air can be sterilized by many methods namely:
(i) filtration (ii) heat (iii) electrostatic precipitation (iv) U.V. light or (v) chemical agents

The sterilization of air in fermentation industries is widely carried out by the filtration method. Other methods, for some reasons, are not employed. For example, sterilization of air by heat is not carried out owing to the high cost involved. Therefore, the only method for sterilizing the air for industrial use is filtration, which has been briefly discussed.

Filtration method (Figure 13.6) for sterilizing very large volumes of air was studied by Terjesen and Cherry. These scholars used a preformed slab of slag wool 3 inches thick, packed to a density of at least

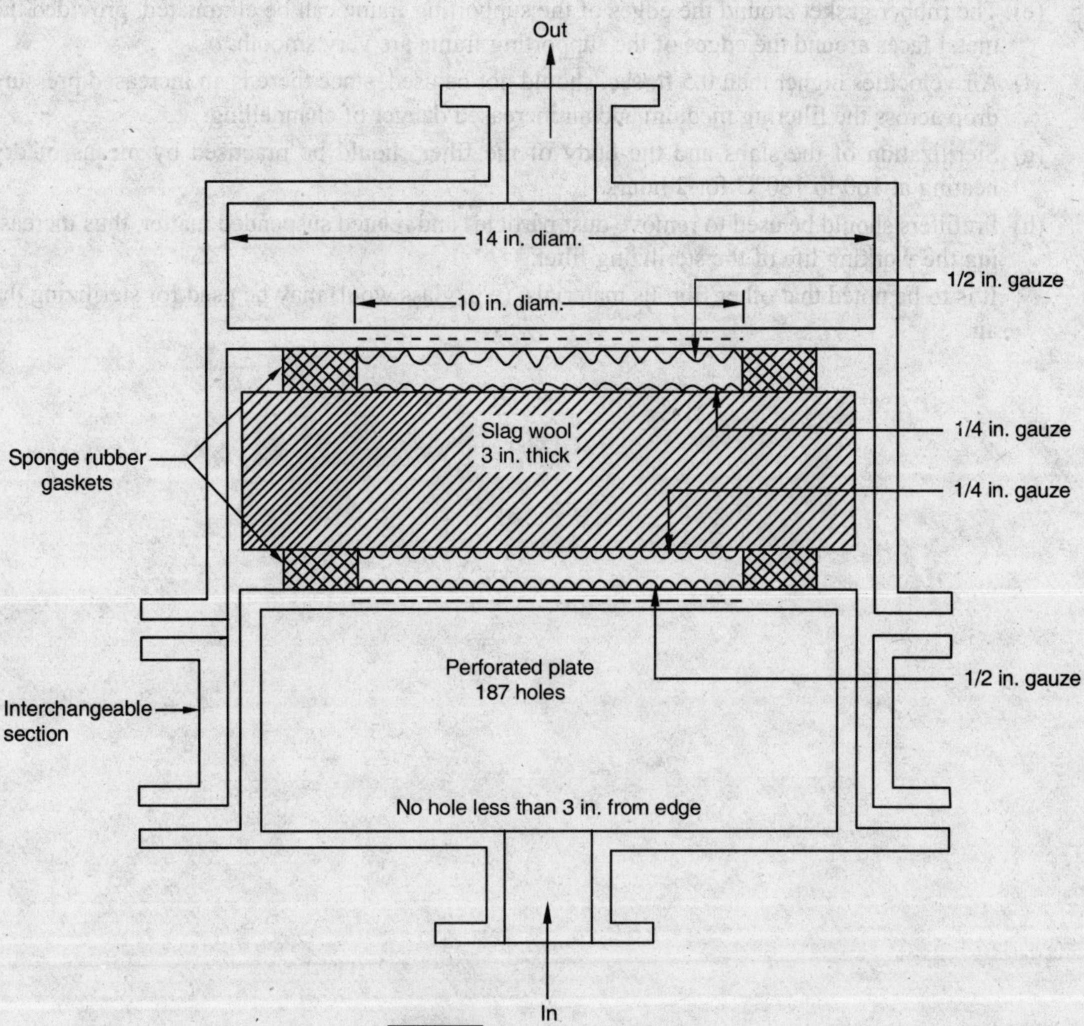

Fig. 13.6 *Diagram of air-filter.*

17lb./ft^3., and of mean fibre diameter less than 6μ. The air velocity through the slab was kept below 1 ft./sec. to avoid channelling through the slag wool material. Techniques for evaluating the air sterility developed by Bourdillon *et al.* were used by Terjesen and Cherry for assuring the efficiency of each method under experimental conditions. After many experiments, they made the following suggestions:

(a) The slab to be installed in the filtration system must follow certain specifications, (e.g. fibre diameter, uniformity, density of the slab, etc.).

(b) Attention should be paid to the air which may escape by passing around the edges of the filtering medium.

(c) No holes should be drilled less than 3 inches from the edges of the perforated supporting plates.

(d) A non-continuous sponge-rubber gasket should be inserted between the slab and the 3 inches blank annulus of the supporting plates.

(e) The rubber gasket around the edges of the supporting frame can be eliminated, provided the metal faces around the edges of the supporting frame are very smooth.

(f) Air velocities higher than 0.5 ft./sec. should not be used, since there is an increased pressure drop across the filtering medium and an increased danger of channelling.

(g) Sterilization of the slabs and the body of the filter should be practised by means of dry heating at 160 to 180°C for 2 hours.

(h) Prefilters should be used to remove dust particles and related suspended matter, thus increasing the working life of the sterilizing filter.

It is to be noted that other fibrous materials, (e.g. glass wool) may be used for sterilizing the air.

14 Principal Types of Fermentors

1 INTRODUCTION

In fermentation industries, microbes are to be grown in specially designed vessels loaded with particular types of nutritive media. These vessels are referred to as fermentors. Usually, they are complicated in design, since they must provide for the control and observation of many facets of microbial growth and biosynthesis. The design of the fermentor depends upon the purpose for which it is to be utilized. Many fermentors have been designed for use in submerged fermentation on a laboratory, semi-pilot-plant and pilot-plant scale. This chapter deals with the biochemical engineering aspects of fermentors.

2 FACTORS INVOLVED IN FERMENTOR DESIGN

According to Levenspiel (1971), there are two problems involved in the reactor design. They are:

(i) the selection of the best type of reactor for the particular reaction

(ii) the determination of the best operating conditions

Usually, the scale of operation, (i.e. the required daily output) and the kinetics of the given reaction are determined beforehand. Apart from these features, there is considerable freedom of choice. For example, a batch process or one of several kinds of continuous processes can be adopted and, within limits, any initial concentration of the reactants, operating temperature and pH can be used. Controlled changes in these variables can be made during the course of the reaction.

The objective of design is to be able to describe completely the effect of operating conditions on the performance of a bioreactor, and to compare alternative designs with economic criteria. Unfortunately, such situations are rare. Often, there is a dearth of information; the mathematics may be intractable or too costly to undertake, or it may be difficult to justify isolating the design of the fermentor from the overall process. It is a matter of compromise between the time and the effort required to design the fermentor and the necessity to maximise the probability of success of the process as a whole. Thus, simplifying the assumptions and short cuts are necessary, and some factors have to be ignored. In these circumstances, good engineering judgement, which comes only with experience, is at a premium.

Although, the design of a fermentor in which a given conversion and throughout can be achieved, is not a simple matter, the procedure involved may be broken down into a number of steps:

(1) the selection of an appropriate strain of a particular species of micro-organism. This largely determines the phase of growth in which the product is formed, the ranges of temperature and pH that can be considered, the degree of aerobiocity required and the probable effect of contamination

(2) the selection of an appropriate fermentor configuration, i.e. a form of batch stirred-tank fermentor, a continuous stirred tank fermentor or a tubular fermentor, etc.

(3) the determination of the fermentor dimensions, e.g. volume and diameter, and the values of the operating variables, notably concentration, temperature and pH, as well as the process time for batch fermentation and flowrate with continuous fermentation

(4) the extent of the heat transfer surface and the mixing devices required

(5) the power and aeration requirements

(6) the mechanical design, including selection of materials of construction and, if required devices for the maintenance of aseptic conditions

(7) facilities for monitoring and control

(8) safety factors

It is only on completion of this whole design exercise that a reasonable estimate of the process economics can be attempted. For this, knowledge and experience of microbiology, biochemistry, thermodynamics, microbial and biochemical kinetics, fluid mechanics, mass and heat transfer and economics are necessary. Two differing types of information are required, notably that associated with the biochemical changes which take place, and that which is concerned with the rate at which they occur. In a fermentor design it is the latter problem, (i.e. the rate processes) that largely concerns us, as at this stage, the biochemical changes are largely known from the basic research programme.

2.1. Differences between biochemical and chemical processes

Biochemical conversions, with the help of microbes, differ from purely chemical processes in a number of ways, particularly:

(1) the complexity of the reactant mixture

(2) the increase in the mass of microbes together with the accomplishment of the biochemical conversion

(3) the ability of microbes to synthesise their own catalysts, (i.e. enzymes)

(4) the mild conditions of temperature and pH (usually)

(5) the difficulty in maintaining the required biochemical conversion (stability)

(6) the restriction to the aqueous phase

(7) the relatively low concentrations of both substrate and products

2.2. Classification of biochemical reactions

There are many ways of classifying biochemical reactions. In biochemical engineering, probably the most useful scheme is the breakdown according to the number and types of phases involved, the main division being between homogeneous and heterogeneous systems. A reaction is called homogeneous if it takes place in one phase alone, and heterogeneous if it requires the presence of at least two phases. It is immaterial for the purpose of definition whether the reaction takes place in one, two or more phases, or at an interphase between phases, or whether the reactants and products are distributed among the phases, or are all contained within a single phase. What matters is the number of phases necessary for the reaction to proceed.

According to these definitions all microbiological reactions are heterogeneous, since they involve, at the very least, a 'solid' phase, (i.e. microbes) and an aqueous phase. Occasionally, a gas phase and further liquid and solid phases may be involved. Cell-free enzymatic reactions may be either homogeneous or heterogeneous. In the latter case either the substrate or the enzyme can occur as the 'solid' phase.

2.3. Rate processes

The overall rates of reaction are influenced both by the reactants and the products, (i.e. they are autocatalytic). This leads to rather unusual optimisation problems.

Essentially, all configurations of microbiological reactors are concerned with microbes dispersed in aqueous nutrient media. For aerobic reactions there is usually an additional dispersed phase consisting of air bubbles as shown in Figure 14.1.

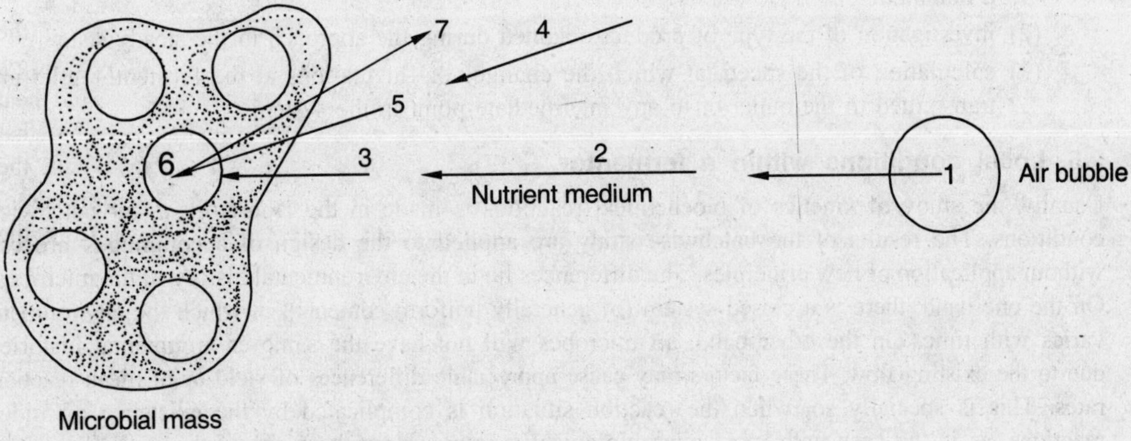

1. Oxygen absorbed by aqueous phase
2. Oxygen transferred through the aqueous phase
3. Oxygen absorbed by and transferred through the intercellular gel to reaction zone
4. Organic substrate transferred through aqueous phase
5. Organic substrate absorbed by and transferred through intercellular gel to reaction zone
6. Microbial reaction zone
7. Products transferred from reaction site into aqueous phase

Fig. 14.1 *Transport processes in aerobic fermentation.*

In this case the overall rate of reaction depends upon the absorption of oxygen from the gas phase, and its subsequent diffusion through the liquid phase to the microbes, where it reacts with the substrates in the nutrient media. These latter reactants also have to diffuse through the liquid phase to reach the reacting site, (i.e. the microbes). Some of the products of the reaction, (e.g. secondary metabolites, CO_2 and other oxidation products) are returned to the aqueous phase, where the less-soluble components come out of solution. CO_2, as a relatively insoluble gas, falls into this category. It forms bubbles of pure CO_2, and also enters the air bubbles by desorption and coalescence.

2.4. Operational considerations

The difference between a batch and a continuous operation lies in the nature of the flow in the continuous process. In particular, all the molecules passing through the flow system will not necessarily have the same residence times, nor will all these molecules undergo the same history of concentration or temperature changes.

The statement that a system usually tends towards a time-invariant state applies, of course, only if there are constant feed conditions, constant heat removal, etc. However, even in continuous operation, transients, (e.g. of concentration or of temperature) are important especially in connection with the start-up of the process, and with its automatic control. Typical areas, in which transient behavior is of importance, include:

(1) the calculation of the time required for a system to approach to within, say, 1 per cent of the steady state, and the determination of the start-up procedure that will reduce this time to a minimum

(2) investigation of the type of product obtained during the approach to the steady state

(3) calculation of the speed, at which the changes or fluctuations at the fermentor inlet are transmitted to the outlet or to any intermediate point in the system

2.5. Local conditions within a fermentor

Usually, the study of kinetics of biochemical reactions is made in the laboratory under batchwise conditions. The results of the batchwise study are applied to the design of a continuous process without application of new principles. The differences lie in the environmental history of the microbes. On the one hand, there is a closed system (of generally uniform contents), in which the environment varies with time. On the other hand, all microbes will not have the same environmental histories due to the existing flow. These factors may cause appreciable differences of yield or of mean reaction rates. This is specially so, when the reaction situation is complicated by the existence of 'side' reactions, as is the case with most microbiological systems. Here the yield of the desired product may differ considerably from batch and continuous operation, and also from the two main types of continuous processes. Fermentor yield is not necessarily lower in continuous processing; in some instances it may be higher. However, where it is lower, this factor may so outweigh the normal advantages of continuous operation as to argue in favour of a batchwise system.

In a batchwise operation the change in the composition of broth coordinates with time. Whether or not a batch system is uniform throughout its spatial coordinates, it always changes from moment to moment. Moreover, it changes as long as is needed to reach thermodynamic equilibrium, or until the process is brought to an end, as is the usual industrial practice. The significant difference in composition is shifted to a space coordinate, and any part of the system usually tends towards a time-invariant state, and the variation in the composition shows itself between one region of the system and another, (e.g. between adjacent vessels in a train of continuous stirred-tank fermentors or between neighbouring cross-sections of a tubular fermentor).

For any reaction the conversion depends upon the residence time, and the concentrations and temperatures that occur within the fermentor. It follows that differences in performance between the various fermentor configurations can be identified with differences of residence time distribution and the concentration and temperature histories of the fermentor contents.

Residence time Distributions

Whereas in a batch fermentor all the fluid elements undergo reaction for the same period of time, this is not usually the case in continuous processes. In a continuous stirred-tank fermentor, the mixing action results in all the elements of the fluid, having virtually the same composition, and this is equal to that of the outflow. Thus, a given molecule entering the fermentor has a significant possibility of finding its way to the outflow almost immediately. With the tubular fermentor (except in the limiting case of plug-flow) the residence times of individual fluid elements are likewise distributed over a range of values. The important contributing factors are:

(1) the variation of fluid velocity over the cross-section of the fermentor, (i.e. the fact that some elements of fluid move faster through the fermentor than others).

(2) diffusion (molecular or turbulent) that occurs in the direction of flow, as well as at right angles to it. The effect of longitudinal diffusion is to reduce the mean residence time of the reagent molecules, and also to increase the spread of residence times of individual molecules. Lateral diffusion works in the opposite direction, and tends to bring the performance of the system closer to that which would be expected on the basis of plug-flow.

Whenever these variations of residence time occur, there are several important results, in particular, an increase in the size of fermentor is necessary for a given output of product at a required efficiency of conversion.

Concentration Distributions

The significance of the concentration distribution is not quite the same as that of the residence time distribution, although the two factors are related.

On one side the concentration of the reagents in a batch or tubular fermentor changes continuously. On the other side, the concentration of the reagents in a train of continuous stirred-tank fermentors changes discontinuously, (i.e. there is an abrupt step change from one such tank to the next). Moreover, the height of the steps is determined in part by the reaction itself, as well as by the size of the tanks. Hence, it is possible for two different continuous processes to have the same residence time distribution (relative to an inert tracer substance), and yet to have different concentration distributions and, therefore, different performances.

Temperature Distributions

The temperature distribution usually has a larger effect than the concentration distribution, and is particularly significant in tubular fermentors. A fluid moving close to the axis of a tubular fermentor may pass through a sequence of temperatures quite different from that of the fluid moving nearer to the wall. The overall composition of the fluid at the outflow is, therefore, a function of what is virtually an infinite number of different temperature distributions.

3 FERMENTOR CONFIGURATIONS

Before discussing the different types of fermentors individually, it is essential to consider the general features of an ideal industrial fermentor. They are:

(1) Material used in the fabrication of a fermentor should be strong enough to withstand the interior pressures due to the fermentation medium. It should be resistant to corrosion, and free from any toxic effect for the microbial culture.

The nature of the construction material to be selected depends upon the particular fermentation that is to be carried out in the vessel. For example, wooden tanks are used in carrying out some fermentations, such as those producing ale and lactic acid. On the other hand, materials in the fabrication of or lining of fermentation vessels include copper, stainless steel, iron and glass. Stainless steel or glass is not used for lining of fermentors except in certain cases, since they are costly materials.

(2) A fermentor should permit easy control of contaminating microbes.

(3) The fermentation tank should be provided with the inoculation point for aseptic transfer of inoculum.

(4) In case of aerobic submerged fermentations, the tank should be equipped with the aerating device.

(5) A fermentor should be provided with a stirring device for the uniform distribution of air, nutrients and microbes.

(6) Baffles should be present to avoid vortex formation.

(7) The fermentation vessel should be equipped with a sampling valve for withdrawing a sample for different laboratory tests.

(8) There should be a provision for controlling temperature and pH of the fermentation medium.

(9) There should be a facility for the intermittent addition of an antifoam agent. Similarly, there should be provision for feeding certain medium components during the progress of fermentation, (e.g. penicillin fermentation).

(10) A drain at the bottom is essential for the removal of the completed fermentation broth for further processing.

(11) A man-hole should be provided at the top for access inside the fermentor for different purposes, (e.g. repairing and thorough cleaning between fermentation runs).

The mechanical features of fermentors and the auxiliary equipment involved are shown in Figure 14.2.

3.1. The Batch Fermentor (BF) (Figure 14.3)

Batch fermentors are used to carry out microbiological processes on a batch basis. There are a number of steps involved. These are associated with the development of micro-organisms from a stock culture, and include agarslope and shake-flask stages. Thereafter, this is followed by 'seed' and production stages.

Size: Batch fermentors are available with varying capacities. The capacity of the tank may range from a few hundred to several thousand gallons. The capacity of the fermentor is usually stated on the basis of the total volume capacity of the same. Thus, small laboratory fermentors, pilot-plant fermentors and larger or production fermentors may be available. Small laboratory fermentors are in the size range of 1 to 2 litres with a maximum up to 12 to 15 litres. Pilot plant fermentors have a

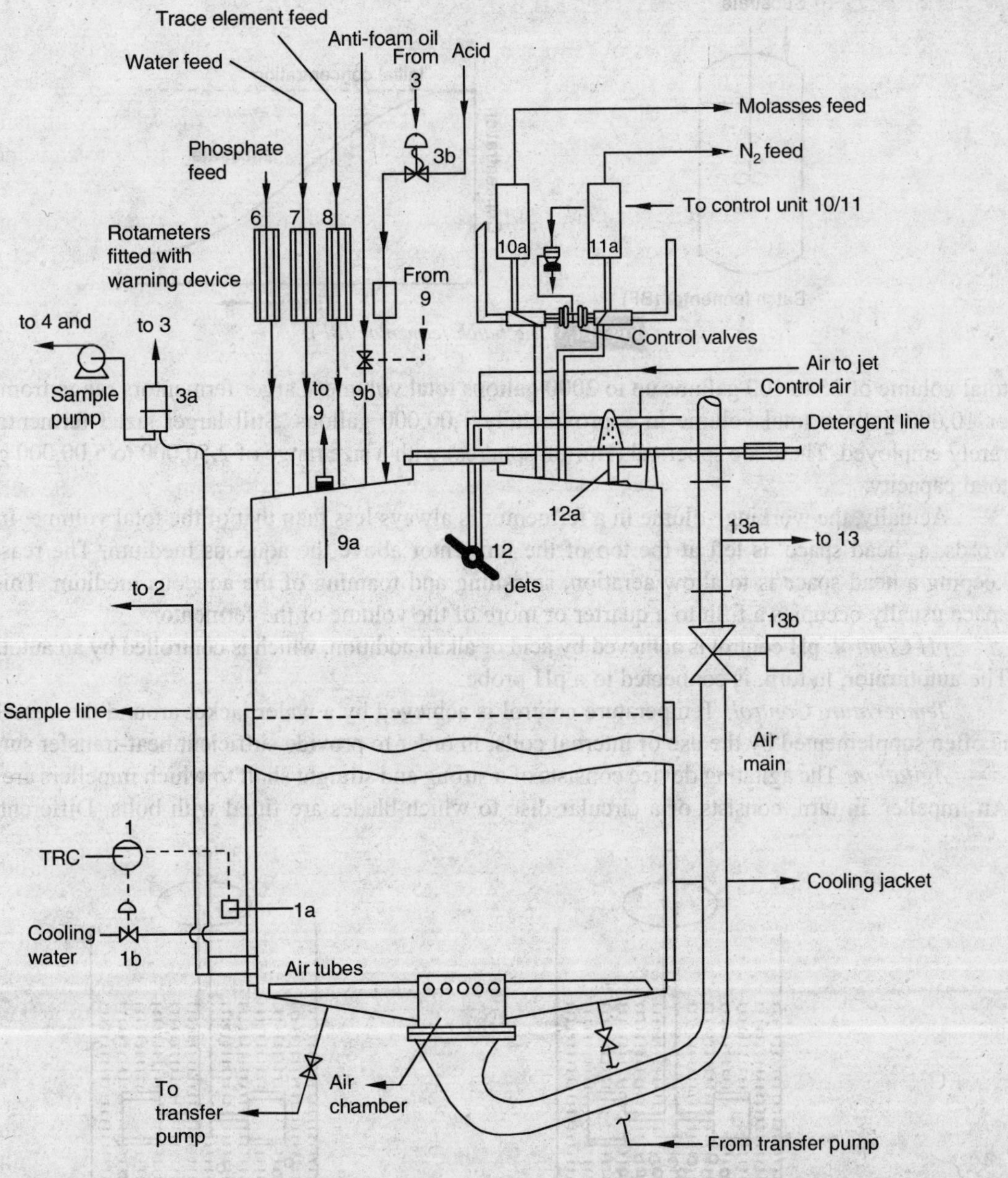

*1. Temperature controller and recorder 1a. Resistance thermometer lb. Control valve *2. Fermentor level
*3. pH controller and recorder 3a. pH electrode system 3b. Control valve *4. Yeast concentration recorder
*5. Recorder controller 6. Phosphate feed rotameter 7. Water feed rotameter 8. Trace element feed rotameter
*9. Foam controller 9a. Foam detector 9b. Control valve 10/11. Dosage control unit 10a. Molasses feed
11a. Nitrogen feed 12. Rotor jet 12a. Power unit *13. Air controller and recorder 13a. Venturi
13b. Power operated air control valve 14. Cooling jacket 15. Air chamber

***Indicates panel-mounted instruments.**

Fig. 14.2 *Typical fermentor with auxiliary equipment (Olsen, 1960).*

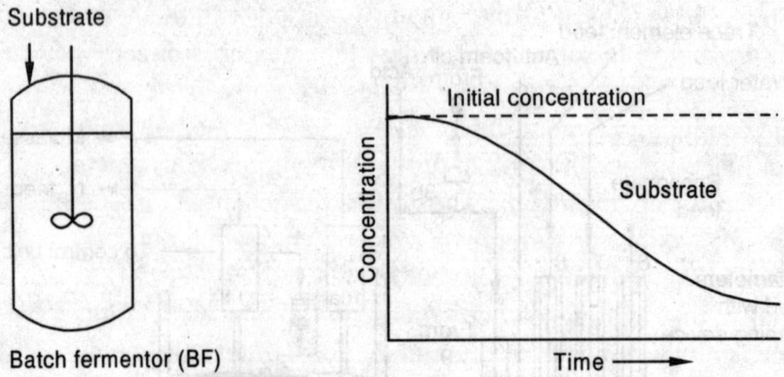

Fig. 14.3 *The batch fermentor (BF).*

total volume of 25 to 100 gallons up to 2000 gallons total volume. Larger fermentors range from 5,000 or 10,000 gallons total volume to approximately 1,00,000 gallons. Still larger sized fermentors are rarely employed. These are spherical (Horton spheres) with a size range of 2,50,000 to 5,00,000 gallons total capacity.

Actually, the working volume in a fermentor is always less than that of the total volume. In other words, a 'head space' is left at the top of the fermentor above the aqueous medium. The reason for keeping a head space is to allow aeration, splashing and foaming of the aqueous medium. This head space usually occupies a fifth to a quarter or more of the volume of the fermentor.

pH Control: pH control is achieved by acid or alkali addition, which is controlled by an autotitrator. The autotitrator, in turn, is connected to a pH probe.

Temperature Control: Temperature control is achieved by a water jacket around the vessel. This is often supplemented by the use of internal coils, in order to provide sufficient heat-transfer surface.

Agitation: The agitating device consists of a strong and straight shaft to which impellers are fitted. An impeller, in turn, consists of a circular disc to which blades are fitted with bolts. Different types

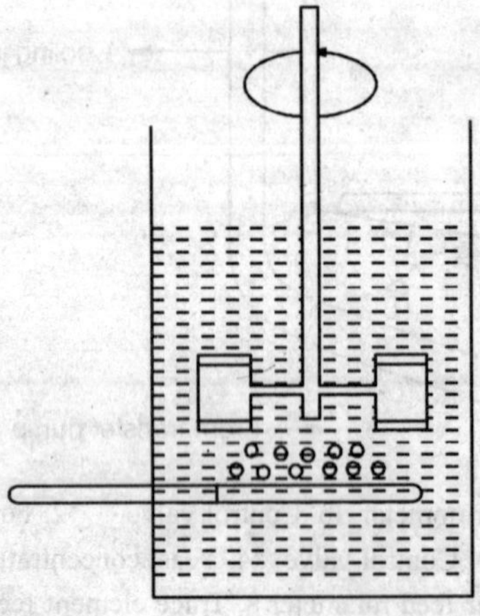

Fig. 14.4 (a) *Conventional ring sparger–orifices facing upwards.*

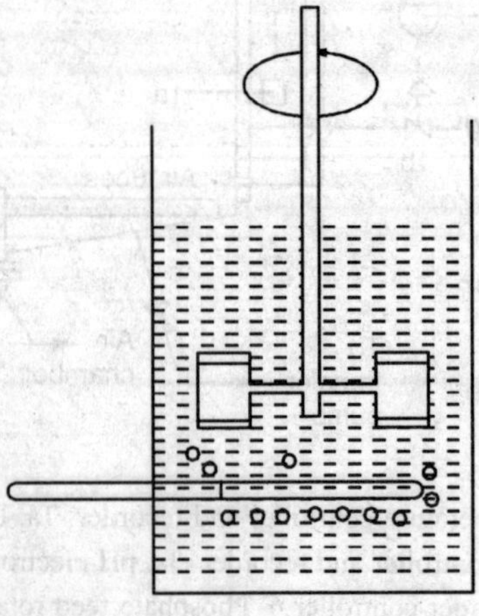

Fig. 14.4 (b) *Suggested modification–orifices facing downwards.*

of blades are available, and are used according to the requirements. The shaft passes through a bearing in the lid of the fermentation tank. It is rotated with the help of an electric motor mounted externally at the top of the tank. Usually, the speed of the agitator is varied with the help of adjustable pulleys and belts connected with the motor. In some cases, where the agitator is directly driven, impeller action is varied by the use of different types of impeller blades. Recently, the impeller of small fermentors is moved by a magnetic coupling to a motor mounted externally at the bottom of the tank. Moreover, the height of the impeller blades above the bottom of the fermentor is adjustable according to one's desire. The liquid medium is thrown up towards the walls of the fermentor while rotating the impeller blades at a high speed. This results in the formation of a vortex which is eliminated, usually, by four equally spaced baffles attached to the walls of the fermentor.

Aeration: Usually, the aerating device consists of a pipe with minute holes, through which pressurised air escapes into the aqueous medium in the form of tiny air bubbles. This aeration device is called a sparger. The size of the holes in a sparger ranges from 1/64 to 1/32 of an inch or larger. Holes smaller than this require too high an air pressure for economical bubble formation. One should always remember that the smaller the air bubbles, the greater is the bubble surface area. Subsequently, it is

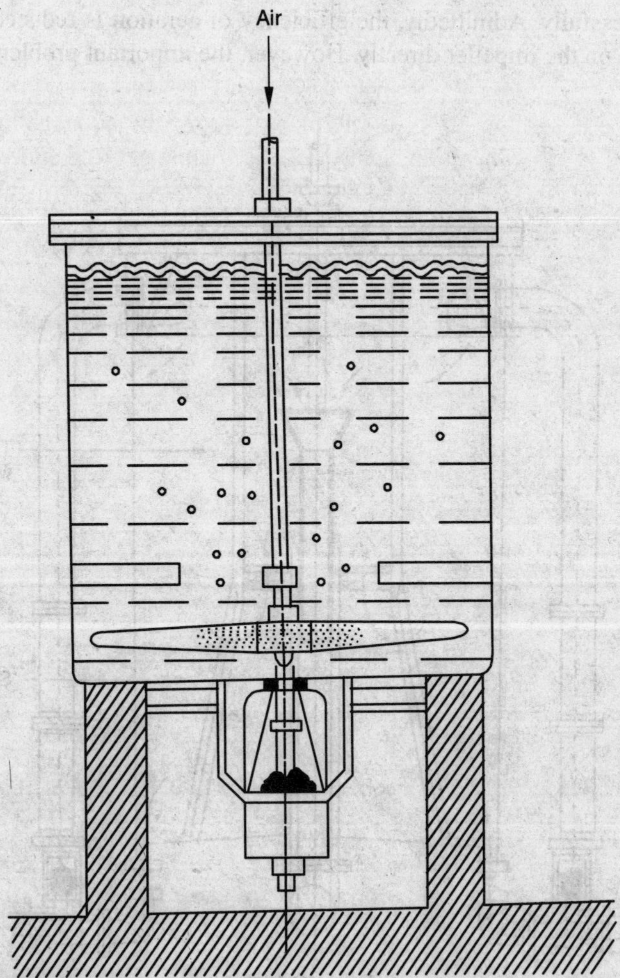

Fig. 14.5 *Fermentor with impeller air injection (Hospodka, 1966).*

more likely that the oxygen of that air would pass across the bubble boundary and dissolve in the aqueous medium. However, small air bubbles require higher air pressure for their formation through the fine holes. It is desirable to adjust the size of the air bubbles to give the greatest possible aeration without greatly increasing the overall cost of the fermentation process. The reason for this is that sterile air is a costly item for large scale fermentation.

Figure 14.4 (a) shows the conventional type of ring sparger where the orifices face upwards, just below the impeller. When working with mycelium-forming microbial fermentations with heavy mycelia and involving long fermentation times, clogging problems can become serious. This renders aeration completely ineffective. Such a problem has been encountered by some workers in the laboratory in the case of cellulose fermentation to produce cellulases with an LKB Ultroferm laboratory fermentor. The second reason for the clogging of the orifices is occasional power failures, when the cellulose particles settle down by gravity to plug the orifices.

These problems have been overcome by a simple modification by changing the direction of the orifices by a full 180° as shown in Figure 14.4(b). As a result, the orifices now face downwards, and fermentations run successfully. Admittedly, the efficiency of aeration is reduced, because now the air bubbles do not impinge on the impeller directly. However, the important problem of clogging has been solved.

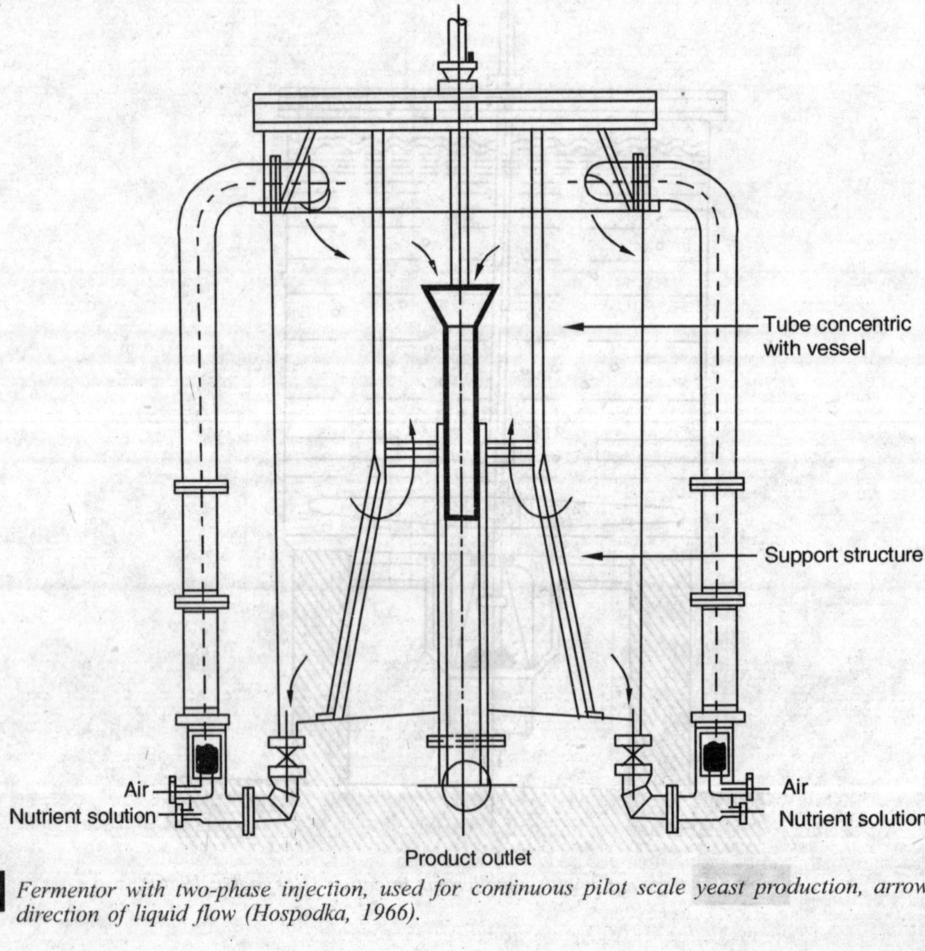

Fig. 14.6 *Fermentor with two-phase injection, used for continuous pilot scale yeast production, arrows indicate direction of liquid flow (Hospodka, 1966).*

The application of the latter sparger arrangement is more likely in the small laboratory fermentors where relatively narrow orifices are used.

There are various ways for introducing air into the fermentation vessel. The principal modes of injecting air are:

1. impeller air injection; air is fed to the impeller by means of a hollow drive shaft, and injected into the medium through holes in the impeller (Figure 14.5).
2. two-phase injection; a two-phase mixture of air and nutrient medium is fed into the fermentor in the form of a foam or suspension (Figure 14.6).
3. air-lift fermentor; air is used to circulate the contents of the fermentor, either through tubes external to the vessel or internally by using a draft tube (Figure 14.7).
4. sparger air injection; the most widely used fermentor is the stirred, baffled, upright cylinder with air sparger, the main features of which are illustrated in Figure 14.8.

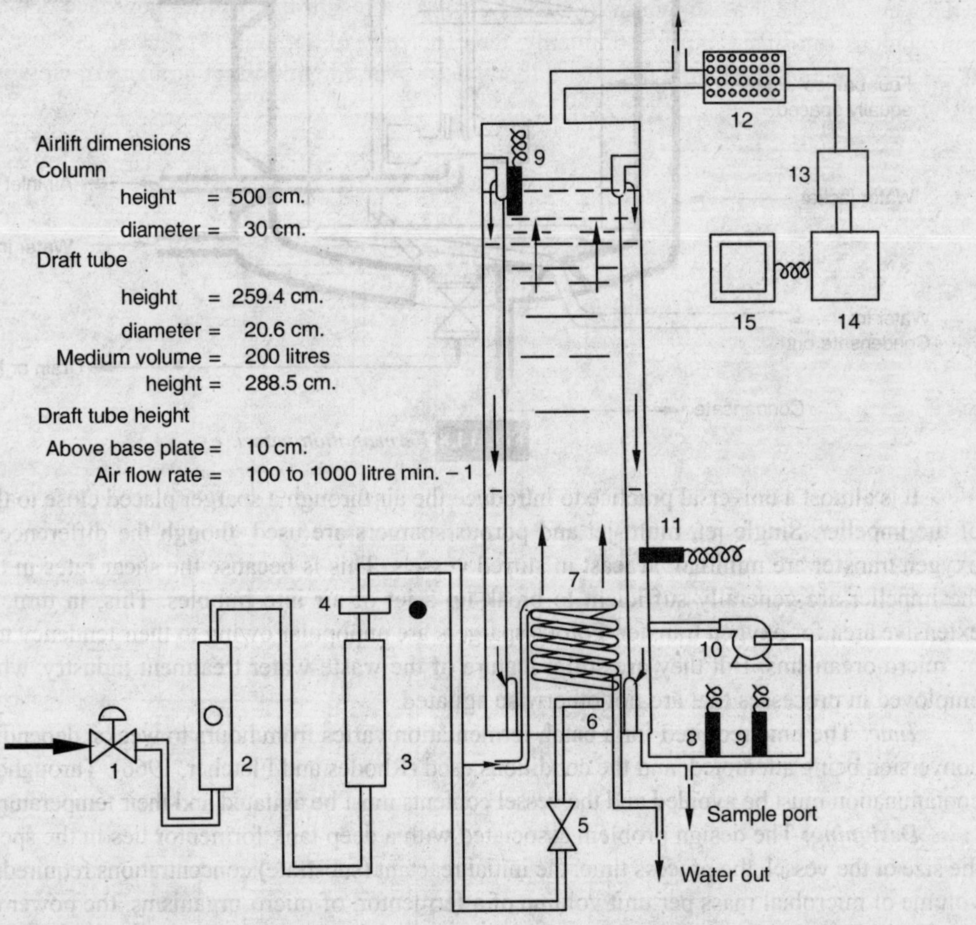

1. Air-pressure regulator 2. Rotameter 3. Glass-wool air filter 4. Millipore air filter 5. Check valve 6. Air sparger 7. Heating and cooling coil 8. pH electrode 9. Oxygen probe 10. Centrifugal pump 11. Thermistor probe 12. Silica-gel bed 13. Diaphragm pump 14. Paramagnetic oxygen analyser 15. Recorder

Fig. 14.7 *Air-lift fermentor, arrows indicate direction of liquid flow (Wang and Humphrey, 1969).*

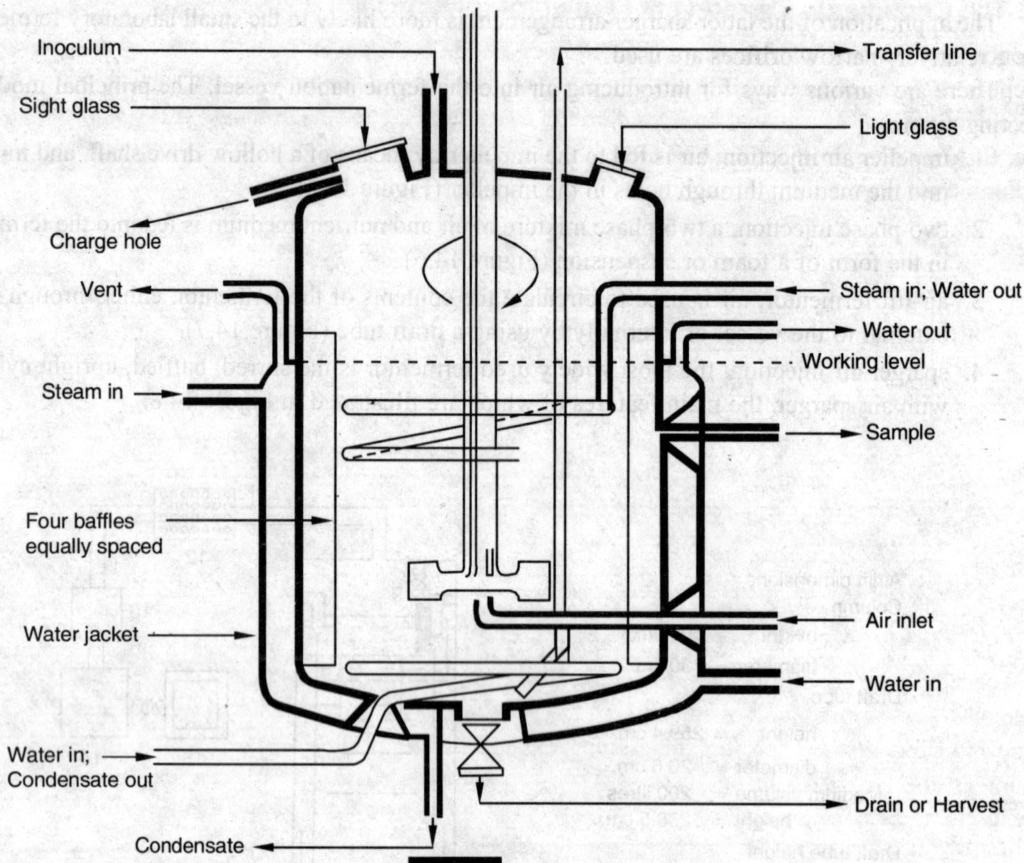

Fig. 14.8 *Fermentation vessel.*

It is almost a universal practice to introduce the air through a sparger placed close to the underside of the impeller. Single jet, multi-jet and porous spargers are used, though the differences as regards oxygen transfer are minimal, at least in stirred vessels. This is because the shear rates in the region of the impeller are generally sufficient to break up a jet of air into bubbles. This, in turn, provides an extensive area for oxygen transfer. Porous spargers are unpopular owing to their tendency to be blocked by micro-organisms. But they are still a feature of the waste-water treatment industry, where they are employed in processes that are not otherwise agitated.

Time: The time required for a batch fermentation varies from hours to weeks, depending upon the conversion being attempted, and the conditions used (Rhodes and Fletcher, 1966). Throughout this time, contamination must be avoided and the vessel contents must be agitated and their temperature controlled.

Designing: The design problem associated with a deep tank formentor lies in the specification of the size of the vessel, the process time, the initial reactant (substrate) concentrations required, the hold-up volume of microbial mass per unit volume of a fermentor, of micro-organisms, the power and aeration requirements and the area of heat-transfer surface.

The process of 'sludge digestion' in the waste-water treatment industry is an example of a batch fermentation.

3.2. The Continuous Stirred-tank Fermentor (Figure 14.9)

A continuous stirred-tank fermentor (CSTF) need not be basically different from the batch fermentor, except in sofar as feed and overflow devices are added. The fundamental difference lies in the fact that the contents of the vessel are at a *steady state*. In other words, contents of the vessel no longer vary with time, this applies to the hold-up of micro-organisms and the concentration of the components of the medium in the fermentor.

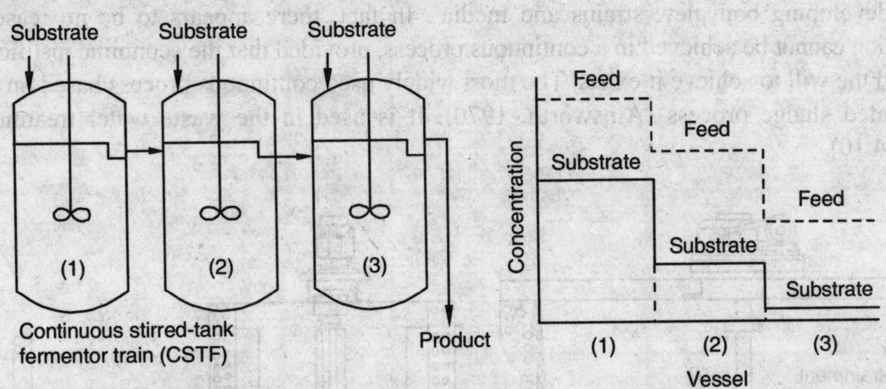

Fig. 14.9 *Continuous stirred-tank fermentor (CSTF) train.*

Steady-state conditions can be achieved by operation on either 'chemostatic' or 'turbidostatic' principles. The former involves adjusting of the flow-rate to the fermentor to an appropriate and constant value and allowing the micro-organism, substrate and biochemical product concentrations to attain their natural levels. The 'Turbidostat' requires an experimental determination of the turbidity, (i.e. an indirect measurement of the microbial concentration). This is then used to control the flow-rate. Both these methods have been employed in practice, though the former is obviously the simpler from every view point. Consequently, it is the one used in operations other than those on a laboratory scale.

The immediate consequence of the steady state condition is, that for satisfactory economic operation the environmental conditions selected for the fermentor have to lead to acceptable yields of microbial and biochemical products. This single feature demands depth of knowledge of the physiological and biochemical factors (which influence the microbial activity) far in excess of that demanded by batch fermentations (where detailed empirical procedures for fermentor operation are usually developed over a period of years). Once a small yield of product has been achieved in a batch fermentation, this can be enhanced on a developmental basis (Demain, 1971). In contrast, for continuous operation detailed knowledge is demanded of all the relationships between the rate of reaction and the operating variables. A further complication arises in that 'natural' media, media obtained from complex natural sources, in particular farm products or other fermentations, e.g. molasses and corn steep-liquor, as opposed to specially prepared chemical substrate solutions, are the norms in the fermentation industries. Therefore, the determination of the effects of all the concentration variables is scarcely feasible. The use of such media results in a further restraint, since the concentrations, on a relative basis, are fixed and this makes it difficult to consider a variety of inlet compositions as part of the design exercise. Detailed application of batch data to a continuous system that presents significant difficulties, and these are scarcely made easier by the physiological and biochemical changes known to occur in the micro-organisms themselves during the period of a batch fermentation (Herbert, 1961).

The most successful continuous systems to date have been those employing yeasts and bacteria, in which the desired products are the cells or primary metabolites, compounds that form the chemical 'inventory' of a microbe, (e.g. enzymes and amino acids), or some product clearly associated with growth or energy producing mechanisms, (e.g. the production of alcohol). A comment made by Righelato and Elsworth (1970) is worth noting, '...if they wish to adopt continuous methods, industrial researchers must be prepared to carry out their own empirical research as they have done with batch cultures, perhaps developing both new strains and media'. In fact, there appears to be no reason why any fermentation cannot be achieved in a continuous process, provided that the economic justification can be made, and the will to achieve it exists. The most widely used continuous process based on the CSTF is the activated sludge process (Ainsworth, 1970). It is used in the waste water treatment industry (Figure 14.10).

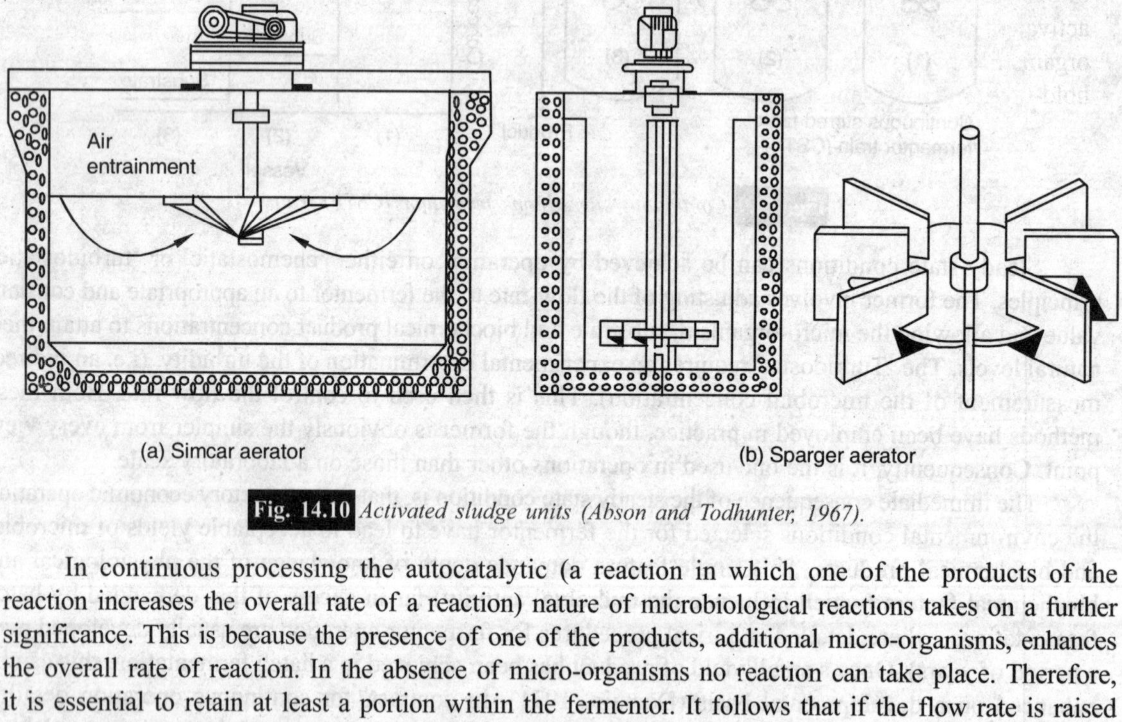

Fig. 14.10 *Activated sludge units (Abson and Todhunter, 1967).*

In continuous processing the autocatalytic (a reaction in which one of the products of the reaction increases the overall rate of a reaction) nature of microbiological reactions takes on a further significance. This is because the presence of one of the products, additional micro-organisms, enhances the overall rate of reaction. In the absence of micro-organisms no reaction can take place. Therefore, it is essential to retain at least a portion within the fermentor. It follows that if the flow rate is raised to a high value, then all the micro-organisms will be swept from the fermentor, and the conversion will cease. This phenomenon is commonly known as 'Wash-out'. Obviously, if micro-organisms are fed to the fermentor simultaneously with the substrate feed, the problems associated with wash-out are abated, and the reaction proceeds normally. Operation under such conditions requires a continuous flow of micro-organisms identical with those within the fermentor. The logical source of these is the effluent stream, as this contains micro-organisms and nutrients in the same biochemical and physiological conditions, as those within the fermentor. The effluent may be passed through either a centrifuge or a sedimentation tank. This operation produces a concentrated microbial suspension for recycling to the fermentor (Figure 14.11).

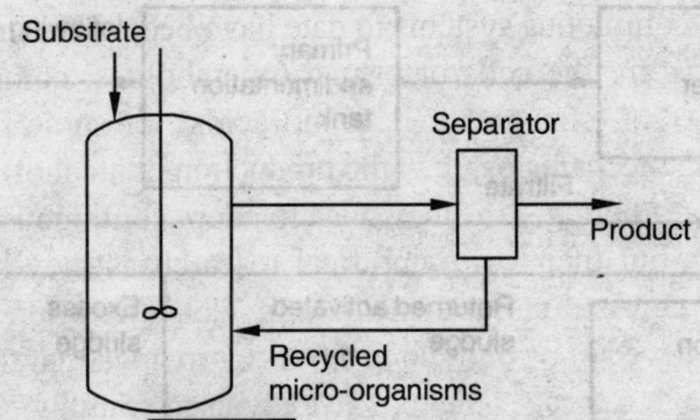

Fig. 14.11 *CSTF with recycle.*

Both these methods have been successfully applied under industrial conditions for the production of beer (Coutts, 1958, 1961) as illustrated in Figure 14.12. The latter method is a normal feature of the activated sludge process (Ainsworth, 1970) as shown in Figure 14.13. A constant inflow of micro-organisms obviates the wash-out problem. It also results in increased productivity from the higher total hold-up of micro-organisms within the fermentor.

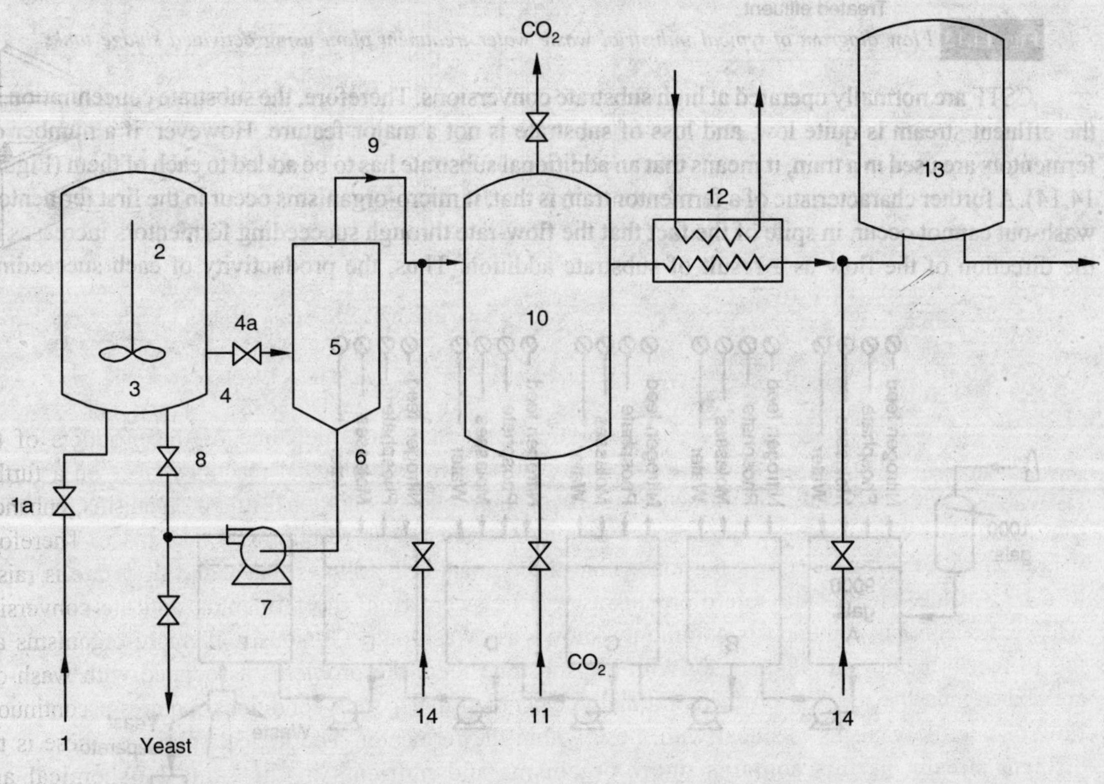

1. Wort inflow la. Control valve 2. Fermentor 3. Impeller 4. Beer outflow 4a. Control valve
5. Sedimentation vessel or Centrifuge. 6. Yeast recirculation 7. Pump 8. Recirculation control valve
9. Clearbeer outflow 10. Washing vessel 11. Carbon dioxide inlet 12. Heat-exchanger (cooler)
13. Reservoir 14. Finings inlet

Fig. 14.12 *Flow sheet of the continuous, single-stage fermentation for the production of beer (Coutts, 1958,1961).*

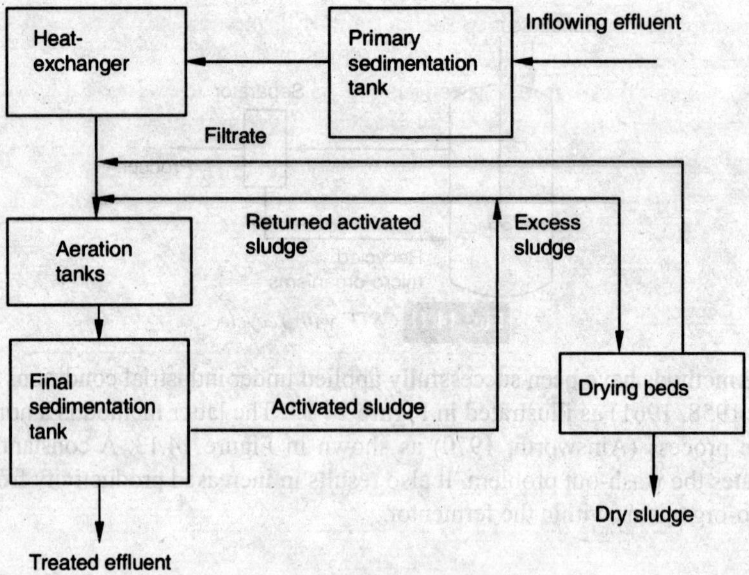

Fig. 14.13 *Flow diagram of typical industrial waste water treatment plant using activated sludge tanks.*

CSTF are normally operated at high substrate conversions. Therefore, the substrate concentration in the effluent stream is quite low, and loss of substrate is not a major feature. However, if a number of fermentors are used in a train, it means that an additional substrate has to be added to each of them (Figure 14.14). A further characteristic of a fermentor train is that, if micro-organisms occur in the first fermentor, wash-out cannot occur, in spite of the fact that the flow-rate through succeeding fermentors increases in the direction of the flow as a result of substrate addition. Thus, the productivity of each succeeding

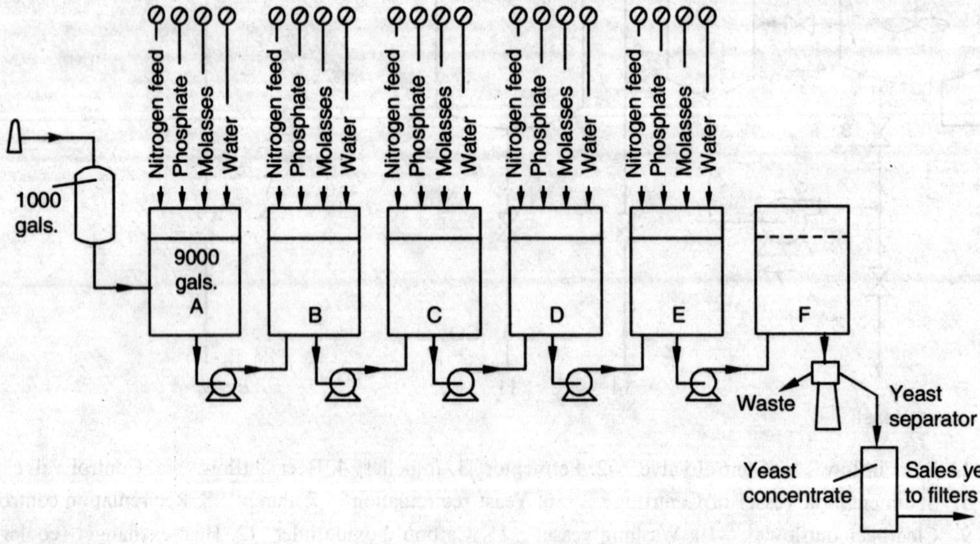

Fig. 14.14 *Flow sheet of continuous production of bakers' yeast (Oslen, 1960). Fermentors are from A to F.*

fermentor is greater than the previous one. Such a train of fermentors has been used successfully for the production of bakers' yeast (Olsen, 1960) as shown in Figure 14.14.

Multi-stage systems for microbial reactions are particularly useful when, in the equivalent batch process, it is necessary to vary the environmental conditions over the course of the reaction. In a multi-stage system this is equivalent to providing different environmental conditions at the various stages.

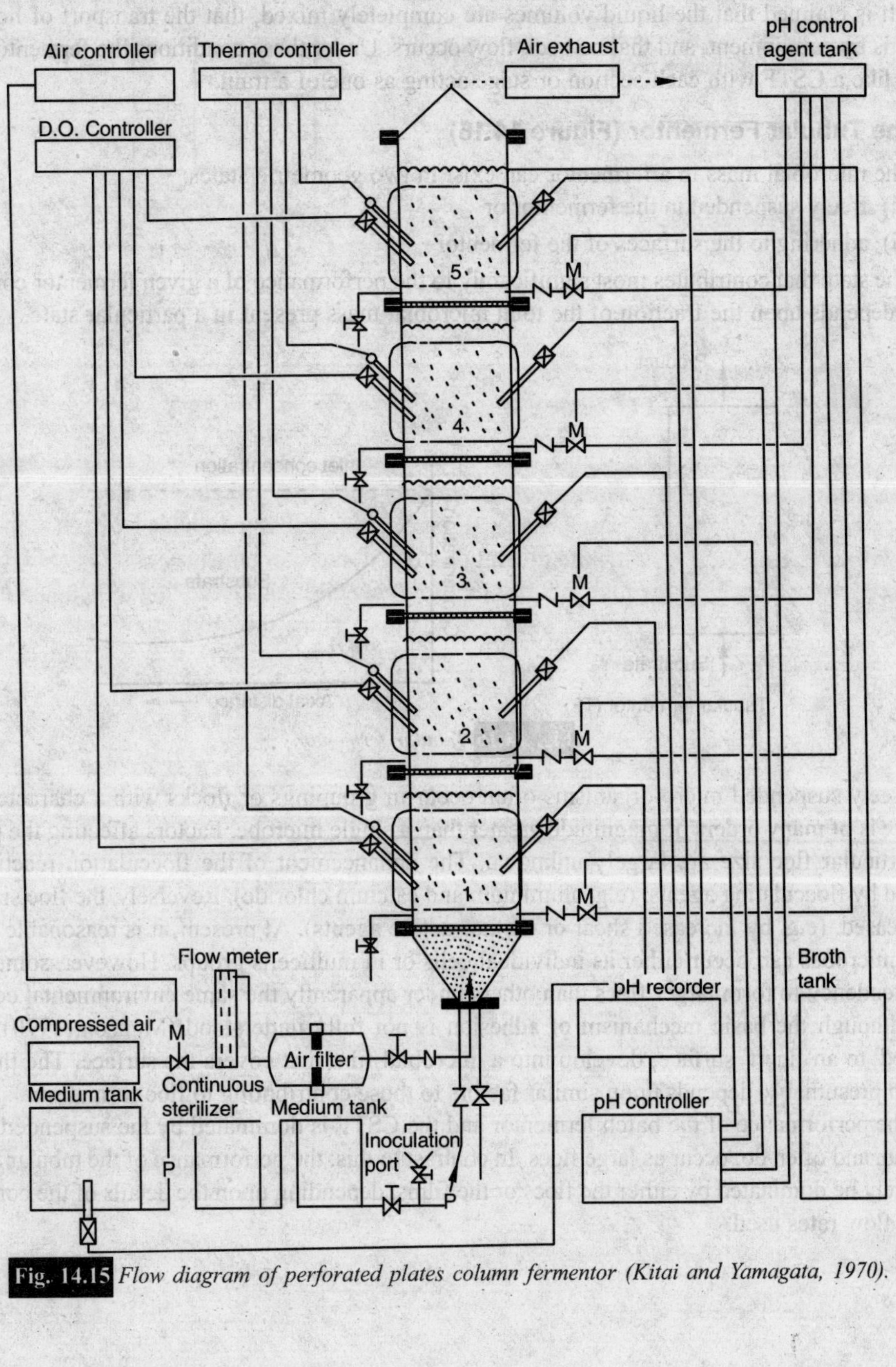

Fig. 14.15 *Flow diagram of perforated plates column fermentor (Kitai and Yamagata, 1970).*

An alternative form of CSTF train to that illustrated in Figure 14.14 has been described by Kitai and Yamagata (1970). This fermentor is shown in Figure 14.15. It consists basically of a tube divided into equal sections by horizontal perforated plates. The fermentor is intended for use under aerobic conditions with the medium and air fed concurrently to the base of the column. In operation the gas liquid ratio can be adjusted, so that a pool of liquid may be maintained on each plate with an air space above. It is claimed that the liquid volumes are completely mixed, that the transport of liquid up the column is by entrainment, and that no back flow occurs. Under these conditions the fermentor performs exactly like a CSTF with each section or stage acting as one of a train.

3.3. The Tubular Fermentor (Figure 14.16)

The microbial mass in a fermentor can exist in two geometric states:

(i) freely suspended in the fermentor or

(ii) adhering to the surfaces of the fermentor

The state that contributes most significantly to the performance of a given fermentor configuration largely depends upon the fraction of the total microbial mass present in a particular state.

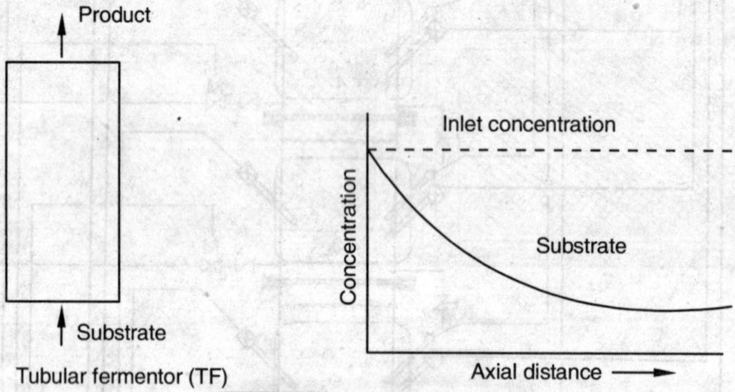

Fig. 14.16 *Tubular fermentor.*

Freely suspended micro-organisms often occur in groupings or flocks with a characteristic size. This size is of many orders of magnitude greater than a single microbe. Factors affecting the production of a particular floc size are largely unknown. The enhancement of the flocculation reaction can be achieved by flocculating agents, (e.g. aluminium and calcium chloride). Reversely, the floc size can also be decreased, (e.g. by increased shear or surface active agents). At present, it is reasonable to suggest that all microbes can occur either as individual cells or in multicells groups. However, some microbes have a tendency to form larger flocs than others under apparently the same environmental conditions.

Although the basic mechanism of adhesion is not fully understood (Munson, 1970), microbes 'attached' to an 'inert' surface, develop into a microbial film that covers the surface. The thickness of this film presumably depends upon similar factors to those contributing to floc formation.

The performance of the batch fermentor and the CSTF is dominated by the suspended microbes. They can, and often do, occur as large flocs. In contrast to this, the performance of the tubular fermentors can largely be dominated by either the flocs or the films, depending upon the details of the configuration and the flow rates used.

In a tubular fermentor containing microbial flocs, individual microbes are exposed to different environments. This may be an advantage in that the conditions can be compared with the batch fermentation, if the distance along the tubular fermentor is related to the time under batch conditions. These fermentors are analogous, since the conditions at a particular position in a tubular fermentor are characterized by the time the fluid passing that position has been in the fermentor.

In the case of the CSTF, the substrate concentration in the outflow from a CSTF is generally very small, (i.e. the bioconversion is large). Because of the progressive fall in substrate concentration along the length of a tubular fermentor (Figure 14.17) the outlet concentration is likely to be of measurable proportions, resulting in some loss of the raw material.

Processes with micro-organism adhering to a support surface are chiefly used in the waste water treatment industry under the name, *'Trickling Filter'* (Jenkins, 1970) as shown in Figure 14.17.

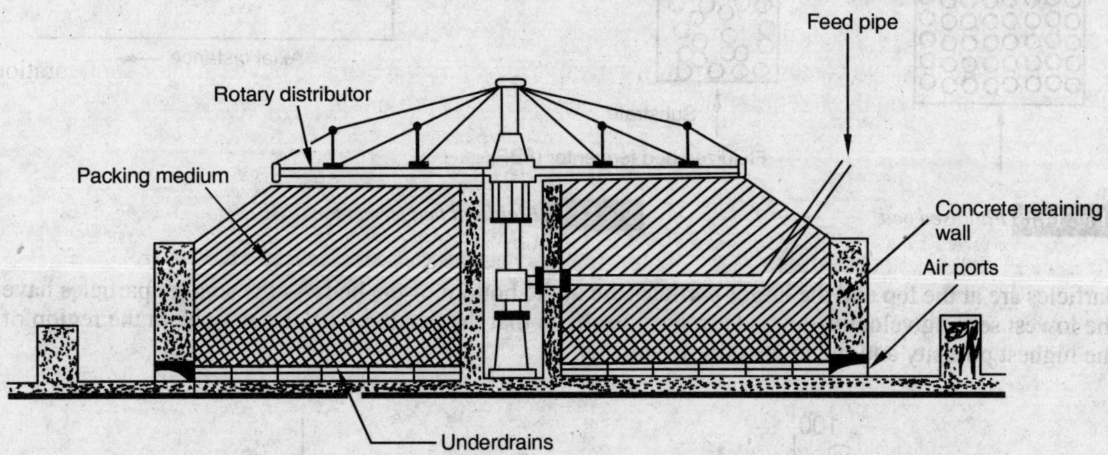

Fig. 14.17 *Trickling filter.*

3.4. The Fluidised Bed Fermentor (Figure 14.19)

The condition shown in Figure 14.18 represents 'particulate' fluidisation. This is a uniform mode of fluidisation. This is a characteristic of beds of regular particles suspended in an upflowing liquid stream.

If an additional gas phase is involved, there is a tendency for the particles in the bed to become less evenly distributed.

There are two important features of the beds of mixed particle sizes:

(i) the increase in porosity from the bottom to the top of the bed

(ii) the decreased particle movement when compared with beds containing particles of constant size

Since porosity or voidage is a measure of the local free space within a bed, it also represents a measure of the microbial hold-up when expressed as wet volume per unit bed volume. Thus, a variation in microbial hold-up is to be expected within a 'fluidised bed' fermentor (Figure 14.20) on fluidisation, the smaller particles rise relative to the larger particles, and produce a situation where the smaller

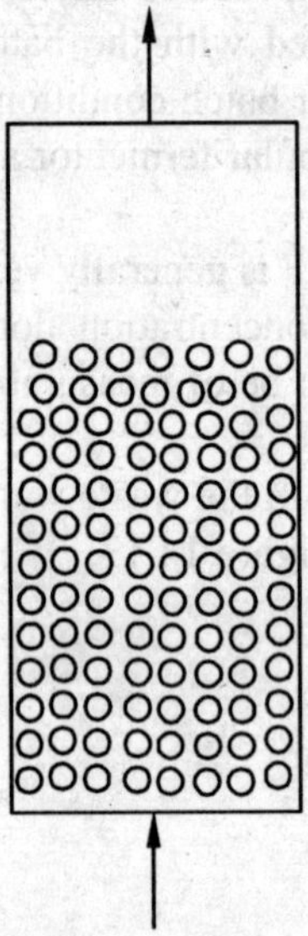

Fig. 14.18 *Fluidised bed.*

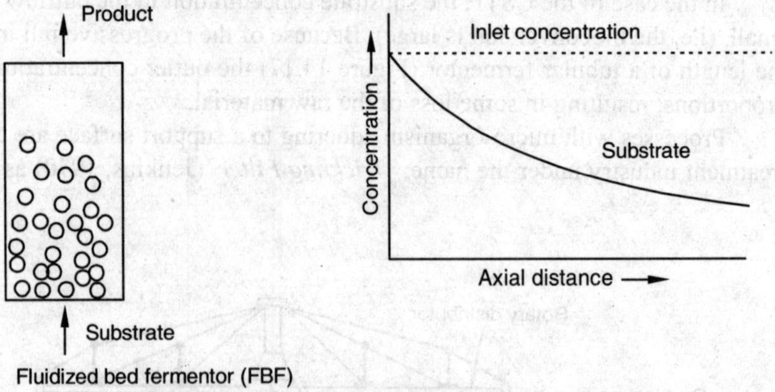

Fig.14.19 *Fluidised bed fermentor (FBF).*

particles are at the top and the larger particles are at the bottom of the bed. As the smaller particles have the lowest settling velocity, the bed arranges itself, so that the smaller particles may be in the region of the highest porosity and the lowest linear velocity.

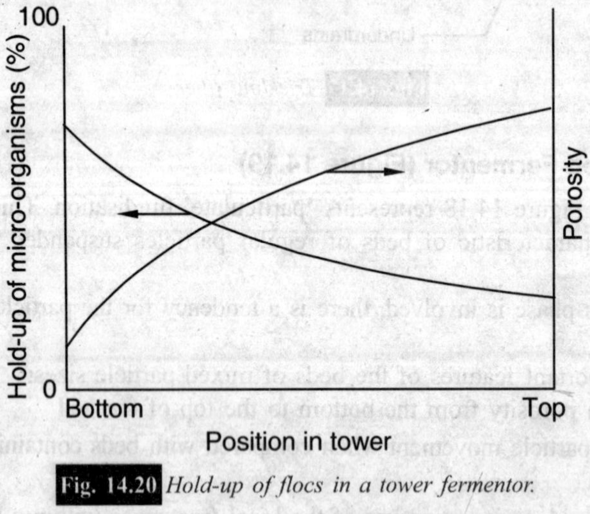

Fig. 14.20 *Hold-up of flocs in a tower fermentor.*

The **tower fermentor** (developed for the continuous production of beer) is based upon these general principles (Ault et al. 1969). In this process yeast flocs are maintained in suspension by the upward movement of the nutrient medium. Moreover, any entrained particles are returned by means of a sedimentation device at the top of the tower. Essentially, the fermentor consists of a vertical cylinder with an aspect ratio (length to diameter) of approximately 10:1 (Figure 14.21).

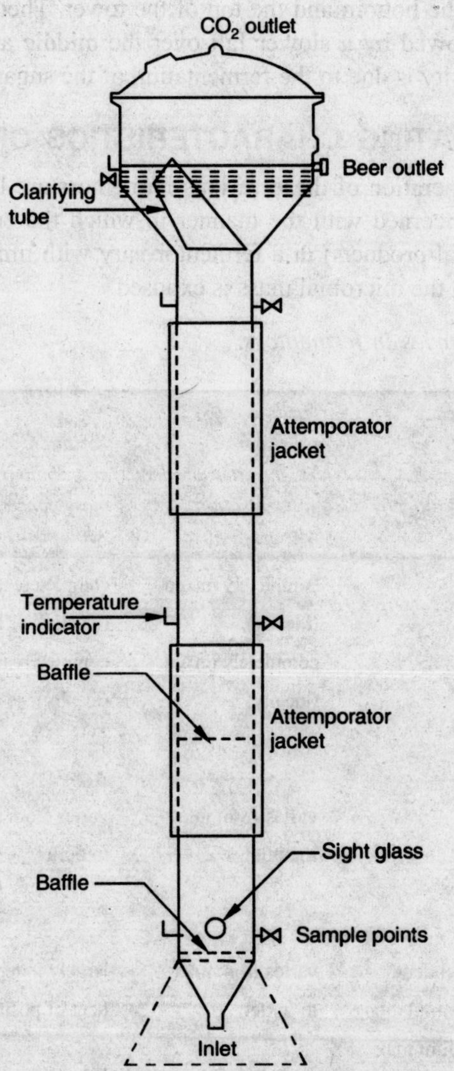

Fig. 14.21 *Diagram of the tower fermentor (Shore and Royston, 1968).*

At the top of the tower a separator is provided to induce the gas bubbles produced by the reaction, to coalesce and escape from the liquid phase. Within the separator there is a quiescent zone, free of the rising gas, so that the yeast may settle and return to the main body of the tower, and clear beer can be removed.

A flocculent yeast, (i.e. a yeast capable of achieving relatively large floc sizes) is essential for an alcoholic fermentation in a FBF at acceptable flow rates, otherwise a large proportion of the yeast would be washed out. As a result of this, an insufficient yeast concentration is maintained. A mean yeast concentration of 25 per cent by weight (expressed as centrifuged wet weight) is typical with values as high as 30-35 per cent by weight at the bottom of the tower, and as low as 5-10 per cent by weight at the top. A significant feature of the tower is the progressive and continuous fall in the specific gravity of the nutrient medium between the bottom and the top of the tower. There is an initial rapid fall at the bottom of the tower. It is followed by a slower fall over the middle and the top of the tower. This gradual fall in the specific gravity is due to the fermentation of the sugars.

4 PRINCIPAL OPERATING CHARACTERISTICS OF FERMENTORS

The principal features in the operation of these various fermentors are listed in Tables 14.1 and 14.2. The first of these tables is concerned with the manner in which the concentrations, (i.e. substrates, microbial mass and biochemical products) in a fermentor vary with time and position, as well as the environmental history to which the microbial mass is exposed.

Table 14.1 *Concentration histories in fermentors.*

Fermentor	*Time dependence of substrate, microbial mass & biochemical product concentration*	*Variation of substrate & biochemical product concentration with position*	*Variation of microbialr mass-concentration with position*	*Environmental history of microbial mass*
(A) Batch	dependent	completely mixed (ideally)	completely mixed (ideally)	varies over course of the fermentation.
(B) CSTF	independent	completely mixed (ideally)	completely mixed (ideally)	constant (all flocs exposed to the same environment at all times).
(C) Tubular fermentor containing microbial flocs	independent	varies from inlet to outlet	varies from inlet to outlet	varies as flocs travel through the fermentor.
(D) Tubular fermentor containing microbial films	largely independent, but some accumulation of microbial film may occur, this will also affect the substrate and product concentrations	varies from inlet to outlet	largely independent of position	constant (however, the microbial film in different parts of the fermentor is exposed to different environments).
(E) Fluidised bed fermentor	independent	varies from inlet to outlet	varies from inlet to outlet	largely constant, but some movement of flocs does take place (flocs in different parts of the fermentor are exposed to different environments).

From an industrial viewpoint the ability to control pH and temperature may be of considerable importance, as are any restrictions that might be placed on the flow-rate. These features are contained in Table 14.2.

Table 14.2 *Performance features of fermentors.*

Fermentor	*pH control*	*Temperature control*	*Features of industrial importance*	*Chief industrial applications*
(A) Batch	if required	if required	labour-intensive	most commercial fermentations.
(B) CSTF	if required	if required	flow rate limited by wash-out	waste-water treatment, microbial protein production.
(C) Tubular fermentor containing microbial flocs	difficult to control, except at high re-circulation rates	if required	requires a constant feed of microbes	—
(D) Tubular fermentor containing microbial films	difficult to control	if required	difficult to control microbial hold-up	waste-water treatment, vinegar production.
(E) Fluidised bed fermentor	difficult to control	if required	flow-rate limit by wash-out	beer and cider production.

15 Computer Control of Fermentation Processes

1 INTRODUCTION

New developments in digital electronics and information processing have permitted a level of monitoring and control of fermentation processes. It could only previously be achieved by more expensive analogue equipment and the dedicated diligence of process operators. Small, inexpensive units are now available which will render the empirical, uncontrolled operation of fermentors a thing of the past. Information on what happened during a fermentation will be available in an ever increasing quantity, eliminating dead times (overnight). This will enable a more refined analysis of the cause and effect relationship existing between the microbes in the fermentor and the environmental conditions to which we subject them. Table 15.1 shows commonly measured and controlled parameters.

Table 15.1 *Commonly measured and controlled parameters.*

Measured only	*Measured/controlled*	*Normally controlled*
—outlet O_2	—substrate feed rate	—temperature
—outlet CO_2	—substrate concentration	—pH
—NH_3	—biomass concentration	—dissolved O_2
—pressure(s)	—level	—stirrer speed
	—power to stirrer	—antifoam addition
	—gas flows	
	—pH control fluids	

The use of computers for control of fermentation processes has increased significantly in the last decade. Those engaged in both industrial fermentation and fermentation research are fast realizing that the computer can be a vital instrument for process optimization and control. The decrease in cost and the improved reliability of computer hardware in recent years have made the use of computers even more alluring. The advent of micro-processors has had an enormous impact in the area of real-time computer applications to fermentation processes. Micro-computers can provide a relatively inexpensive and reliable means of integrating computer control into the fermentation processes.

The first project to computerize a fermentation facility at Lilly began in 1966 at Dista Products Ltd. (a Lilly subsidiary), located in Liverpool, England. The project consisted of interfacing an 8K

Elliott Automation Arch 102 computer to production fermentation vessels for the purpose of data logging, alarming, and control. Control (114 loops) was accomplished by direct digital control (DDC) and included control of temperature, pH, air-flow, and foam level. The computer software was designed and coded by 'in-house' professional personnel. Assembly language was used to implement modular building blocks which could be easily assembled to form a working control strategy.

In early 1970, Lilly personnel, after careful study of process control and information management requirements, resident technical expertise, and systems then available in the market, decided to develop a new generation computer system to be used with fermentation. The system, which is now fully developed, consists of a hierarchy of computers.

The first descriptions of computer-controlled fermentation were given in 1969 by Grayson and Yamashita *et al.* In 1971, a philosophy of data acquisition, analysis and computer control of fermentation processes was presented by Nyiri. The use of multi-level control and on-line process optimization was also suggested. Since then, many reports have dealt with applications involving varying degrees of computer control. The first international conference on computer applications in fermentation technology was held in 1973 in Dijon, France. It helped sustain interest in this topic and spurred subsequent conferences in Philadelphia, U.S.A., in 1978 and Manchester, England, in 1981.

For a fermentation production plant, the primary objective of computer control is to produce a product as economically as possible. The computer is generally used to provide quality control, save the operational time, furnish automatic documentation and decrease the per loop control costs. Computer control has found its way not only in plants, where the economic advantages are more obvious, but also in pilot plants and research laboratories. In the latter cases, the computer provides fast and efficient data acquisition, the ability to monitor and control experimental conditions, and flexibility in the operation of the fermentor.

One aspect in which the capabilities of the computer can truly be realized is in the implementation of advanced control and the optimization strategies. It was not until 1970s that much of the modern control theory, developed in the 1950s and 1960s, could be utilized practically. This change was sparked primarily by tremendous economic and technological advances in computer hardware, especially with the advent of micro-computers. The fermentation industry has lagged somewhat behind the chemical and petro-chemical industries in the area of advanced computer control. However, a real interest has been generated for using modern control techniques for fermentation.

The author predicts a rapid growth of fermentation industries with the computerization of the fermentation process and the possibility of competing with chemical and petro-chemical industries.

2 COMPUTER HARDWARE AND SOFTWARE

2.1. Instrumentation

The basis of any system used in monitoring and controlling a fermentation process is instrumentation. Proper measurements provide the means for evaluating the microbial system and its dynamics. The selection and subsequent performance of basic measurement instrumentation is important in the development of a computer-controlled fermentation system.

It has been stated that the primary measurement problem in fermentation is to decide what is to be measured, and in what way. The complexity of the biological process makes measurement problems significantly different from those associated with typical chemical processes. Hatch divided the measurable process output variables into three categories:

(i) environmental
(ii) physiological
(iii) cell culture composition

Common environmental variables include temperature, pH, dissolved oxygen, agitation speed, aeration rate and nutrient concentrations. Physiological variables can be subdivided into two groups:

(a) products of metabolism, such as cell mass and extra-cellular metabolic products
(b) variables describing the state of metabolism

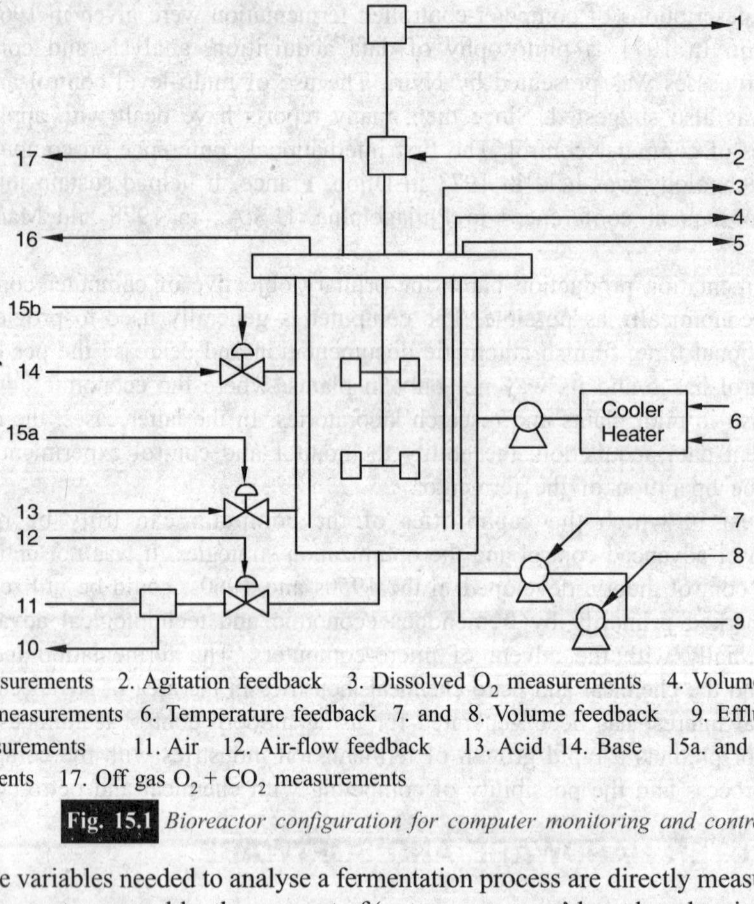

1. Agitation measurements 2. Agitation feedback 3. Dissolved O_2 measurements 4. Volume measurements 5. Temperature measurements 6. Temperature feedback 7. and 8. Volume feedback 9. Effluent 10. Air-flow measurements 11. Air 12. Air-flow feedback 13. Acid 14. Base 15a. and 15b. pH feedback 16. pH measurements 17. Off gas O_2 + CO_2 measurements

Fig. 15.1 *Bioreactor configuration for computer monitoring and control.*

Many of the variables needed to analyse a fermentation process are directly measurable. For those variables which are not measurable, the concept of 'gateway sensors' has played an important role. By this technique, certain biologically significant parameters can be extracted from a combination of available sensors. The computer, in this case, is very useful in making the necessary correlations, thus providing a 'gateway' to calculate unmeasurable quantities. Measurable quantities which are frequently used for correlation purposes include the concentration of O_2 and CO_2 in the exit gas, rates of nutrient or energy addition, and redox potential. A bioreactor configuration which depicts key physical variables used in computer monitoring and control is given in Figure 15.1.

2.2. Interfacing Techniques

The interfacing techniques used in coupling the computer with measurement and control instrumentation for fermentation processes do not differ significantly from interfacing techniques used in other types of processes.

The basic components of a typical computer interface unit are shown in Figure 15.2.

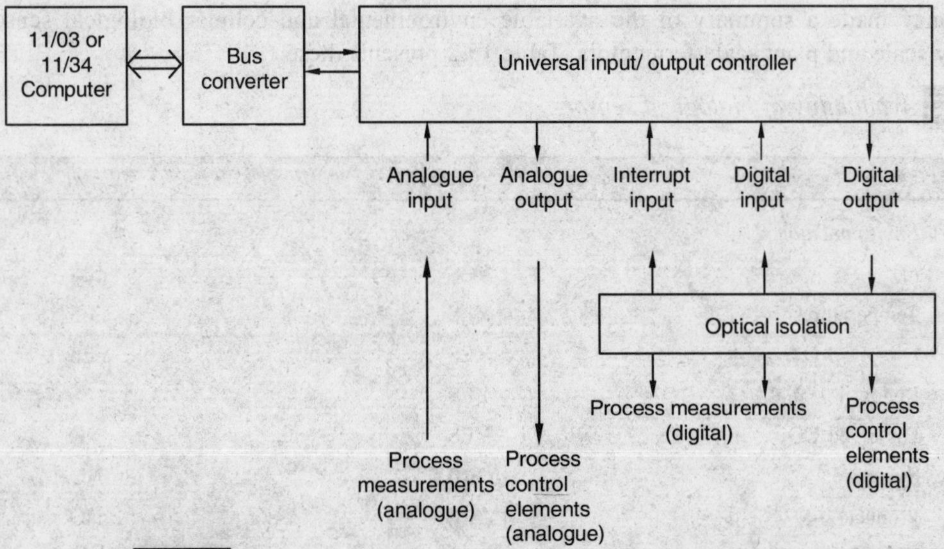

Fig. 15.2 *The basic components of a typical computer interface unit.*

The unit includes modules providing analogue inputs, analogue outputs, digital inputs, digital outputs and interrupt inputs. In this unit, sets of eight analogue input signals are multiplexed into each analogue to digital (A/D) converter. This is a common practice with analogue input signals due to the high cost of A/D converters, which is acceptable by virtue of the generally low data acquisition rates associated with the fermentation processes. Electronic interfacing units can be purchased as standard equipment on most computer systems, or they are readily available from computer accessory manufacturers.

3 COMPUTER APPLICATIONS IN FERMENTATION TECHNOLOGY

3.1. Automation in penicillin fermentation

The primary objective of penicillin fermentation is productivity or the conversion yield of the product from the carbon-energy substrate. To achieve effective control, one must continuously assess the product and residual substrate concentrations. Herein lies one of the major bottle-necks in the development of computer-aided fermentation. The sensors for direct monitoring of biomass, product (other than biomass) concentration and substrate concentration are not available for use in commercial-scale fermentation.

Automatic process control in industrial fermentation is currently limited to the control of temperature, pH, and in a few cases, dissolved oxygen. Many investigators are now examining more sophisticated control systems that monitor and analyse data while directly coupled to the fermentation. The importance of biomass as a parameter in the penicillin fermentation process cannot be overemphasized. But unfortunately, there are no commercially available sensors for its measurement. Jefferies and Humphery and Cooney *et al.* have suggested indirect methods of analysis to derive the biomass concentration through O_2 and CO_2 material balancing. The approach to the indirect monitoring of biomass concentration, as well as other biological parameters is generally applicable to fermentation processes. It is particularly well suited to on-line computer-aided fermentation processes and is useful as a signal for process control of substrate, nutrient and precursor feed.

Cooney made a summary of the available environmental and cellular biological sensors for laboratory scale and plant scale fermentors. Table 15.2 presents these.

Table 15.2 *Availability of biological sensor.*

Sensor	*Laboratory*	*Plant*
(i) *Environmental*		
pH	YES	YES
Temperature	YES	YES
Dissolved O_2	YES	YES
Redox potential	YES	YES
Dissolved CO_2	YES	NO
Substrate	(YES)*	NO
Product	(YES)*	NO
Mass flow rate	YES	YES
(ii) *Cellular*		
Gaseous O_2	YES	YES
Gaseous CO_2	YES	YES
Heat evolution	YES	(YES)**
ATP	YES	NO
NAD/NADH	YES	NO
Product	(YES)**	NO
Cell Mass	(YES)**	NO
Enzyme level	(YES)**	NO

*Specialized analytical schemes for laboratory use are available for only a limited number of substrates and products.
**Several specialized sensors are under development.

He also showed that it was possible to develop an adaptive control strategy for continuous addition of glucose in order to meet the demands of the growing mycelium of *Penicillium chrysogenum*. The control strategy which was based on the integration of the total CO_2 evolved provided an accurate measurement of the amount of mycelium present. This would add sufficient carbon source to maintain the maximum growth rate desired during the growth phase and any desired present slow growth rate required during the production phase. The control thus achieved would not overfeed the glucose which would otherwise lead to a wasteful side-product formation or the repression of penicillin synthesis. Figure 15.3 shows the results.

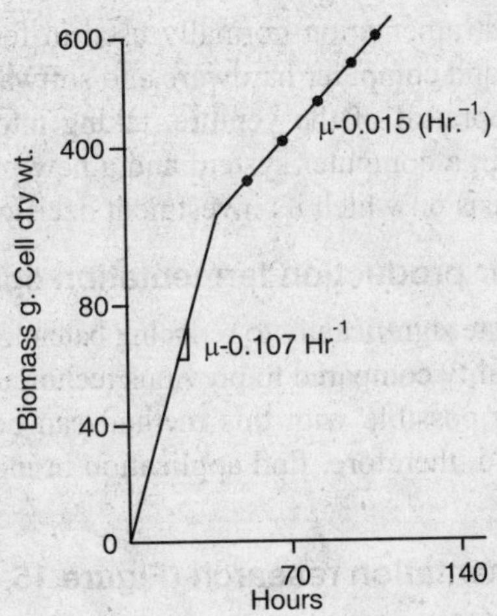

Fig. 15.3 *Growth of Penicillium chrysogenum under computer control of a fed-batch fermentation.*

During the production phase, it is desirable to maintain carbon-limited growth of *P. chrysogenum* in order to avoid the carbon catabolite repression of penicillin synthesis.

Ohase *et al.* have described a cell mass sensor for measuring optical density (O.D.) in situ. This is a submersible colorimeter probe that permits continuous measurement of O.D. of the broth by removing air bubbles which otherwise pose a problem in O.D. measurement. Though this may be suitable for yeast or bacterial growth, it is unsuitable for mycelial growth. Hewetson *et al.* have shown that it was possible to measure penicillin accretion in a fermentation lasting four days using a penicillin sensing electrode constructed by retaining a layer of penicillinase covalently linked to controlled pore glass (CPG) in front of a flat surface pH electrode. The main problems experienced were pH variation and electrical interference, upsetting the readings from the electrode, and an increase in response time throughout the fermentation.

The production of a secondary metabolite has generally been considered with respect to biomass growth. Bajpai and Rau proposed for penicillin production, a substrate inhibition model based upon evidences relating carbon catabolite regulation to secondary metabolite production and a model relating product formation rate with growth rate for the production of secondary metabolite.

3.2. Practical implementation of basic computer control strategies for enzyme production

The computer control strategy for fungal enzyme fermentations reduces the production costs as follows:

- — The microbe's enzyme formation capacity can be optimally utilized.
- — The fermentation equipment can be more efficiently used, which means increased productivity, (i.e. More enzyme is produced per time unit).
- — The final enzyme concentration in the broth is higher, thus reducing recovery costs.
- — Less energy is required.
- — Labour costs can be reduced to a minimum governed by the consideration of safety (accidents, illness) in the plant.

Furthermore, the application of the control strategy requires only a restricted amount of 'new', but industrially used and reliable instrumentation—gas-analysis and enzyme auto-analysis equipment in addition to the physical process instrumentation normally used in fermentation. This means that the investment in new instrumentation and computer hardware and software can be kept reasonably low.

A technical and economic appraisal of the venture, taking into account all other benefits and drawbacks involved in investment in a computer system and a new processing strategy is required to give the fermentation company a basis on which its investment decision could be made.

3.3. Fault analysis methods for production fermentation control

Automatic fault analysis can contribute significantly to reducing batch losses. The examples given should serve to show its utility and universality compared to previous techniques, which were limited to single variables. Furthermore, the savings possible with this method can be realized with micro-computer technology available today. It should, therefore, find application in industrial fermentation in the near future.

3.4. Computer system for fermentation research (Figure 15.4)

The micro-minicomputer hierarchical control system meets its objectives by providing an extremely versatile environment for fermentation research. With its help the microbiologist can control the numerous

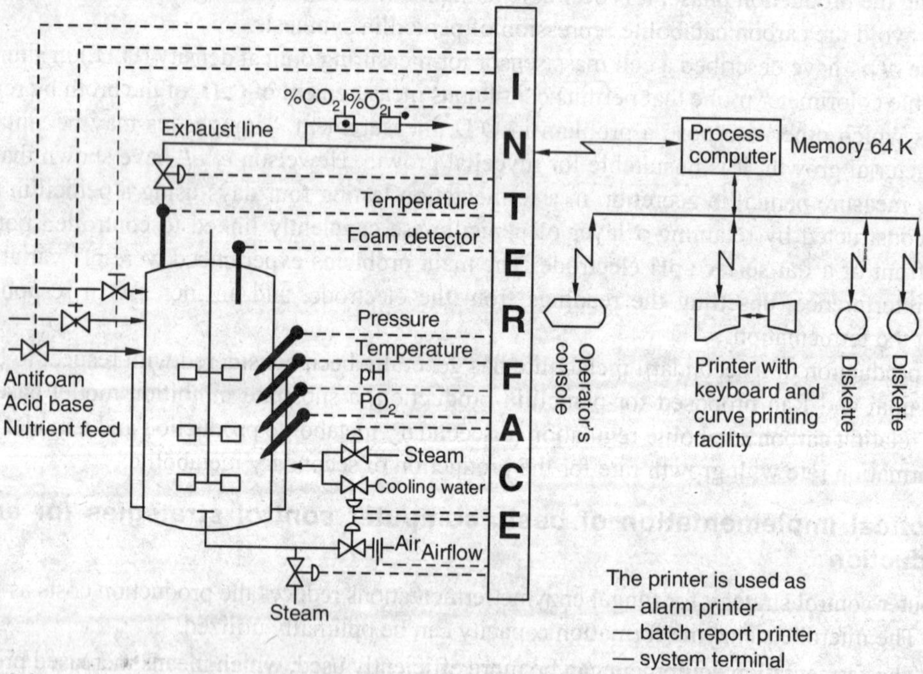

Fig. 15.4 *Scheme of automation of pilot fermentors (Gas analysis and foam control were not included in the first phase of the project).*

bioparameters and easily obtain documentation for the study. The future evolution of such a system especially for the wider use of micro-processors, will deal with the technical realization. But the objectives, the fulfillment criteria and the functions will remain the same. The use of such a system will help the researcher in the complex elucidation of the behavior of micro-organisms, saving cost and time. The mathematical representation of this behaviour, (i.e. the model) is the basis for actuating the dynamic optimization.

Incorporation of the micro-computer has given more versatility, generated an intelligent real-time interface, and supplied a convenient structure for control, especially in the case of the fermentors which uses DDC. With much of the routine work being done by the micro-computer, the mini-computer performs other tasks more efficiently. These tasks include supervisory control, advanced control and optimization strategy, data analysis, and programme development.

The application of the various fermentors in this system is further evidence of the versatility of the system and ease of operation for the researcher. Fermentors have been operated in batch, continuous, and fed-batch modes, as well as in series configuration. Various micro-organisms have also been studied. This research includes modelling, process optimization, and investigation of the stability of certain bio-reactor configurations.

The limitations of the system are in the basic process measurements of the fermentors. There is always a need for developing more accurate and sophisticated measuring devices, as well as instrumentation to measure previously unmeasurable variables.

4 JUSTIFICATION AND PLANNING

Beaverstock and Trearchis analysed the present status of control functions which is illustrated in Table 15.3.

Table 15.3 *Control function comparison matrix.*

Function	*USE*			
	Pilot Plant		*Industrial Plant*	
	Existing Processes	*New Processes*	*Existing Processes*	*New Processes*
Feedback control			0	0
Data acquisition	0	0	0	0
Scheduling and sequencing			0	0
Computation of non-measured variables	0			
Advanced control techniques	0	0	Limited use	
Optimization	0	Challenge work	Opportunity	Critical areas
Modelling, parameter estimate	0	in progress	for use	Work needed

The present interest in computer applications for fermentation processes and micro-processor techniques indicates that a computer-based control system will be a critical part of future plants. The time has come for those with a fermentation process background and those with an industrial control background to work together for the successful application of a control system in fermentation process plants.

In an industrial environment, a computer-based automation project can only be justified if a rigid investment plan and a tight time schedule can be adhered to.

In order to fulfil these demands, the following rules should be observed:

(a) Begin small: Choose a low-risk project, (i.e. a project with a clearly defined problem tested transmitters and actuators and safe control strategies).

(b) Plan well: This includes

- — detailed specification
- — no changes during designing
- — follow-up on contractors' progress
- — test of hardware, software and field equipment in separate modules

When one project is successfully concluded, it is time to plan the next careful step towards more advanced applications of computers.

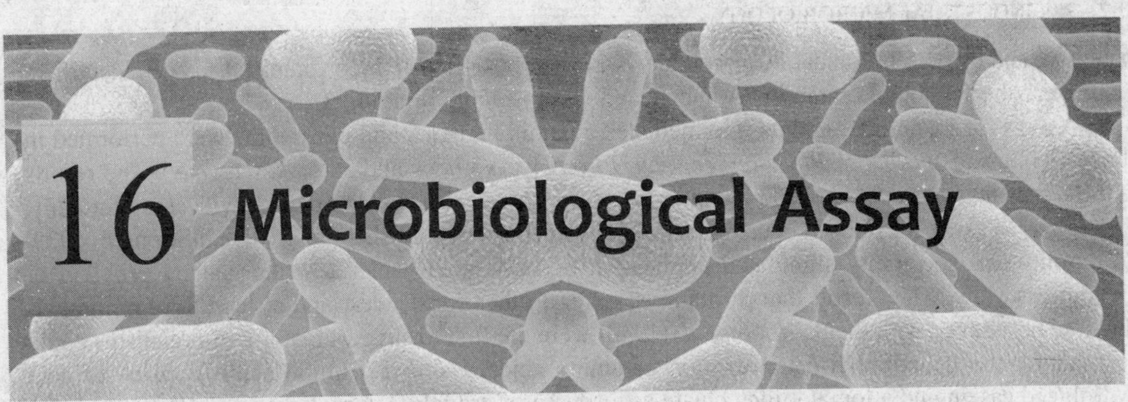

16 Microbiological Assay

1 INTRODUCTION

Microbiological assay may be defined as qualitative and/or quantitative determination of any chemical compound from a simple or even complex material with the use of microbes. Thus, microbes can be and are exploited as analytical tools in an analytical laboratory. Today, a new branch of microbiology, analytical microbiology, has been developed wherein various analytical techniques involving microbes are dealt with. Microbes which are important in analytical work include bacteria, (e.g. lactic acid bacteria), fungi, (e.g. mutant of *Neurospora crassa),* and protozoa, (e.g. *Tetrahymena pyriformis*).

It is not possible in the limited space available, to give details of microbial assay methods for all types of antibiotics, vitamins, amino acids and trace elements, and the literature should be consulted for details of individual methods.

2 HISTORY

The principles of microbiological assay were discovered and used by many early workers, including Fleming, in the 1920s for the assay of the enzyme, lysozyme, and penicillin. But it was not until the development of penicillin manufacture in 1940 that the method had any widespread application. Large-scale manufacture of antibiotics made it imperative that analytical procedures capable of detecting the amount of antibiotic produced during fermentation were further developed and factors affecting their sensitivity and reproducibility elucidated.

Microbiological assay of the vitamin B-complex and amino acids, basically evolved and developed from the purely academic studies on the nutritional requirements of bacteria and fungi. The first landmark in the history of this subject was the discovery in 1901 by Wildiers that certain species of yeasts require substances which he called 'bios' to give full growth. Eighteen years later R.J. Williams noticed the great similarity between the properties of 'bios' and the 'water-soluble vitamins' needed by the animal organism. He was sufficiently ahead of his time to suggest that yeast might be enlisted as an assay micro-organism for the quantitative determination of this substance.

Microbial vitamin assays were not created *de novo*; they arose in response to the needs of plant and animal physiologists.

Animal physiologists studied the nutrition of animals (rats, pigeons, chicken, cattle), by feeding them purified or restricted diets. Responses of the animals indicated the presence or absence of accessory factors in the diet. Such responses could be used as a form of bio-assay to measure the amount of growth factor in the diet. The assays were slow (an answer might take 6 weeks or more to obtain), expensive and by present standards, inaccurate. Nonetheless, by 1924, three vitamins had been identified. They were the fat-soluble vitamins A and D and water-soluble vitamin B.

Assays for vitamin B_1 and its intermediates, vitamin thiazole, and pyrimidine were performed in about 1932 using species of *Phycomyces,* (e.g. *Phycomyces nitens* and *P. blakesleeanus*). The results of these fungal assays could be reported in terms of quantities of known compounds with some certainty. Thereafter, an assay for riboflavin using lactic acid bacteria was published by Snell and Strong in 1939. Its rapidity and specificity stimulated application of lactic acid bacteria to the general problem of vitamin assays with the result that 8 vitamins, several other substances, and at least 18 amino acids can now be assayed with them. Wood *et al.* (1940) were the first to suggest that micro-organisms could be used for the quantitative determination of amino acids. Using 17 amino acids in place of acid hydrolyzed casein and a tomato juice eluate as a source of growth factors, Kuiken *et al.* reported that 9 amino acids were essential for the growth of *Lactobacillus arabinosus* 17-5 and that 3 amino acids, (i.e. leucine, isoleucine and valine) could be quantitatively determined by the microbiological techniques similar to those previously exploited for vitamin assays. Because of this, it has grown the ability to conveniently determine, with a good degree of precision, accuracy and specificity, 18 amino acids.

3 MICROBIOLOGICAL ASSAY OF VITAMINS AND AMINO ACIDS

Vitamins and amino acids are important growth factors needed for growth and multiplication of micro-organisms. If a specific growth factor is absent in the growth medium, the specific microbe that is dependent on a specific growth factor fails to grow. Then the growth factor dependent microbe shows signs of growth if a specific growth factor is added to the growth medium. The degree of growth occurs as per the concentration of the growth factor. In other words, more the concentration of the specific growth factor in the growth medium, more is the growth of the particular micro-organism. Therefore, the graph plotted as in Figure 16.1, (i.e. the amount of growth factor against growth response) remains linear only up to a certain extent. Thus, growth response is sensitive only up to the limiting concentration of the growth factor (Figure 16.2).

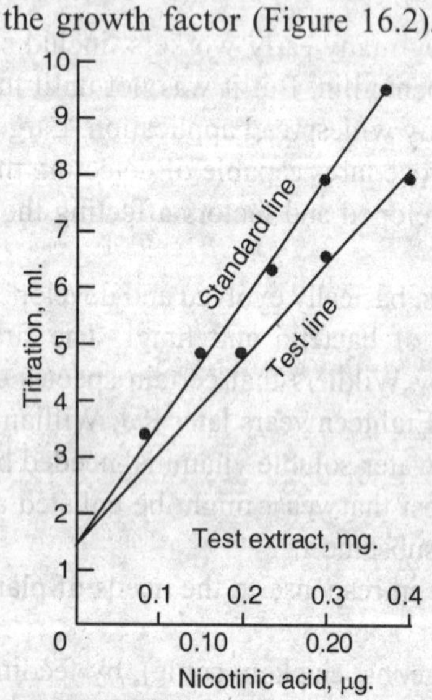

Fig. 16.1 *Assay of nicotinic acid.*

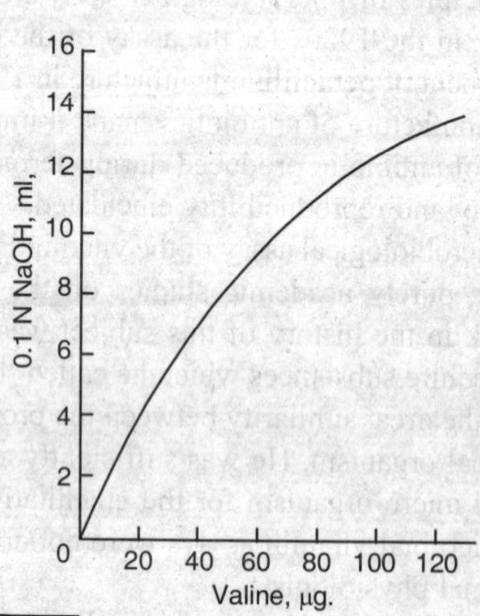

Fig.16.2 *Typical standard curve for valine using L. arabinosus 17/5.*

The bioparameters controlling microbiological assay for vitamins and amino acids have been dealt with in general, in the following discussion:

(A) *Test micro-organisms:* The lactic acid bacteria were and are extensively employed for assays of vitamins and amino acids. These include species of *Lactobacillus, Streptococcus,* and *Leuconostoc.* Over the years, lactic acid bacteria have remained the micro-organisms of choice because of their many advantageous characteristics:

1. They are better lactic acid producers. Therefore, they are considered to be more suitable for acidimetric methods involved in the measurement of growth response. Usually titration of lactic acid is done against 0.1N NaOH solution using bromothymol as an indicator. In the case of microbiological assays using *Streptococcus faecalis,* 0.05N NaOH is used owing to low acid production.

2. They possess uniform biochemical characteristics. Compared to many bacteria, such as *Escherichia coli,* lactic acid bacteria are extremely stable and apparently have a very low mutation rate. It is indeed fortunate for both vitamin and amino acid assayists who use these micro-organisms that mutations are rare.

3. All of the lactic acid bacteria have relatively complex and exact nutritional requirements. Most of these bacteria need one or more of the vitamins, and in some cases, a specific form of a particular vitamin. All lactic acid bacteria require at least a few amino acids and for many, growth is greatly stimulated by the presence of many amino acids. Thus, this characteristic allows the use of a single micro-organism for assaying more than one growth factor. For example, *Lactobacillus arabinosus* can be used for the microbiological assays of nicotinic acid and riboflavin. Therefore, the number of assay micro-organisms needed for the routine microbiological assays of vitamins and amino acids is less, thus reducing the culture maintenance expenditure.

4. None of the *Lactobacillus* or *Leuconostoc* species are known to be pathogenic or to cause food poisoning in man. A few *Streptococcus* species, particularly *S. pyogenes*, are human pathogens.

Some species of lactic acid bacteria, that can be and are exploited in the microbiological assay techniques of the vitamins and amino acids, are listed in Table 16.1.

Table 16.1 *Some species of lactic acid bacteria and corresponding vitamins/amino acids.*

Micro-organism	*Vitamin/Amino acid*	*Growth-limiting concentration mug./ml.*
Streptococcus faecalis ATCC 8043	Folic acid	0 – 0.8
Lactobacillus plantarum ATCC 8014	Biotin	0 – 0.2
Lactobacillus plantarum ATCC 8014	Nicotinic acid	0 – 40
Lactobacillus leichmannii ATCC 7830	Cyanocobalamin	0 – 0.025
		mcg./ml.
Leuconostoc mesenteroides ATCC 8042	L-Lysine	0 – 9
Leuconostoc mesenteroides ATCC 8042	Methionine	0 – 6

Certain vitamins of the B-complex, notably pyridoxine, pyridoxal and pyridoxamine, inositol and the nitrogen base choline cannot be assayed by means of the lactic acid bacteria. In such cases X-ray induced mutants of the mould *Neurospora* and certain strains of yeasts, (e.g. *Saccharomyces carlsbergensis*) may be resorted to. A large number of mutant strains of the species *Neurospora crassa* have been produced. Only three of these X-ray induced mutants are of practical importance as assay micro-organisms and have been used with success. The growth response is measured by harvesting the mycelium, drying to constant weight and weighing on a suitable balance. From the practical stand-point, there are a number of disadvantages connected with the use of these X-ray induced mutants of the species *Neurospora crassa* as assay micro-organisms:

(i) A separate strain must be used for each individual assay, which leads to the necessity of maintaining a large number of micro-organisms as stock cultures.

(ii) In the actual carrying out of an assay, these mutants must be grown in flasks, test-tubes being unsuitable for this purpose. Incubator space, therefore, becomes a critical problem.

(B) *Preparation of inoculum*: It has been found to be most suitable to keep all assay micro-orgainsms as lyophilized cultures. When obtained from the ATCC the cultures are grown and immediately a portion is lyophilized for future reference. Depending on the micro-organism, the working subcultures are kept in different ways. *Streptococcus faecalis* has been found to be stable when carried as a liquid culture in the complete assay medium for long periods of time. It can be routinely carried in this manner. Working cultures of other micro-organisms such as *Leuconostoc mesenteroides* p-60 and *Lactobacillus delbrueckii* (9649) can be carried as agar stab cultures in any one of a variety of rich organic media that contain glucose, meat and/or yeast extract, and a suitably hydrolyzed protein.

Inocula can be taken either directly from lyophilized cultures to the inoculum medium or indirectly through a preliminary growth in a rich organic medium. In many cases, for example, where *Streptococcus faecalis* or *Leuconostoc mesenteroides* is the test-organism, inocula can be successfully grown in the synthetic assay medium.

Usually, inoculum for the assay can be grown in the assay medium. Two inoculum tubes are grown: one for inoculating the assay and one for future cultures. With *Lactobacillus plantarum*, ATCC 8014-biotin assay, the micro-organism is transferred from the stab culture into a liquid medium, and incubated at 30°C for 24 hours. A suitable medium is 1% glucose, 1% yeast extract, tubed and sterilized in 7 to 8 ml. quantities in 5/8 × 5 inch culture tubes. A 24-hour culture (cotton plug secured with a rubber band) is centrifuged at about 3000 r.p.m. for 15 minutes. The spent medium is discarded and the cells are resuspended in approximately 10 ml. of sterile physiological saline. The cell suspension is diluted aseptically 1 to 100 with sterile saline and one drop is added to each sterilized assay tube.

Inoculum grown in the assay medium containing a growth limiting quantity of the growth factor, (e.g. amino acid) to be assayed eliminate the need to wash the inoculum before use in the assay.

(C) *Preparation of sample for assay*: This is the most important bioparameter that is to be carefully controlled for obtaining valid and reliable results. Majority of the materials suspected to contain vitamin/amino acids need some sort of treatment, since interfering substances, if present, are responsible for producing erroneous assay results. Therefore, material to be tested is subjected to treatment by a suitable method. The method to be used in preparing the sample for bio-assay depends upon:

(i) the nature of the material under question

(ii) the nature of the vitamin/amino acid to be estimated

(iii) the test micro-organism to be used

(iv) the nature of interfering substances, if present

With microbiological assays of the vitamin B-complex, interfering substances must be eliminated for securing successful assay results. For example, the microbiological assay of riboflavin is undoubtedly the most difficult of all to carry out, and unless the utmost care is taken over details, invalid and useless results will be secured. The stimulatory substances, (e.g. free starch, certain free fatty acids, notably stearic and oleic acid, etc.) should be removed by the correct extraction procedure for the successful assay of riboflavin with *Lactobacillus casei.* Materials such as blood, cooked foods and malt often contain other stimulatory substances, possibly of a lipid nature, which must be eliminated before an assay is practised. Such fatty compounds may be eliminated from materials under assay by simple filtration through paper at pH 4.5. When the concentration of such interfering fatty compounds is high, however, materials should be subjected to extraction with equal volumes of ether prior to dilution and assay for the riboflavin.

With microbiological assay of amino acids, it is necessary to bring about hydrolysis of proteinic samples to make the amino acids free. This may be accomplished by one of the three following methods where choice of method depends upon the nature of the amino acid to be estimated:

1. acid hydrolysis
2. alkaline hydrolysis
3. enzymatic hydrolysis

1. Acid hydrolysis: This method is not suitable for releasing two amino acids, tryptophan and tyrosine. The former is completely destroyed by prolonged hydrolysis with strong acid, while tyrosine is partially decomposed. Other amino acids may also suffer losses during acid hydrolysis, and in some instances corrections can be made. The acids used for protein hydrolysis include hydrochloric, sulphuric, hydriodic, and hydrochloric: formic acids. There is an advantage in using sulfuric acid, since excess acid may be eliminated with $Ba(OH)_2$ as the insoluble precipitate. However, almost invariably HCl or HC1: formic acid hydrolyses have been employed for microbiological assays. For microbiological assays, only a degree of hydrolysis sufficient to liberate in a free state all of the amino acids being estimated is needed.

With microbiological assay of amino acids such as leucine, isoleucine, valine, etc., the hydrolysates may be prepared as under:

(i) The material for assay is finely ground and the requisite amount (usually 2-20 mg.) accurately weighed out and added in a 16 x 120 to 18 x 150 mm. heavy walled pyrex tube.

(ii) A minimum volume of 0.5 ml. 6N HC1, 1 ml. 3N HC1 or 0.3 ml. 3N HC1 in 40% formic acid is added in a tube containing a sample.

(iii) The tubes are then subjected to freezing in dry ice-ethanol.

(iv) Such frozen tubes are evacuated with an oil pump and are sealed in a flame.

(v) The suspension is hydrolysed for not less than 6 hours, or better still for 10 hours, in an autoclave at 121°C (Hydrolysis in 3N and 6N HC1 are usually carried out for 16 to 24 hours and in 3N HC1 in 40% formic acid for 8 to 24 hours).

(vi) After hydrolysis the tubes are cooled in dry ice-ethanol, and opened with caution.

(vii) The opened tubes are placed in a desiccator over NaOH and evacuated with care to avoid bumping of the hydrolysates (In most instances it is equally satisfactory to neutralize with alkali).

(viii) Hydrolysate is dissolved in water and filtered to remove humin.

(ix) A portion of the filtrate is now shaken with ethyl ether in a separating funnel and the pH of an aliquot brought to 6.8 to 7.0 with sodium hydroxide solution and made up to a fresh suitable volume.

It is to be kept in mind that hydrolysate solutions must be protected from the microbial contamination. Hydrolysate solutions may be preferably stored at subzero temperatures.

2. Alkaline hydrolysis: Alkaline hydrolysis is to be used in the preparation of hydrolysates for assay of tyrosine and tryptophan because these two amino acids are destroyed during acid hydrolysis. With microbiological assay of tyrosine, alkaline hydrolysis is carried out as under:

(i) Dry material is finely ground, and 1 gm. of the sample is accurately weighed out and added to a tube with a loose cover (Tubes, which may be sealed as in acid hydrolysis, can also be used).

(ii) 20 to 50 ml. of 5N NaOH solution are added to the tube containing 1 gm. of the sample.

(iii) The tube is closed with a loose cover and autoclaved at 121°C for 8 to 10 hours to bring about hydrolysis.

(iv) The hydrolysate is adjusted to pH 4.0 with HC1.

(v) Any precipitate that is formed is removed by centrifuging the hydrolysed sample.

(vi) The solution is adjusted to pH 6.8 and made up to volume.

During alkaline hydrolysis amino acids are racemized so that care must be taken to be sure that the growth response of the test micro-organism is specific for the L-isomer and not interfered with by the D-isomer. Therefore, the figure obtained by the assay will have to be doubled to give the correct value for the L-amino acid in the sample.

Alkali may cause not only the partial and erratic destruction of tryptophan, but racemization is often not complete. Therefore, modified alkaline hydrolysis has been recommended by Kuiken *et al*. to overcome the difficulty in liberating tryptophan without damage.

3. Enzymatic hydrolysis: Wooley and Sebrell developed an enzymatic method for releasing tryptophan from proteinic materials because this amino acid was found to be sensitive to destruction during the alkaline hydrolysis using NaOH. Enzymatic hydrolysis of different biological materials for the liberation of tryptophan is carried out as under:

(i) Sufficient assay material estimated to contain 1 gm. of protein is accurately weighed into a 100 ml. Erlenmeyer flask.

(ii) Fifty ml. of 0.1N sulfuric acid and 10 mg. of pepsin (1 : 10,000 potency) are added.

(iii) The flask is subjected to incubation overnight at 37°C.

(iv) Three grams of $K_2HPO_4.12H_2O$ are then added as buffer and the pH is adjusted to 8.4.

(v) Ten milligrams of trypsin (1 : 300 potency) are added and the flask is subjected to incubation at 40°C for a further period of 12 to 24 hours.

(vi) The pH is readjusted to 7.8 with dilute acid and 100 mg. of erepsin are added, and the flask is incubated at 40°C for 2 more days.

(vii) After this final digestion period, the pH is brought to 6.8 and the volume brought to 100 ml. with water.

The flask should be occasionally shaken during each incubation period and the digestions should be carried out under sulphur-free toluene. The recommended test micro-organisms, *Streptococcus faecalis,* does not show a response to indole and anthranillic acid so that extraction with ether is not needed. The digest may be filtered before use, if necessary.

Since there will be some liberation of tryptophan, as well as stimulatory peptides, as a result of the autodigestion of the added enzymes for hydrolysis, it is essential to determine the degree of stimulation and the amount of tryptophan entering the assay via this route.

In actual assays, evidence of complete digestion of the materials is indicated by the lack of 'drift' in the assay values calculated from the results at different concentration levels. An upward 'drift' in the assay will indicate either incomplete digestion of the protein sample or the presence of stimulatory substances. A downward 'drift' will indicate the presence of inhibitory substances.

A comparison of the quantitative values for tryptophan in different biological materials assayed microbiologically and also estimated by the chemical method of Horn and D.B. Jones is given in Table 16.2.

Table 16.2 *Tryptophan values for proteins and other biological materials on moisture-free basis expressed in percentages.*

Material	*Digested by pepsin trypsin and erepsin*		*Refluxed with Sodium Hydrate Over Electric Hot Plate*	
	Microbiological	*Chemical*	*Microbiological*	*Chemical*
Labco casein	1·20	1·20	0·47	1·10
Lactalbumin	1·79	1·79	0·75	1·74
Ox-muscle protein	1·17	1·18	0·45	1·24
Phaseolin	0·38	0·38	0·25	0·70
Velvet-bean globulin	0·73	0·73	0·34	0·69
Corn	0·24	0·26	0·22	0·33

(D) Assay media: The principle upon which all microbiological assays, whether of vitamins or amino acids, is based, is the preparation of a suitable basal medium which shall contain an ample supply of all the specific nutrients required by the particular micro-organisms to be used for the assay, except the substance to be estimated. If that substance is added to the medium in limiting concentrations, the growth of the test micro-organism will be proportional to the quantity added. A wide variety of basal media have been devised by different investigators for the assay of vitamins and amino acids.

In general, completely synthetic media have been exploited for the assay of vitamins and amino acids. For example, of the recommended methods for microbiological assays of amino acids, only the assay for cystine is carried out on a non-synthetic media. With microbiological assays of the amino acids, one precaution must be observed, that is that the amounts of nutrients are not so far in excess as to be toxic to the assay micro-organism or, in the case of the amino acid nutrients, to lead to competitive antagonisms.

All basal media contain an assortment of amino acids and vitamins, purine and pyrimidine bases, a buffer which may be one or a combination of sodium acetate, sodium or potassium citrate, or sodium or potassium phosphate, small amounts of inorganic salts and glucose as an energy source. At present, there is only one assay medium, (i.e. modification of the oxidized peptone medium of Riesen *et al.*) which contains oxidized peptone in place of some of the amino acids used.

The glucose concentrations used are either 10 or 20 mg./ml. The quantity used is apparently dependent on the assay micro-organisms used, and to a lesser extent on the method of growth measurement used. Ammonium salts are included in only a few media.

In the various microbiological assays, the basal medium is always made up in double concentration. Thus, if one litre of medium is required (which is sufficient for 100 tubes), then the constituents of the assay medium are dissolved in 500 ml. of water. If 500 ml. of final medium are needed then the constituents are dissolved in 250 ml. of water, and so on.

It is not within the scope of this book to give chemical composition formulae of different specific assay media for assaying various vitamins and amino acids. Interested readers are requested to refer 'Analytical Microbiology' Volume I by F. Kavanagh (1963).

(E) Apparatus and instruments: It is necessary to have lots of clean glassware for microbiological assays of the vitamins and amino acids using lactic acid bacteria. Large numbers of volumetric and graduated pipettes and culture tubes are needed (Calibrated Culture Tubes are used if growth is to be measured by turbidimetric method). The usual supplement of volumetric flasks, graduated cylinders, funnels, beakers, and Erlenmeyer flasks are required. In the preparation of the assay medium, glass-stoppered graduated cylinders are needed. Pipettes with delivering capacity of 6, 7, 8 and 9 ml. are very useful in such types of practical work. The size of culture tubes to be used depends upon: (i) the volume of the assay and (ii) the method of measurement. Assay fermentations are carried out in rimless tubes (18-20 x 160-180 mm.). Noncorrosive borosilicate glassware is usually recommended.

The glassware can be washed by automatic glass washing equipment. The glassware washing problem for amino acid assays is not as strict as it is for vitamin estimations because the quantities of a substance to be estimated are larger. Satisfactory cleaning of culture tubes is essential to avoid the adhering of liquid in large amounts to the inner wall of the tube after mixing. Therefore, thorough rinsing of all glassware in high quality distilled water is imperative. Due to the possible resin content of de-ionized water, it is advisable to avoid its use for the purpose of a final rinse.

Sterilization of pipettes and other glassware may be done by conventional dry heat methods. Pipette cans and long glassine paper bags are useful. Moreover, some sort of cover is needed for sterile cultures and assays. Cotton plugs, glass, polypropylene, aluminium and stainless steel caps, and screw-capped tubes have all been used. Screw-capped tubes and flasks are particularly useful for stock cultures as they reduce drying and spilling. The stainless steel caps are preferably used for assay tubes.

The use of a constant temperature water bath that can maintain ± 0.1°C at 30° to 39°C is recommended for incubating assay tubes. Stainless steel test tube racks are required to avoid corrosion problems during incubation in the water bath. It is advisable not to mount the stirrer directly on the water bath in order to avoid vibration.

(F) Design of assay: In practice, two sets of assay tubes are prepared. One set is prepared for standard growth factor, (i.e. pure vitamin or amino acid) and another set for test extract suspected to contain a particular substance. In all cases, the final volume of the solution in each assay tube is always 10 ml. In other words, uniform condition in all assay tubes is to be maintained by the addition of the required quantity of water to make the final 10 ml. volume. The level at 5 ml. of the double strength basal medium plus standard vitamin solution, or solution of test extract is adjusted to 10 ml. with water. These assay tubes are plugged, autoclaved (at 120°C for 10 minutes), inoculated and incubated.

With microbiological assays using a nutritional mutant of *Neurospora crassa*, sample and standard are pipetted into 125 ml. Erlenmeyer flasks at increasing concentrations to a maximum of 0.5 ml. Then, 25 ml. medium is added to each flask, sterilized by autoclaving for 5 minutes at 120°C, cooled and aseptically inoculated into each sterile assay flask with 1 drop of prepared inoculum. The inoculated assay flasks are incubated at room temperature or 28°C for 4 or 5 days.

(G) Assay conditions: Assay conditions influencing the microbiological assay results include temperature and time. The temperature to be used for incubation of assay tubes depends upon the assay micro-organism involved in a particular estimation. For example, temperatures between 30° and 37°C are recommended for microbiological assays of vitamins and amino acids using lactic acid bacteria. Approaching the optimal temperature for the assay, micro-organisms may somewhat reduce the time required for an assay. Constant incubation temperature, however, is the more important factor for precise microbiological assays.

(H) Measurement of response: There are three principal methods of response measurement. They are:

(a) titrimetric (or acidimetric) (b) turbidimetric (c) gravimetric (These methods have been briefly outlined here).

Incubation time for which assay tubes are to be incubated largely depends on the type of method for growth measurement involved. For example, long periods of incubation time are required in assays using the titrimetric method.

Acidimetric methods are dependent upon the quantity of acid excreted as an end product during the fermentative metabolism of the micro-organisms. Rate of growth and acid production do not run exactly parallel courses. The maximum growth will have been reached considerably before maximum acid production has occurred. In fact, the total quantity of acid accumulated would be a function of the cellular mass and the time that this cellular mass has had to produce the acid. If the incubation is for a sufficiently long period of time, the quantity of acid excreted is closely related to the cellular mass. This is the reason why these methods often require long incubation periods (72 hours or more) as compared to the turbidimetric method which requires a shorter incubation period (16 to 48 hours). At the end of the incubation period, the total amount of lactic acid produced is measured by a suitable method and will be found, over a certain range, to be proportional to the concentration of the added essential factor. Acidity may be measured by an automatic titrating and recording apparatus. Alternatively, acidity may be measured by a simple method using a comparator.

In turbidimetric methods, absorption of light by the cellular mass is usually determined. This may be accomplished by a suitable instrument such as photoelectric colorimeter or spectrophotometer. Alternatively, the light-scattering of the number of cells present may be measured with nephelometers. Nephelometers have not been used as extensively as spectrophotometers and colorimeters. Difficulties in standardization and extreme sensitivity of nephelometers limit their use. The photometers used have ranged from home-made ones to expensive recording spectrophotometers. Any sufficiently sensitive photoelectric densitometer is suitable. For rapid, precise, and critical work, the instrument should hold at least two cuvettes and have a sufficiently long (over 6 inches minimum) direct reading or expandable scale. Null-point type instruments are too slow and burdensome for estimation work. A system for using a particular wavelength of light is highly desirable, particularly to remove colour effects of culture media or of the assay sample. Ordinary light filters are usually satisfactory for this purpose. A narrower spectrum is often desirable and either interference filters or a spectrophotometer may be used. A properly selected narrow light band reduces colour interference. Various wavelengths have been used ranging from 400 to 760 $m\mu$, depending some cases on instruments and filters available. The use of lower wavelengths increases the sensitivity of the assay, since turbidity readings increase as the wavelength of light used is decreased. The observed turbidity readings (either transmission percentage or optical density) are adjusted to compensate for deviations from Beer's law. By such adjustments, linear relationships that are not apparent from unadjusted readings would be disclosed. This method of response measurement is preferably used in assays that make use of micro-organisms with a tendency to produce smaller amounts of acid. On the other hand, the turbidimetric method is not preferred if assay micro-organisms have a tendency to clump during growth, making uniform suspension difficult.

In gravimetric methods, growth is determined by harvesting the mycelium, drying in an oven at 90°C for at least 2 hours or overnight to constant weight and weighing on a suitable balance. Dry weight measurements are laborious and of limited precision. Either a micro-balance or a large sample of cells is needed.

Among the above outlined methods, the titrimetric method has been widely used.

(I) Computation of Answers: In all microbiological assays a standard curve is obtained in which the growth response of the micro-organism to the standard substance, either vitamin or amino acid, is plotted against its concentration. It is essential that a new standard curve should be obtained for each fresh assay. The sample is also assayed at several different concentration levels, the concentrations of the test substance being so adjusted that the amount of growth at each level falls upon the standard curve. There are two methods for calculating an assay result. They are: 'slope-ratio' and 'log-log'. These have been outlined below:

It has been suggested by E.C. Wood that, for those microbiological assays which show a strictly linear portion to the standard curve, the so-called 'slope-ratio' method should be used for calculating the results. This method, of course, has the merit of being statistically sound, and it also has the valuable advantage that it is possible to calculate the fiducial limits of an assay. The successive steps for calculating an assay result by the 'slope-ratio' method are as follows:

1. On a piece of squared paper the mean response (usually the mean titration figure in ml. of standard NaOH solution) of the standard preparation against the dose in μg is plotted. The response of the zero-dose (which is the 'blank') should be included (Figure 16.1). The most even straight line is now drawn between the plotted points.

2. If the point corresponding to the zero-dose or 'blank' or that corresponding to the maximum dose, does not lie on the 'best' line through the points, it should be ignored.

3. On the same graph, the points for the various doses of the 'test' preparations are plotted in the same way. The scale used is immaterial, so long as the zero corresponds with that of the standard preparation graph. The 'best' line is now drawn through these points.

4. The slopes of the two lines, (i.e. for each line in turn) are now calculated. From any two convenient points on the graph, the values for dose and response are read off at opposite ends of the line. In the next steps of the calculation the lower dose is subtracted from the higher dose and the lower response from the higher response, in order to obtain the increase in response for a given increase in the dose. From the slope is calculated the increase in response per unit of standard preparation, or test preparation as the case may be.

5. Whether or not an assay is valid can at once be detected by this method of calculation. If the assay is valid, the two lines should intersect exactly on the vertical axis of the response. In actual practice, a small divergence may often be found, but this is of no significance, so long as the distance between the intercepts of the two lines is no more than a few tenths of a milliliter. Unfortunately, the 'slope-ratio' method of calculation has only a very restricted use in the computation of microbiological assay values because there is no linear portion to the standard curve in the vast majority of assays, especially in the case of amino acids.

It is rare that the standard curves for amino acid assays show a linear portion; they are almost invariably curvilinear in shape (Figure 16.2). Any transformation that could be effected for such curves, so that they would give linear graphs would have manifest advantages, because it would give a sound mathematical basis for computation which is a desired consideration for any analytical method. E. C. Wood has been able to establish such a transformation by plotting the logarithms of both the doses and the responses. The resulting points were found to lie in a reasonably straight line, at least over a certain range. Therefore, this method of calculating the microbiological assay results is called 'log-log' procedure.

The method is best illustrated by the actual example of an assay of lysine with *Leuconostoc mesenteroids* P-60 (Figure 16.3). It will be seen that the curves obtained by plotting dose against response, both for the standard curve and for the test preparation, are transformed into two parallel straight lines when log dose is plotted against log response. In this case, the range of linearity is at least from 20 to 80 μg in terms of dose, and from 3.55 to 9.05 ml. of 0.05 N sodium hydroxide in terms of response. In a valid assay, the test line must be parallel to the standard line. Any significant lack of parallelism is the evidence of invalidity. Similarly, any curvature of the test line indicates that the assay is suspect.

Assuming that the lines are straight and parallel between approximately the same limits of response, the best and quickest way of calculating the result of an assay is to measure the horizontal distance from the test line to the standard line. Taking any point on the log-response axis, draw a line parallel to the log-dose axis and read off the values for log-dose corresponding to the points of intersection made by this line and the test and standard lines. Thus, in the example given (Figure 16.3), the horizontal distance between the standard and test lines along the line meeting the log-response axis at 0.75 is 0.765–0.421=+0.344. This figure represents log (ml. standard solution equivalent to 1 ml. test extract). On adding the value for the logarithm of the potency of the standard solution (μg./ml.), this figure then represents log (μg. vitamin or amino acid equivalent to 1 ml. test extract). On taking the anti-logarithm,

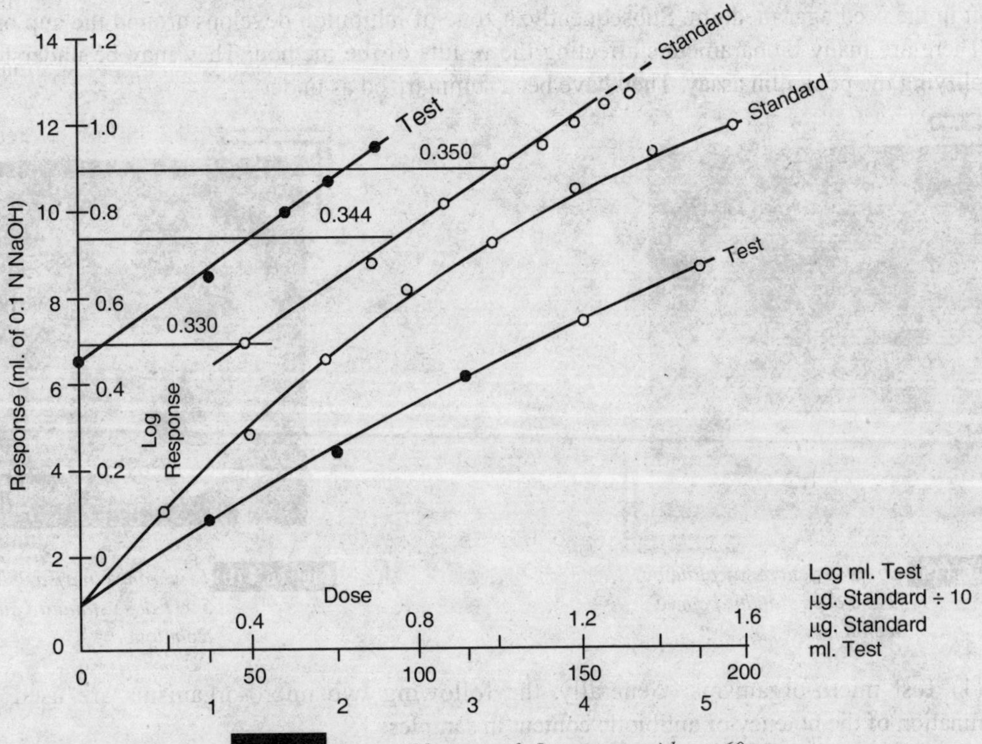

Fig. 16.3 *Assay of lysine with L. mesenteroides p-60.*

and knowing the concentration of the test extract, the vitamin or amino acid content of the test material is readily obtained. The standard preparation in this assay was pure lysine, and the unit 10 μg. and the logarithm of this is 1.00. From this we obtain that 1 unit of the test extract, which is here 1 ml., contains antilogarithm 1.344 = 22.1 μg. of lysine.

This 'log-log' relationship is not general for all these curvilinear curves but can frequently be used.

4 MICROBIOLOGICAL ASSAY OF ANTIBIOTICS

The principle of the microbiological assay of antibiotics is different from that of the microbiological assay of growth factors, (e.g. vitamins and amino acids). The growth of a sensitive test micro-organism does not occur in the presence of an antibiotic, provided a sufficient quantity of the antibiotic is present. There are several methods for quantitative determination of antibiotics in test samples. The important methods may be listed as under:

(A) agar diffusion
(B) serial dilution technique
(C) photometric assay

(A) *Agar diffusion* [Figures 16.4(a) and 16.4(b)] : In this method, the diameter of the inhibition zone is directly proportional to the quantity of an antibiotic. The antibiotic present in the solution *diffuses* radially through the agar medium layers and exerts its toxic effect on the sensitive test micro-organism present in the seed agar medium. Subsequently, a zone of inhibition develops around the cup or paper disc. There are many bioparameters affecting the results of the method. They may be understood by exemplifying the penicillin assay. They have been summarized as under:

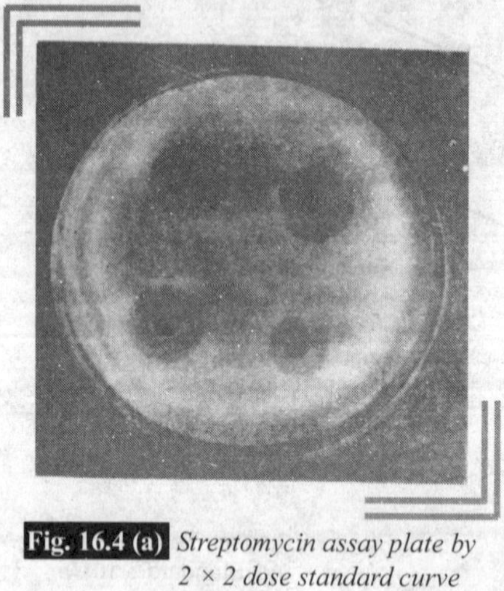

Fig. 16.4 (a) *Streptomycin assay plate by 2 × 2 dose standard curve technique.*

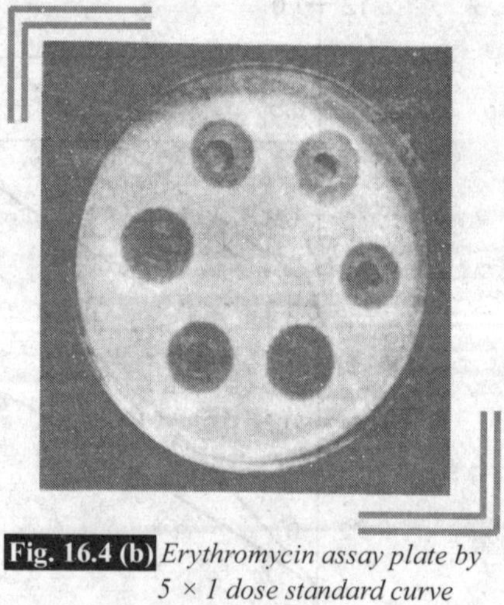

Fig. 16.4 (b) *Erythromycin assay plate by 5 × 1 dose standard curve technique.*

(1) Test micro-organisms: Generally, the following two micro-organisms are used in the determination of the potency of antibiotic content in samples:

(i) *Staphylococcus aureus* (ATCC 6538P)
(ii) *Sarcina lutea* (ATCC 9341)

The stock cultures of these micro-organisms are maintained by weekly transfers to fresh sterile slants of Grove and Randall agar medium No. 1. In the preparation of the inoculum suspension, the freshly prepared slants are incubated at 32 to 35°C for 16 to 24 hours.

(2) Standard Solutions: A stock of solution of penicillin working standard is prepared by careful weighing of the dried standard and then diluted to a suitable concentration (usually 100 u./ml.). The dilution is made with a sterile pH 6 phosphate buffer. In some penicillins (not G), it is necessary to solubilize them in a small quantity of solvent, (e.g. formamide) before diluting in a buffer. Such stock solution is stored in the refrigerator. The stock solutions can be used for a maximum of 2 days. It is to be kept in mind that all penicillins or penicillin derivatives do not exhibit identical dose-response curves even under identical assay conditions. Therefore, it is extremely important that one makes use of a standard identical to the compound being assayed. Two important precautions to be taken are:

(i) A stock solution of penicillin working standard should be prepared from the solid penicillin that has been preserved in tightly stoppered vials over a desiccant, (e.g. silica gel) in a desiccator.

(ii) Weighing of the solid penicillin should be done in an area with controlled humidity (preferably a relative humidity of 50 per cent or less).

(3) Sample preparation: There are many different materials to be handled by the assayist for testing purposes. Therefore, it is impossible to narrate here all the methods of preparing samples for estimation. A description of the method of preparing the samples for assay in a fermentation industry may be given as under:

Penicillin solutions and bulk powders to be tested may be diluted directly with pH 6 phosphate buffer before assay. In the case of some penicillins, it is essential to bring about the solubilization of the sample using a small quantity of solvent, (e.g. methanol or formamide) before dilution. Certain conditions, (e.g. pH) of the sample solution are kept identical with that of standard solution.

(4) Mechanics of the assay: In the routine work, a 100-mm. flat bottom petri plate is used. Such a plate can accommodate four to six cylinders or discs per plate. Usually, the standard dose-response curve is limited to four or five dose levels. Thus, the entire curve may be accommodated on a single plate. The sample or test plates are replicated five times, each plate containing four cylinders. Two alternate cylinders on each sample plate are used for the application of the unknown solution and the remaining two cylinders are used for the application of one of the concentrations used in the dose-response curve itself. With samples likely to contain contaminants with microbial activity (either stimulatory or inhibitory), at least two dilutions of the sample should be assayed. If this is done, and parallelism demonstrated, the assayist may have considerably more faith in the validity of the results.

Generally, such assay methods make use of the two-layer agar system. In this system, the first layer called 'base' layer is prepared by pouring melted agar into the plate and then allowed to harden on a levelled surface. The next layer called 'seed' agar is prepared by pouring melted agar that has been cooled to approximately 48°C, and inoculated with the appropriate test micro-organism. There are two reasons for the use of the two-layer system. They are:

(i) to improve the definition of the resulting zones of inhibition

(ii) to provide the optical contrast necessary for the precise measurement of the zones

The selection and application of agars in this phase of the assay is of utmost importance to the success of the method. It has been found that 21-ml. 'base' and 4-ml. 'seed' layers for the *Staphylococcus aureus* method, and 10-ml. 'base' and 4-ml. 'seed' layers for the *Sarcina lutea* method are satisfactory in assay techniques using 100-mm. assay petri plates. Inoculum to be added in the seed agar is an important factor so far as validity of the assay results is concerned. The cellular mass is produced in a Roux bottle containing appropriate medium, (e.g. G and R agar medium No. 1 for *Staphylococcus aureus*). Then the resultant growth is harvested by washing the agar surface with a suitable quantity of sterile 0.85 per cent sodium chloride solution. Finally, the inoculum suspension is standardized by a suitable method. In the case of *S. aureus*, a quantity of 1.0-1.5 ml. of standardized inoculum suspension (1-10^8 viable cells/ml.) is added to 100 ml. of liquefied and cooled (48°C) agar.

The standard stock solution is diluted according to the requirements to obtain a linear response. This dilution operation is carried out immediately before use on the day of the assay. The number and spacing of doses in the standard response curve is a matter which must be largely decided upon by the assayist in light of his knowledge of the materials which he is estimating. Like standard penicillin, the sample to be assayed is diluted in the range of the standard curve, giving the importance of the diluent that one chooses. Thus, this is an extremely important phase of the assay.

There are many ways to apply the standards and samples to the seeded agar surface. For example, the use of cylinders for penicillin assays is widespread. Dilutions of standard and sample are carefully added to sterile glass or porcelain cylinders kept on flood plates of the test micro-organism. With this method, considerable care must be taken in transporting plates to the incubator, and in placing the plates on the shelves. Spillage from the cylinders causes irregularly shaped zones which cannot be interpreted. Various trays and holding devices have been developed to overcome this problem. In any event, it is probably advisable to hold all assay plates under refrigeration until immediately before use and then to apply all solutions as rapidly as is consistent with good technique. The plates are incubated at 28 to 30°C for 16 to 18 hours with subsequent development of satisfactory inhibition zones.

(5) Measuring the response: After incubation under optimum conditions, circular zones surrounding the cylinders containing the antibiotic develop. The size of the circular zone depends upon the potency of the antibiotic present in the cylinder. In other words, the diameter of the inhibition zone is proportional to the potency of the antibiotic. Therefore, response in such assays is measured in terms of the diameter of the inhibition zone. This task of measuring the diameter may be accomplished by one of the following listed devices:

(i) calipers
(ii) millimeter rules
(iii) scale projectors
(iv) automatic and semi-automatic measuring devices

The selection of any of the above measuring devices is governed by the number of assays being run, the degree of accuracy required, and the allowable cost. A valid assay is possible only when standard and sample give the same type of zones.

(6) Computation of answers: The first requirement in the computation of the answer is of the standard dose-response curve. The standard curve is plotted by plotting the average response value (millimeters) for each of the standard levels against the logarithm of the respective doses, and by joining these plotted points. By the use of such a standard curve, the potency of the unknown antibiotic, (e.g. penicillin G) may be determined.

(B) *Serial dilution technique* (Figure 16.5): This assay method is simple in principle. It is easy to perform and is of considerable utility. No elaborate equipment is needed. Several dilutions of the antibiotic in small tubes (12 or 13 × 75 mm. Kahn tubes) are inoculated with a test micro-organism (any micro-organism which can grow in liquid medium), incubated for 18 to 24 hours. And the highest dilution of the substance which apparently inhibits the growth of the micro-organism is taken to be the minimum inhibitory concentration (Figure 16.5). The potency of the unknown sample is computed from the minimum inhibitory concentration and may be reported as micro-grammes of substance per milliliter.

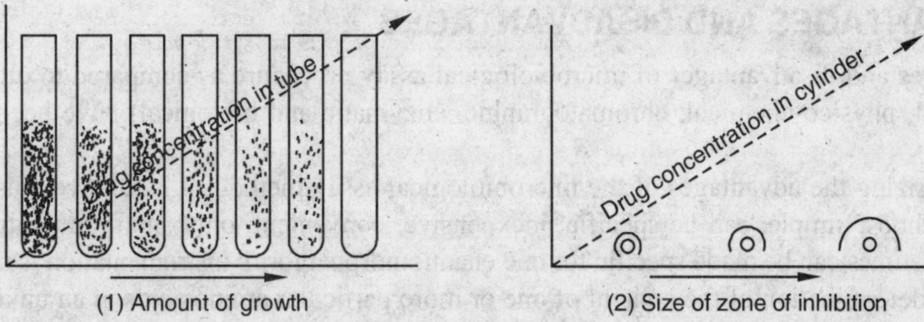

(1) In the tube-dilution technique, the inhibition of growth (decrease in turbidity) produced by the unknown sample is compared with that produced by the known or standard sample. The amount of antibiotic present in the unknown sample can then be calculated.

(2) The cylinder-plate technique follows much the same procedure for determining antibiotic potency, except that inhibition of growth is measured in terms of the size of the zones of inhibition.

Fig. 16.5 (1), (2) *Microbiological assay of antibiotics and some other chemotherapeutic agents is accomplished by either the tube dilution or the cylinder-plate method (a variation of the paper-disk plate technique).*

The activity of an antibiotic substance measured by this method is reproducible over a long period of time if attention is paid to certain details which include sensitivity of the test micro-organisms, constancy of composition of medium, inoculum size, temperature of incubation, and pH of the medium. All these factors can be controlled.

(C) *Photometric assay:* This method is very simple in principle. The test substance is added to a suspension of the test micro-organism in a nutrient medium, the mixture incubated, and the response (in terms of turbidity) of the test micro-organism measured. The factors governing this assay method include concentration of cells, size of cells, opacity of cells, medium, incubation, and response of the test micro-organism to the test substance. The advantages of this method are rapidity, ease of operations, objective measurement of response to the antibiotic, absence of diffusion effects, and accuracy. The disadvantages of the method are minor ones, including complexity of equipment and assuring that the sample is not grossly contaminated and that it does not contribute measurable colour to the assay medium.

5 MICROBIOLOGICAL ASSAY OF TRACE ELEMENTS

Exploitation of micro-organisms in the field of analytical microbiology has been extended to the estimation of simple mineral elements, as well as complex organic compounds. For example, trace elements in soil samples and biological materials can be estimated with the help of *Aspergillus niger* as shown in Table 16.3.

Table 16.3 *Bioassay of some trace elements with the help of Aspergillus niger.*

Trace elements	*Range μg./50 ml. medium*
Fe	0.1 — 10
Cu	0.05 — 4
Zn	0.5 — 10
Mo	0.0005 — 0.01
Mn	0.01 — 5

6 ADVANTAGES AND DISADVANTAGES

The advantages and disadvantages of microbiological assay procedure as compared to other methods, (e.g. chemical, physico-chemical, chromatographic, enzymatic and biological) have been outlined as under:

Summarizing the advantages of the microbiological assay method, it is relatively as accurate as chemical methods; simple, can be specific, inexpensive, convenient, only small amounts of material required; sometimes can be made specific for one enantiomorphic form; instrumentation is not elaborate; excellent for determination of the content of one or more particular amino acids in an unknown.

The disadvantages of the microbiological assay method are that somewhat larger amounts of material are required than for chromatography when a number of amino acids are to be determined; continuous checks are necessary for consistency, specificity, and interferences.

7 AUTOMATION OF MICROBIOLOGICAL ASSAY

The mechanical and biological principles for automating microbiological assays have been known for many years. It is only recently, however, that instrumental systems have been applied to the automation of these assays. There are two objectives of automation:

(i) to increase accuracy of answers

(ii) to increase sample handling capacity of the laboratory

Gerke *et al*. introduced the concept of employing a continuous flow analysis, (e.g. Autoanalyzer instrumental system) to automate microbiological assays and described both turbidimetric and respirometric methods.

It is beyond the scope of this text to narrate automation operations involved in different microbiological assay methods. Automations of microbiological assays may be understood by exemplifying automatic systems of photometric assay. The instrument automatically measures a present amount of antibiotic solution, adds inoculum and nutrient medium, incubates the mixture for 100 minutes, and transfers the incubated mixture to a photometer where the results are automatically read and recorded. Fifty complete assays are carried out during each hour of operation. Before the instrument can operate by itself, inoculum and reagents must be prepared and test solutions placed in a predesigned order on a sample table.

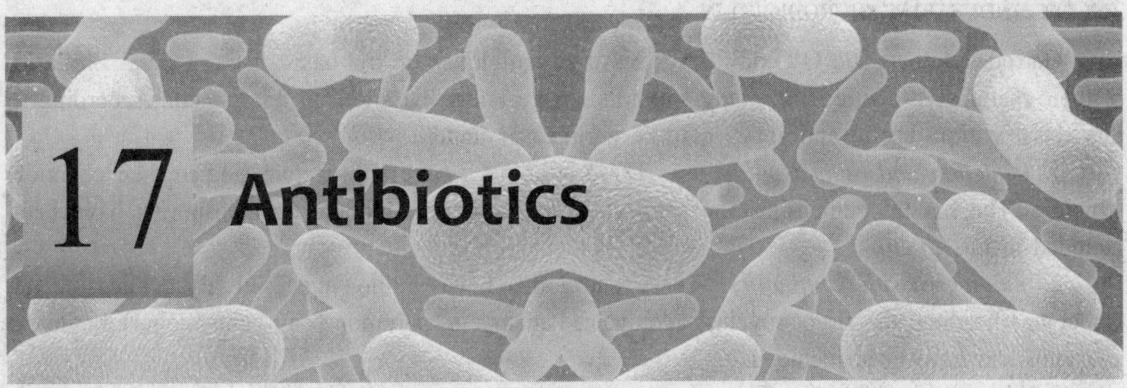

17 Antibiotics

1 INTRODUCTION

A chemical compound acting against life is called an antibiotic. It can be obtained either from natural sources, (e.g. microbes) or by a synthetic method. Those antibiotics which fulfil the requirements for their use in a specific field are manufactured on a large scale. Antibiotics are widely used in different fields, such as medical, agricultural and others.

The science of antibiotic fermentations is still imperfectly developed. Antibiotics are produced by the *submerged culture* method. This method of production is to date the most widely used method because of its simplicity and the advantages of batchwise operation.

Since 1945, thousands of different antibiotics produced by fungi, actinomycetes (closely related to bacteria) or bacteria have been isolated and studied. A few of these are of therapeutic importance. At present about 50 antibiotics are produced on a commercial scale, since they are useful in medical and veterinary fields. This chapter deals with the manufacture of some important antibiotics: penicillin, streptomycin, tetracycline and griseofulvin.

2 CLASSIFICATION OF ANTIBIOTICS

Antibiotics have been extensively studied. Generally, they are low molecular weight compounds. They have a variety of chemical structures, elemental composition and physical-chemical properties. Although the chemical characterization of antibiotics has been extensive, a good chemical classification is not yet possible.

Classification of antibiotics has been made on the basis of the following two criteria, in addition to their chemical structure:

(i) microbial source

(ii) mode of action

The first classification, based only on the microbial source, suffers from the drawback that it is much too broad. Another classification system, based only on mode of action, is also imperfect in that some antibiotics may have more than one primary site of attack.

Garrod, Lambert and O'Grady have classified antibiotics on the basis of the general similarity of their chemical structure. But, many antibiotics were left unclassified in this system. According to this classification, useful antibiotics may be grouped as under:

1. *Penicillin and related antibiotics:* These antibiotics have a β-lactum ring in their chemical structure. The natural penicillins, the semisynthetic penicillins and the cephalosporins have been included in this group.

2. *Aminoglycoside antibiotics:* These antibiotics have amino sugars in glycosidic linkage. Under this heading, the streptomycins, neomycin, kanamycin, paromomycin, gentamicin, tobramycin and amikacin have been included.

3. *Macrolide antibiotics:* These antibiotics have a macrocyclic lactone ring to which sugars are linked. This group includes erythromycin, oleandomycin and spiromycin.

4. *Tetracycline antibiotics:* These antibiotics are derivatives of the polycyclic naphthalene carboxamide. Antibiotics included in this group are tetracycline, chlortetracycline, oxytetracycline, demeclocycline and minocycline.

5. *Peptide antibiotics:* These antibiotics have peptide linked amino acids which commonly include both D- and L- forms. This group includes bacitracin, gramicidin and the polymyxins.

6. *Antifungal antibiotics:* This group has been divided into two main sub-groups: (i) polyenes containing a large ring with a conjugated double bond system. The important antibiotics in this category are nystatin and amphotericin B and (ii) other antifungal antibiotics. This category includes 5-fluorocytosine, clotrimazole and griseofulvin.

7. *Chloramphenicol:* This antibiotic is in a class by itself. It is a nitrobenzene derivative of dichloroacetic acid.

8. *Unclassified antibiotics:* These have varied structures and have not been grouped among the main groups listed above. This group includes cycloserine, fusidic acid, novobiocin, prasinomycin, spectinomycin and vancomycin. This group may be further divided to form additional antibiotic groups. Rehm (1967) has also classified antibiotics (Materials and Technology Vol. 5). But, this arbitrary classification is open to criticism.

3 PENICILLIN

3.1. Production of Penicillin

The Mould: It is highly desirable to use a high-yielding strain in the manufacture of an antibiotic. This can be achieved by *Sequential genetic selection.* In other words, such a strain can be obtained by step-wise development with the help of a series of mutagenic treatments, each followed by the selection of improved mutants. This selected mutant possesses a far greater capacity for antibiotic production than the 'wild' strain. From *Penicillium chrysogenum* (ancestral fungus), high-yielding strains have been developed for the commercial production of penicillin (Table 17.1).

Table 17.1 *Some high-yielding strains of Penicillium chrysogenum.*

Designation of mould	*History*	*Use*
(i) *Penicillium chrysogenum,* NRRL 1951. B25	Substrain of NRRL 1951. Yields in excess of 200 units/ml.	Submerged culture production in industry during the latter half of 1944 and the first half of 1945.
(ii) *P. chrysogenum,* Standford 25099	Substrain of NRRL 1951. B25 resulting from U.V. light irradiation.	Superior submerged culture production.
(iii) *P. chrysogenum,* X-1612	Substrain of NRRL 1951. B25 resulting from X-rays irradiation. Yields in excess of 500 units/ml.	Superior submerged culture production.
(iv) *P. chrysogenum,* Q-176 (Wisconsin)	Substrain of *P. chrysogenum,* X-1612 resulting from treatment of conidia with U.V. light. Yields in excess of 761 units/ml.	Submerged culture production.

It has been found that high-yielding strains of *Penicillium chrysogenum* are genetically unstable (in fact, the higher the yield capacity, the more unstable the strain is). Therefore, they are carefully maintained. Production strains are stored in a dormant form by any of the following standard culture preservation techniques:

(i) A spore suspension may be mixed with a sterile, finely divided inert support (soil or sand is traditional) and desiccated,

(ii) Spore suspensions can be lyophilised in appropriate media,

(iii) Spore suspensions can be stored under liquid nitrogen, i.e. in a frozen state.

All manipulations are carried out in laminar flow cabinets in rooms in which filtered air is maintained at a slight positive pressure in relation to the outer environment. Personnel wear sterilized clothing and they are careful in their aseptic technique. There is a strict necessity for pure culture in antibiotic fermentations. This fact was first revealed during penicillin production. Many bacteria, (e.g. *Staphylococcus aureus*) produce an enzyme, penicillinase. This catalyzes the hydrolytic cleavage of the four membered β-lactum ring of penicillin, with the formation of penicilloic acid (a compound without antibiotic activity). Therefore, contamination of a fermentation medium by penicillinase producing micro-organisms can result in the complete destruction of the accumulated penicillin.

Inoculum preparation: The aim is to develop a pure inoculum in sufficient volume and in the fast-growing phase for the production stage fermentor. Figure 17.1 shows a typical route by which the inoculum is prepared. The time taken for each stage is measured in days, and it decreases as the sequence progresses.

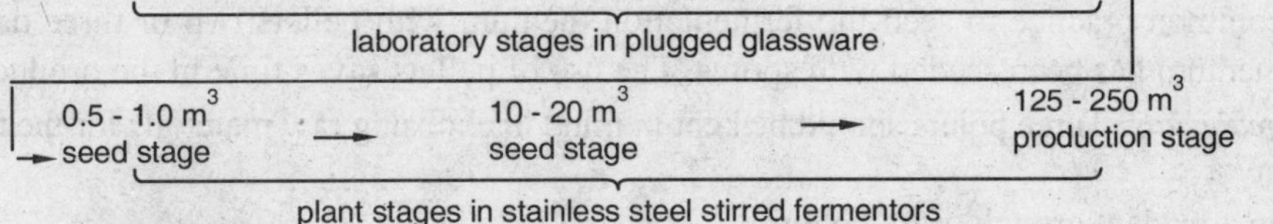

Fig. 17.1 *Stages in the preparation of inoculum for penicillin production.*

The medium is designed to provide the organism with all the nutrients that it requires. Many sporulation media have been developed to obtain large numbers of spores. For example, Moyer and Coghill (1946a) suggested the following sporulation medium:

Component	*(gm./ltr.)*
Glycerol	7.5
Cane molasses (edible quality as commonly sold at retail)	7.5
Corn-steep liquor	2.5
$MgSO_4 . 7H_2O$	0.050
KH_2PO_4	0.060
Peptone	5.00
NaCl	4.00
Fe-tartrate	0.005
$CuSO_4 .5H_2O$	0.004
Agar	2.50
Distilled water to make	1.0 litre

Note: The flasks containing the seeded medium are incubated on a rotary shaker to effect aeration and agitation.

The above medium may be solidified by making the total amount of agar at least 15-25 gm. per litre. Adequate oxygen is supplied in the form of sterile air. In addition, the temperature (24°C) is controlled. Principal criteria for the transfer to the next stage in the progression are:

(i) freedom from contamination

(ii) growth to a predetermined cell density

Contamination in the inoculum tanks is tested both by microscopic observations and by subcultures to broth medium.

The volume of culture increases approximately ten-fold with each successive stage in the progression. The mould, *P. chrysogenum,* grows in a filamentous form.

Inoculation methods: Any one of the following inoculation methods may be used to inoculate the fermentation medium in the submerged culture production of penicillin:

(i) Dry spores may be used to seed the fermentation medium.

(ii) Inoculation of the fermentation medium may be accomplished by a suspension of ungerminated mould spores. A non-toxic wetting agent, (e.g. 1 : 10,000 sodium lauryl sulfonate) in sterile water is used for preparing a more uniform suspension than can be obtained with sterile water alone. Such suspension may be applied by means of a suitable technique, (e.g. spray guns, pipettes and other means). This is followed by agitation and aeration of the fermentation medium for the equal distribution of the spore suspension.

(iii) The fermentation medium may be seeded by pellet inocula which are obtained by the germination of spores, with the formation of mycelial growth, under submerged conditions. It is a common practice to seed the fermentation medium with pellets two or three days after the medium has been seeded with spores. The use of pellets saves time in the productive cycle.

Raw materials: Three points should be kept in mind in choosing raw materials for the manufacture of penicillin:

(i) an abundant growth of mycelium

(ii) maximum accumulation of penicillin

(iii) ease of extraction and purification of the antibiotic

Carbon sources: Lactose acts as a very satisfactory carbon compound, provided that it is used in a concentration of 6 per cent (Moyer and Coghill, 1946a). Other carbohydrates, for example, glucose and sucrose, may be used as satisfactory substitutes for lactose, if slow feeding rates are used. Glycerol and sorbitol are less suited for this purpose.

Nitrogen sources: Ammonium sulphate, ammonium acetate, ammonium lactate or other materials can be used as nitrogenous compounds. Nitrogen is often supplied as ammonia gas. Corn-steep liquor also supplies cheap and readily available nitrogen.

Mineral sources: Many experts have studied the mineral requirements for mould growth and subsequent penicillin accumulation. Among them are the reports of Pratt (1945), Moyer and Coghill (1946a, b), Hosler and Johnson (1953), and others.

Elements, namely potassium, phosphorus, magnesium, sulphur, zinc and copper are essential for penicillin production. Some of these are supplied by corn-steep liquor. Potassium and phosphorus are customarily supplied as potassium dihydrogen phosphate, whereas magnesium and sulphur, as magnesium sulphate ($MgSO_4 . 7H_2O$). Iron and copper, if required in the medium, are usually supplied as sulphates. Moyer and Coghill (1946b) reported that a fermentation medium containing corn-steep liquor does not require addition of zinc or manganese ions. They are, if necessary, supplied as the sulphates.

Corn-steep liquor seems to contain sufficient amounts of potassium dihydrogen phosphate and magnesium sulphate ($MgSO_4.7H_2O$). However, small amounts of these salts may be incorporated in the fermentation medium.

Corn-steep liquor: Our readers are requested to refer to Chapter 12 for details of corn-steep liquor. According to Moyer and Coghill (1946a, b), the presence of this material in the fermentation medium gives higher yields of penicillin. Foster and his associates (1946) reported that the stimulatory effects of corn-steep liquor were due, in part to arginine, histidine, glutamic acid and possibly to phenylacetic acid derivatives.

Precursors: Chapter 5 may also be referred to for details of precursors. There are many related penicillins, (e.g. penicillin G, penicillin V, penicillin X, etc.). The most important naturally occurring penicillins are penicillin G (benzyl-penicillin) and penicillin V (phenoxymethyl-penicillin). The formation of a desirable penicillin can be stimulated by the addition of phenylacetic acid derivatives. For example, the mould, *Penicillium chrysogenum,* synthesizes large quantities of penicillin G, if phenylacetic acid ($C_6H_5 . CH_2 . COOH$) is present in the fermentation medium. Phenylacetic acid (PAA) supplies the side-chain of penicillin G. Fortunately, corn-steep liquor, employed in the formulation of the fermentation medium, is a source of phenylacetic acid derivatives. High levels of PAA are, however, toxic to the mould, and so all of it cannot be added at a time. The feed aims at providing a reserve of PAA in the medium at all times, without approaching toxic limits.

The penicillin nucleus, to which the 6-amino-penicillanic acid (6-APA) side-chain is attached, is believed to be synthesized from the amino acids L-cysteine and valine, as shown in the following diagram:

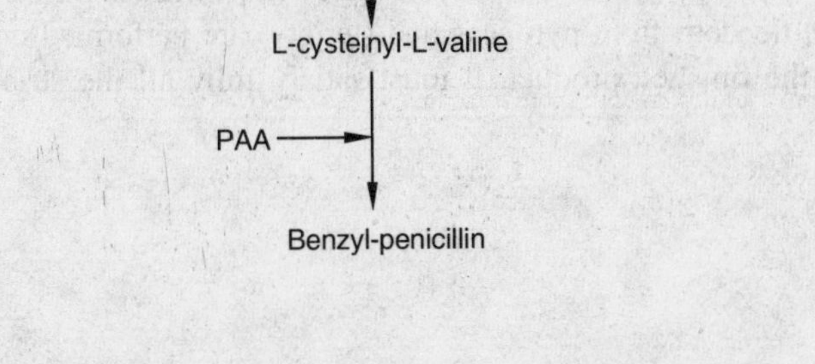

A typical medium reported by Jackson (1958) has the following chemical composition:

Component	*Per cent*
Corn-steep liquor solids	3.5
Lactose	3.5
Glucose	1.0
Calcium carbonate	1.0
Potassium dihydrogen phosphate (KH_2PO_4)	0.4
Edible oil	0.25
Penicillin precursor with slow Feeding rate	

Extraction and *Purification:* This operation is carried out in three different stages:

(a) *Removal of mycelium:* At harvest, the penicillin is in solution exocellularly, together with a range of other metabolites and medium components. The first step is to remove the mycelium by filtration. A rotary vacuum filter is employed for the filtration of fermented production medium. This stage is carried out under such conditions as would avoid contamination of the filtrate with penicillinase producing micro-organisms which may cause serious or total loss of an antibiotic.

(b) *Countercurrent solvent extraction of penicillin:* The next stage is to extract the penicillin. Countercurrent extraction is the generally accepted process. The pH of the filtrate [from stage (a)] is adjusted to 2-2.5 with the help of phosphoric or sulphuric acids. In aqueous solutions, at such low pH values there is a high partition coefficient in favor of certain organic solvents, (e.g. amyl acetate, butyl acetate or methyl iso-butyl ketone). This step has to be carried out quickly, for penicillin is very unstable at low pH values. Podbielniak countercurrent solvent extractor is used for this purpose. The antibiotic is then extracted back into an aqueous buffer at pH 7-7.5, the partition coefficient now being strongly in favour of the aqueous phase. The resulting aqueous solution is again acidified and re-extracted with an organic solvent. These shifts between the water and the solvent help in the purification of the penicillin. The spent solvent is recovered by distillation for reuse.

(c) *Treatment of crude extract:* Penicillin is manufactured as various salts according to its intended use, (e.g. for administration to humans and/or animals).

The treatment of the crude penicillin extract varies according to the objective, but involves the formation of an appropriate penicillin salt (usually sodium penicillin). The solvent extract as finally secured in stage (b) is carefully extracted back with aqueous sodium hydroxide. This is followed probably by charcoal treatment to eliminate pyrogens, (i.e. fever-producing substances), and by sterilization, (e.g. by the use of Seitz filter to remove any bacteria). Pure metal salts of penicillin can be safely sterilized by dry heat, if desired. Thereafter, the aqueous solution of penicillin is subjected to crystallisation.

For parental use, the antibiotic is packed in sterile vials as a powder (for reconstitution before use) or suspension. For oral use, it is tabletted, usually now with a film coating. Searching tests, (e.g. for potency, purity, freedom from pyrogens and sterility) are performed on an appreciable number of random samples of the finished product. It must satisfy fully all the strict government standards before being marketed.

Flow sheet: Figure 17.2 presents a flow sheet of the industrial production of penicillin G.

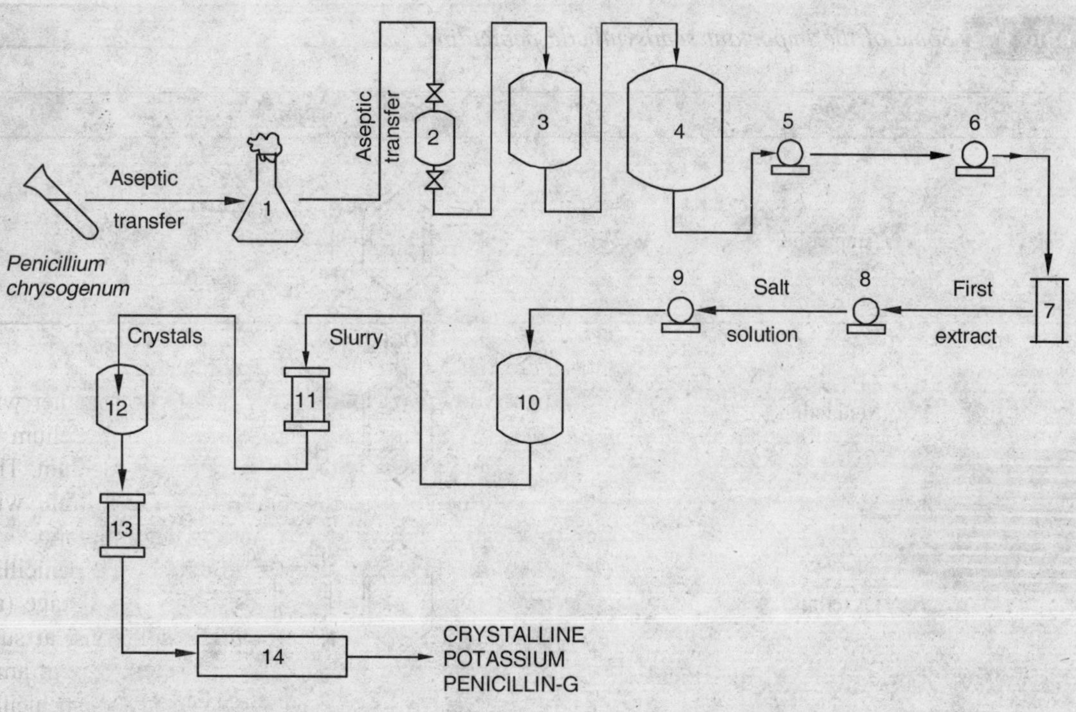

1. Shaker flask
2. Transfer vessel
3. Seed tank
4. Fermentation tank
5. Rotary filter
6. First extractor
7. Purification column
8. Second extractor
9. Third extractor
10. Vacuum still
11. Ceramic filter
12. Crystal wash
13. Ceramic filter
14. Vacuum drier

Fig. 17.2 *Flow sheet for large-scale production of penicillin G.*

3.2. Semisynthetic Penicillins

The semisynthetic penicillins are prepared from 6-aminopenicillanic acid (6-APA). 6-APA is a penicillin without an acyl group. Acylation of the 6-APA is easily and efficiently brought about by chemical means. In this way, thousands of semisynthetic penicillins have been obtained. A number of them have proved to be markedly superior to penicillin G in clinical application.

Acylation of the 6-APA is accomplished by a chemical reaction with a suitably activated derivative of a carboxylic acid (most frequently an acid chloride or a mixed anhydride). The coupling reaction can be carried out in organic solvents under anhydrous conditions or in an aqueous solution. Organic amines, (e.g. triethyl amine) are added to the anhydrous reactions. Alternatively, the aqueous reactions are frequently carried out in acetone-water mixtures in the presence of sodium bicarbonate. The formed semisynthetic penicillins are usually isolated by distributing the product as its free acid into a water-immiscible organic solvent and the salts and other impurities into water. Table 17.2 shows structural formulae of some of the more important semisynthetic penicillins.

Table 17.2 *Some of the important semisynthetic penicillins.*

Name	*Nature of acyl group*
Ampicillin	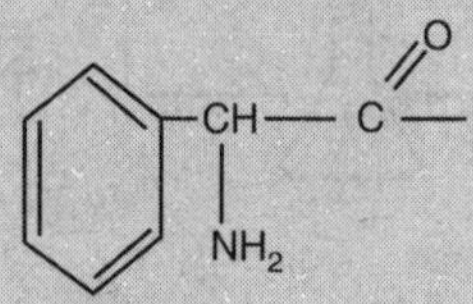
Methicillin	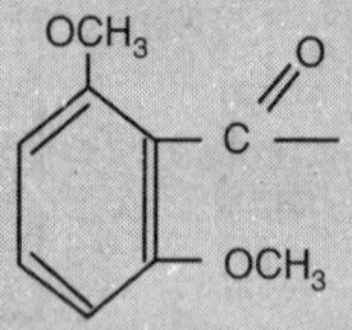
Oxacillin	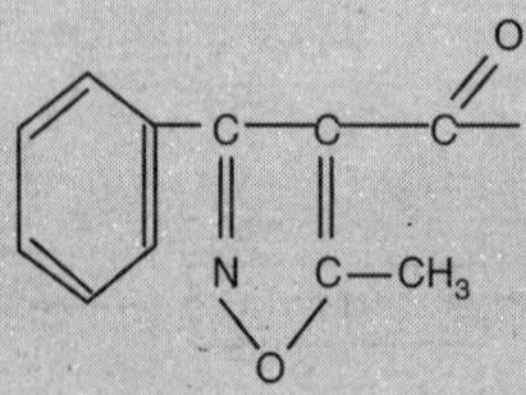
Propicillin	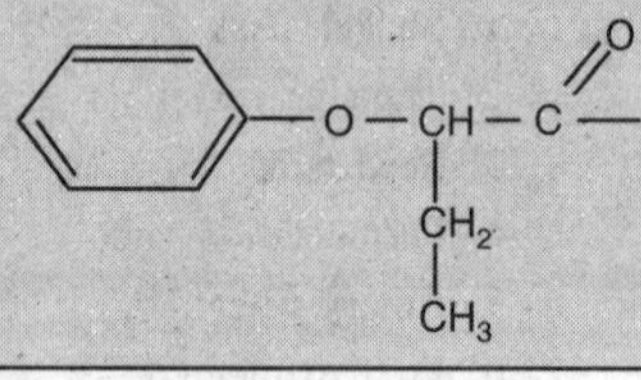

4 STREPTOMYCIN

The culture used in Streptomycin production is *Streptomyces griseus.* Glucose, fructose, galactose, xylose, mannitol, maltose, lactose or starch can be used as substrates in the fermentation medium. Other carbohydrates, like arabinose, rhamnose, sorbitol, dulcitol, inositol, sucrose and raffinose are not so well suited for this purpose. Oligo- and poly-saccharides give low yields, so that glucose must be regarded as one of the best carbon sources. Fats, oils and fatty acids may also be used in the presence of glucose. As a source of nitrogen, peptone, meat extract, soy extract, the residues from alcohol distillation and dried penicillium mycelium, ammonium salts, nitrates or glycine may be used. There is no need to supply any preformed structure analogous to the PAA of the penicillin G fermentation. It is important that mycelial growth is not over-stimulated by compounds such as glutamate. It is better to use proline which gives a slow growth and a high streptomycin production. When necessary, Mg^{++}, Ca^{++}, K^{+}, B^{+++}, Mo^{+++}, SO_4^{--}, Cl^{-}, HPO_4^{--} and growth stimulating compounds such as phenylacetic acid and α -naphthalene acetic acid may be added. Antioxidants such as starch, agar or sodium sulphite may also be added, if necessary.

The streptomycin production by submerged culture method has been presented in Figure 17.3.

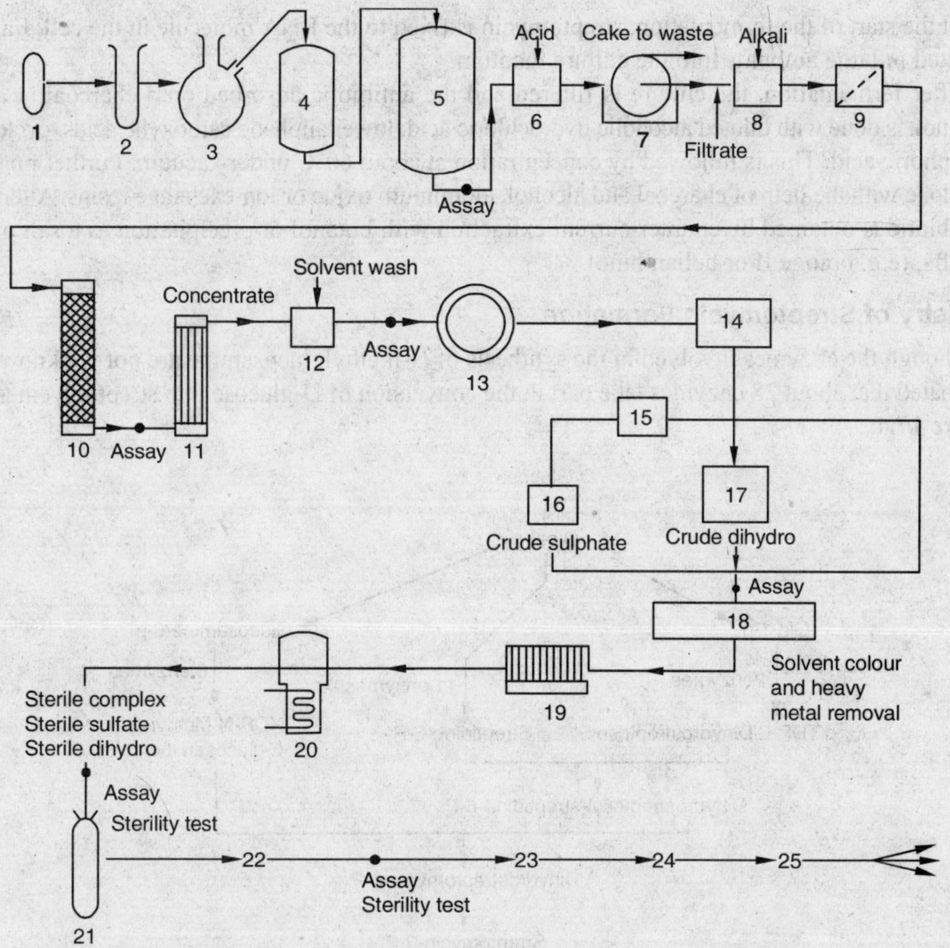

1. Master culture 2. Agar slopes 3. Shaker flask 4. Seed vessel 5. Fermentor 6. Acidification 7. Filtration 8. Neutralization 9. Filter clarification 10. Ion-exchange reagent 11. Evaporator 12. Crystallization 13. Vacuum oven 14. Calcium chloride crude complex 15. Calcium chloride removal 16. Crystallization 17. Catalytic hydrogenation 18. Finishing 19. Seitz filter 20. Freeze drying 21. Vial filling machine 22. Capping 23. Labelling 24. Packing 25. Dispatch

Fig. 17.3 *Flow sheet of streptomycin production by submerged-culture method.*

There are three phases in the fermentation process:

(A) Mycelial growth, with a large requirement for oxygen, glucose, nitrogen and phosphorus. The pH will rise to about 8.0 due to the production of ammonia.

(B) Streptomycin production without production of new mycelium. The ammonia is utilised and the pH falls to about 7.0. In this phase, glucose and oxygen are required in a large quantity.

(C) The mycelium undergoes autolysis with the rise in the pH value. After a period of continued streptomycin production, the pH falls again. There is a reduction in the oxygen requirement and the content of the medium is exhausted.

The first phase takes about 2 days, the second 1 to 2 days, while the total time required is about 5 days.

At the start of the fermentation, streptomycin is fixed to the RNA molecule in the cell. Later on, it is excreted in large amounts into the culture medium.

After fermentation, the culture is filtered and the antibiotic adsorbed onto charcoal at a neutral pH. Elution is done with diluted alcoholic hydrochloric acid, lower aliphatic carboxylic acids, cyclohexanol or phosphoric acid. This is followed by concentration at about 60°C under vacuum. Further purification can be done with the help of charcoal and alcohol, aluminium oxide or ion-exchange resins. Alternatively, the antibiotic is obtained by counter-current extraction with butanol or precipitation as a salt of certain dye-stuffs, (e.g. orange II or helianthine).

Chemistry of Streptomycin Formation

Though the enzymes involved in the synthesis of N-methylglucosamine are not yet known, it can be estimated that about 28 enzymes take part in the conversion of D-glucose into streptomycin as shown in Figure 17.4.

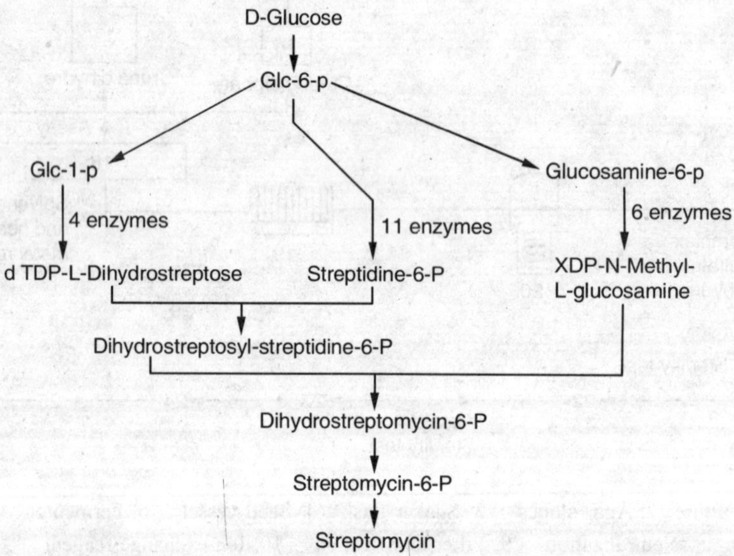

Fig. 17.4 *Postulated biosynthetic pathway from D-glucose to streptomycin involving about 28 enzymes.*

5 TETRACYCLINES

Tetracyclines are produced exactly like penicillin and streptomycin. Cultures used in the manufacture of tetracyclines are *Streptomyces aureofaciens, S. rimosus* or other species. The method used in the commercial production of these antibiotics is aerobic submerged fermentation at about 28°C and at the initial pH of about 6.0.

The culture medium used in this case is similar to that of penicillin. The nitrogen source may be ammonium salts, urea, groundnut flour and corn-steep liquor. The carbon source may be molasses, starch and even oils and fats. The fermentation vessels must be well-constructed so that no iron ions contaminate the culture medium, thus preventing a reaction with the chlortetracycline (CTC) produced.

There are two phases in the fermentation process:

(A) There is an abundant growth of mycelium, with very restricted antibiotic production.

(B) The bulk of the antibiotic is produced.

After about 2½ days of fermentation at 27°C, it is best to harvest it for CTC production. On the other hand, it is best to harvest after 5 days at 28°C for oxytetracycline (OTC) production.

CTC and OTC can be obtained from the substrate by adsorption with active charcoal at a low pH from the centrifuged culture fluid and elution with methanol, ethanol or acetone. The other way to obtain these antibiotics is by extraction of the filtered culture with suitable organic solvents, such as amyl alcohol. Further purification can be achieved in the usual way.

6 GRISEOFULVIN

Griseofulvin is produced by *Penicillium patulum*. It is formed from acetate building blocks. It contains neither sulphur nor nitrogen, but does contain chlorine. Chlorine is supplied as potassium chloride throughout the fermentation, otherwise the product formed is dechlorogriseofulvin. This is identical to griseofulvin, except that the chlorine is missing.

The principle in griseofulvin fermentation is to encourage early growth and then, by nutrient limitation, to maintain a cell population just within the aeration potential of the fermentor.

Griseofulvin, unlike penicillin and streptomycin, remains intracellular. Consequently at harvest, it is the cells that are significant and the filtrate is of no use. The cells can be easily removed by filtration. The first step in the recovery of griseofulvin is to extract the cells with acetone. The resulting extract is subjected to concentration, and thus a crude product is obtained. This crude product is washed with further organic solvents to remove acetone-soluble impurities, (e.g. the natural fats of the cells). The washed griseofulvin is redissolved in acetone and the solution is mixed carefully with water. This causes the antibiotic to precipitate. The large volumes of acetone used are recovered for further use.

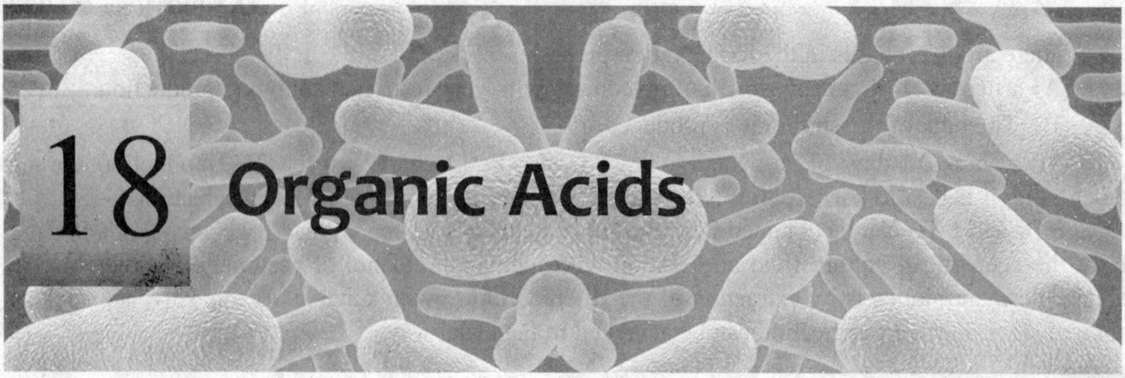

18 Organic Acids

1 INTRODUCTION

About sixty kinds of organic acids are known as microbial products. The major ones are shown in Table 18.1:

Table 18.1 *Main organic acids produced by micro-organisms.*

Acid	*Micro-organism(s)*
Citric acid	*Aspergillus niger*
	Candida lipolytica
Lactic acid	*Lactobacillus delbrueckii*
Itaconic acid	*Aspergillus terreus*
Acetic acid	*Acetobacter aceti*
Gluconic acid	*Aspergillus niger*
Kojic acid	*Aspergillus oryzae*
2-Keto-gluconic acid	*Pseudomonas fluorescens*
5-Keto-gluconic acid	*Gluconobacter suboxydans*
D-Araboascorbic acid	*Penicillium notatum*
Propionic acid	*Propionibacterium shermanii*
Pyruvic acid	*Pseudomonas aeruginosa*
Succinic acid	*Bacterium succinicum*
Tartaric acid	*Gluconobacter suboxydans*
Fumaric acid	*Rhizopus delemar*
Malic acid	*Lactobacillus brevis*
α-Keto-glutaric acid	*Candida hydrocarbofumarica*
L(+) Isocitric acid	*Candida brumptii*
L(+) Allocitric acid	*Penicillium purpurogenum*

Only a few of the above listed organic acids are commercially produced by fermentation processes.

2 CITRIC ACID

Citric acid is used in many fields. Its industrial uses may be listed as under:

(i) an acidulant in food, (e.g. jams, preserved fruits, fruit drinks, etc.) and pharmaceutical industries
(ii) a chelating and sequestering agent, (e.g. in the tanning of animal skins)
(iii) production of carbonated beverages
(iv) citrate and citrate esters as plasticizers

Before the development of fermentation technology, citric acid was obtained by extraction from the juices of certain citrus fruits, (e.g. lemon), and later from pineapple wastes. Today most of the commercial citric acid is obtained by a microbial fermentation process. As a result, the process of extracting this acid from the juices of certain citrus fruits became obsolete.

There are various fermentation processes used in the manufacture of citric acid. These processes have been briefly dealt with in this chapter.

(a) **Surface-culture process:** In this process, the culture is inoculated across the surface of the production medium. In addition, culture remains on the surface throughout the fermentation, since this is a stationary fermentation. This is an old process and it is still practised. The details of the process are not easily available, since they are treated as trade secrets.

The various bioparameters affecting the fermentation process are briefly discussed below:

Fungi: It is very important to choose a suitable fungal strain for producing citric acid successfully. Many fungi have been reported to possess the ability to produce citric acid. For example, *Aspergillus niger*, *A. clavatus*, *Penicillium luteum*, *P. citrinum*, *Paecilomyces divaricatum*, *Mucor piriformis*, *Ustulina vulgaris*, and other species of *Mucor* have been employed to produce citric acid in the laboratory or on a commercial basis. But only *A. niger* strains among these fungal strains are used for the following reasons:

(i) They are efficient, (i.e. high-yielding) strains.
(ii) They possess fairly uniform biochemical properties.
(iii) They produce a negligible or small amount of oxalic acid, provided fermentation conditions, (i.e. pH and salts) have been adjusted to favour the formation of citric acid.
(iv) They can easily be cultivated.

Preparation of inoculum: The spores of *Aspergillus niger* strain required to inoculate shallow pans are produced by growing the fungus from a stock culture on a suitable solid sporulation medium at 25°C for 4 to 14 days. In the composition formula of inoculum medium, the presence of trace amounts of manganese salts, unless balanced against proper amounts of zinc or iron salts, may lower yields of citric acid in the actual fermentation receiving these spores. Suspension of spores is obtained by suspending the grown spores in a suitable diluent, such as water containing a wetting agent.

The spores of inoculum are added to the production medium so as to keep them floating on the surface, since this is a surface-culture process. This may be accomplished by a suitable special inoculating device.

Source of carbon: Sucrose, according to laboratory studies, is the best source of carbon among various tested organic substances, particularly sugars, in producing high yields of citric acid. It was also reported that sucrose concentration exceeding 15 per cent should not be used, since the excess amount of sugar (less than 3 per cent) remained unconverted to citric acid. When a part of sucrose was substituted by fructose or glucose, it resulted in lower yields of citric acid than when controls containing sucrose alone were used. Therefore, precaution is to be taken while using sucrose to avoid partial hydrolysis.

Beet molasses is extensively used as a carbon substrate in the fungal production of citric acid on a commercial basis. Beet molasses requires pre-treatment, since it contains excessive amounts of trace metals. Therefore, ferrocyanide or ferricyanide may be added to the production medium before sterilization. The metals of the beet molasses form a complex with the added chemical agent, thus eliminating themselves as the precipitate. Alternatively, the clarified molasses may be passed through a cation-exchange resin. Under laboratory conditions, the highest yields of citric acid are secured, provided sucrose has been passed through an ion-exchange resin.

Inorganic salts: Apart from the carbon, hydrogen and oxygen supplied by the added carbohydrate, the trace metals, namely nitrogen, potassium, phosphorus, sulphur and magnesium are needed in the fermentation medium used for citric acid production. According to Currie (1917), the fermentation medium with the following chemical composition is the most favourable medium in the manufacture of citric acid:

Component	*Gms./Litre*
Sucrose	125 – 150
NH_4NO_3	2.0 –2.5
KH_2PO_4	0.75–1.0
$MgSO_4.7H_2O$	0.20–0.25
HC1 to pH 3.4 to 3.5 (5-4 ml. of N/5HC1)	

Doelger and Prescott (1934) reported the following medium to be most satisfactory:

Component	*Gms./Litre*
Sucrose	140
NH_4NO_3	2.23
K_2HPO_4	1.0
$MgSO_4.7H_2O$	0.23

Note: Salts and sugars are dissolved and made up to 1 litre with distilled water. The pH of the medium is adjusted in the range of 2.20 to 1.60 with N/1 HCl. The sterilization of the medium is done at 8 to 10 lbs. steam pressure per square inch for 30 minutes.

It is highly essential to add NH_4N0_3, KH_2PO_4 or K_2HPO_4 and $MgS0_4.\ 7H_2O$ only in a minimum quantity, since the presence of these metallic salts beyond their limits affects the fermentation adversely. For example, more than 2.50 gms. of NH_4NO_3, 1.50 gms. of potassium monohydrogen phosphate, and 0.30 gms. of $MgSO_4.7H_2O$ favour the increased yield of oxalic acid and the citric acid yield decreases. It is highly desirable that mycelial mats should be thin and the sporulation light or nearly absent because high yields of citric acid may be secured.

Many scholars studied the effect of different metallic ions in both directions: qualitative and quantitative. Conclusively, the type and quantity of other metallic ions, (e.g. iron and zinc) required for citric acid—*Aspergillus niger* fermentation largely depend on the type of fungal strain to be used in the process. For example, *Aspergillus niger* 62 required 0.1 mg. of iron per litre to produce optimum results when Perlman, Dorrell and Johnson (1946) conducted an experiment using highly purified media. On the other hand, *A. niger* 59 required 10 mg. of iron per litre to produce optimum results. In addition, the effects of the metallic ions are interrelated. Therefore, the proper level of one metal may depend on the concentration of other metals present in the medium. The actual required quantity, (e.g. a fraction of a part per million) of trace metals may be added to the medium because improved techniques of analysis are now available.

pH: According to Currie, the pH of the medium should be adjusted to 3.4 to 3.5 with HCl. Also, low pH values (1.60 to 2.20) have been found to be the most favourable by Doelger and Prescott. These scholars also advocated the use of HCI, which has been found superior to nitric, sulphuric, and acetic acids in producing higher yields of citric acid.

Generally, calcium carbonate is not added to the production medium to bring about the neutralization of citric acid produced because its presence favours contamination. Moreover, its absence favours higher yields of citric acid and a shorter fermentation time.

Low pH values in *Aspergillus niger* - citric acid fermentation are desirable for the following reasons:

(i) Sterilization of medium is more readily effected.
(ii) Formation of citric acid is favoured.
(iii) Formation of oxalic acid is suppressed.
(iv) The danger of contamination is minimized.

Lastly, stationary pans or trays must be protected from corrosion or they should be made of stainless steel or aluminium because production media with low pH values are used, and because of the sensitivity of the fermentation process to excessive iron.

Temperature: The exact required temperature of incubation depends in part on the fungal strain and the fermentation conditions. According to Doelger and Prescott, a temperature range of 26 to 28°C is considered to be satisfactory.

Ratio of surface area to volume: This is an important bioparameter in this process, since the rate of bioconversion of sugar to citric acid depends on the ratio of surface area to the volume of the medium. The lower the ratio of the volume to surface area, the higher the citric acid yield is. In this shallow-pan method, a large surface area of the mycelial mat is exposed to a relatively shallow layer of a production medium. Under such conditions, high-level bioconversions of sugar occur. In this respect, this process is superior to the submerged-culture process. Doelger and Prescott studied the effect of varying the ratio of the volume to the surface area of the medium. The results of their research is shown in Table 18.2.

Table 18.2 *The effect of varying the ratio of the volume to the surface area on the yields of citric acid.*

Volume/Surface area ml./sq. cm.	Original volume ml.	Sucrose per pan gm.	Final volume ml.	Citric acid /100 ml. gm.	Total citric acid per pan gm.	Yield of citric acid to sugar per cent
2.45	2,000	280	1,810	6.35	114.9	41.0
2.20	1,800	252	1,620	7.05	114.2	45.3
2.08	1,700	238	1,490	7.40	110.3	46.3
1.83	1,500	210	1,310	7.75	110.2	48.3
1.22	1,000	140	780	8.80	68.6	49.0

Usually, the production medium is placed in shallow pans in such a way that a shallow layer of medium with a depth of 1 to 2.5 cm. (but sometimes up to 8 cms.) is formed.

Aeration: It is necessary to supply air to the surface of the seeded medium. From laboratory studies it has been found that the rate of aeration, either higher or lower than the optimum rate, gives lower yields of citric acid. Therefore, it is essential to determine the rate of the air supply required for each new apparatus installed.

Time: With the shallow-pan method, (i.e. surface-culture process) the period required for the progress and termination of the fermentation process is in the range of 7 to 10 days.

Yields: It has been reported that yields of citric acid may be in the range of 60 to 80 gms. (or higher) of anhydrous citric acid per 100 gms. of incorporated sugar. Wells, Moyer and May obtained a maximum yield of 90.7 per cent of citric acid from glucose on the basis of the sugar consumed.

Recovery of citric acid: The recovery operation for citric acid from the harvested fermentation broth is difficult for the following important reasons:

(i) the presence of unconverted sugars
(ii) the presence of other acid fermentation products, (e.g. oxalic acid)
(iii) the presence of trace salts as impurities

Usually, recovery of citric acid is practised as under:

(i) The fermentation liquor is drained off to separate the mycelium. In addition, any intracellular citric acid present in the mycelium is obtained by pressing the mycelial mat.
(ii) The recovered fermentation liquor, as in step (i), is treated with milk of lime forming the precipitate of calcium citrate.
(iii) The precipitate of calcium citrate obtained from hot neutral aqueous solution, as in step (ii), is filtered and washed.
(iv) Calcium citrate is treated with an equivalent of sulphuric acid to liberate citric acid, leaving behind the precipitate of calcium sulphate.
(v) The precipitated solution, as in step (iv), is filtered and washed.
(vi) An impure solution of citric acid is subjected to decolorization by the activated carbon. Then, it is also demineralized.
(vii) Finally, the pure citric acid solution is evaporated and is crystallized from the solution.

The alternative method is the counter current extraction method. Organic solvents (100 parts of tri-*n*-butyl phosphate and 5 to 30 parts of *n*-butyl acetate) are employed for the purpose of extracting citric acid. Then citric acid is extracted from the organic phase with lime water. Lastly, the precipitate of calcium citrate is treated with sulphuric acid as previously discussed.

(b) **Submerged-culture process using** *Aspergillus niger*: In this process the fungus is grown dispersed through a liquid production medium. Usually, fermentation is carried out in the fermentation vessel consisting of a sterilizable tank of the capacity of thousands of gallons. The fermentation tank is equipped with a mechanical agitator and a sparger.

The origins of the submerged-culture method using a small flask to simulate deep culture conditions date back to the work of Amelung. Perquin established that phosphate limitation plays a significant role in the production of citric acid by the submerged-culture process using *Aspergillus niger.* Martin and Waters established the relationship of the effect of *A. niger* morphology with citric acid production from beet molasses. Subsequently, Clark found that, specific quantities of ferrocyanide lead to the formation of an optimum mycelial form.

In practice the mould spores are produced under controlled aseptic conditions. When harvested, they are used in specific quantities to seed the inoculum tank containing a medium designed to develop cellular mass and to control morphology rather than to produce citric acid. Then it is transferred to the production tank under aseptic conditions. In contrast to the composition of the inoculum medium, production medium contains constituents so as to favour citric acid production rather than growth. High aeration rates, (i.e. 0.5 to 1.5 volumes per minute) producing much foam are necessary. Samples are regularly withdrawn to determine the citric acid and sugar content revealing the progress of the fermentation process. In addition to this, pH, dissolved oxygen and solids content are determined. The initial sugar concentrations and the citric acid yield compare favourably with those of the surface-culture process. This method also utilizes pretreated molasses medium like the surface-culture process. The selection of raw materials for the submerged-culture process suffers from the same constraints as the surface-culture process, except that it allows a wider choice of raw materials.

A patent was issued in 1951 for the use of *Aspergillus niger* in the submerged-culture process employing a production medium that consists of pure sucrose or a solution of cation-depleted molasses. Subsequent patents cover production by *A. niger* on a substrate produced from starch treated with amylase and amyloglucosidase. Very recently, glucose solutions derived directly from corn have been employed. It has been reported that yields of citric acid from such sugar substrates may exceed 90 per cent.

(c) **Submerged-culture process using a yeast:** The belief that industrial production of citric acid could be possible only by *Aspergillus niger,* was disproved around 1969 or 1970. A challenging patent was issued in 1970 by demonstrating the production of citric acid by the species of yeast, (e.g. *Candida guilliermondii)* grown submerged in a medium containing either glucose or black-strap molasses with an equivalent amount of sugar. With this fermentation, the duration of the fermentation period was shorter than that of the submerged-culture process involving *A. niger.* A subsequent improvement patent quotes citric acid concentrations of 110 gms./1tr.

Using certain strains of *Candida lipolytica*, the possibility of using hydrocarbons as a raw material in citric acid fermentation has been shown. A patent was issued in 1970 for the bioconversion of C_9 to C_{20} normal paraffins to citric acid by *C. lipolytica.* Thus, citric acid weight yields exceeding 100 per cent were claimed. In 1972 other patents were issued describing a method to select high-yielding mutant strains of *C. lipolytica*. By using a chemical agent, monofluoroacetate, in media, Japanese workers have succeeded in isolating mutants with only one hundredth of the *cis*-aconitase activity exhibited by the parent strain (*C. lipolytica* ATCC 20114). These strains are able to accumulate 112 gms. of citric acid per litre after 3 days of incubation at 30°C. On the basis of the substrate consumed, the yield is 145 per cent. In 1974 Pfizer patented a continuous process for *C. lipolytica*-citric acid fermentation. The process utilizes a single vessel. Continuous feeding of paraffin and continuous withdrawal of fermented broth are to be practised.

3 LACTIC ACID

Lactic acid is commercially produced by the synthetic method as well as by the fermentation process. It is manufactured in Japan and the United States by the synthetic method. On the other hand, European manufacturers produce lactic acid by the fermentation process. Half of the estimated world capacity for lactic acid production (28,000-30,000 metric tons per annum) is by European manufacturers. Lactic acid is largely used in pharmaceutical industries, (e.g. iron lactate) and some plastic industries.

The following bioparameters are to be controlled during lactic acid manufacture by the fermentation process:

The micro-organism: There are numerous microbial species capable of producing large quantities of lactic acid. These include *Lactobacillus bulgaricus, L. delbrueckii, L. leichmannii, L. casei, L. pentosus, Streptococcus lactis* and *Rhizopus oryzae.* The selection of a micro-organism largely depends on the nature of the carbon source being used in the fermentation process. For example, *L. bulgaricus* is preferred to other microbial species when whey is used as a source of carbohydrate. The inoculum volume is usually about 5 per cent of the fermentation broth volume.

Carbon sources: Glucose, maltose, sucrose or lactose may be used as a satisfactory source of carbon. Crude substrates such as corn starch, potato starch, rice starch, whey, molasses or sulfite lyes have been used to supply carbon in the production medium. Pre-treatment to starchy materials by enzymes or by acid, (e.g. sulfuric acid) is necessary to bring about hydrolysis to maltose and glucose. It is essential to steam strip the sulfite lyes to eliminate sulphur dioxide and to treat them with alkali to remove lignin during the pre-treatment operation. Jerusalem artichokes are a potential raw material for lactic acid production. The sugar concentration in the production medium is initially adjusted to 5-20 per cent, but usually does not exceed 12 per cent.

Nitrogen sources: Ammonium salts have been used, giving satisfactory results. For example, ammonium hydrogen phosphate in the concentration of 0.25 per cent is used for the production of lactic acid at Hammond, Ind., in America.

Growth factors and *mineral sources*: Lactobacilli require complex nutrients, particularly vitamin B-complexes. These may be supplied by enriching the medium with crude vegetable sources, (e.g. malt sprouts, i.e. the rootlets of germinating barley grains). This explains why care is to be taken not to overheat the enrichment materials during drying, avoiding any nutritional defects. Minerals, based on the requirement of a selected micro-organism are added to the production medium. Calcium carbonate in high concentration, (e.g. 10 per cent) is used as a buffer.

pH: The pH of the broth in the range of 5.5-6.5 is considered satisfactory. It is essential to control the pH, since the yields and the rates can be increased by doing so. Neutralization of the accumulated acid is carried out by the continuous addition of a slurry of calcium hydroxide. If neutralization of the acid is not done, it will result in high acidity. Such high concentrations of acid are not tolerated by lactobacilli. Therefore, fermentation is not completed. Such a fermentation medium with a low pH value is advantageous, since most contamination problems are eliminated and it allows sterilization at low temperatures.

Temperature: The temperature maintained during fermentation depends on the type of micro-organism involved. For example, a temperature in the range of 45 to 50°C has been found satisfactory in fermentation involving *Lactobacillus bulgaricus* or *L. delbrueckii.* On the other hand, a temperature of about 30°C is considered satisfactory using *L. casei, L. pentosus* or *Streptococcus lactis.*

Aeration and *agitation:* The supply of sterile air is not necessary in the fermentation process using lactobacilli, since this is anaerobic fermentation. Moreover, the complete removal of air is not necessary because these bacteria are facultative aerobes. The broth is usually gently stirred to keep the calcium carbonate in suspension.

Time: The duration of a plant fermentation is commonly 5-10 days. It is possible to complete the fermentation in 72 hours, using 12 to 13 per cent glucose in the medium, provided the pH is controlled at 6.3-6.5 by the continuous neutralization of the acid produced.

Yield: Commercial fermentation yields are in the range of 93-95 per cent of the weight of glucose supplied.

Recovery: A major problem in lactic acid production is that of product recovery and not of fermentation. The method to be employed for recovery depends on the type of grade required. There are different types of grades needed for different purposes. They are: food-grade, technical-grade and plastic-grade.

It is necessary to use a medium with a higher-grade sugar source and a mineral protein content for food-grade lactic acid production. The very first step is filtration by conventional industrial precoat filters. The filtrate is acidified with sulphuric acid for regenerating the lactic acid. Calcium is precipitated as calcium sulfate and it is washed. This wash is combined with the filtrate and is then treated with activated vegetable carbon to remove organic impurities. Thereafter, concentration of crude lactic acid to 25 per cent solids by evaporation is carried out. The refining and evaporation steps are repeated with a result of 50-65 per cent total acidity. If the final product is discoloured, sodium ferrocyanide can be used to remove heavy metals, (e.g. copper) by precipitation. The last traces of contamination are removed by passing it through ion-exchange resins.

For technical-grade products, calcium present is removed as calcium sulfate dihydrate by precipitation. Thereafter, the filtration step is carried out to separate the precipitate of calcium sulphate dihydrate. Finally, the filtrate is subjected to concentration to 35-40 per cent lactic acid by evaporation. If more calcium sulfate precipitates, it is also removed by filtration.

Plastic-grade lactic acid can be obtained by esterification with methanol after concentrating it.

4 ITACONIC ACID

Itaconic acid is a valuable compound for the manufacturers of plastics. It may also be used as a material for artificial dentures and certain synthetic fibres, (e.g. Courtelle).

Industrial production of itaconic acid is carried out by the fermentation process (surface and submerged methods). Numerous chemical syntheses, (e.g. thermolysis of citric or aconitic acids) have been reported, but none of them are practised on a commercial scale. Itaconic acid fermentation is to be carefully controlled, since some typical problems are bound to arise. The factors are briefly discussed here:

The micro-organism: Kinoshita (1929) was the first to obtain itaconic acid as a metabolic product of *Aspergillus itaconicus*. Unfortunately, the amounts obtained were too poor for successful commercial development. A suitable strain, called *Aspergillus terreus* (NRRL 1960), was isolated from Texan soil for the industrial production of itaconic acid. Thereafter, mutants of *A. terreus* were developed by irradiation with ultraviolet rays. These mutants produce enhanced yields. Some industrially important strains are: *A. terreus* ATCC 10020, *A. terreus* ATCC 10029 and *A. terreus* NRRL 265.

Carbon Sources: Glucose and sucrose are satisfactory carbon substrates. Due to the low cost of cane-molasses, it is used for the large-scale production of itaconic acid. Sugar concentrations, of the order of 15-25 per cent, are fermented efficiently, provided beet-molasses is used in the spore-germinating medium.

Nitrogen sources: Ammonium salts, [e.g. NH_4NO_3 and $(NH_4)_2SO_4$] are a satisfactory source of nitrogen.

Mineral sources: The production of itaconic acid, like citric acid, is sensitive to minerals such as iron, copper, magnesium and zinc. For example, the effect of varying concentrations of iron on the itaconic acid production by a mutant of *Aspergillus terreus is* illustrated as shown in Table 18.3.

Table 18.3 *Effect of iron concentration on itaconic acid production by a mutant of Aspergillus terreus.*

Iron (mg./L)	*Yield* $\left(\frac{\text{gm. of itaconic acid}}{\text{gm. of hexose used}} \times 100\right)$
0	57
1	25
2	17
4	17

pH: Initial pH of the production medium is usually adjusted in the range of 1.8 to 2.0. pH may be adjusted by the use of sulfuric acid, nitric acid, hydrochloric acid or itaconic acid. Of these, sulfuric acid is preferred. It is necessary to neutralize the accumulated itaconic acid. This may be accomplished by step-wise addition of ammonia.

Temperature: The optimum temperature is in the range of 30 to 35°C.

Aeratton and agitation: It is essential to supply sterile air with constant agitation of the broth in submerged culture method. The rate of aeration and the speed of the agitator are determined for each new process.

Duration of fermentation: For the completion of the fermentation process, 3 to 7 days are needed.

Yield: Kobayashi (1967) reported yields of over 50 per cent of sugar consumed in three days on a medium containing 6 per cent glucose. Stoichiometric yields of over 80 per cent have been reported.

Recovery: Itaconic acid must be separated from broth, which contains mycelium, residual sugar, fermentation products including other organic acids and inorganic salts. Acidification, clarification and other purification treatments are followed by concentration, crystallization, and collection of itaconic acid. Solvent extraction of the broth may also be employed for recovering itaconic acid.

Biochemistry of itaconic acid formation: First of all, citric acid is produced via the tricarboxylic acid cycle. Then, citric acid undergoes dehydration by aconitate hydratase with the formation of *cis*-aconitic acid. Finally, decarboxylation of *cis*-aconitic acid by aconitate decarboxylase leads to the formation of itaconic acid. The sequence of these reactions may be schematized as under:

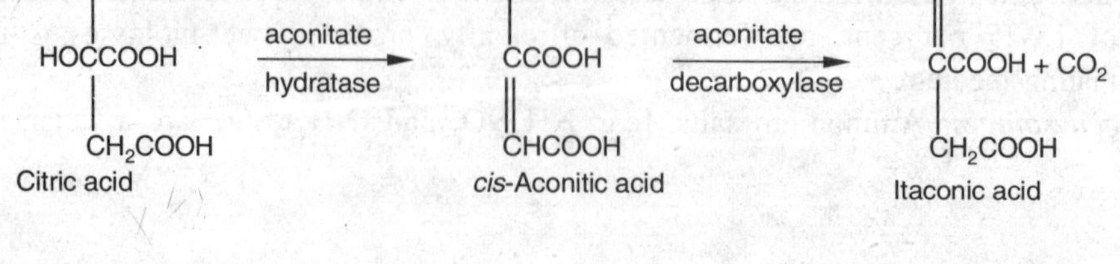

5 ACETIC ACID (VINEGAR)

The term 'vinegar' has originated from the French word *vinaigre* (sour wine). Formation of vinegar involves souring of wine under specific conditions. Thus, there are two stages in the formation of vinegar using sugar-containing media:

(a) alcoholic fermentation

(b) acetic acid fermentation

(a) **Alcoholic fermentation:** During this stage sugar-containing materials, (e.g. fruit juices, honey or hydrolyzed starchy materials) are fermented to ethyl alcohol. This bioconversion is brought about by the agency of enzymes of yeast, (i.e. a selected strain of *Saccharomyces cerevisiae)*. According to the Gay-Lussac equation as shown in Figure 18.1.

$$C_6H_{12}O_6 \longrightarrow 2CO_2 + 2C_2H_5OH$$

Fig. 18.1 *Gay-Lussac equation.*

One gram of glucose should yield 0.5114 gms. of ethyl alcohol. It is necessary to adjust the fermentable sugar concentration in the raw material undergoing fermentation in the range of 8-20 per cent. With certain fermentations, (e.g. using honey as a fermentable substrate) it is necessary to add potassium and ammonium phosphates to the production medium. On the other hand, these mineral salts are not needed in case of fermentations using crushed grapes and grape juice or with apple juice. A temperature of 75 to 80°F has been found to be satisfactory.

This stage, in turn, is carried out in two stages: (i) Primary stage—It lasts from 3 to 7 days. It is usually conducted in open fermentation vats. (ii) Secondary stage—It lasts over a period of several weeks. It is carried out in closed fermentors. Sulfur dioxide is added to the production medium at the start of fermentation and prior to the addition of the starter culture of yeast. This inhibits the growth and activity of unwanted yeasts and bacteria. An initial level of 50-100 mg. of sulfur dioxide per litre is enough. The yield of alcohol is usually only about 85-90 per cent of the stoichiometrical yield. The clarified wine is obtained by racking (drawing off) it from the sediment.

(b) **Acetic acid fermentation:** There are three main processes for the microbial conversion of dilute ethanolic solutions to acetic acid. They are:

(a) The Orleans process in which acetic acid bacteria grow very slowly as a film over the surface of still liquid. This is said to produce the finest culinary vinegar.

(b) The 'Quick' vinegar process which had been introduced in the last century.

(c) The deep fermentation processes which have had commercial success only recently. Despite the attractions of modern deep fermentations, the 'Quick' vinegar process retains considerable industrial importance.

'Quick' vinegar process: This method makes use of Frings generator (Figure 18.2). It consists of an airtight tank usually 14 ft. in diameter and 15 ft. high. There is a wooden grating near the bottom. In addition, the sprayer or distributing arm is fitted near the top of the tank. At the top, a 4-in. vent with a damper connected to a vapour-liquid separator serves to control the air supply. There are ten air intakes (containing air filters) situated around the tank near the level of the wooden grate.

In order to control the temperature, the tank is equipped with tubular cooling coils situated at the bottom of the tank in the collection chamber. The coils are connected by a centrifugal pump to a feed line supplying the sprayer. The generator is packed with beechwood shavings (supported by the wooden grating) up to about 1-5 ft. from the top of the generator.

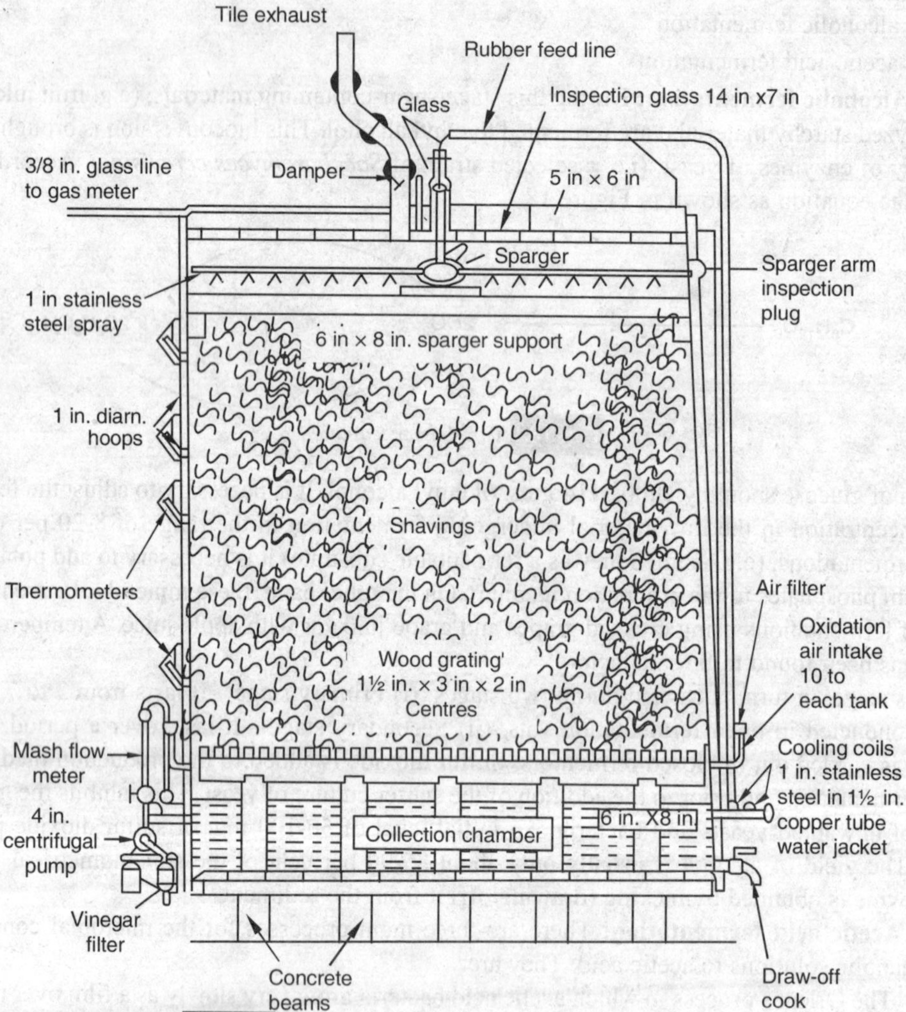

Fig. 18.2 *Trickling (Frings) generator (Hansen, 1935).*

The vinegar stock or mix possesses the following composition:

Compound	*Amount* (%)
Glucose	0.9
Ammonium phosphate	0.4
Magnesium sulfate	0.1
Potassium citrate	0.1
Pantothenic acid	0.00005

Acetic acid bacteria (a selected species or strain of the genus *Acetobacter*) are inoculated onto the beechwood shavings. The vinegar stock, or mix is allowed to undergo circulation repeatedly through the beechwood shavings. Thus, alcoholic solution undergoes more and more oxidation. This operation continues until vinegar of the desired strength has been obtained.

There are several factors to be controlled during vinegar manufacture. For example, nature of packing material, nature of vinegar stock used, rate of flow of ethanolic solution, rate of air flow and temperature are important factors for the successful operation of the acetic acid fermentation stage.

Packing materials that may be used for the packing of generators include beechwood shavings, coke, pumice, rattan, grape and other twigs, corn cobs and Berl saddles. Among these, beechwood shavings are preferred for packing purposes. Allgeier *et al.* reported the comparative performances of generators that were packed with beechwood shavings, coke, rattan and Berl saddles. The limiting factors of coke were reported and they are: surface area and impurities in coke. Removal of impurities by acid washing and increasing the surface area by using smaller pieces ($\frac{1}{4}$ to $\frac{1}{2}$ in. in diameter) increased the operating efficiency of the coke-packed generator from 76 to 98 per cent of that obtained with expensive beechwood shavings. A higher concentration and a different type of nutrient were needed with coke than with beechwood packing. Thus, it is to be kept in mind that the type and concentration of nutrients vary with the type of packing.

In re-circulating generators, difficulty is experienced with wine or cider vinegar due to sliming. Sliming is caused by the development of the so-called 'mother of vinegar' (actually exocellular bacterial cellulose produced by *Acetobacter xylinum*). Slime formation may plug the Frings-type generators. Sliming is common in case of beechwood shavings-packed generators. Also, it occurs in case of coke-packed generators after 6 to 12 months of use, necessitating repacking. Maintenance of proper balance of nutrients is also essential in the avoidance of plugging. Slime formation decreases considerably as the concentration of acetic acid in the finished vinegar increases. For example, in cider or wine vinegar stock which has an adequate quantity of alcohol to produce 10 per cent vinegar, sliming occurs less frequently.

In one experiment using the Frings-type pilot plant vinegar generator, Allgeier *et al* adjusted ethanolic solution flow rate to 12 gal./(hr.) (100 ft.^{3}) of packing and an airflow rate of 55 ft.^{3}/(hr.) (100 ft.^{3}). It is necessary to supply adequate amounts of air, since this is an oxygen-requiring process:

$$C_2H_5OH + O_2 \longrightarrow CH_3COOH + H_2O$$

A typical air flow rate is about 80 ft.^{3}/(hr.) (100 ft.^{3}) of packing. Therefore, the rate of air flow is regulated. And the outlet oxygen concentration is maintained slightly above 12 per cent.

The optimum temperature range for the operation of Frings generator is 80 to 90°F. Heat generation occurs during the bioconversion of ethanol to acetic acid, since it is an exothermic reaction. Therefore, it is necessary to control the temperature within the desired range. Deviation in temperature out of the desired range exerts an adverse effect on acetic acid bacteria. Also, unwanted microbes may grow, producing objectionable substances in the finished vinegar.

Deep fermentation process: Recently, Hromatka and Ebner and Haeseler have introduced this method for acetifying vinegar stock. The 'Quick' vinegar process is now being replaced by the deep fermentation process.

They found *Acetobacter acetigenum* and *A. pasteurianurn* to be better species than other species tested. During their studies, fermentors were operated at 76-82°F with 20 litres of air introduced per hour. A satisfactory pH range for acetification was 3.9-5.0. High yields of acetic acid were produced if the basal medium was added daily to the fermentors than when it was continuously diluted. Optimisation of vinegar process using waste pineapple juice was made by Richardson. According to him, a satisfactory medium has the following components:

Corn sugar
Diammonium hydrogen phosphate
Autolysed yeast
Citric acid
Powdered whey
Potassium chloride

There is difficulty in completing the acetification of vinegar stock in the case of a large-scale fermentation process. It is necessary to have an alcohol content of 0.25-0.30 per cent in the fermenting wash. The vinegar prepared by submerged fermentation process possesses a characteristic flavour and aroma. It has met with favourable customer acceptance. The main defect of vinegar produced by submerged process is its cloudy appearance. Therefore, it is necessary to filter for obtaining clear vinegar.

6 GLUCONIC ACID

Gluconic acid is used as a mild acidulant in metal processing, leather tanning and foods. Sodium gluconate is widely used as a sequestering agent to prevent the precipitation of lime soap scums on cleaned products. Calcium gluconate is widely used in calcium therapy.

Process brief: Gluconic acid is manufactured only by submerged fermentation process: The fungus used is a strain of *Aspergillus niger.* Inoculum consists of either a sporulated culture or spores germinated in seed tanks. Each type of inoculum has special advantages not seen in others. For example, direct spore inoculation avoids the cost of installation and operation of the seed tanks. On the other hand, use of germinated spores reduces the operating cycle for the main fermentor. The acid is quickly neutralized (using $CaCO_3$ or NaOH) upon excretion by the fungus. Therefore, the fermented broth contains either calcium or sodium gluconate.

The raw materials are D-glucose and oxygen. The glucose substrate may be supplied as a solution of crystalline glucose. Alternatively, it may be added as a syrup obtained from starch or crude starchy materials by the action of the α-amylase (from *Bacillus subtilis)* followed by the amyloglucosidase (from *Aspergillus foetidus, A. phoenicis* or *Endomycopsis fibuliger*). The glucose concentration that can be handled depends on whether the sodium or the calcium salt is to be prepared. For example, if sodium gluconate is to be prepared, it is possible to use initial glucose concentrations as high as 28-30 per cent. In addition to carbohydrate substrate, other ingredients, usually added, are (Gastrock, Proges, Wells and Moyer, 1938):

Ingredient	*Amount (gm./litre)*
Corn-steep liquor (Commercial concentrate grade)	3.70
$MgSO_4.7H_2O$	0.17
KH_2PO_4	0.20
Urea	0.10
$(NH_4)_2HPO_4$	0.40
H_2SO_4 to adjust pH at 4.5	
Tap water to make one litre after inoculation; Antifoam agent as needed.	

It is necessary to maintain the maximum concentration of oxygen dissolved in the solution. This is achieved by the application of vigorous agitation with a turbomixer or by the action of a cavitator. The greater the consumption of power by the agitator motor, the more dissolution of oxygen. Table 18.4 gives an idea about the yield of gluconate in relation to air pressure (May *et. al.*, 1934).

Table 18.4 *Yield of gluconate in relation to air pressure.*

Pressure (Atmospheres)	*Weight Yield* (%)*
1	42.5
3	80.4
4	82.4
5	81.3
6	86.1

Operating optimum fermentation conditions in the manufacture of calcium or sodium gluconate are:

Temperature = 28–30°C
Aeration = 1–1.5 volumes air/volume solution/minute
Agitation = Vigorous
Initial pH = ~6.5

The pH is maintained at about 5.5-6.0 by the excess $CaCO_3$ in the production of the calcium salt, or at about 6.8-7.2 by NaOH in the production of the sodium salt.

Recovery: At the end of the fermentation, fungal mycelium is filtered off from the solution. The mycelium is used for the recovery of glucose aerodehydrogenase. On the other hand, the filtrate is used to recover calcium or sodium gluconate, gluconic acid, and glucono-δ-lactone. Recovery of calcium gluconate is achieved by heating the filtrate with a slight excess of $Ca(OH)_2$, then decolourizing with carbon and filtering. On cooling to a temperature below 20°C and seeding with calcium gluconate crystals, the compound readily gets crystallized. A second crop of crystals of calcium gluconate is obtained by the evaporation of mother liquors through heating to about 10-15 per cent volume followed by treatment with carbon, filtration and chilling. On the other hand, sodium gluconate is recovered by concentrating the filtrate to about 42-45 per cent solids. Thereafter, pH of the concentrate is adjusted to 7.5 with NaOH and the salt is drum-dried. To recover gluconic acid and its delta lactone, calcium is precipitated by the addition of a stoichiometric quantity of sulfuric acid. Calcium sulfate is filtered off and the filtrate is decolourized with carbon. After the removal of carbon, the acid solution is subjected to a concentration of 50 per cent acid strength. The product, thus obtained, is a mixture of free gluconic acid and its γ– and δ–lactones. Crystals separating from a gluconic acid solution at a temperature below 30°C and preferably near 0°C, are principally of gluconic acid. Crystals separating from solution at a temperature between 30-70°C are of δ-lactone. Finally, γ-lactone crystallizes out at a temperature above 70°C.

---

* calculated as gluconic acid. Initial glucose 200 gm. $dm.^{-3}$, $CaCO_3$ 50 gm., aeration rate 40 cm^3 200 cm^3 culture, temperature 30°C.

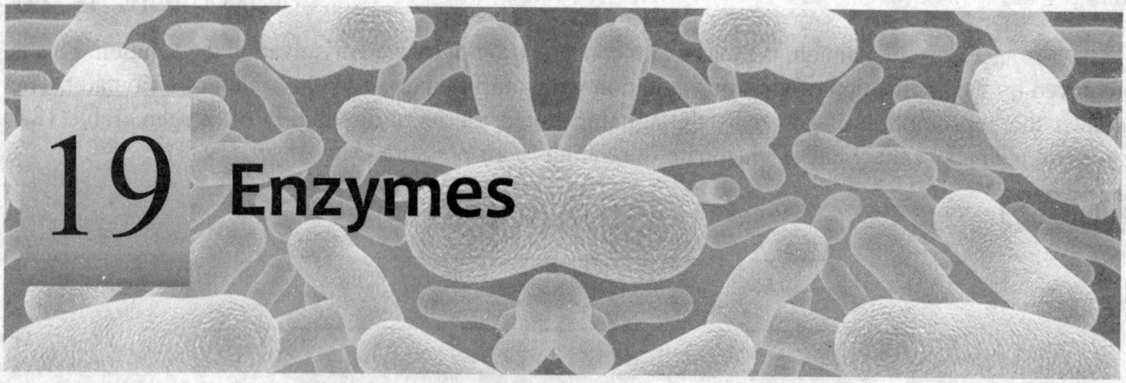

19 Enzymes

1 INTRODUCTION

Uses of enzymes

Enzymes are employed for a variety of purposes, which rely on the specificity and rapid catalysis achieved by these molecules. This description can be divided into three classes: (A) laboratory (B) industrial (C) clinical

(A) Laboratory uses (Table 19.1)

Table 19.1 *Major research and analytical enzymes.*

Name	*Use(s)*
– Alcohol dehydrogenase	– Ethanol estimation
– Lactate dehydrogenase	– Used as a coupling enzyme in alanine aminotransferase assay – Isoenzyme studies
– Malate dehydrogenase	– Used as a coupling enzyme in aspartate aminotransferase assay
– Glucose-6-phosphate dehydrogenase	– Generation of reduced NADP
– Glucose oxidase	– Glucose estimation (Coupled with peroxidase)
– Galactose oxidase	– Galactose estimation
– L–aminoacid oxidase	– Amino acid estimation
– Uricase	– Uric acid estimation
– Peroxidase	– Used as an indicator enzyme for glucose estimation with glucose oxidase
– Pyruvate kinase	– ATP generation for protein biosynthesis

Contd.

Name	*Use(s)*
– Creatine kinase	– ATP generation for protein biosynthesis
– DNA polymerase	– Formation of DNA from mono- and oligo-nucleotides
– Firefly luciferase	– ATP estimation
– Acetyl cholinesterase	– Estimation of neuro active pesticides and neurochemical research
– Ribonuclease T1	– Sequential degradation of RNA for structural determinations.
– DNAase I	– Degradation of DNA for structural determination
– α –Amylase	– Research on possible commercial applications
– β –Amylase	– Research on utilization of natural polymers
– Cellulase	– Research on utilization of cellulose
– Lysozyme	– Structural analysis of microbial cell-walls
– Hyaluronidase	– Tissue disintegration
– Aminopeptidase, Trypsin, chymotrypsin, and carboxypeptidase	– Structural analysis of proteins
– Elastase	– Gerontology research
– Collagenase	– Histology
– Asparaginase	– Possible treatment of acute lymphatic anaemia
– Urease	– Urea estimation. Removal of urea from blood by immobilized enzyme and ion-exchanger

Some enzymes play an important role in analytical techniques. They are often employed on the estimation of a substance which is the substrate of the particular enzyme. Thus, the ethanol, found in one's urine and blood after imbibing alcoholic beverages can be measured using the enzyme, alcohol : NAD oxidoreductase. This enzyme converts ethanol to acetaldehyde, using the coenzyme NAD. This can be shown in the following reaction :

$$\underset{\text{Ethyl alcohol}}{CH_3CH_2OH} + NAD \xrightleftharpoons[\text{oxidoreductase}]{\text{Alcohol : NAD}} \underset{\text{Acetaldehyde}}{CH_3CHO} + NADH_2$$

The consequent increase in ultra-violet absorption by $NADH_2$ at a wavelength of 3400 Å using a spectrophotometer is measured. NAD itself does not absorb ultra–violet light of this wavelength.

Many other biochemicals can be assayed through the action of one enzyme, or alternatively by the use of a series of linked enzyme reactions. For example, the substrate of the first enzyme reaction can itself be assayed[+] or the product modified by having a second enzyme present. Thus, glucose oxidase and peroxidase are used for estimation of glucose in blood. The H_2O_2 produced in the oxidation of glucose by glucose oxidase is used to oxidise a colourless organic compound added to form another compound which can be estimated by measurement of its intense colour using a colorimeter at appropriate wavelength. This reaction of H_2O_2 is catalyzed by peroxidase.

Frequent use of enzymic assay methods is made by hospital laboratories. Here, hundreds of specific estimations on, e.g. blood or urine may be needed daily to help in diagnosis, and control of treatment of patients. Automated techniques are, therefore, extensively employed in this work. One such enzymic assay method is used to estimate the level of certain enzymes in the bloodstream, the transaminases. These enzymes are released by damaged heart muscle following a coronory thrombosis,* and their build-up in the bloodstream is measured to give an indication of the extent of any such damage. The two transaminase enzymes which are assayed are called (a) alanine aminotransferase and (b) asparate aminotransferase. The amount of pyruvate formed by (a), is estimated using another enzyme, lactate dehydrogenase. By the action of this reversible enzyme, pyruvate is converted to lactate and the resulting loss of $NADH_2$ is measured. Similarly, the amount of oxaloacetate formed by (b), is assayed using malate dehydrogenase. Thus, oxaloacetate is converted to malate with loss of $NADH_2$.

It is of interest that lactate dehydrogenase itself is measured in blood-serum, using measurement of $NADH_2$ formed, following damage to the liver. Various tissues in the body have characteristically different enzymically active forms of this enzyme which can exist in five different forms referred to as its *isoenzymes*. The isoenzymes can be resolved by electrophoresis of blood serum. Thus, liver damage can be detected, due to such damage allowing the release into the blood stream of liver lactate dehydrogenase, which is mainly one of the five normal forms. Very little of this form is present in normal blood serum, so that the increase is easily detected. Isoenzymes of lactate dehydrogenase have a quaternary structure which consists of four subunits each a single protein chain. Each of these subunits may be one of two possible types, e.g. A or B, so that in all, five possible combinations are formed, i.e. A_4, A_3B, A_2B_2, AB_3, B_4.

Attempts to automate enzyme assays of substrate is facilitated by using enzymes in insolubilised form, prepared by binding them to a solid support such as a cellulose derivative. Columns of supporting material with enzyme bound to it can be used continuously in the automated systems, so that often very expensive enzymes are reused several times instead of being used only once and then being destroyed by subsequent processing of the products. Enzymes bound to solid supports can be more stable than the corresponding enzyme in solution, provided that a favourable environment is thereby provided, and large-scale industrial uses for these products are being developed.

+ Conditions should be such that substrate of the first enzyme is completely used in the formation of products finally measured.

* Following the blockage of the coronary artery by a blood clot.

(B) Enzymes in industry (Table 19.2)

Table 19.2 *Summary of some common industrial uses of enzymes.*

Industry	*Enzyme*	*Use(s)*
– Baking	– Bromatin, Maltase	– As a baking agent
	– Fungal α - amylase	– General improvement (starch → glucose)
	– Fungal protease	– Soft doughs
	– Lipoxidase	– Whitens bread
– Brewing	– Bacterial α-amylase and bacterial protease	– Use of unmalted barley and liquefaction of cereal adjuncts
	– Fungal amyloglucosidase	– Removes dextrans from wort or sweetens beer
	– Ficin	– To liquefy starch and proteins during the treatment of beer
	– Papain or β-glucanase	– Chill haze stability lowers viscosity of wort or beer
	– Maltase	– To convert starch into sugar like maltose
	– Cellulase	– Hydrolysis of cellulose or celluloid material
– Syrups	– Bacterial α -amylase or Fungal amyloglucosidase	– To solubilize starch at high temperatures to complete the conversion of starch to defined syrups
	– Invertase	– Sucrose ⟶ invert sugar
	– Glucose isomerase	– Glucose ⟶ invert sugar
– Cheese	– Rennin or Marzyme	– To precipitate the protein (casein) from milk
	– Penicillinase and Nicinase	– Removal of penicillin and nicin from milk if present
	– Catalse	– Removal of added H_2O_2 from milk
– Ice-cream	– Lactase	– Removal of lactose that might crystallize out and prevents grittiness
– Margarine	– Lipases	– Flavour improvement
– Dried milk	– Proteases	– Slight degradation improves dispensibility and inhibits coagulation. Also some liberation of tyrosine from casein
– Fruits	–Pectinase	– Many improvements in processing and products, e.g. clarification or colour and cloud replacement
– Vegetables	– Cellulases	– Softening and flavour promotion
– Eggs	– Glucose oxidase and Lipase	– To remove glucose before drying. Whipping properties of egg white

Contd.

Industry	*Enzyme*	*Use(s)*
– Meat	– Papain	– To tenderize meat : the action of this fairly heat resistant enzyme can continue while the meat is warming. The enzyme is, of course, inactivated as cooking proceeds.
	– Fungal protease	– Also canning industry
– Textile	– Bacterial α-amylase or Rhozyme	– Removal of starch (sizes) added during processing, i.e. desizing
	– Bacterial proteases	– Recovery of wool from sheep skin pieces
– Paper	– Bacterial α-amylase or Rhozyme	– Used to degrade starch to lower-viscosity product needed for sizing and coating processes
– Laundering	– Bacterial proteases	– Enzyme-containing washing powders (protein stain removal)
– Rubber	– Catalase	– To generate oxygen from peroxide to convert latex to foam rubber
		– To decompose fat
– Tanning industries	– Bacterial proteases	– Loosening of hair without damage to hide or hair. Leather treatment (bating)
	– Papain	– As a hair remover (dehairing hide)
– Photographic	– Ficin	– To dissolve gelatin of scrap film allowing recovery of silver present
– Sewage disposal plant	– Lipase	– To split up of fat molecules present in sludge like material
– Canning industries	– Pectinase	– As a hydrolyzing agent and as a depolymerizer of pectin molecules present within the substrate
	– Glucose oxidase	– As an oxidizing agent
– Pharmaceuticals	– Cellulase	– To hydrolyze cellulose or celluloid material
	– Diastase	– To liquefy and hydrolyze starch, i.e. as a digestive aid.
	– Pancreatine	– To hydrolyze fat, starch and proteins i.e. as a digestive aid
	– Papain	– Used in tooth powders to remove protein deposited on teeth
	– Bromatin	– Hydrolysis of organic complex made up of protein components other than protein of enzyme molecules in meat

Contd.

Industry	Enzyme	Use(s)
– Food industries	– Trypsin	– To predigest baby-foods
	– Pepsin	– Hydrolysis of protein components present in cereals
	– Maltase	– To convert starch or polysaccharides into sugar like maltose (in sweet drinks)
	– Bacterial proteases	– Preparation of hydrolysed proteins, e.g. for animal feedstuffs
	– Catalase	– Removal of H_2O_2
	– Pectinase	– In the treatment of wine and preparation of jellies
	– Cellulase	– Used in the treatment of citrus material, oil, and fat to hydrolyze cellulose or celluloid material

Recently food manufacturers have started taking advantage of commercially available enzymes. Enzymes are relatively non-toxic. They have great specificity of action and work excellently under conditions of moderate temperature and near neutral pH. They act rapidly and their action could easily be arrested when desired. A variety of beneficial changes can be achieved in the taste, texture, colour, clarity and solubility of various products, by the use of enzymes to modify the constituent molecules of these foods.

Many of the enzymes used fall into the category of hydrolases, which are enzymes that add the elements of water to particular chemical bonds, e.g. in macromolecules, causing degradation to smaller fragments. Amylase is an example of this class of enzyme. Enzymes with amylase activity may be isolated from bacteria and fungi. They degrade starch, which is ploymer of glucose, mainly to maltose the corresponding dimer, and to glucose itself. Amylases are used in the baking industry to treat flour, prior to bread-making. The resulting loaf of bread has many desirable properties. These include an improved taste, as well as a larger volume. The latter is caused by the extra CO_2 released by fermentation, by the baker's yeast, of the additional sugars produced from the starch owing to amylase action. Amylases are also employed in the production of sweet syrups from various raw starches. The process is very economical. Similarly, pectic enzymes are used for the degradation of pectic substances and help in increasing the yield of juice extracted from various fruits. When added to the juice itself, they reduce its viscosity and help in clarifying juices, particularly wines.

Proteases, the protein hydrolysing enzymes, are employed in dairy industry for preparing cheese, in bakery for bread manufacture, in brewery for chill-proofing, and most commonly as a tenderizing agent for meat. An example of tenderizing action of enzymes in meat is provided in the household use of crushed raw papaya fruits containing papain. The latest development in meat-tenderizing is the injection of proteolytic enzyme solutions into the vascular system of animals before slaughter which leads to uniform distribution of enzymes throughout the body tissues.

Papain is also employed in brewing industry. First, barley is converted to malt by the traditional malting process. A mash is prepared, which is suitable for yeast growth and fermentation of the sugars present, such as maltose, to ethanol plus carbon dioxide. The product is, of course, beer. The public prefer this beverage to be clear, not hazy. Unfortunately beer, especially when cooled, can develop a cloudiness, known as chill-haze. Prevention of haze formation is possible by the addition of papain to the beer. This proteolytic enzyme breaks down a protein constituent of beer, so preventing the formation of the haze, which is a proteintannin complex.

One of the proteolytic enzymes used in the food industry (dairy) is also used in the home. This is rennin, and it is isolated from calf stomach. This enzyme attacks the major protein component of milk, which is casein. The attack causes the molecules of casein to aggregate, due to a small change in structure, and the casein curd is precipitated from the milk. Rennin is the active principle of junket-making preparations used in the home. Its industrial use is in the manufacture of cheese from milk, and the rennin (also called rennet) is used to produce the necessary curd.

Other enzymes used in the food industry are glucose oxidase and catalase. Here glucose oxidase can be used to remove glucose from, for example, eggs before the manufacture of dried egg powder. If the glucose is not removed, brown colouration is produced by reaction of the glucose with other components of the egg, such as amino acids. It is of interest that this enzyme requires one molecule of oxygen to oxidise one molecule of the glucose. Thus glucose oxidase will remove traces of oxygen from foods if required, provided that sufficient glucose is present. The H_2O_2 produced in this reaction is degraded to water and oxygen by the action of the enzyme, catalase that is used in conjunction with glucose oxidase. Eventually, all the oxygen present in the system can be used up with no accumulation of H_2O_2. Many products lose their sales potential upon storage, if oxygen is present. For this reason, dried foods and mayonnaise may be treated with glucose oxidase plus catalase, for the removal of oxygen.

Another enzyme is of considerable interest because it solved major problem in the confectionery industry-how to put a chocolate coating around a liquid sugar core to produce soft-centered chocolates. Of course, it is easier to coat a hard sugar core and this is done—but an enzyme is added. This enzyme is called invertase, and it slowly hydrolyses the sucrose molecules into its two parts, one molecule of each of glucose and fructose. These two very soluble products can dissolve in the small amount of water present to give the liquid center. The name of the enzyme, invertase, derives from the name of the equimolecular mixture of glucose and fructose it produces. The mixture is known as invert sugar because the conversion of sucrose to glucose and fructose results in 'inversion'. This is the reversal which occurs in the direction of rotation of the plane of polarized light passed through the system. It is due to the difference, in optical activity, between sucrose and its products. Invertase is isolated from yeast, on an industrial scale, by the use of a brief treatment with papain to release invertase from debris produced by disruption of the yeast cells with hot toluene.

Lipases from another important class of lipolytic enzymes employed in diary industry. Controlled lipolysis is necessary in cheese-making to develop characteristic flavour and aroma.

There are several other enzymes which are finding increasing use in food industry. It will, therefore, be worthwhile if new, cheap sources for large-scale commercial production of these enzymes are explored.

Many other industries in addition to the food industry, employ enzymes. For example, enzymes are used in the pharmaceutical, leather, paper, textile, rubber, canning and photographic industries. Most of the uses to date have been relatively simple ones. Much fundamental knowledge about an enzyme is needed to harness it for more sophisticated purposes, and further studies can provide this information. Obviously an understanding is required of the specificity, stability, pH and temperature optima, and possible inhibition, if successful use of an enzyme is to be achieved.

(C) Clinical uses of enzymes

Some enzymes are employed in the actual treatment of diseases of man. As an often-quoted example, one could envisage the treatment of the small proportion of the population who are allergic to penicillin. The bacterial enzyme, penicillinase, could be injected to destroy the penicillin, if allergy were indicated. It is of interest that it is the production of precisely this enzyme by certain penicillin-resistant strains of bacteria which allows them to destroy the penicillin being used against them. Penicillin normally stops the bacterium from growing (and dividing) by interfering with its ability to finish building its cell wall. Penicillinase attacks the penicillin molecule and breaks one of the two rings of atoms it contains, thus destroying its anti-bacterial action. The enzyme lysozyme is itself an antibiotic, which attacks the carbohydrate of the cell-walls of certain bacteria. It is an antibiotic found in human tears, and is used in the therapy of eye infections. Large quantities of lysozyme can be isolated from hen egg white.

Certain proteolytic enzymes are employed for therapy, one of the most widely used of which is trypsin. This is used to clean wounds by dissolving body secretions including blood clots. Let us end with mention of an enzyme which has been used in the treatment of patients with some types of leukemia, a form of cancer. The therapeutic value of this enzyme can be explained, in simple terms, by the following mechanism. It was discovered that the leukaemic white cells found in the blood stream were unusual in needing a continuous supply of asparagine, from the blood, for survival. Asparagine is the amide form of the amino acid, aspartic acid, and has the formula:

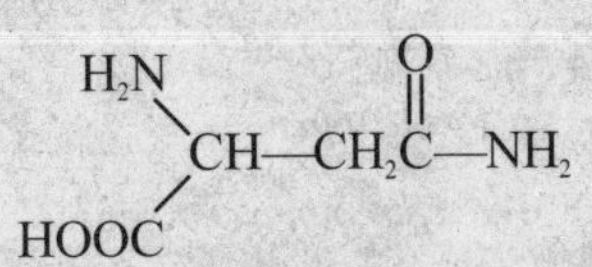

This amino acid is usually made inside the cell, from aspartic acid, and is incorporated into proteins by the usual mechanisms. This enzyme may be isolated from the bacterium *E.coli.* Injection of the enzyme, asparaginase, into the blood stream, hydrolyses the asparagine to aspartic acid. But due to a metabolic derangement the leukaemic white cells cannot use this-unlike normal cells.

Microcapsules containing L-asparaginase have been found to suppress the growth of certain asparagine-dependent tumors in mice (Chang, 1971). When the soluble enzyme was injected into the mice, the tumors were not suppressed because of the organisms' immunological rejection of foreign protein. A similar procedure was employed to treat acatalusaemic mice with encapsulated catalase (Chang *et al.* 1968). Chang suggests the use of encapsulated enzymes in cases of clinical abnormalities associated with enzyme deficiencies due to in-born errors of metabolism, and for removing uremic metabolites. This technique is still in its infancy, but holds great possibilities for the future.

It must be admitted that there are many problems, even in the apparently simple cases described above, of therapy with enzymes. Nevertheless, the expectation is that enzymes will be used in treatment of many diseases in the future. In some cases, they may be used in their crude forms. But at other times, they are used in highly purified states (e.g. urease for urea estimtaion).

Microbial enzymes and their corresponding producing micro-organism(s) are listed in Table 19.3.

Table 19.3 *Microbial enzymes and their corresponding producing micro-organism(s).*

Enzyme(s)	*Micro-organism(s)*
α-Amylase	*Aspergillus oryzae, A. niger, Bacillus subtilis, B. amyloliquefaciens.*
Glucoamylase	*A. niger, Rhizopus niveus, R. delemar, Endomycopsis* spp.
Protease	*A. oryzae, A. niger, A. saitoi, Mucor pusillus, B. subtilis, Streptomyces griseus*
Lipase	*A. niger, Rhizopus* spp. *Candida cylindracea*
Cellulase	*A. niger, Trichoderma viride*
Pectinase	*Coniosthyrium diplodiella*
Glucose oxidase and others	*Penicillium amagasakiense*
Invertase	*Saccharomyces cerevisiae*
Naringinase	*A. niger*
Melibiase	*Mortierella vinacea*
Aspartase	*Escherichia coli*
Aminoacylase	*A. oryzae*
Antocyanase	*A. niger*
Glucose isomerase	*Streptomyces* spp.
Catalase	*A. niger*

The future is extremely promising for further uses. Entirely new enzymes will be tailor-made for particular uses in all three fields.

2 GENERAL ASPECTS OF ENZYMES' PRODUCTION

Enzymes are commercially produced by two methods:

(1) semisolid culture

(2) submerged culture

There is great competition among enzymes' manufacturers. Therefore, manufacturers are reluctant to reveal the process details. It is not possible to undertake a survey of the production methods being used. But, it is evident that the submerged culture method has been gaining ground during the past three to four decades. These two methods have been briefly dealt with here.

(1) *Semisolid culture*: The enzyme producing culture is grown on the surface of a suitable semi-solid substrate. The substrate usually consists of moistened wheat or rice bran, supplemented with nutrient salts. The production medium is prepared by mixing bran with a solution containing any desired nutrient salts. The desired pH for optimum growth of the mould is adjusted with acid. Then, the medium is steam-sterilized in an autoclave while stirring. This sterilized medium is spread on metal trays up to a depth of 1-10 cms., the total quantity being of the order of thousands of kilogrames. Such a transfer is performed under aseptic conditions. Alternatively, the cultivation may be carried out in rotating drums.

The fungal spores are inoculated, either in the autoclave after cooling or in trays. A series of trays are enclosed in a large vessel. Aeration is ensured by the circulation of suitably humidified air over the surface of the culture. It is necessary to keep the temperature within narrow limits. Moreover, heat generation occurs during fermentation. Therefore, the trays should be equipped with a cooling system. It should be borne in mind that, direct air cooling is not practical, since drying of the culture takes place. Subsequently, extraction of the enzyme is performed with water. Enzymes produced by a semisolid culture, or a surface culture process are:

Enzyme	*Micro-organism(s)*
α-Amylase	*Aspergillus oryzae*
Glucoamylase	*Rhizopus* spp.
Lactase	*A. oryzae*
Pectinase	*A. niger*
Protease	*A. niger* and *A. oryzae*
Rennet	*Mucor pusillus*

Precautions to be taken:

(i) It is necessary to keep aseptic conditions during the fermentation, since contaminations are considered a major problem in a semisolid culture.

(ii) Large numbers of spores should be prevented from escaping into the environment.

Advantages:

(i) This method involves comparatively low investment.

(ii) It allows the use of substrates with a high dry matter content. It, therefore, yields a high enzyme concentration in the crude fermented material.

(iii) It has to be used for cultivating some moulds which are very difficult to grow in fermentors due to wall growth.

(iv) This method allows the moulds to develop in their 'natural' state. This results in the normal differentiation in mycelium, conidophores, conidia, etc. and results in a broader variety of enzymes being formed than in submerged fermentation. Such a phenomenon is desirable in some cases (e.g. in the production of pectic enzymes).

Disadvantages:

(i) This method requires more space.

(ii) It requires more labour.

(iii) It involves a greater risk of infection.

(iv) Modern control methods are not readily applicable to this fermentation system.

(v) It is difficult to introduce automation in such systems.

(2) *Submerged culture:* Nowadays submerged culture methods are widely used in the production of enzymes. The fermentation equipment used is the same as in the manufacture of antibiotics. It is a cylindrical tank of stainless steel. The tank is equipped with an agitator, an aerating device, a cooling system and various ancillary equipment, (e.g. means of foam control, monitoring of pH, temperature, oxygen tension, etc.). The quantity of production medium taken in the fermentation tank is in the range of 1000-30,000 gallons or more.

With enzyme fermentations, the formulation of the production medium and to a lesser extent, control of fermentation conditions play major roles in the success of the process. In other words, the chemical composition of production media makes these processes tricky. The production medium should basically contain an energy source, carbon and nitrogen sources and any special growth requirements, (e.g. essential amino acids or vitamins). However, good growth is not enough to obtain a high enzyme yield. Certain compounds present in the production medium may induce or inhibit enzyme formation. For example, the presence of lactose (inducer) in the medium induces β-galactosidase. Other examples of inducers are given in Table 19.4.

Table 19.4 *Effect of various inducers on yields of selected enzymes.*

Enzyme	*Culture*	*Inducers*	*Yield in international units*
Cellulase	*Trichoderma viride*	Cellulose	22.5
		Cellobiose	0.2
		Cellobiose diplamitate	4.8
Dextranase	*Penicillium funiculosum*	Dextrans	1080
		Isomaltose	2
		Isomaltose dipalmitate	1098
Invertase	*Aureobasidium pullulans*	Sucrose	1.3
		Sucrose monopalmitate	108

As the inducers are rather expensive, it is preferable to use constitutive mutants which do not require the inducer. Such mutants have been developed for several enzymes. On the other hand, certain compounds present in the medium act as repressors for specific enzymes. For example, glucose represses the formation of some enzymes, (e.g. α-amylases). Under such conditions, the glucose concentration should be kept low. This can be achieved either by incremental feeding of glucose or by using a slowly metabolizable sugar, (e.g. lactose or partly hydrolyzed starch).

The presence of certain surfactants in the production medium increases the yields of certain enzymes. Non-ionic detergents, (e.g. Tween 80) are frequently used. Table 19.5 shows the effect of 0.1% Tween 80 on the yield of various enzymes.

Table 19.5 *Effect of 0.1% Tween 80 on the yield of various enzymes.*

Enzyme	*Culture*	*Yield in medium + Tween 80 / Yield in medium (ratio)*
Cellulase	*Trichoderma viride*	20
Invertase	*Aureobasidium pullulans*	16
Amylase	Numerous fungi	5
Pullulanase	*Enterobacter aerogenes*	1.5

Occasionally Triton (Rohm and Haas) produces better results. But the mode of action of these surfactants is not understood.

Most enzyme fermentations are carried out near neutral pH. Therefore, it is necessary to control the pH within the desired limits during fermentation. This can be achieved by adding a buffer system, (e.g. phosphates or calcium carbonate to the medium). The alternative to this is to add certain compounds, which upon metabolism, bring about a change in pH in the desired direction. This occurs due to the formation of either acid or base, as the case may be. Thus, salts of organic acids and nitrates will tend to raise the pH, while ammonium salts will tend to lower the pH.

Economy is very important in a medium formulation. The raw material costs account for 60-80% of the variable costs in a typical fermentation process. Common agricultural raw materials, which are available in a good and consistent quality, are generally preferred. For this purpose, defined or synthetic media may be used to elucidate the nutritional requirements of the culture for the formation of a desired enzyme. Once what is indispensable is understood, a search can be made for cheaper materials, giving the same result. Moreover, empirical determination of the optimum concentration of each component of the medium involves a lot of work. The enzyme recovery steps should also be taken into consideration when formulating the production medium. The composition should be such, that at the end of the fermentation the suspended solids and viscosity are as low as possible. Moreover, materials which could confuse separation of solid and liquid phases, should not be present.

Most production media are still steam sterilized batchwise in the fermentor. Complete sterilization of the medium is ensured at approximately 120°C for 1-2 hours. Alternatively, the production medium can be sterilized separately by continuous sterilization. The latter method offers certain advantages: less colour formation due to Maillard reactions between proteinaceous materials and carboyhydrates and superior fuel economy when heat recovery is practised. In addition, dilution of the production medium due to the condensate is avoided.

After the completion of fermentation, the fermented liquor is subjected to rapid cooling to about 5°C to reduce deterioration. Separation of micro-organisms is accomplished, either by filtration or centrifugation of refrigerated broth with adjusted pH. The colloidal particles present in the filtrate are eliminated with coagulating or flocculating agents, (e.g. calcium phosphate). Diatomaceous earth (2-4%) may be added to the fermented broth as a body feed before filtration operation. Removal of the suspended solids is carried out by vacuum drum filtration or by a disc-type centrifuge equipped with a self-cleaning bowl. In order to obtain a higher degree of purity, the enzyme is precipitated with acetone, alcohols or inorganic salts, (e.g. ammonium or sodium sulfate). Fractional precipitation gives purities higher than one-step precipitation. In the case of large-scale operations, salts are preferred to solvents because of explosion hazards.

Extraction of endocellular enzymes involves the disintegration of the microbial cells. This can be accomplished by a homogenizer or a bead mill. Thereafter, purification methods being employed for exoenzymes are used for purification. The different steps involved in the purification of endocellular enzymes are schematically shown in Figure 19.1.

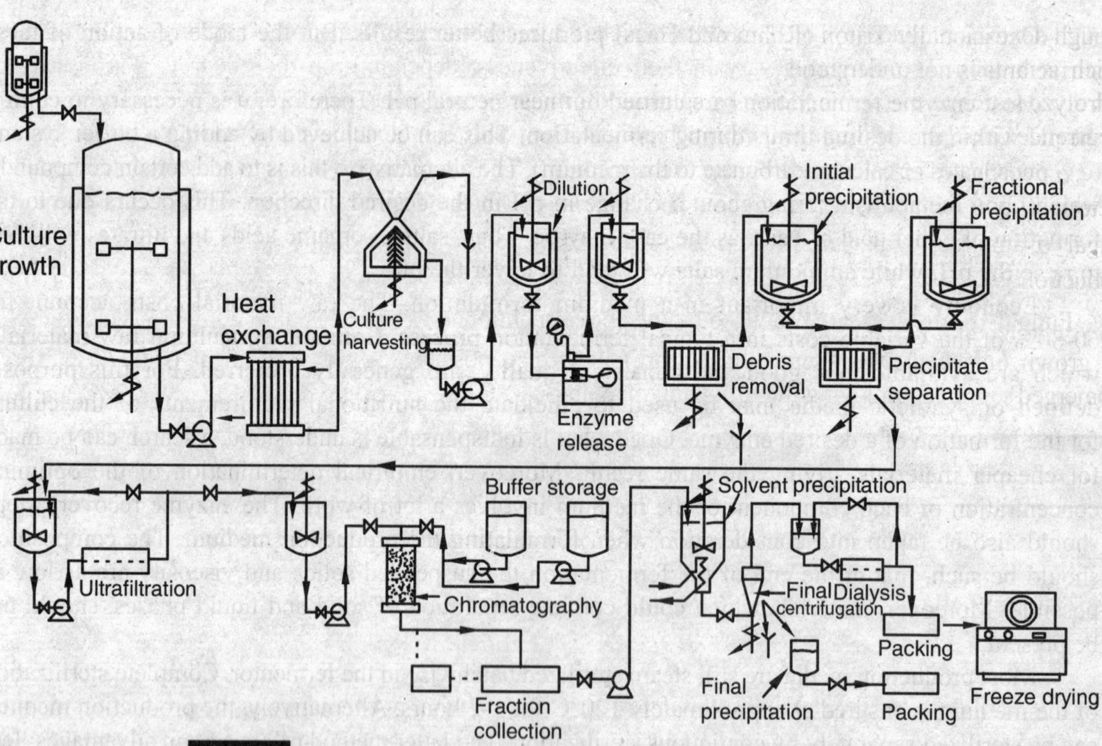

Fig. 19.1 Flow sheet for production of purified endocellular enzymes

Advantages:

(i) This method requires less space and labour.
(ii) The method involves lower risk of infection.
(iii) Modern control methods can easily be used.
(iv) Automation in such methods is easier.
(v) The yields are also generally higher on a dry matter basis.

Disadvantage:

(i) Initial investment cost is high.

3 AMYLASES

Amylases play the most important part in food technology, (e.g. bread-making, beer-making, etc.). Therefore, concentrates of α - and β -amylases are prepared and used in a variety of ways. These enzyme preparations must be carefully standardized for activity, according to the purpose for which they are to be used.

The amylases constitute a large group of enzymes. They are characterized by their ability to hydrolyze 1, 4-glucosidic linkages in polysaccharides, (e.g. starch and glycogen). There are two main subgroups: α-amylases and β-amylases. α-Amylases are endo-enzymes. They attack all the linkages between glucose units in the starch molecule. The bond hydrolyzed is that between C-1 and the oxygen atom linked to the adjacent glucose group. The process eventually results in complete degradation, through dextrins to glucose. Thus, amylose, which is a linear starch, is degraded faster than amylopectin, which is branched. α-amylases vary in their effectiveness, depending on their source. β-Amylases hydrolyze starch and other amyloses by splitting off maltose molecules until the action is blocked by the occurrence of either 1, 3 linkages or branch points. The residual molecule is then called a limit dextrin.

α-Amylases are produced by the use of fungi, (i.e. *Aspergillus niger* and *A. oryzae),* as well as bacteria, (i.e. *Bacillus amyloliquefaciens* and *B. licheniformis*). Therefore, α-amylases are called either fungal α-amylases or bacterial α-amylases according to the nature of the microbes used for their production.

Fungal α-amylases are produced by the above-mentioned two species of the fungus. These fungi are grown on wheat bran (semisolid culture). It is also possible to produce fungal α-amylases by submerged-culture, employing the following medium:

Component	*Amount (gm./litre)*
Corn starch	24
Corn-steep liquor	36
KCl	0.2
Na_2HPO_4	47
$CaCl_2$	1
$MgCl_2.6H_2O$	0.2

There is a problem for aeration and agitation because of a very high viscosity of the medium due to the presence of mycelial body. Amylases biosynthesis is inhibited when the medium contains glucose.

Bacterial α-amylases are produced by the above-mentioned two bacterial species. Bacterial amylase is produced only by submerged-culture using the following medium:

Component	*Amount (%)*
Ground soya bean meal	1.85
Amber BYF (autolyzed brewers' yeast fractions, Amber Laboratories)	1.50
Distillers' dried solubles	0.76
N-Z amine (enzymic casein hydrolyzate, Sheffield Chemical Co.)	0.65
Lactose	4.75
$MgSO_4.7H_2O$	0.04
Hodag KG-1 antifoam (Hodag Chemical Corp.)	0.05
Water	90.40

A temperature in the range of 30° to 40°C is satisfactory. The optimum pH for the fermentation medium is 7.0. It is necessary to maintain the pH near neutrality, since the amylase is denatured below 6. Calcium carbonate is used as the buffer to maintain neutral pH. The production of α-amylase begins when the bacterial count reaches 10^9–10^{10} cells per milliliter after about 10-20 hours, and continues for another 100-150 hours. Preservation of liquid preparations of bacterial α-amylases is done by 20% sodium chloride. The most active preparations contain 2 per cent active amylase protein. On the other hand, the most active solid preparations contain 5 per cent active amylase protein.

4 PROTEASES

Complex mixtures of true proteinases and peptidases are usually called proteases. The concentrations of the peptidases in the production medium is low, since they are endoenzymes. Proteases, like α-amylases, are produced by bacteria, (i.e. *Bacillus subtilis* and *B. licheniformis*) as well as by fungi, (i.e. *Aspergillus niger* and *A. oryzae*). It is essential to take care during their production, because they are relatively unstable, and tend to lose their activity during dehydration. There are two types of proteases: (a) alkaline serine proteases and (b) acid proteases. The production methods for both these proteases are briefly discussed here:

(a) *Alkaline serine proteases:*

Subtilisin Carlsberg is the most widely used detergent protease. It is obtained from *Bacillus licheniformis* by submerged-culture method. The bacterium is grown on the medium with the following composition:

Component	*Amount (gm./litre)*
Starch hydrolyzate	50
Soya bean meal	20
Casein	20
Na_2HPO_4	3.3

Other media are available for this purpose. The temperature of fermentation in the range of 30° to 40°C has been found to be satisfactory. The pH of the production medium is kept 7.0 for satisfactory results. The production of the enzyme begins when maximum cell growth is reached after 10 to 20 hours. And this continues at an almost constant rate till the completion of fermentation. At the end of the productive fermentation, protease is only the protein present in the production medium. The reason for this is the occurrence of hydrolysis of all the proteins present in the medium by protease. The yield may be 10 per cent of the initial protein content of the medium. The enzyme is marketed primarily in the form of dust-free granules. These granules contain 1-5 per cent enzyme protein. The enzyme remains stable in liquid preparations. Liquid preparations contain about 2 per cent of the enzyme.

(b) *Acid proteases:*

These enzymes are mostly produced by fungi. The fungi employed for producing these enzymes are: *Mucor pusillus, M. miehei, Aspergillus oryzae, A. phoenicis, A. niger* var. *macropus.*

Acid proteases may be produced by either semisolid-culture or submerged-culture, depending upon the fungal species employed. For example, *Mucor pusillus* is cultivated on a semisolid medium. The medium consists of 60 per cent wheat bran with water. The optimum temperature of the fermentation is 30°C. The fermentation requires 3 days for completion. The yield is 3,200 Soxhlet units per gram of wheat bran. The yield of the enzyme can be increased by the addition of ammonium salts. Finally, the enzyme is extracted with water. On the other hand, *Mucor miehei* is grown by submerged-culture method. The production medium used for this purpose has the following composition:

Component	*Amount (%)*
Starch	4.0
Soya bean meal	3.0
Ground barley	10.0
Calcium carbonate	0.5

The optimum temperature of the fermentation is 30°C. The fermentation is completed in 7 days. The yield is about 3,500 Soxhlet units per millilitre of the fermented broth.

The enzyme preparations are marketed at concentrations of 10,000-1,50,000 Soxhlet units per milliliter. These preparations contain 0.2-3.0 per cent of active enzyme.

5 PECTINASES

Pectinases are produced by species of *Aspergillus,* (*e.g. A. niger*) and *Penicillium.* These fungi are grown on beet sugar molasses because this material contains pectins. Semisolid culture can be used to produce pectinases employing *A. niger.* At the end of the fermentation, drying and pulverization of the mycelial material is carried out. Subsequently, a mixture of pectin esterase, polygalacturonase, polymethylgalacturonase and pectin transeliminase is extracted with the help of cold water. After the extraction is completed, the crude enzyme mixture is subjected to concentration and refrigeration without solidifying.

6 OTHER ENZYMES

In addition to α-amylases, proteases and pectinases, many other enzymes, (e.g. naringinase, glucoamylase, lipases, etc.) are commercially produced. It is beyond the scope of this book to discuss the details of the production of these enzymes.

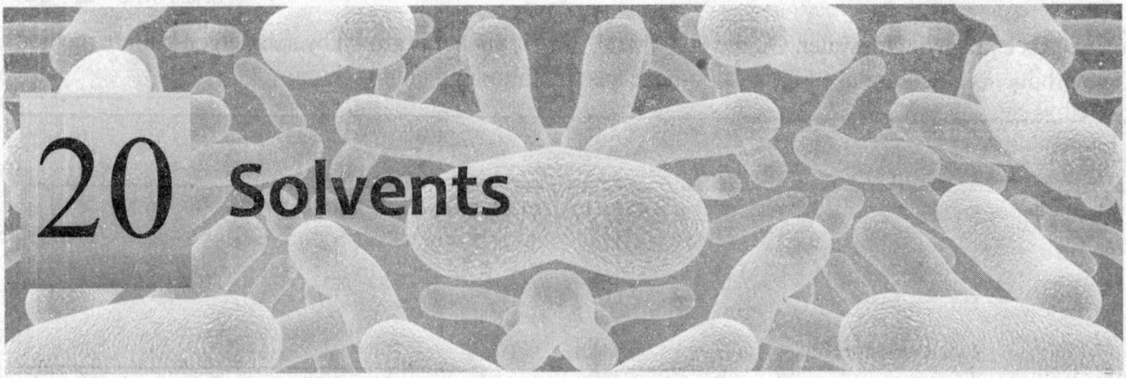

20 Solvents

1 INTRODUCTION

Industrial solvents, (e.g. ethanol, glycerol, acetone, butanol and 2, 3-butanediol) are widely used in chemical as well as fermentation industries. For example, ethanol is used as a solvent and a precursor of many synthetic chemicals. On the other hand, extraction procedures of certain antibiotics from fermented broth, (e.g. penicillin) need suitable solvents. These chemical compounds have been produced by fermentation processes. Today, fermentation processes have virtually disappeared as a result of the development of chemical ones. This has happened for the following reasons:

(i) high prices of the fermentable substrates

(ii) increasing economic competition from the petrochemical industry

These chemical compounds are produced as by-products during the refining of petroleum.

2 ETHYL ALCOHOL PRODUCTION

Alcohol fermentation was the first fermentation known to mankind. Distilleries began to appear in Europe in the middle of the seventeenth century. At first, alcohol was used only for human consumption. Later, the demand for alcohol as a universal solvent and chemical raw material developed. The distilling industry, therefore, grew very rapidly. The annual production in the United States alone has now reached one billion gallons.

Micro-organisms: The micro-organisms responsible for alcoholic fermentation of molasses are yeast species of *Saccharomyces, Torulopsis, Kloekera, Candida,* and certain species of *Mucor* and bacteria. The most important among them are the yeast species belonging to genus *Saccharomyces.* There are two types of *Saccharomyces* species, one is a top fermenter and the other a bottom fermenter. Top fermenter yeasts, (e.g. *S. cerevisiae)* are active at the top surface of the fermenting liquid. They remain either in an evenly distributed layer, or as a ring along the walls of the fermentation vat. Bottom fermenters, (e.g. *S. carlsbergensis)* develop only at the bottom of the fermentation vats. They are whirled up in the liquid owing to carbon dioxide evolution. The two classes, however, vary in their morphological and biochemical characteristics.

The strain of yeast to be used for alcohol fermentation should possess the following selective features:

(1) It should be an efficient strain. In other words, it should produce a large quantity of alcohol.

(2) It should be a fast growing strain.

(3) It should have a high tolerance to alcohol, as well as to osmotic pressure. The reason for this is that the yeast should continue its activities and accumulate large amounts of alcohol in a concentrated medium.

(4) It should possess uniform and stable biochemical properties.

Usually the strains of yeast, *Saccharomyces cerevisiae,* are employed for commercial production of ethyl alcohol. Special strains of this yeast have been developed for industrial ethyl alcohol fermentation by genetic selection.

However, the choice of micro-organism depends upon the type of raw material to be used for alcohol fermentation. Generally, strains of *Saccharomyces cerevisiae* are selected when starchy material, as well as saccharide material like molasses are used as raw material. Strains of *Candida* are preferred when whey and waste sulphite liquor are employed as raw material in alcohol fermentation.

Preparation of inoculum: A large amount of yeast culture is required in the industrial production of ethyl alcohol. The size of inoculum ranges from 8 - 10 per cent with an average of 4 per cent volume by volume. The medium employed in the preparation of inoculum is the same as that of the actual fermentation process. A high degree of aeration and agitation is necessary, since this favours rapid growth of yeast, producing a large amount of cell-mass. The optimum pH is 4.8 to 5. The optimum temperature is 28° to 30°C. Steps involved in the building up of inoculum are:

(a) 10 ml. of sterile medium in a test-tube is inoculated with pure yeast culture. The tube is incubated at 28° to 30°C for a suitable period.

(b) The tube culture is transferred to a flask, containing 200 ml. of sterile medium. The flask is incubated at 28° to 30°C until pre-determined cell-mass is obtained.

(c) The flask culture is, in turn, transferred to a suitable glass container, containing about 4 litres of sterile medium and incubated for an appropriate length of time.

(d) Finally, the contents of the glass container are equally transferred into small tanks (yeast vessels) installed closely to a fermentor. Each small tank contains 10 to 40 gallons of sterile medium. After incubation the inoculum is ready for inoculating the fermentor. The inoculum is added to the fermentor by means of pumping or gravity. This operation of addition of inoculum to the fermentation medium is called 'pitching'.

Raw materials: A variety of carbohydrates available as waste-products of agricultural industries are used for the preparation of fermentation medium. On the basis of their chemical nature, they may be grouped as:

(1) saccharide materials—molasses, hydrol, whey, technical glucose, fruit juices

(2) starchy materials—potato starch, corn starch, wheat flour, cereals, (e.g. oats)

(3) cellulosic material—sulphite waste liquor and others

Preparation of fermentation medium: The raw material used for the preparation of fermentation medium depends upon the agricultural produce of the country. For example, cane-molasses is used in India. Generally, molasses is used as the main raw material in the preparation of a fermentation medium. However, starchy materials and cellulosic materials may also be used for the preparation of fermentation medium. These materials require treatment before their use. For example, starchy material is subjected to saccharification, where starch gets hydrolysed, thus forming fermentable sugars, maltose and glucose.

The optimum sugar concentration ranges between 10 to 18 per cent in the fermentation medium. Also, a suitable source of nitrogen, such as ammonium sulphate, is added in optimum concentration. When beet-molasses is used, usually a small quantity of black-strap molasses is also added to meet the biotin deficiency in the beet-molasses. The pH of the medium is adjusted to 4.8 to 5. The pH can be adjusted by the use of H_2SO_4 or lactic acid. Alternatively, it can be adjusted by growing lactic acid bacteria prior to fermentation. It is not necessary to sterilize the fermentation medium. This is mainly owing to highly acidic pH which hinders the growth of undesirable bacteria. However, pasteurization of the fermentation medium may be practised.

Fermentation conditions: For the satisfactory progress and completion of the alcohol fermentation, the following conditions are carefully controlled:

Carbon sources: Sugar concentration in the range of 10 to 18 per cent is found satisfactory. Sugar concentration of 12 per cent is often used. For example, cane molasses contains sucrose (cane sugar) as a carbon source. The sucrose content of this raw material is about 48 to 55 per cent. Wort (molasses diluted with water) is prepared in such a way as to contain about 10 per cent sugars. High concentration affects yeast adversely while low concentration makes the process uneconomical.

Nitrogen sources: Many organic, as well as inorganic nitrogenous compounds may be used in this fermentation. Ammonium sulphate (0.15 gms. per 2.5 gallons of molasses) is generally used with success to supply nitrogen for growing yeast cells. The level of nitrogen in the medium is limited. Excess nitrogen is objectionable, since it promotes the growth of yeast, inhibiting the fermentation.

Growth factors: It is not necessary to add the growth factors separately in the preparation of the fermentatión medium, since the raw materials, (e.g. cane-molasses) used satisfy the requirements of the culture.

pH: pH of the fermentation medium is adjusted to 4.8 to 5. The higher values of pH increase the chances of contamination, whereas lower pH values inhibit the yield of ethyl alcohol.

Temperature: Usually, the temperature range of 70° to 80°F is preferred. As there is heat evolution during fermentation, the temperature in the fermentation tank rises gradually and is controlled by means of cooling coils. Temperature can also be controlled by cold water spray around the fermentation tank. Intermittent agitation is required for the uniform cooling of the medium. Higher temperatures favour the growth of bacteria and also cause loss of ethyl alcohol due to evaporation.

Time: Fermentation starts within a few hours after the addition of yeast. The process becomes very rapid after 24 hours to complete the fermentation. Duration for ethyl alcohol fermentation is usually 30 to 72 hours when the specific gravity of the fermented liquid (wash) becomes constant. At this stage it contains 6 to 8 per cent of ethyl alcohol.

Yield: On an average 0.4 gallons of ethyl alcohol is obtained from one gallon of molasses. Strict technical control is essential to ensure satisfactory yield. About 90 per cent of carbohydrate is converted into alcohol.

Recovery: The fermented liquid (wash) is allowed to settle for a few hours, and then, distilled in analyser and rectifier columns to obtain rectified spirit (ethyl alcohol) and fusel oil (higher alcohols). The highest concentration that can be secured by fractional distillation is the constant boiling mixture. This mixture contains 95.6 per cent ethyl alcohol and remaining 4.4 per cent water. The spent wash and bottom sludge are drained as distillery wastes. The product is marketed as rectified spirit, denatured spirit, or special denatured spirit. For the manufacture of rum, gin, whisky and brandy, matured rectified spirit is used according to the ISI specifications. Different techniques of ageing, dilution and blending are employed for the manufacture of these potable drinks.

Large distilleries make use of patent stills, (e.g. Coffey's still or Ilgis or Borbet's still). Coffey's still is composed of two columns:

(i) the analyser at the top follows a zigzag course down the column—steam enters at the bottom and travels counter-currently depriving the wash of its alcohol content. The exhausted wash slop is run out. The vapours (steam and alcohol) issuing from the top of the analyser enter the rectifier at the bottom.

(ii) The rectifier is composed of a specially designed fractioning column with a number of chambers. The less volatile constituents (slop and fusel oil) gradually condense and are drawn off from a higher point of the still. The temperature at the higher point of the still is roughly equal to the boiling point. The head products which are very small consist of aldehydes, formic esters, etc. which owing to their greater volatility pass out through the top of the column, along with a small quantity of uncondensed alcohol. The higher alcohols are removed from the receiver every two or three days.

Flow-sheet: Figure 20.1.

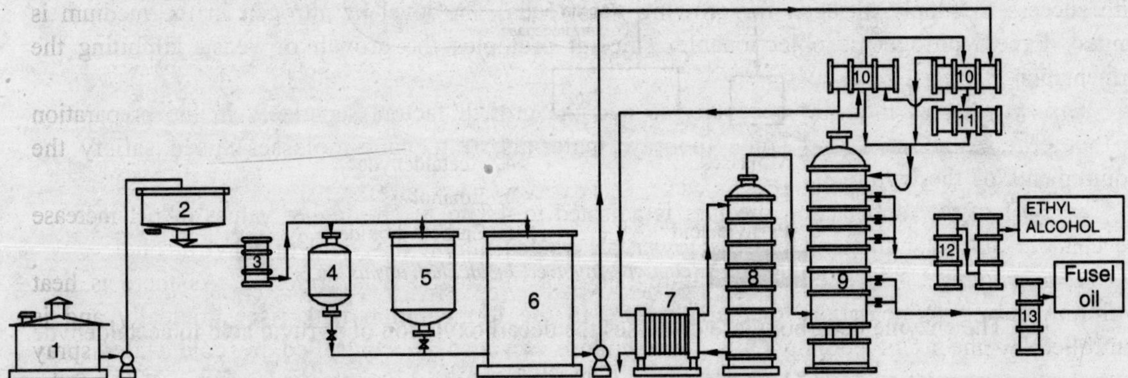

1. Molasses tank
2. Overhead tank
3. Diluter
4. Yeast vessel
5. Buffer tank
6. Fermentor
7. Heat-exchanger
8. Analyzer
9. Rectifier
10. Beer heater
11. Total condenser
12. Cooler
13. Fusel oil separator

Fig. 20.1 *Flow diagram of industrial production of alcohol from molasses.*

Bio-chemistry: Alcohol fermentation is an anaerobic process, (i.e. takes place in the absence of air). Micro-organisms utilize the carbohydrate present in the raw material, (e.g. molasses) to obtain energy for growth and metabolic activities, leading to the formation of alcohol. Monosaccharides (hexoses) are directly fermented. Disaccharides are first hydrolyzed to hexoses, and then fermented to form carbon dioxide and ethyl alcohol. The flow of carbon in the ethyl alcohol formation takes place via the well known Embden-Meyerhof-Parnas (EMP) pathway:

$$C_6H_{12}O_6 \rightarrow 2CH_3CH_2OH + 2CO_2$$

A series of intermediate steps involving enzymes result in the breakdown of glucose molecules into pyruvic acid. The pyruvic acid is then further converted into ethyl alcohol and carbon dioxide by the following enzymatic reactions (Figure 20.2):

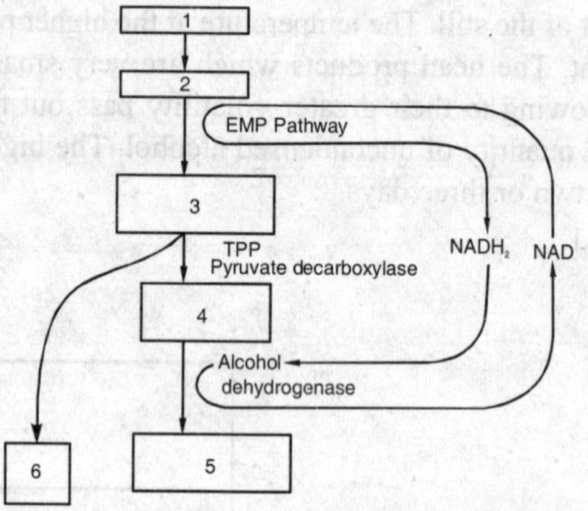

1. Molasses
2. Hexoses
3. Pyruvic acid
4. Acetaldehyde
5. Ethanol
6. Carbon dioxide

Fig. 20.2 *Reactions involved in alcohol formation.*

(a) The enzyme decarboxylase catalyses the decarboxylation of pyruvic acid to acetaldehyde.

$$CH_3COCOOH \xrightarrow{\text{pyruvate decarboxylase}} CH_3CHO + CO_2$$

(b) The enzyme alcohol dehydrogenase further catalyses the reduction of acetaldehyde to ethyl alcohol at the expense of $NADH_2$:

$$CH_3CHO + NADH_2 \xrightarrow{\text{alcohol dehydrogenase}} CH_3CH_2OH + NAD$$

By-products: The three main by-products of the alcohol industry are: carbon dioxide, the yeast and the distillery effluent (Figure 20.3). Carbon dioxide is produced in large quantities. This gas after purification can be liquified or converted into dry ice. The latter finds wide commercial use. For example, large quantities of liquified carbon dioxide are used in the soft drink industry. The yeast withdrawn from alcoholic mash is a valuable fodder due to its higher content of proteins, vitamins and essential trace elements.

During the production of alcohol some quantities of higher alcohols (isoamyl alcohol, butyl alcohol), collectively called fusel oil, are also formed. After refining they are used in small quantities in the manufacture of perfumes and potable liquors, and the bulk portion is processed into various acetates by esterification. Among them *n*-propyl acetate, isopropyl acetate and amyl acetate are mainly used in the paint and lacquer industry. Distillery-spent wash or effluent is a good source of potassium salts and can be used for the manufacture of potassium fertilizer.

The main product, alcohol, offers many possibilities for further processing, besides its use for drinking purposes. The most important industrial uses of alcohol are in the manufacture of ether, acetaldehyde, acetic acid, butanol, etc.

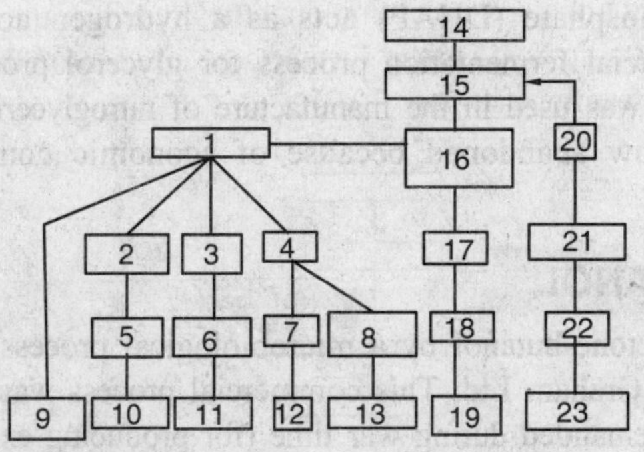

1. Distillation
2. Ethyl alcohol
3. Fusel oil
4. Slop
5. Methyl spirit
6. Higher alcohols
7. Yeast
8. KCl recovery
9. **Fine alcohol**
10. **Solvent**
11. **Solvent**
12. **Food yeast**
13. **Fertilizer**
14. Molasses
15. Fermentation
16. Separation of yeast
17. Yeast
18. Drying
19. **Fodder yeast**
20. Carbon dioxide
21. Liquid carbon dioxide
22. Press
23. **Carbon dioxide dry ice**

Fig. 20.3 *By-products of alcohol fermentation (indicated by bold types).*

In India the alcohol fermentation industry is still not much developed. Much research and development work is required to develop it further. Fermentation efficiency and overall efficiency could be increased multifold if preclarification of molasses, recycling of yeast and continuous fermentation processes are adopted. There is a lot of scope, and all we need is to carry out more research to make this industry profitable in India.

3 GLYCEROL

An interesting aspect of the alcoholic fermentation is the formation of glycerol. Research was carried out during World War I for accumulating more glycerol than the usual amount formed during the alcoholic fermentation. Subsequently, investigators discovered that yeast cultures could be forced to produce five to ten times more glycerol than they normally produce. This was achieved by the addition of sodium sulphite during the fermentation process. Formation of glycerol occurs due to the blocking of an enzymatic pathway by sodium sulphite as below:

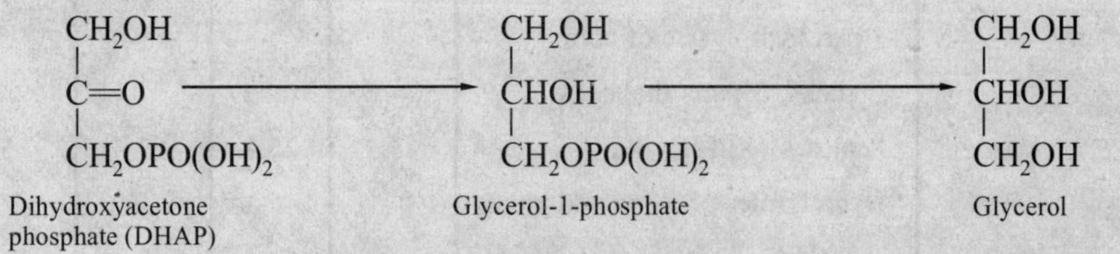

Biochemistry of glycerol formation.

Dihydroxyacetone phosphate (DHAP) acts as a hydrogen acceptor and gets reduced to glycerol. Thus, the commercial fermentation process for glycerol production was established. The glycerol thus manufactured, was used in the manufacture of nitroglycerine and other explosives. This fermentation process is now abandoned because of economic competition from the synthetic production of glycerol.

4 ACETONE-BUTANOL

Industrial production of acetone-butanol by a microbiological process was first started in England by the firm of Strange and Graham Ltd. This commercial process was developed just before World War I. Since acetone was demanded during war time (for producing explosives), the acetone-butanol industry was expanded rapidly. In addition to the wide use of acetone, butanol was needed as a solvent for the rapid drying of nitrocellulose paints in the developing automobile industry. Thus, this commercial process of acetone-butanol survived even after a lack of demand for acetone after World War I. Riboflavin (a by-product of the fermentation) also helped to sustain the large-scale practicability of the process.

Clostridium acetobutylicum acetone-butanol fermentation made significant technological contributions to industrial microbiology. This was the first industrial microbiological process where the removal of contaminating microbes from the fermentor was found to be of paramount importance in the success of the process. The establishment of this fermentation process led to the first successful use of pure culture techniques on a large-scale. Later, these pure culture techniques were improved and refined in connection with the manufacture of antibiotics.

Process brief: Tables 20.1 and 20.2 provide information regarding the micro-organisms and substrates used in acetone-butanol fermentations together with the yields and relative concentrations obtainable.

Table 20.1 *Acetone/butanol/ethanol fermentation of corn cobs, corn stalk and wood [Walton and Martin (1979)].*

Culture	*Substrate(s)*	*Time (h)*	*Yield (%)*	*Solvent composition (%)*		
				Butanol	*Acetone*	*Ethanol*
Clostridium acetobutylicum S_{25}	Hydrolyzate of corn cobs (7%)		26.2			
	Hydrolyzate of sawdust (7 %)		22.2			
C. acetobutylicum 314	Hydrolyzate of corn cobs (40-60% , 3.7% total), molasses (40-60% sugar)	48 - 72	40	67.5		
Butyl culture	1 part hydrolyzate of corn stalks, 3 parts molasses	50 - 55	31 - 37			
	Pentoses (13.5%)		25.4	67	33	
	Hydrolyzates of wood and plants (8%)		35.5	62	32	6

Table 20.2 *Production of acetone/butanol in countries other than the United Kingdom and the United States [Walton and Martin (1979)].*

Culture	Substrate	Time (h)	Yield (%)	Butanol	Solvent Composition (%)		
					Acetone	Ethanol	i-Propanol
Clostridium acetobutylicum	Sugar solution (12%) in presence of active carbon (6%): carbon can be reused	36	66	33			
	Molasses plus rye or wheat meal. Addition of animal charcoal (1.0%) increases solvent yields	26.5	66	31	3		
C. acetobutylicum S_{25}	Millet (5%)		34.4				
	Corn (5%)		33.9				
	Sweet potatoes		33.7				
	Corn cobs		26.2				
	Molasses		24.9				
	Corn (6%) molasses (4%)		24.3				
	Sawdust (7%)		22.2				
C. acetobutylicum isolated from sugarcane roots	Mash in 1 *l* containing sugarcane juice (250 ml, 20 Brix) ground *Vicia sativa* (1 g) KH_2PO_4 (2.5g) $CaCO_3$(4 g)	32	73	19 - 23	3 - 4		
C. aurianticum	Glucose, fructose, sucrose and starch used with $(NH_4)_2SO_4$, $NaNO_3$, peptone or urea as nitrogen source		30 - 40	60.5	0.5	7.4	25.6
C. saccharoperbutylacetonicum	Blackstrap-molasses (51.4 g sugar 55g per *l* of mash)		34.2	75.4	20.2	4.4	
C. saccharoperbutylacetonicum ATCC 13564	Sugar solution (4%), $(NH_4)_2SO_4$(0.2%), calcium superphosphate (0.1%), calcium carbonate (0.3%)	60	28.5				
C. saccharoperbutylacetonicum N-1-41	Molasses with $(NH_4)_2SO_4$, $CaCO_3$ and Ca-superphosphate	60		82.1	15	2.9	
C. saccharobutylacetonicum liquefaciens	Molasses [sugar (5%)], protein nutrients, alkaline buffer added to mash	36 - 48	30	60	30	10	
Butyl culture	Calcium acetate and acetic acid increase fermentation yield						
	Rye-flour with sterile air blown in for 2-3 min. per hour, butanol output increased 3.4-9.1%	40					
Butyl culture	Starch (100%)		37.1				
	Starch (90%), beet-molasses (10%)		36.7				
	Starch (50%), beet-molasses (50%)		38.6				
	Starch (37%), beet-molasses (63%)		37.9				

Contd. Table 20.2

Culture	Substrate	Time (h)	Yield (%)	Butanol	Solvent Composition (%) Acetone	Ethanol	i-Propanol
Culture 1 (1,00,000-*l* tank)	Molasses [sugar (6.37%)] +rice bran (0.3%), $(NH_4)_2SO_4$(0.3%), $CaCO_3$ (0.4%). Temperature, 27-37.5°C	104	33.6				
SES-4	Molasses [sugar (5-6%)]		33	56.8	33.5	9.6	
SES-5	Temperature, 33°C		33	56.2	29.9	13.8	
SES-5 (1,00,000-*l* tank)	Molasses [sugar (6.28%)] + rice bran (0.3%) $(NH_4)_2SO_4$(0.3%), $CaCO_3$(0.4%) Temperature, 35-37°C	90	32				

Schematic diagram for the acetone-butanol fermentations is illustrated in Figure 20.4.

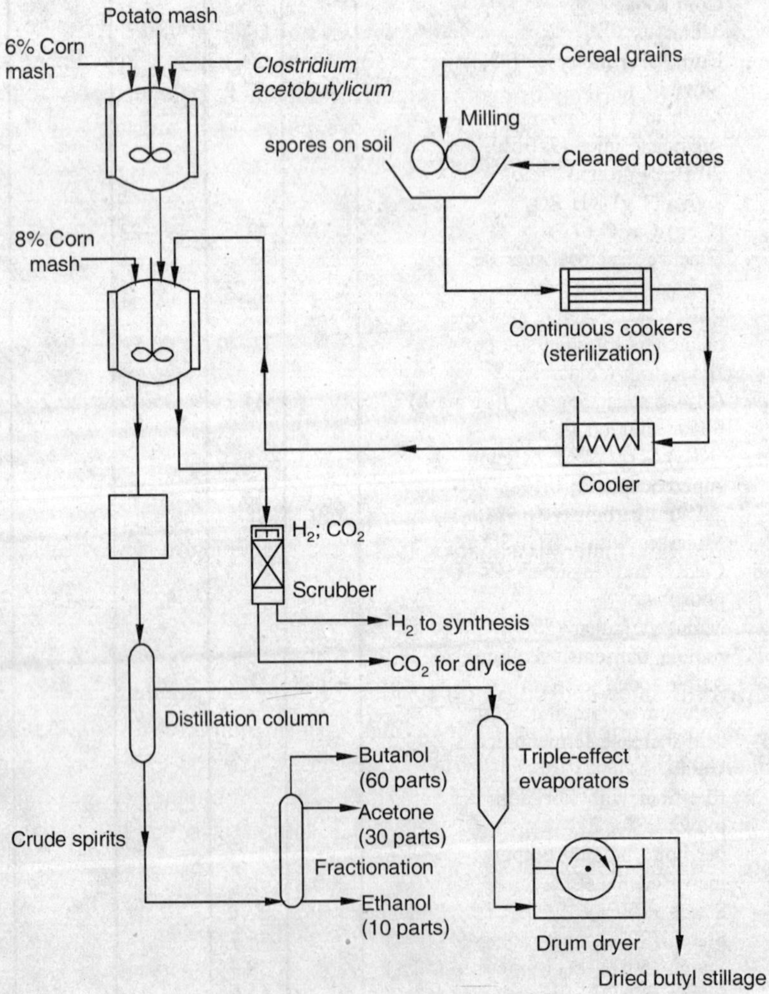

Fig. 20.4 *Flow diagram of a typical acetone-butanol fermentation using starch products (Beesch, 1953).*

Moreover, fermentation conditions, (i.e. pH and temperature) are optimum for the fermentation. For *Clostridium acetobutylicum* acetone-butanol fermentation, the optimum ranges of pH and temperature are 5.0 - 7.0 and 30 - 37°C respectively. The kinetics and regulation of acetone/butanol production by *Cl. acetobutylicum* is receiving increasing attention (Prescott and Dunn, 1959, Petitdemange *et al.* 1976, Spivey, 1978, Gottschal and Morris, 1982). A study has been carried out on production of acetone and butanol by the batch and continuous culture of *Clostridium acetobutylicum* under nitrogen limitation by F. Monot and J.M. Engasser in 1982. Continuous production of acetone and butanol has been recently reported in a single stage chemostat at sufficiently low dilution rates. The highest productivities in continuous cultures have been attained on complex media with high glucose and nitrogen concentrations (Wang, 1981). Significant levels of solvents were also reported in phosphate limited chemostats (Bahl *et al* 1982). On the other hand, only weak or no solvent production was observed in continuous cultures with ammonium limitations (Gottschal and Morris, 1981, Andersch *et al* 1982).

5. 2, 3-BUTANEDIOL

2, 3-Butanediol (also called diol or 2, 3-butylene glycol) is a possible source for l, 3-butadiene (used in the synthesis of rubber). It is also a source for a permanent-type antifreeze. Moreover, other compounds, readily derived from this mother compound, include methyl ethyl ketone, methyl vinyl carbinol and methyl vinyl ketone. The last two compounds are used in the plastics industry. 2, 3-Butanediol exists in three stereoisomeric forms:

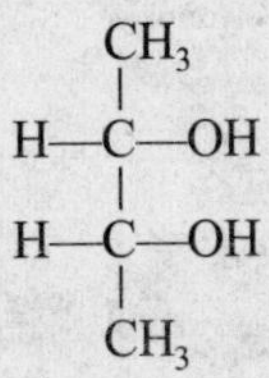

meso

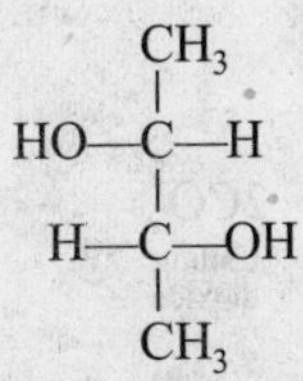

D(–)

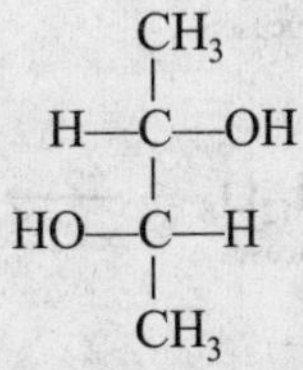

L(+)

Most bacteria, except *Bacillus polymyxa*, produce the meso isomer with only traces of the D(–) or L(+) forms. *B. polymyxa* produces more than 98% of the D(–) form.

Raw materials: Raw materials satisfactory for the manufacture of 2, 3-butanediol are grouped in three major classes:

(i) starchy substrates, (e.g. corn, wheat, barley, sweet potatoes and potatoes)

(ii) sugars and sugar residues, (e.g. beet and cane molasses)

(iii) wood hydrolysates and pulp-mill residues, (e.g. waste sulphite liquor)

From the above-mentioned list of materials as fermentation substrates, it appears that ethyl alcohol fermentation also uses the same raw materials.

Process brief: Bacterial species able to produce 2, 3-butanediol have been studied using 5% glucose under anaerobic conditions. Table 20.3 summarizes the list of bacterial species and yields of various fermentation products.

Table 20.3 *Comparison of yields of various fermentation products.*

Bacteria*	*Sugar utilized in anaerobic production of following compounds(%)*				
	Diol-hydrogen 1	*Diol-formic acid* 2	*Diol-glycerol* 3	*Ethyl alcohol* 4	*Lactic acid* 5
Enterobacter aerogenes	52	2	...	23	2
Bacillus polymyxa	68	...	...	33	2
Pseudomonas hydrophila	56	...	...	26	12
Serratia marcescens	1	55	1	27	6
Bacillus subtilis (Ford's type)	...	...	84	3	8

*The difference between the percentage of sugar utilized by each bacterium and 100% is accounted for by the cell material produced.

(1) $\underset{\text{Glucose}}{C_6H_{12}O_6} \longrightarrow \underset{\text{2, 3-Butanediol}}{CH_3 \cdot CHOH \cdot CHOH \cdot CH_3} + \underset{\text{Hydrogen}}{H_2} + \underset{\text{Carbon dioxide}}{2CO_2}$

(2) $\underset{\text{Glucose}}{C_6H_{12}O_6} \longrightarrow \underset{\text{2, 3-Butanediol}}{CH_3 \cdot CHOH \cdot CHOH \cdot CH_3} + \underset{\text{Formic acid}}{HCOOH} + \underset{\text{Carbon dioxide}}{CO_2}$

(3) $\underset{\text{Glucose}}{3C_6H_{12}O_6} \longrightarrow \underset{\text{2, 3-Butanediol}}{2CH_3CHOH \cdot CHOH \cdot CH_3} + \underset{\text{Glycerol}}{2CH_2OH \cdot CHOH \cdot CH_2OH} + \underset{\text{Carbon dioxide}}{4CO_2}$

(4) $\underset{\text{Glucose}}{C_6H_{12}O_6} \longrightarrow \underset{\text{Ethyl alcohol}}{2C_2H_5OH} + \underset{\text{Carbon dioxide}}{2CO_2}$

(5) $\underset{\text{Glucose}}{C_6H_{12}O_6} \longrightarrow \underset{\text{Lactic acid}}{2CH_3 \cdot CHOH \cdot COOH}$

Comparison of yields by different bacterial species only under anaerobic conditions does not give a true picture of their real potentialities. All 2, 3-butanediol fermentations are sensitive to changes in fermentation conditions, (i.e. air-supply, pH, or sugar concentrations). For example, dissimilation of glucose under aerobic conditions by these bacterial species takes place as under:

$$\underset{\text{Glucose}}{C_6H_{12}O_6} + \tfrac{1}{2}O_2 \longrightarrow \underset{\text{2, 3-butanediol}}{CH_3 \cdot CHOH \cdot CHOH \cdot CH_3} + \underset{\text{Carbon dioxide}}{2CO_2} + \underset{\text{Water}}{H_2O}$$

Under aerobic conditions, free hydrogen production is lowered for the first three bacterial species listed in Table 20.3. *Serratia marcescens* gives little formic acid, and *Bacillus subtilis* (Ford's type) gives scarcely any glycerol. Generally, the yields of 2, 3-butanediol or its oxidation product, acetoin, are increased as compared to the other products under aerobic conditions. Oxygen retards ethyl alcohol formation by both *Bacillus polymyxa* and *Enterobacter aerogenes.* Automatic pH control is necessary to obtain the highest yield of 2, 3-butanediol in the shortest possible fermentation time. The optimum fermentation conditions for an aerobic dissimilation of sucrose by *Bacillus polymyxa* are in a growth medium of 8% sugar concentration at 33°C maintained with pH at 6.2 by automatic additions of ammonium hydroxide. Fermentation time is 30 hours, giving 130 moles of 2, 3-butanediol and 122 moles of ethyl alcohol per 100 moles of sucrose used.

Bacillus polymyxa is a unique bacterial species in that it produces its own amylolytic enzymes, hydrolyzing whole grain mashes during fermentation.

Recovery: Recovery of 2, 3-butanediol from fermented mashes is rather difficult because of its high boiling point (179-180°C). Various procedures have been studied. The procedures studied are: spray-drying, drum-drying, chemical conversion, dialysis, solvent extraction and steam stripping. In most of these procedures the first step consists of elimination of ethyl alcohol. Thereafter, liquor is subjected to filtration or screening, followed by concentration by evaporation. Subsequently, one of the above-mentioned procedures is used to recover 2, 3-butanediol. With grain fermentations the most feasible procedure of recovery was found to be vapour-phase steam extraction.

Remarks: Commercial plants for the manufacture of 2, 3-butanediol have not been established till today. Only pilot plants are in operation for the purpose of study.

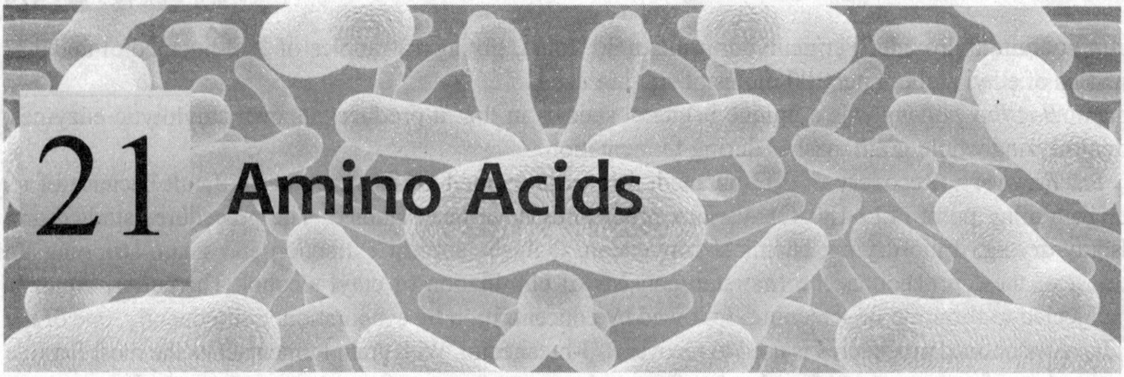

21 Amino Acids

1 INTRODUCTION

Amino acids, like vitamins, form an important class of compounds. They are required by living cells as growth factors. The production of amino acids through the fermentation process was initiated by Kinoshita and his co-workers in 1956. They produced L-glutamic acid as the first amino acid by the fermentation method. Thereafter, various L-amino acids were economically produced by the use of suitable micro-organisms. Thus, L-α-amino acids were made available at low cost to be used in many areas as economical raw materials.

However, the production of surplus amino acids is a problem for the micro-organisms. Over-production of the amino acids by micro-organisms may be accomplished by controlling the complex

Table 21.1 *Productivities of amino acids using various auxotrophic and analogue resistant mutants.*

Amino acid	*Micro-organism*	*Mutation specificity Amino acid requirement*	*Concentration of carbo-hydrate, %*	*Yield, (gm./litre)*
L-lysine	*C. glutamicum*	homoserine	15	45
L-threonine	*E. coli*	lysine	15	24
L-leucine	*C. glutamicum*	phenylalanine & histidine	12	16
L-homoserine	*C. glutamicum*	threonine	10	15
L-valine	*C. glutamicum*	isoleucine	7.5	11
L-ornithine	*C. glutamicum*	arginine	10	36
L-citrulline	*C. glutamicum*	arginine	10	10.7
L-proline	*C. glutamicum*	base	15	31
		Analogue resistance[b]		
L-isoleucine	*Serratia marcescens*	isoleucinehydroxamate & α-aminobutylic acid	12	12
L-arginine	*Bacillus subtilis*	argininehydroxamate & 6-azauracil	8	28
L-histidine	*C. glutamicum*	TRA, PUA & PYA	15	15
L-tyrosine	*C. glutamicum*	PheA and TyrA	10	17.6
L-phenylalanine	*C. glutamicum*	PheA	10	9.5
L-tryptophan	*C. glutamicum*	PheA, TyrA & TryA	10	12

[b] TRA, 1, 2, 4-triazol-3-alanine; PUA, purine analogues; PYA, pyrimidine analogues; PheA, phenylalanine analogues; TyrA, tyrosine analogues; TryA, tryptophan analogues.

regulatory systems. Artificially derived auxotrophic or regulatory mutants as the strains resistant to growth inhibition by the structural analogues of the desired amino acid may be used for fermentative production of these compounds. Table 21.1 lists various auxotrophic and analogue-resistant mutants used in the production of amino acids.

Patents have been produced, particularly in Japan, describing the production of most of the naturally occurring amino acids. But only glutamic acid and lysine are produced in large quantities. These amino acids are used principally in the food industry.

2 L-GLUTAMIC ACID

Fermentative production of L-glutamic acid

A derivative of glutamic acid, mono-sodium glutamate, is used to develop a flavour in food products, (e.g. soups). The commercial production of this acid in the world exceeds 1,00,000 tons per annum. L-Glutamic acid may be produced in many ways:

(a) by the hydrolysis of wheat gluten, soya bean cake, or other proteinaceous material
(b) by the cleavage of the pyrrolidone carboxylic acid found in Steffen's molasses
(c) by a one-stage fermentation process involving a single micro-organism
(d) by a two-stage fermentation process where α-ketoglutaric acid is produced by one microbe and α-ketoglutaric acid, in turn, is converted to L-glutamic acid by another microbe

Nowadays, a one-stage process utilizing *Corynebacterium glutamicum* is mainly adopted to produce L-glutamic acid on a large-scale. Japanese workers have obtained high-yielding mutants of this microbe.

Glucose and starch hydrolysates are the principal raw materials for the fermentative production of amino acids. More efforts have been made to replace these materials with cheaper and more easily available ones such as molasses, ethanol, pentose, acetic acid and other petrochemicals. The fermentative production of L-glutamic acid from various carbon sources is given in Table 21.2.

Table 21.2 *Fermentative production of L-glutamic acid using various carbon sources.*

Carbon source	*Micro-organism*	*Yield*	
		gm./litre	*per cent*
Glucose	(i) **Corynebacterium glutamicum*	38	38
	(ii) **Brevibacterium flavum*	50	50
Acetic acid	(i) *Brevibacterium flavum*	98	48
	(ii) *Brevibacterium thiogentalis*	51	51
Ethanol	*Brevibacterium* spp.	59	66
Propylene-glycol	*Bacillus megaterium*	27	27
Benzoic acid	*Brevibacterium* spp.	80	80
Xylose	*Brevibacterium pentosaminoacidicum*	5	10

*indicates industrialization in Japan.

Industrial production of L-glutamic acid using glucose as the carbon source was initiated in 1957 by Kyowa Hakko Kogyo Co. It is necessary to add a limited amount of biotin to the production medium. The quantity of biotin needed is in the range of 1 to 5μg./ltr. The accumulated glutamic acid gets excreted through the cell-wall by the mechanism involving biotin. With inexpensive raw materials, fine control of biotin in glutamic acid fermentation is not possible. Glutamic acid formation and its excretion occur even at higher biotin concentrations, provided the fermentation medium is supplemented by penicillin or detergents. Moreover, the use of antibiotics such as penicillin, allows us to use sugar cane waste molasses in place of pure glucose as a carbon substrate. Urea or ammonia may be used as nitrogen sources. Also, phosphates and other ordinary salts are added to the production medium in the manufacture of glutamic acid. The production medium is slightly alkaline in reaction. The temperature to be maintained in *Corynebacterium glutamicum*—L-glutamic acid fermentation is 30°C. The duration of the fermentation period is 40 hours. The carbon source may be converted to glutamic acid in 50 per cent of the yield under optimum aerobic conditions. The general flow-diagram for the industrial production of amino acids is illustrated in Figure 21.1.

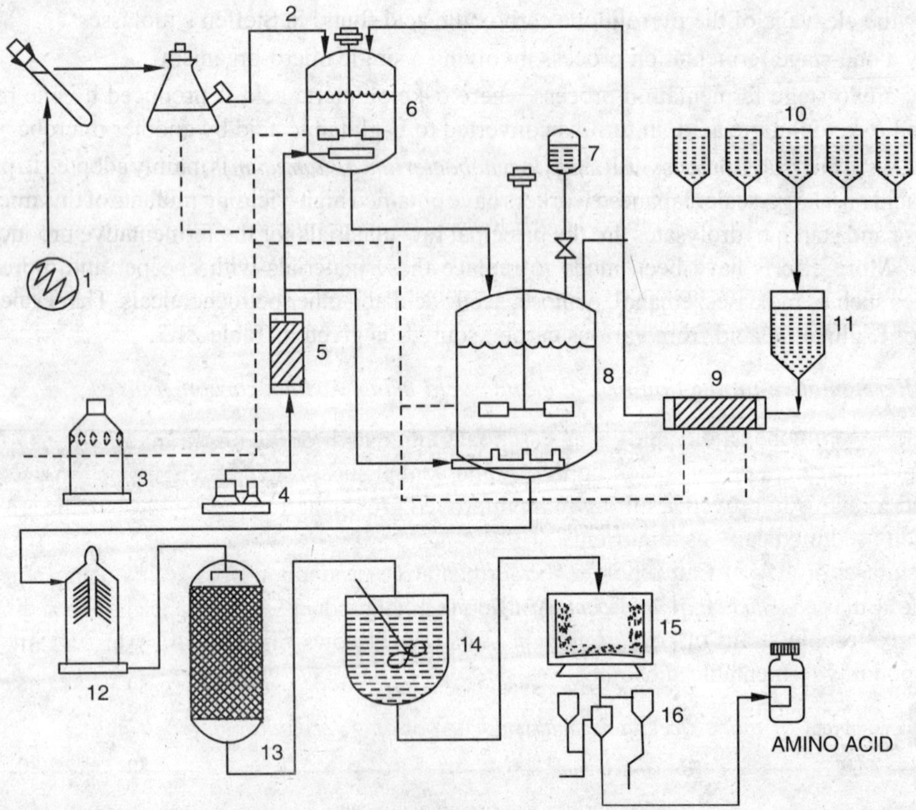

1. Pure culture
2. Inoculation
3. Boiler
4. Air compresser
5. Air filter
6. Seed tank
7. pH control medium
8. Fermentor
9. Sterilizer
10. Culture media
11. Preparation tank
12. Centrifugal separator
13. Ion-exchange column
14. Crystallizing tank
15. Crystal separator
16. Dryer

Fig. 21.1 *Fermentative production of amino acids.*

Biochemistry of L-*glutamic acid fermentation*

In the case of *Corynebacterium glutamicum,* the sugars are metabolized via the glycolysis. Therefore, formation of glutamate does not occur if classical inhibitors of glycolytic pathway, (e.g. fluoride and iodoacetate) are present. Owing to the absence of an α-ketoglutarate dehydrogenase in the mutant, accumulation of glutamate occurs. Thus, the citric acid cycle is not complete (Figure 21.2).

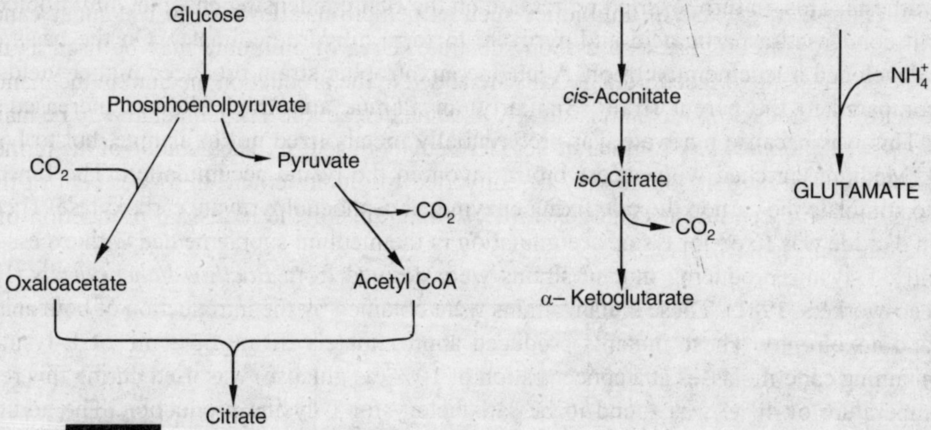

Fig. 21.2 *Principal intermediates in the production of glutamic acid from glucose.*

The energy is provided by a glyoxylate pathway. The accumulation of glutamate continues until a saturation point is reached (50 mg./gm. dry weight of cells). Subsequently, excretion of the glutamate into the medium occurs in the absence of biotin. If molasses is used, addition of penicillin or detergents during the exponential growth phase allows excretion of glutamate.

3 L-LYSINE

L-Lysine (2, 6-diaminohexanoic acid) is an essential amino acid for the nutrition of humans. It is used for supplementing cereal proteins lacking this amino acid. Thus, protein quality of certain foods, (e.g. wheat-based foods) is improved resulting in an improved growth and tissue synthesis. This amino acid has also been used medically as a nutrient.

Kinoshita *et al* (1958) first reported the fermentative production of L-lysine using homoserine auxotrophs of *Corynebacterium glutamicum.* And commercial production of L-lysine has been established using this auxotrophic strain of *C. glutamicum.* Table 21.3 shows microbes of industrial importance with corresponding fermentable substrates.

Table 21.3 *Microbes of industrial interest and corresponding fermentable substrates.*

Micro-organisms	*Fermentable substrate*	*Yield gm./litre*
**Brevibacterium flavum*	Glucose	32
**Corynebacterium glutamicum*	Glucose	39
Brevibacterium flavum	Acetic acid	61
Brevibacterium lactofermentum	Ethanol	66
Nocardia spp. No. 258	*n*-Paraffin	34

*indicates industrialization in Japan.

In the procedures of *Nakayama* and *Hagino;* U.S. patent 3,687,809; August 29, 1972; assigned to Kyowa Hakko Kogyo Company Limited, Japan, one of the main starting materials in the culture medium is L-aspartic acid. Production media containing suitable amount of L-aspartic acid (1 to 10% by weight) are, therefore, used for securing better yields of L-lysine.

Several important factors influencing lysine fermentation were reported by Tosaka *et al.* One of the several important factors was the amount of leucine in the growth medium. Excess of leucine inhibited lysine biosynthesis. This, in turn, exerted repression on the first condensing enzyme, dihydropicolinate synthetase (it condensed aspartic acid and pyruvate to form dihydropicolinate). On the basis of this finding, he developed a leucine auxotroph. A leucine auxotrophic strain produced higher yields of L-lysine as compared to the parent strain. Similarly, an alanine auxotroph showed increased lysine production. This was because pyruvate was preferentially metabolized not to alanine, but to lysine in this mutant. Medium enriched with excess biotin favoured the lysine accumulation. The function of biotin was to stimulate the carbon dioxide fixing enzyme, phosphoenolpyruvate carboxylase. Therefore, more carbon dioxide was fixed for lysine accumulation in the medium supplemented with excess biotin.

Recently, L-lysine producing mutant strains were derived from *Bacillus licheniformis* (Hiroshi Hagino and co-workers, 1981). These mutant strains were obtained by the introduction of both analogue-resistance and auxotrophy. These mutants produced approximately 30 mg. per ml. of L-lysine. The medium containing cane-molasses at a concentration of 10% (as glucose) was used during this research work. A temperature of 40°C was found to be satisfactory for L-lysine production. The addition of antibiotics, (e.g. tetracycline and chloramphenicol) to the growth medium was found to suppress sporulation effectively and to increase L-lysine accumulation.

The success of L-lysine production brought about a new concept in fermentative production of amino acids. The new production method used artificially derived auxotrophic or regulatory mutants as the strains resistant to growth inhibition by the structural analogues of the desired amino acids. The complex regulatory systems controlling the biosynthesis can be artificially bypassed in these mutants to allow overproduction. It has now become possible to produce various amino acids using the corresponding auxotrophic or regulatory mutants.

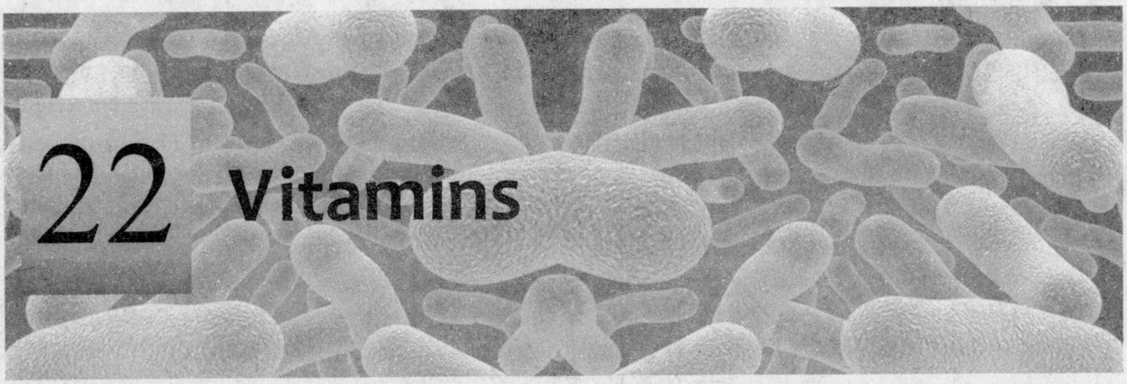

22 Vitamins

1 INTRODUCTION

Vitamins are very important compounds in the diet because they are of great value in the growth and metabolism of the living cell. All the vitamins may be synthesized by prototrophic micro-organisms. These compounds are constituted from simple ingredients supplied by the growth medium. But, they are produced in amounts enough to fulfil the organism's needs. In other words, they do not accumulate in the growth medium. There are certain micro-organisms which excrete vitamins greatly in excess of their own metabolic needs if grown under highly specified and artificial conditions. Thus, such micro-organisms are of industrial interest. This chapter deals with the industrial production of vitamin B_{12}, vitamin B_2 and vitamin C by fermentation method.

2 VITAMIN B_{12}

Vitamin B_{12} is an important dietary component for normal growth in human beings and domesticated animals. Its daily requirement for human beings is 0.001 mg./day.

Looking to the chemistry of vitamin B_{12}, it is a complex chemical compound whose structural formula is as shown in Figure 22.1.

It is cyanocobalamin, consisting of a molecule of cobinamide linked to a nucleotide. The cobinamide molecule has a central atom of cobalt linked to a cyanide group, surrounded by four reduced pyrrole rings joined to form a macro-ring. Moreover, a number of carbon atoms carry methyl or other substituent groups. The nucleotide is atypical in having 5, 6-dimethylbenzimidazole group as its base, instead of a purine or pyrimidine base.

Vitamin B_{12} is entirely produced on a commercial basis by the fermentation. Generally, it is manufactured by primary fermentation. In some instances, it is obtained from broths used primarily for the production of antibiotics. This is largely a matter of economics, since maximum production of the antibiotic and maximum production of cobalamin generally do not take place under the same fermentation conditions. Its production has become one of the important fermentation industries in the United States.

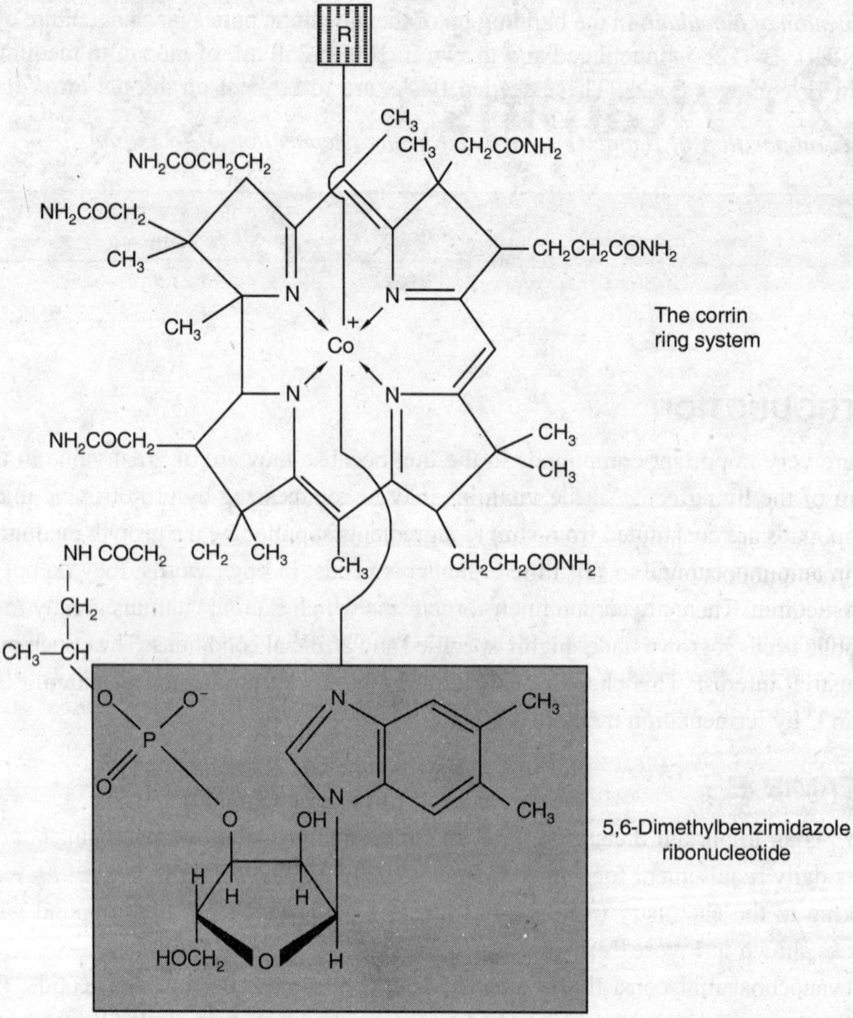

Fig. 22.1 *Structural formula of vitamin B_{12} (R = cyanide).*

Micro-organisms that may be employed in the industrial production of vitamin B_{12} are: *Streptomyces griseus, S. olivaceus, Bacillus megaterium, B. coagulans, Pseudomonas denitrificans, Propionibacterium freudenreichii, P. shermanii,* and a mixed fermentation of a *Proteus* spp. and a *Pseudomonas* spp.

Generally, Vitamin B_{12} is manufactured by the submerged-culture process. Aeration and agitation of the production medium are essential. Usually, such fermentations are completed in 3 to 5 days.

Vitamin B_{12} production using Streptomyces olivaceus NRRL B-1125

The factors affecting *Streptomyces olivaceus* NRRL B-1125–vitamin B_{12} fermentation have been briefly discussed here.

Preparation of inoculum: In the building up of the inoculum, pure agar slant culture of *Streptomyces olivaceus* NRRL B-1125 is inoculated and grown in 100 to 250 ml. of inoculum medium (Table 22.1) contained in Erlenmeyer flasks. These seeded flasks are to be kept on the platform of a mechanical shaker during incubation in order to aerate the medium. Subsequently, the flask cultures are used to inoculate larger amounts of inoculum media contained in inoculum tanks in series. Usually, two or three such successive transfers are made to obtain the required amount of inoculum culture. It has been shown that inoculation of the production tank with about 5% of the volume of production medium is satisfactory.

Table 22.1 *Composition of Bennett's agar used in the preparation of inoculum.*

Component	*Amount* (gm./litre)
Yeast extract	1.0
Beef extract	1.0
N-Z-Amine A (Enzymatic hydrolysate of casein)	2.0
Glucose	10.0
Agar	15.0
Distilled water	1,000 ml.
(pH is adjusted to 7.3 with NaOH)	

Production medium: Production media used in this fermentation, usually, consist of carbohydrate, proteinaceous material, and a source of cobalt and other salts. A typical production medium is presented in Table 22.2.

Table 22.2 *Composition of a typical production medium used in vitamin B_{12} production.*

Component	*Amount* (per cent)
Distillers' solubles	4.0
Dextrose	0.5 to 1
$CaCO_3$	0.5
$COCl_2.6H_2O$	1.5 to 10 p.p.m
(pH adjusted to about 7 with NaOH)	

Most forms of distillers' solubles, soya bean meal, yeast, casein, etc. are found satisfactory under suitable conditions. It is necessary to add cobalt to the medium for the maximum yields of cobalamin. It is to be noted that cobalt has nothing to do with the growth of the *S. olivaceus*. In some cases, it is essential to add cyanide to the medium to aid the conversion of other cobalamins to vitamin B_{12}.

Sterilization of the medium may be practised batchwise or continuously. In batchwise sterilization, the medium contained in the production tank is heated at 250°F for 1 hour. In the latter method, the production tank is charged, (e.g. at 330°F for 13 minutes) by mixing it directly with live steam. During sterilization, steam is blown into openings and all transfer lines are kept filled with live steam when not in use to ensure sterility.

Temperature: A temperature of 80°F in the production tanks is satisfactory during fermentation.

pH: There are a few tenths of a unit fall in pH and a rapid consumption of sugar in the first 24 hours of the fermentation. After 2 to 4 days, lysis of the mycelium begins resulting in the rise of pH. The stabilization of the mash is practised by reducing the pH to about 5 with sulphuric acid and adding small amounts of a reducing agent, (e.g. sodium sulphite).

Aeration and *agitation:* A proper rate of aeration and a correct speed of agitation are essential because the rate of the growth of the streptomycete strain depends on the rate of aeration and agitation. Aeration rates higher than optimum cause excessive foaming. The optimum rate of aeration is about 0.5 volume air/volume medium/minute. The sterilization of air is, usually, practised by passing it through columns filled with activated charcoal.

Antifoam agents: With this fermentation, foam formation is a serious problem, particularly at the beginning and at the end of the fermentation. There are many antifoam agents that may be used to suppress the foam formation. Important defoaming agents are: soya bean oil, corn oil, lard oil and silicones. A defoaming agent in its sterile form is added to the medium during foaming according to the requirement. In some instances, an antifoam agent is added at the time of the production medium make up.

Prevention of contamination: It is very essential to maintain sterility until the fermentation ends because contamination invariably results in very low yields. Therefore, all equipment must be sterile. In addition, transfers are carried out under aseptic conditions.

Yields: The yields of cobalamin are usually in the range of 1 to 2 mg. per litre in the fermented broth.

Recovery: During the major part of the fermentation period, most of the cobalamin is associated with the mycelium. But, a considerable portion of the cobalamin is in the solution at the end of the fermentation period. Heating the mixture to boiling at pH 5 or below, liberates the cobalamin quantitatively from the mycelium. Broth containing cobalamin is subjected to further work-up depending on the type of the product to be produced.

In order to obtain crystalline vitamin B_{12}, the very first operation is of filtration to remove mycelium. Then the filtered broth is treated with cyanide to bring about the conversion of cobalamin to cyanocobalamin. Alternatively, this conversion may also be done after some concentration has been secured while fermentation is in progress. The adsorption of the cyanocobalamin from the solution is practised by passing it through an adsorbing agent packed in columns. Several adsorbents such as activated charcoal, bentonite, fuller's earth and ion-exchange resins are available for this purpose. Finally, elution of cyanocobalamin from adsorbent is accomplished by the use of an aqueous solution of materials ranging from organic bases to hydrochloric acid. Several materials have been reported for this purpose: water, water-acetone, and solutions of sodium cyanide or sodium thiocyanate.

Thereafter, some type of extraction is usually carried out. It may be countercurrent distribution between cresol, amylphenol, or benzyl alcohol and water or a single extraction into an organic solvent, (e.g. phenol). Precipitation as a copper or zinc cyanide-cyanocobalamin complex has been reported.

Another purification step applicable to aqueous concentrates consists of dissolving a zinc salt in the slight acid solution and then raising the pH to bring about precipitation of zinc hydroxide, which eliminates many impurities.

Chromatography on alumina and final crystallization from methanol-acetone, ethanol-acetone, or acetone-water, usually complete the process.

In order to obtain a concentrate of vitamin B_{12} to use as feed-supplement, the final fermented broth is evaporated to dryness. These final broths containing about 3% solids are first subjected to evaporation *in vacuo* to a solids content of 15 to 20%. Then the syrups are drum-dried or spray-dried. The concentrates obtained in this way by cobalt-supplemented fermentation may contain 10 to 30 mg./lb. of cobalamin.

Flow-sheet: see Figure 22.2.

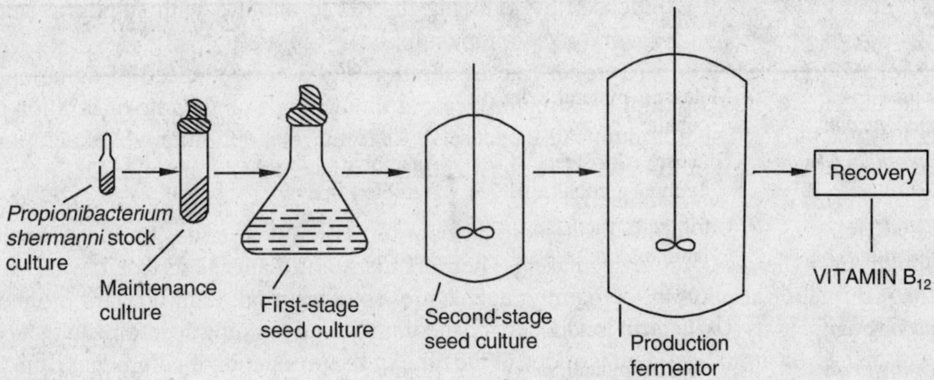

Fig. 22.2 *Semipilot plant scale fermentation process of vitamin B_{12} using Propionibacterium shermanni.*

Vitamin B_{12} production using *Propionibacterium freudenreichii:*

Species of the genus *Propionibacterium* have been reported to produce the highest yields of vitamin B_{12}. For example, *P. freudenreichii* has been reported to produce 20 mg. per litre.

The production medium containing glucose, corn-steep, betaine, and cobalt has been shown to be satisfactory. Betaine is added to the medium to furnish methyl groups. Betaine, probably, does this by aiding in the synthesis of methionine acting as a methyl donor. The concentration of betaine in the medium is 0.5%. In connection with the use of cobalt, its concentration in the medium is critical. It has been shown that concentration of 5μg./ml. of cobalt is optimum. Excess of cobalt suppresses the formation of cobinamide. The pH of the production medium is adjusted to 7.5. A temperature of 30°C has been reported to be satisfactory.

This fermentation is divided into two fermentation cycles: anaerobic fermentation cycle of 70 hours and aerobic fermentation cycle of 50 hours. The formation of cobinamide occurs only in the anaerobic cycle because oxygen inhibits the biosynthesis of cobinamide. In addition, the pH falls from 7.5 to 6.5 and then rises up to 8.5. It is necessary to add 0.1 per cent of 5, 6-dimethylbenzimidazole to the production medium, since *P. freudenreichii* cannot readily synthesize this base. The formation of the nucleotide occurs during the next aerobic fermentation cycle. Then, this nucleotide links with the previously formed cobinamide to give cobalamin. Thus, the yield obtained has been less than 20μg/ml. In a continuous culture version of the process, two large fermentation tanks in series are used, each having a retention period of 60 hours; the first is anaerobic and the second aerobic, with 5, 6-dimethylbenzimidazole added continuously. Thus, the yield of 15μg./ml. is obtained.

Vitamin B_{12} production using other micro-organisms

In addition to the two above discussed micro-organisms, *Streptomyces olivaceus* and *Propionibacterium freudenreichii,* in the industrial production of vitamin B_{12}, other commercially important micro-organisms with brief process details have been listed in Table 22.3.

Japanese workers have reported that various strains of *Corynebacterium, Mycobacterium* and *Nocardia* can produce vitamin B_{12} on hydrocarbon media. But, the yields produced by these micro-organisms are poor, being in the order of a few milligrammes per litre.

Table 22.3 *Some vitamin B_{12} producing micro-organisms.*

Species	*Medium*	*Aeration*	*Temperature (°C)*	*Time of process (hour)*	*Yield (mg./ltr.)*
Bacillus megaterium	Molasses, mineral salts, cobalt	Aerobic	30	18	0.45
Propionibacterium shermanii	Glucose, corn-steep, ammonia, cobalt, pH 7.0	Anaerobic (3 days) Aerobic (4 days)	30	150	23
Bacillus coagulans	Citric acid, triethanolamine, corn-steep, cobalt	Aerobic	55	18	6.0
Pseudomonas denitrificans	Oxalic acid, betaine, cobalt, mineral salts	Aerobic	...	...	10

3 RIBOFLAVIN (VITAMIN B_2)

Riboflavin is an important vitamin required by man and animals. Therefore, its presence in food and animal feed is essential. Riboflavin allowances of the National Research Council have been set at 0.07-0.1 mg./0.75 kg. of body weight in human beings.

Process brief: Commercial production of riboflavin is practised by a fermentation process as well as by a synthetic method. The fermentation process may be outlined as under:

Micro-organisms of interest for the manufacture of riboflavin include bacteria, yeasts, or fungi. The outstanding two closely related species of ascomycetes are: (i) *Ereomothecium ashbyii* and (ii) *Ashbya gossypii*. Preparation of inoculum is started from either slants or spores dried on sand. After one or two flask stages, further propagation is practised through one or two tank inoculum stages.

The size of the fermentors used in riboflavin production ranges from 10,000 to 1,00,000 gals. Selection of a suitable production medium is made according to the type of micro-organism to be used in the process. For example, in the case of *Ashbya gossypii* the production medium contains glucose, collagen, soya oil and glycine. For *Ereomothecium ashbyii,* stillage (still slops) from the alcohol industry with skim milk, soya bean meal or casein added as a proteinaceous source is used while the carbohydrate source may be maltose, sucrose or glucose.

Aeration and agitation of the production medium are essential operations, since fungi are employed in these commercial processes. Fermentation time is 96-120 hours. Optimum titres may be 3-6 gm. of riboflavin or more per litre.

Recovery: Recovery of riboflavin meant for animal feed supplements involves two steps. They are: (i) evaporation of the whole broth in multiple effect evaporators (ii) drying by drum- or spray-drier

For drug and fine food uses, it is necessary to obtain riboflavin in pure crystalline form. In this case, recovery steps are:

(i) heating of the fermentation broth at 120°C for 1 hour
(ii) filtration
(iii) precipitation of the riboflavin by addition of dithionite (hydrosulfite)
(iv) subjecting the riboflavin to several purification steps, including crystallization

4 L-ASCORBIC ACID (VITAMIN C)

L-Ascorbic acid was discovered in 1919. It was isolated in pure form from lemon juice for the first time in 1932 by American biochemists, C.G. King and W.A. Waugh. Svirbely and Szent-Gyorgi showed that this vitamin is a carbohydrate derivative. L-Ascorbic acid is one of the simplest vitamins in structure (Figure 22.3).

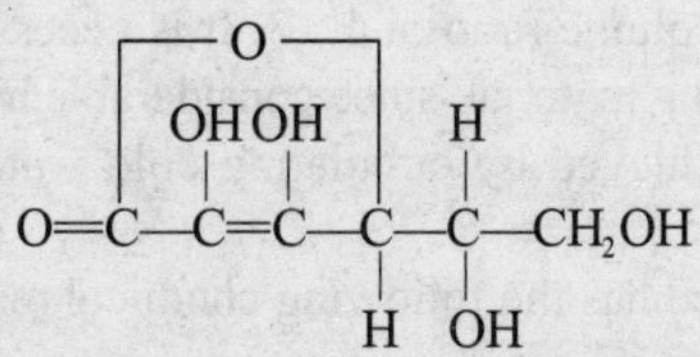

Fig. 22.3 *Structural formula of L-ascorbic acid (vitamin C).*

It is a strong reducing agent and plays a part in cellular oxidation-reduction reactions. It appears to be particularly important for the formation of bone, teeth and collagen tissues. On the basis of studies made by the Food and Nutrition Board of the National Academy of Sciences (U.S.A.), a daily intake of 60 mg. of vitamin C has been recommended for persons aged 15 and above, 45 mg. for children under 11 and 50 mg. for those 11 to 14.

The most important natural sources of vitamin C are fresh or frozen citrus fruits (lime, oranges, grape-fruit, lemon, citron, etc.) amla, fresh strawberries, cantaloupe, guava, tomatoes, green chillies, broccoli, brussels sprouts, spinach, kale cabbage, turnips and many other green vegetables. Rose-hip syrup is specially rich in vitamin C (200 mg. per 100 ml.).

Process brief: The raw material used in the synthesis of L-ascorbic acid is L-sorbose. L-Sorbose, in turn, is obtained from D-sorbitol by fermentation reaction. The bacterium used for bringing about bioconversion is *Acetobacter suboxydans:*

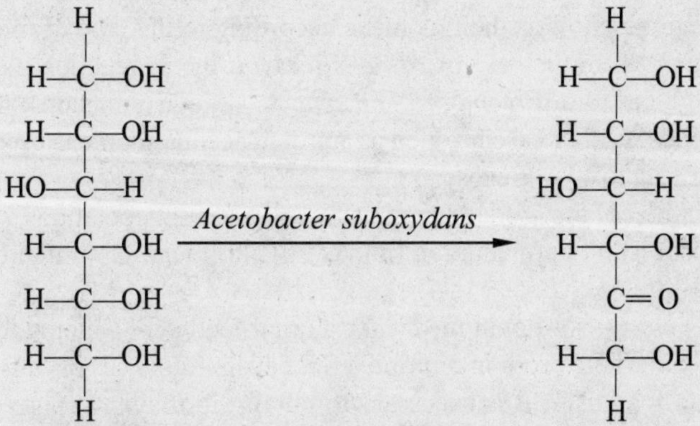

D-Sorbitol L-Sorbose

A requirement of a very high level of aeration of the fermenting solution is a specific characteristic of the fermentation. L-Sorbose has been produced by fermentation since 1934.

Vertical fermentors used for this fermentation process range from 5,000 to 10,000 gal. capacity. They are designed so that large volumes of air may efficiently be supplied. Therefore, fermentors are equipped with a turbomixer-type agitator and baffles along the side walls to ensure maximal turbulence of the solution. There is also an air inlet underneath the agitator. Air inlets may be open pipe ends, small holes or jets through which the air passes at high linear velocity or porous stones. The high-velocity jets or porous stones produce very small bubbles. Thus, this results in the maximum liquid-gas interface for oxygen transfer. Therefore, bioconversion of sorbitol to sorbose occurs at faster rates. At the most, fermentors may be operated at atmospheric pressure or with a head pressure up to 30 p.s.i. In commercial operations with high velocity jets and 30 lb. head pressure, the maximal rate of fermentation may be obtained with an air to solution volume ratio of 1 : 3. It is necessary to have provision for cooling the fermentor along with the fermenting material, since considerable heat is liberated during the fermentative reaction. The cooling may be achieved by circulating cold water either through coils, or preferably through jackets.

The production medium used has the following chemical make-up:

Ingredient	*Amount*
D-Sorbitol	200-250 gm.
Corn-steep liquor	5 ml.
Tap water	1 litre
Antifoam is added as required.	

With higher sorbitol concentrations, the fermentation is slower in the beginning.

The complete fermentation cycle takes 25-30 hours. To obtain the shortest fermentations, a solution containing 100-150 gms. of sorbitol per litre is inoculated with a culture of *A. suboxydans* amounting to 3% of the total volume, and later enough concentrated sorbitol syrup is added to give the final sorbitol concentration.

Analytical yields are about 95-98%. Recovery yields should be about 85%.

23 Yeast and Its Uses

1 INTRODUCTION

Yeasts constitute an interesting group from a technical and industrial standpoint in the microbial world. They can be isolated from many source materials, (e.g. soil, grape and curd) by the use of suitable media. There are some species and strains of yeast which are used commercially. The well-known commercial uses of yeasts are in alcoholic beverage production, baking, food stuffs, animal feeds and the production of biochemicals. Strains of *Saccharomyces cerevisiae,* among known genera and species of yeast, are of the greatest technical importance.

Yeasts can be grown on media containing essential nutrients for the development of yeast. Essential nutrients include assimilable carbohydrate, nitrogen, trace elements, phosphate, sulphate and vitamins (usually biotin). In addition to essential nutrients, other growth stimulators, such as amino acids, inositol, thiamin, pantothenate, riboflavin and PABA are added to the medium to accelerate yeast growth. Media, such as brewers' wort, grape musts and apple juice contain all the essential and non-essential growth stimulating nutrients. These media serve to propagate those yeasts of the genus *Saccharomyces* which are used to produce alcoholic beverages. Moreover, they are employed as the substrates for the production of beverages.

Commercial production of those yeasts used for baking, distilling and as food and animal feeds and biochemicals is carried out on media based on molasses, cheese whey, sulfite liquor and gas oil fractions. But these media usually need to be supplemented by essential or/and non-essential growth stimulating nutrients, as the case may be. In other words, media based on these raw materials are incomplete. Therefore, it is necessary to supplement:

(i) molasses-based media with biotin and ammonium ions
(ii) cheese whey-based media with ammonium ions, sulphate and phosphate
(iii) sulfite liquor-based media with ammonium ions, yeast extract, occasionally corn-steep liquor, sometimes cellulose or starch hydrolysates
(iv) petroleum-based media with minerals, yeast extract and ammonium ions

These media are widely used because they are considerably less expensive and are not so susceptible to seasonal variations in supply as worts, musts and apple juice are.

Thus, molasses-based media are employed for the large-scale production of baking yeasts, food yeasts and distillers' yeasts. Cheese whey-media are used to propagate *Kluyveromyces fragilis* (formerly *Saccharomyces fragilis*) and *Candida pseudotropicalis* food and fodder yeasts. These two species of yeast are capable of assimilating lactose, a characteristic not exhibited by other commercially grown yeasts. Sulphite-liquor based-media are useful in propagating *Candida utilis.* This yeast assimilates a wider range of sugars than other commercially produced yeasts. Consequently this property of yeast allows the producer to vary the supplementary carbohydrate in the medium. Molasses-based media are also frequently used to propagate *C. utilis.* In contrast to the other yeasts, *Candida lipolytica* assimilates *n*-alkanes and, therefore, it is produced on petroleum fractions. Table 23.1 depicts the partial compositions of some typical yeasts.

Table 23.1 *Partial compositions of unfortified yeast powders.*

Component	*Molasses grown S. cerevisiae*	*Molasses grown C. utilis*	*Sulphite liquor grown C. utilis*	*Debittered brewers' yeast*	*Whey grown K. fragilis*
Moisture, %	5	7	7	6	7
Fat, %	6	3	5	6	1
Protein, %	50	51	50	45	54
Ash, %	7	10	7	8	9
Thiamin, p.p.m.	150	25	125	150	20
Riboflavin, p.p.m.	70	50	50	45	50
Niacin, p.p.m.	500	335	300	400	135
Pyridoxine, p.p.m.	30	20	30	40	40

These yeasts are commercially produced aerobically by batch processes. Although some wine yeasts are cultivated in a similar way to bakers' yeast, they are more often propogated fermentatively, like brewing and cider yeasts.

The sterilization of the production plant is practised by live steam under pressure or chemical sterilants. Production media are seldom sterilized, but are treated to reduce the possibility of infection. Treatments include pasteurization, boiling (brewers' wort), addition of sulphur dioxide (wine and cider industries only) or adjustment to an acid pH(4.5), to inhibit bacteria.

2 PRODUCTION OF BREWERS' YEAST

Brewers' yeast is cultivated on hopped malt wort. Thus, this is a considerably simpler process than the production of bakers' yeast. The sterilization of the production plant is carried out by the use of live steam. On the other hand, the hopped wort medium is pasteurized. The pasteurization of medium is carried out, either by boiling the medium present in the production tank, or separately, before passing the medium into the production tank. A typical system would consist of two propagating closed vessels of 33hl (20 barrels) capacity. Each vessel would be inoculated with 100 gms. pressed yeast (The pressed yeast is produced from laboratory stocks by culture in sterile wort, and aerated for one minute in five vessels at 0.11 m^3 air per vessel). There is the necessity of intermittent aeration without agitation. But it is not necessary to control pH and to feed incrementally. Maintenance of the culture at 20°C is carried out by means of the exterior cooling jackets. The yeast mass increases exponentially to give 17 gms. pressed yeast per litre after 48 hours.

The culture is then used directly to inoculate fermentation vessels. The usual rate of inoculation is of the order of 10 - 20% by volume.

This type of propagation ensures that the yeast develops fermentatively (It is catabolite-repressed by the wort sugars, mainly maltose). This explains why the yield of cells is much lower than that of a bakers' yeast production system. This, however, is desirable since the wort used is duty-paid, and the beer produced can be blended. If the yeast were grown aerobically, the product would not be beer, and would be unsuitable for blending. The empirical equation for the brewers' yeast production is as shown below:

$$\underset{200\text{ gms.}}{\text{Maltose}} + \underset{1\text{ gm.}}{\text{Ammonia}} \longrightarrow \underset{10\text{ gms.}}{\text{Yeast}} + \underset{97.5\text{ gms.}}{\text{Ethanol}} + \underset{93.6\text{ gms.}}{CO_2}$$

There is the liberation of 100 K.Cal. of heat energy per 10 gms. of yeast (18,000 Btu/lb.).

In addition to the direct propagation of brewers' yeast described, breweries routinely crop their production fermentors and re-use part of the yeast so obtained. Therefore, direct propagation is used regularly but intermittently to produce fresh and pure yeast stocks. The surplus brewers' yeast is sold to manufacturers of yeast products for dietary supplementation.

The treatment of yeasts for the production of wines and ciders is similar to that described for brewers' yeast. Musts and apple juices, however, are sterilized by SO_2 addition.

3 PRODUCTION OF BAKERS' YEAST

Bakers' yeast is propagated on beet molasses-, refiners' cane molasses- or black-strap molasses-based media. It is a common practice to mix beet molasses and black-strap molasses in the preparation of the production medium. The reason for blending these two types of molasses is the higher assimilable nitrogen content and lower biotin content of beet molasses than is the case of black-strap molasses.

The molasses are diluted with water, pasteurized, blended and supplemented with the following compounds:

(i) ammonia (basic) and ammonium sulphate (acidic) to a pH of 4.0 - 4.5

(ii) diammonium hydrogen phosphate

(iii) magnesium sulphate

In addition to the above-mentioned compounds, biotin, and often thiamin or yeast extract are added.

A small amount of the production medium is allowed to pass into the sterile, closed production fermentor. The inoculation of the fermentor is then done with either cream yeast or yeast propagated fermentatively in successive seed cultures from laboratory cultures. Sterile air is introduced into the fermentor through a horizontal array of perforated tubes at the bottom of the vessel, or down the centre of a motor-driven impeller and it is let out through the impeller blades.

The cells of the inoculum use the available nutrients, and fresh medium is fed incrementally into the vessel. This Zulauf (fed batch) process is necessary, since high concentrations of sucrose, glucose or fructose (the sugars in molasses) repress respiratory enzymes.

Since the growth of the yeast cell is exponential, exponentially increasing amounts of medium are fed to keep the ratio of the cell mass to sugar constant. The aeration rate is accordingly adjusted. The rapid growing, vigorously aerated culture produces a considerable amount of foam. Either polypropylene glycols or silicone antifoams may be used for suppressing the foam. Heat production occurs as a result of the metabolism of the yeast cells. Therefore, it is essential to maintain the temperature at 30°C. This is accomplished by circulating coolant through internal coils. It is also required to maintain the pH in the range of 4.0 - 4.5. This is accomplished by the addition of suitable amounts of ammonia and ammonium sulphate or ammonia and sulphuric acid. The empirical equation derived by J. S. Harrison for bakers' yeast production is:

$$\underset{200\text{ gms.}}{\text{Sucrose}} + \underset{10.4\text{ gms.}}{\text{Ammonia}} + \underset{102.5\text{ gms.}}{\text{Oxygen}} \longrightarrow \underset{100\text{ gms.}}{\text{Yeast}} + \underset{77.2\text{ gms.}}{\text{Water}} + \underset{145.6\text{ gms.}}{CO_2}$$

Such an equation enables the producer to calculate the amount of sugar, ammonia and air required to produce 100 gms. of yeast dry matter. In addition, the amount of heat liberated during the production of yeast can also be known. For example, 387 K. Cal. of heat is generated during the production of 100 gms. of yeast. Thus, the producer is enabled to calculate the amount of cooling needed.

The pH of the fermentation broth is allowed to rise during the end period of the process. By doing so, yeast cells are made free of the adsorbed coloured material. Finally, the addition of nutrients is stopped and aeration is continued for 30 – 60 minutes. This maturation or ripening process allows cell division to be completed and reduces the number of budding cells to less than 20% of the total population. Bud counts greater than 20% reduce the keeping properties of the yeast. The final culture contains approximately 50 gms. dry matter per litre. The yeast cells are separated from the bulk of the liquid. These cells are collectively called yeast cream. The yeast cream is washed with water and centrifuged. Thus, the product containing 200 gms. dry matter per litre is obtained. This product is further dried on rotary vacuum filters. Salt added prior to filtration shrinks the cells. These cells are subjected to spraying with water on the filter. As a result of this, cells get swollen because they take up extracellular water. This results in the formation of a dried product containing 30 per cent dry matter content. Water and plasticizers may be added to the product scraped from the filter. And this facilitates the formation of blocks of yeast. These blocks must be kept at low temperature until they are used.

Pressed bakers' yeast is gradually being replaced by active dried yeast. The method of production of active dried yeast is similar to that of pressed yeast with the following exceptions:

(i) Specially bread yeasts, which offer stability against drying, are used.
(ii) Nitrogen and sugar feeds are restricted towards the end of the production methods.
(iii) Separation and filtration are designed to give products of some 35 per cent dry matter.
(iv) The yeast mass is extruded through a screen and chopped.
(v) The pellets are dried in a tunnel or rotary driers for 6 - 24 hours.
(vi) The product is packed in airtight containers.

Steps involved in the commercial production of bakers' yeast are shown in Figure 23.1. The advantages of dried yeast are:

(i) It has less bulk than pressed bakers' yeast.
(ii) It may be stored at ambient temperatures.
(iii) It is readily reconstituted by dispersion in warm water.

The major disadvantage is its lower activity, 65 per cent of that of pressed yeast.

Saccharomyces species produced on molasses media for use in distilling are produced in the same way as pressed bakers' yeast.

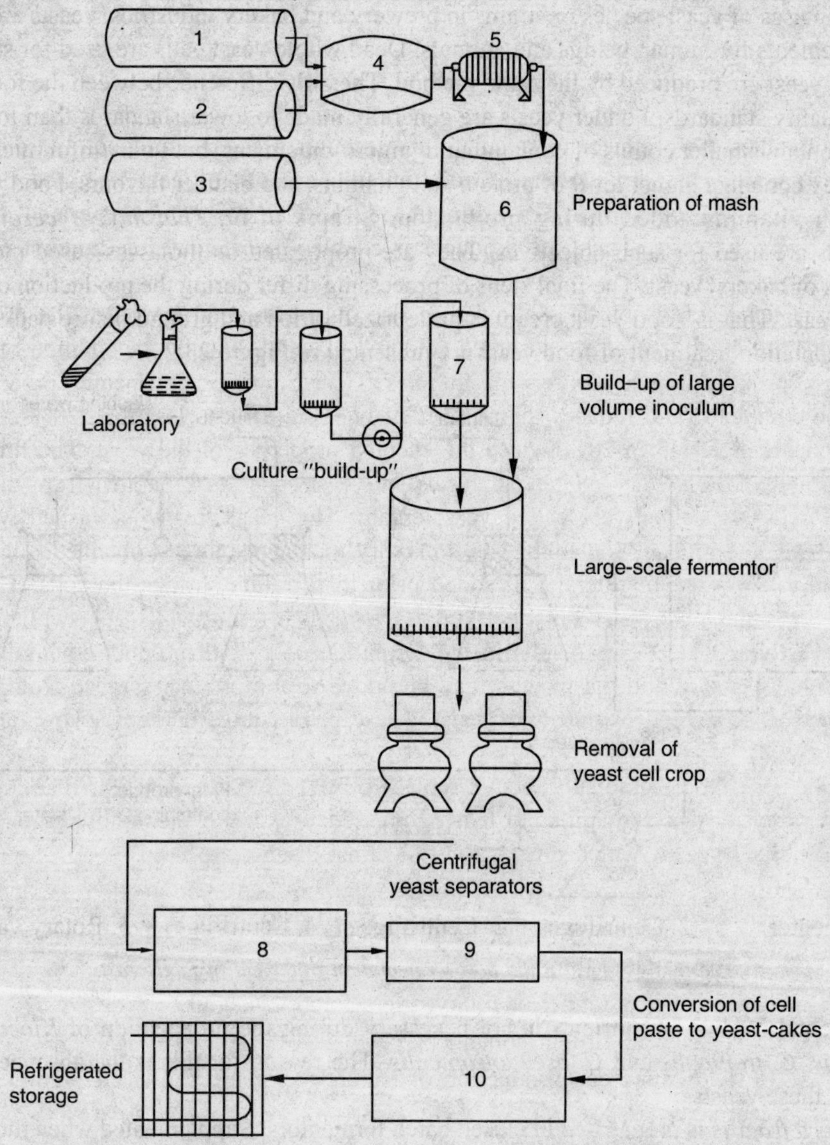

1. Cane-molasses
2. Beet-molasses
3. Additional nutrients
4. Mixing and cooking
5. Filter
6. Finished mash storage
7. Stock inoculum
8. Filter press
9. Mixer and extruder
10. Packing

Fig. 23.1 *Steps in the industrial production of bakers' yeast.*

4 PRODUCTION OF FOOD AND FODDER YEASTS

Apart from the uses of yeast species or strains in brewery and bakery industries, yeasts are also used as dietary supplements for human beings and animals. Dead whole yeast cells are used for such purposes. Both types of yeast are produced by the same method. The only difference between the food and fodder yeasts is in quality standards. Fodder yeasts are generally made to lower standards than food yeasts. So food yeasts contain smaller counts of contaminating micro-organisms, but no harmful micro-organisms. Moreover, they contain a higher level of protein and vitamins, and blander flavours. Food yeasts may be fortified with vitamins added during production. Strains of *Saccharomyces cerevisiae* and *S. carlsbergensis* are used for such objectives. They are propogated on molasses-based media in a way similar to that of bakers' yeast. The final steps of processing differ during the production of these yeasts and bakers' yeast. That is, food yeast cream is pasteurized, dried and ground before packing. Stages in the post-fermentation treatment of food yeast are presented in Figure 23.2.

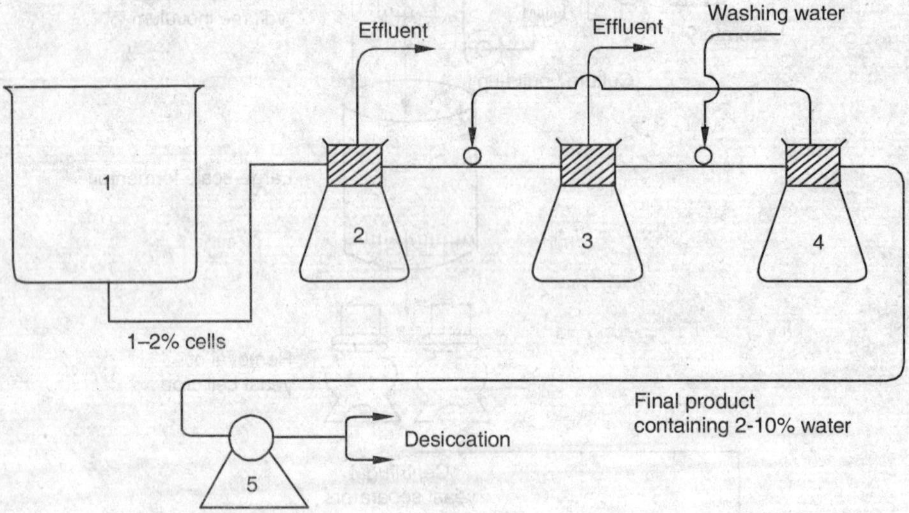

1. Fermentor 2. Centrifuge 3. Centrifuge 4. Centrifuge 5. Rotary vacuum filter

Fig. 23.2 *Stages in the post-fermentation treatment of food yeast.*

Incremental feeding of nutrients is not necessary during the propagation of *Kloeckera fragilis, Candida utilis, C. tropicalis* and *C. pseudotropicalis.* The reason for this is the absence of catabolite repression in these yeasts.

Kloeckera fragilis is propagated in closed batch fermentors. Supplemented whey medium, having pH 5.0 - 5.5, is used for growing this yeast. The optimum temperature is 30°C. It is essential to supply large amounts of sterile air. Finally, the product is subjected to pasteurization before packing.

Candida utilis is produced by a continuous process. Sulphite-liquor-based medium is used for propagation. The gaseous compound, sulphur dioxide, is removed from sulphite-liquor, since it is inhibitory for yeasts. The pH adjustment of the supplemented sulphite-liquor medium is done at 4.5 with the help of ammonia. Continuous feeding of the medium is done into a Waldhof fermentor inoculated with yeast. Vigorous aeration, like *Kloeckera fragilis,* is essential. A temperature of 32°C has been found to be satisfactory. It is maintained by circulating the culture through an external heat exchanger. Sulphuric acid and salts are used to adjust the pH. They are added in proportion to the sugar concentration in the vessels. When the maximum yield in the vessel is reached, the culture is removed at a rate equivalent to the input of fresh medium. The yeast is concentrated, pasteurized, dried and ground prior to packing.

The production of *Candida lipolytica* from petroleum products has not yet reached significant levels.

5 USES OF YEASTS AND YEAST PRODUCTS

Yeasts are, directly or indirectly, very useful micro-organisms to mankind. Species or strains of the genus *Saccharomyces* are widely employed by human beings. This can be exemplified as under:

(i) in the bakery industry (*S. cerevisiae*)

(ii) in the distillery industry (*S. cerevisiae* and *S. diastaticus*)

(iii) in the wine industry (*S. ellipsoideus*)

(iv) in the production of alcoholic beverages (*S. cerevisiae* for ale, *S. carlsbergensis* for lager and cider)

These yeasts are produced in a viable form as pressed yeasts. The bakery industries mainly use active dried yeast. Their enzymatic activities play an important role, because desired bioconversions are brought about by their presence in the yeast cells. In addition to these uses, they are also valuable as human foods and animal feeds. Surplus brewers' yeast and primary food yeasts are exploited for such uses. The reason behind the use of yeasts as food and feed is their high contents of protein and vitamins. Their uses are:

(a) *Treatment of dried yeasts*: Inactive, dried primary food yeast in powdered form is used directly as a dietary supplement. Moreover, it forms an ideal base substance for the addition of flavours, (e.g. cheese, ham, etc.). Yeast extracts are produced from dried yeasts. Acid hydrolysis is the most efficient method to produce yeast extracts. In this method, a 65 - 80 per cent slurry is refluxed with hydrochloric acid for 6 - 12 hours. After hydrolysis, the extract is subjected to neutralization, clarification and concentration, so as to give syrups (45% solids) and pastes (85% solids), or it is spray-dried to give powders (95% solids). Partial destruction of the vitamins of the yeast extract occurs during hydrolysis.

(b) *Treatment of viable yeasts*: Brewers' yeast cannot be directly used as a food additive, since the hop bitter substances are present on the surfaces of the yeast cells. Bitter substances can be removed by treating the yeast cells with alkali and the process is called debittering. Thereafter, viable debittered brewers' yeast, together with primary food yeast as pressed yeast or yeast cream, may be subjected to extraction by any one of the following two methods:

(1) autolysis

(2) plasmolysis

Autolysis involves self-digestion of the viable cells. Self-digestion is carried out at 45°C for 12 - 24 hours in the presence of small amounts of ethyl acetate or chloroform. Self-digestion brings about disintegration of the membrane of the vacuole of the yeast cell. As a result, lytic enzymes are released. Finally, the resulting extract is subjected to clarification and concentration to a syrup. The alternative is oven or spray-drying to yield granules and powders. Such an autolytic procedure is detrimental to the yeast cells.

Plasmolysis involves mixing of viable yeast cells with salts, sugar or certain acetate esters to form a slurry. This process extracts low molecular weight substances, (e.g. vitamins, nucleotides, amino acids, etc.) from the cells. But this method does not extract high molecular weight substances, (e.g. proteins). The resulting plasmolysates are separated from the yeast cells, concentrated and dried. Plamolysis, like autolysis, is lethal to the yeast cells.

Formulations of food yeasts for medicinal purposes invariably contain high levels of vitamins. Vitamins are usually introduced, either during the production of yeast by addition to the fermentor, or by mixing them with the processed product before packing. Either method gives vitamin levels some fifty-fold higher than those of unfortified yeasts.

(c) *Yeast constituents:* Yeast is a major source of both enzymes and biochemicals used in industry and research, respectively.

(i) Invertase: It is obtained from selected strains of *Saccharomyces cerevisiae* grown aerobically on sucrose. It is available in the market as an invertase-rich, dried yeast preparation or as a concentrated syrup obtained following autolysis. Invertase is useful in bringing about the hydrolysis of sucrose:

$$\text{Sucrose} \xrightarrow{\text{Invertase}} \text{Glucose} + \text{Fructose}$$

It is, therefore, used to make soft-centred chocolates. The filling flavour material is mixed with sucrose granules (to make it solid) plus a little invertase. The chocolate coating is then applied, and during storage the enzyme acts upon the sucrose of the filling causing it to undergo liquefaction.

(ii) β-Galactosidase: It is obtained from *Saccharomyces fragilis*, *Candida pseudotropicalis* or *C. utilis* grown on cheese whey. It is available as a vacuum-dried product. It finds application in the manufacture of ice-cream (to remove lactose). In addition to this, it is used to prepare lactose-free milk products to be added to special foods.

(iii) Other enzymes: Enzymes, namely alcohol dehydrogenase, glyceraldehyde-3-phosphate dehydrogenase, hexokinase and many others are obtained from *Saccharomyces cerevisiae*. All these enzymes are useful in research work. These are usually obtained from autolysates and subsequent purification is carried out by precipitation and ion-exchange chromatography.

(iv) Co-factors: Co-factors, viz. Co-enzyme A, oxidized and reduced forms of NAD and NADP, ATP and glutathione are also obtained from *Saccharomyces cerevisiae*. These co-factors, like enzymes, are useful in research work. Isolation and purification of these compounds are the same as those of enzymes discussed in (iii).

(v) Nucleic acids: Nucleic acids, particularly RNA, are obtained from *Candida utilis* and *Saccharomyces*. RNA is used as a raw material for the production of derivatives for biochemical research. In addition to this, RNA is also used in the preparation (by hydrolysis) of flavour compounds, (e.g. guanosine and inosine 5' monophosphates).

24 Production of Single-Cell Proteins

1 INTRODUCTION

We are all aware of the fact that life is dependent on life. Hence, biologists have directed their attention to the use of microbial proteins in animal and human diets. Biomass production has played a significant role in our attempts to make good the protein component of world food shortages. It is alluring to produce biomass to meet the world's protein needs because:

(i) Microbes can synthesize proteins much more rapidly than higher living systems, (i.e. plants and animals).

(ii) Microbes have a short generation time, (e.g. 0.3-2.0 hours in bacteria).

(iii) Microbes can be grown on media containing cheap sources of carbon and nitrogen.

(iv) Microbes have a high protein content (7-12 gms. protein N per 100 gms. dry weight).

For single-cell protein (SCP) production, microbes such as bacteria, yeasts, filamentous fungi and algae have been successfully employed. These have been briefly discussed here separately.

The term 'single-cell protein' (SCP) was coined at Massachusetts Institute of Technology by Professor C.L. Wilson, in 1966, to represent the cells of algae, bacteria, yeasts and fungi grown for their protein contents (Scrimshaw, 1968).

2 BACTERIAL PROTEINS

Some species or strains of bacteria are, and can be used for producing SCP products. The major groups of interest are:

(i) hydrogen-utilizing bacteria

(ii) methane or methanol-utilizing bacteria

(iii) *n*-paraffins-utilizing bacteria

(iv) photosynthetic bacteria

These have been briefly dealt with here:

(i) Hydrogen-utilizing bacteria: These are aerobic bacteria. They obtain energy by the oxidization of hydrogen. They use CO_2 as a source of carbon. The species of interest in this group include:

Pseudomonas facilis, P. saccharophila, P. ruhlandii, P. flava, P. palleronii, Alcaligenes autrophus and *A. paradoxus.*

Hydrogen serves as a substrate for these bacteria. Carbon dioxide is supplied as a source of carbon. Ammonium salts, nitrates or urea serve as a nitrogen source. The trace elements needed include K^+, Mg^{2+}, Ca^{2+}, Fe^{3+}, Ni^{2+}, PO_4^{3-} and SO_4^{2-}. The optimum pH of the production medium is 6.5 - 7.5.

The optimum temperature range for mass cultivation of this group of bacteria is 33-35°C and the oxygen requirement is high. For example, a bacterial culture containing 10 gms./litre of cells (gravimetric determination) requires 300m moles O_2 per hour. In other words, the O_2 supply serves as a limiting factor for growth. A gas mixture of suitable proportions ($H_2/O_2 = 2$) can be furnished by electrolysis of the mineral salts medium within the fermentor, using platinum electrodes. In batch process, yields of 28 gms./litre have been obtained after 10 days. At the end of the process, bacterial cells are recovered by centrifugation. Because to their smaller size, the cost of the centrifugation may be four times as high as that for yeast centrifugation.

These bacteria contain 15 gms. nucleic acid and 20 gms. cell wall material in addition to protein per 100 gms. (dry weight). Table 24.1 shows amino acid profiles of *Alcaligenes eutrophus* and that of casein for comparison.

Table 24.1 *Amino acid profiles of Alcaligenes eutrophus compared with that of casein (gms. amino acid per 16 gms.N).*

Amino acid	*Alcaligenes eutrophus*	*Casein*
Tryptophan	...	1.34
Threonine	4.52	4.30
Lysine	8.61	8.06
Methionine	2.69	3.10
Cystine	...	0.38
Isoleucine	4.58	6.59
Leucine	8.52	10.11
Phenylalanine	3.96	5.42
Tyrosine	3.26	5.86
Valine	7.13	7.44
Histidine	2.48	3.04
Arginine	8.00	4.10
Alanine	8.80	3.38
Aspartic acid	9.57	7.44
Glutamic acid	11.17	2.32
Glycine	5.47	2.00
Proline	3.46	11.82
Serine	3.47	6.69

Many strains of hydrogen-utilizing bacteria accumulate cytoplasmic granules of poly β-hydroxybutyrate. Accumulation of these granules takes place when O_2 is the limiting factor or when the medium is deficient in N or P. Moreover, these granules are not assimilated by animals.

These bacteria are a useful protein source. The noteworthy feature of these bacteria is that, the substrates (CO_2, H_2, O_2 and mineral salts) carry no possibility of contamination of the biomass with extraneous carcinogens.

(ii) Methane or methanol-utilizing bacteria: These bacteria are capable of utilizing one-carbon organic compounds, (i.e. methane and methanol). A number of bacteria utilize methane as a carbon and energy source for growth. These include *Methanomonas methanica, M. methanooxidans, Methylococcus capsulatus, Pseudomonas methanica* and others. On the other hand, methanol-utilizing bacterial species are: *Methylomonas methylovora, M. clara, M. methanolica, Methylophilus (Pseudomonas) methylotrophus, Xanthomonas* spp. and others.

In addition to these bacterial species, members of other bacterial genera, (e.g. *Bacillus*, *Vibrio*, etc.) have been reported as utilizing both methane and methanol.

Apart from the low cost of methane, there are certain difficulties during its use as a substrate. These include :

(i) the risk of explosion
(ii) problems of mass transfer across the cell-membrane
(iii) prolonged generation time of the methylobacteria, (i.e. 3-16 hours)
(iv) high heat generation during growth

When methane is used as a substrate, the oxygen concentration in the gas should be less than 12 % (by volume). On the other hand, some difficulties are eliminated or reduced when methanol is used as a substrate. With this substrate, the problem of mass transfer across the cell-membrane does not arise. In addition, there is a reduction in O_2 requirement and less heat is produced as compared to that of the process using methane as a substrate.

The bacterial cells cultivated on methanol have a protein content of 35-70 gms. / 100 gms. cells (gravimetric). For example, the amino acid composition for *Pseudomonas* (TM 20) cultivated on methanol is:

Amino acid	*Amount (gms. amino acid/16 gms. N)*
Cystine	0.32
Histidine	1.73
Isoleucine	3.90
Leucine	6.96
Lysine	5.30
Methionine	1.81
Phenylalanine	4.18
Threonine	4.52
Tryptophan	—
Tyrosine	2.91
Valine	5.85

The Imperial Chemical Industries (ICI) developed a continuous process for producing biomass using *Methylophilus (Pseudomonas) methylotrophus* (Figure 24.1). The cultivation medium used contains methanol (produced from natural gas), inorganic salts and ammonia. Aeration of the medium is required to be very high. It is not necessary to agitate the medium, since the design of the fermentor (pressure cycle fermentor) facilitates rapid internal circulation of air. Continuous removal of bacterial suspension (containing about 3 per cent dry weight cells) is performed. Finally, the concentration of cells is carried out by a flocculation and flotation process (recovered fluid being recycled) and then drying of cells is done. The product is called 'Pruteen'. It contains 72 per cent protein and 8.6 per cent total lipids.

Fig. 24.1 *ICI's single-cell protein plant.*

Moreover, lysine and methionine are more in amino acid profiles, being comparable to that of fish meal. A process employing a loop-type fermentor has also been developed (Figure 24.2).

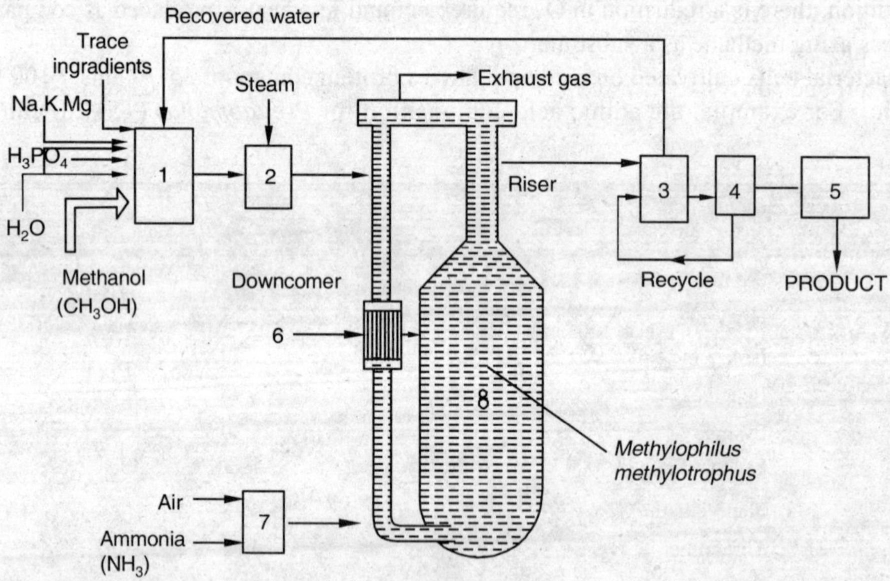

1. Preparation of medium 2. Sterilization of medium 3. Pre-separation 4. Centrifuge 5. Drying 6. Cooling tower 7. Sterilizing filter 8. Pressure cycle fermentor

Fig. 24.2 *Production of bacterial protein using methanol as a carbon substrate.*

(iii) *n*-Paraffins-utilizing bacteria: Hydrocarbon-utilizing bacteria include *Brevibacterium insectiphilium* ATCC 15528, *Corynebacterium* spp. ATCC 15529, *C. paurometabolum* ATCC 15530, *Pseudomonas ligustri* ATCC 15522, *P. pseudomallei* ATCC 15523, *P. orvilla* ATCC 15524, *Alcaligenes* spp. ATCC 15525, *Cellulomonas galba* ATCC 15526 and *Micrococcus cerificans* ATCC 14987. These bacteria were isolated from soil samples by a team of researchers of the ESSO group. Many similar strains were also isolated by the Chinese Petroleum Corporation, Taiwan. Of these isolated soil bacteria, a strain of *Pseudomonas* 5401, was chosen. This becomes more clear from the brief description of the claim given below.

Bacterial protein, Chinese Petroleum Corporation, Taiwan, claim

(a) most promising strain: *Pseudomonas* 5401 (ATCC 21094)

(b) feed stock:

(i) *n*-paraffins—as the sole carbon source

(ii) ammonia or ammonium salts—as the nitrogen source

(iii) air—to provide oxygen

(c) fermentation conditions:

(i) time—24 hours

(ii) temperature—36-38°C

(iii) pH—7.0

(d) maximum cell concentration: 16 gms./litre (dry weight)

(e) with a feeding rate of 12.5 per cent of the volume of the culture per hour, a cell concentration of 10 gms./litre could be maintained

(f) with a feeding rate of 25 per cent of the volume of the culture per hour, a cell concentration of 8 gms./litre could be maintained

(g) yield of dry cells: 105 per cent of original fuel oil

(h) protein content of dry cells: 73.62%

(i) product analysis:

Component	*Unwashed (%)*	*Washed (%)*
Protein	71.21	80.0
Oil	2.07	1.17
Moisture	9.96	—
Ash	9.28	5.63

(j) amino acid pattern comparison (%):

Amino acid	*Pseudomonas 5401 on gas oil*	*Pseudomonas 5401 on fuel oil*	*Unidentified yeast*
Arginine	4.98	4.68	2.37
Aspartic acid	7.24	8.47	5.63
Alanine	6.37	6.41	3.36
Cystine	0.43	—	—
Glutamic acid	8.67	10.05	7.57
Glycine	3.75	4.26	2.45
Histidine	1.21	1.87	1.03
Isoleucine	3.84	4.33	2.35
Lysine	4.29	9.06	3.89
Leucine	6.57	6.74	3.60
Methionine	1.27	0.81	—
Phenylalanine	3.29	3.39	2.17
Proline	2.94	2.36	1.59
Serine	2.96	3.07	2.52
Threonine	4.08	3.77	2.54
Tryptophan	1.16	0.91	0.27
Tyrosine	2.74	2.30	1.56
Valine	3.48	5.08	2.60

Thus, the essential amino acid composition is better than that of yeast, particularly as the content of lysine and methionine is higher. Various B group vitamins are also present in the dried cells:

Vitamin	*Amount* (μ *gms./gm. dry weight*)
Thiamine (chlorhydrate)	2.88
Riboflavin	10.50
Niacin	180
Calcium pantothenate	7.0
Pyridoxine	1.6

(iv) Photosynthetic bacteria: Photosynthetic bacteria may be a potential feed source for aquaculture systems or livestock. At present, processes for biomass production using photosynthetic bacteria as a source of human food have not been developed.

Media with the known and simple composition may be used for the mass-cultivation of photosynthetic bacteria. Previous studies (Wai, 1971 and 1972) have suggested that photosynthetic bacteria may be well-cultivated in extracts of crude carbohydrates prepared from bananas, potato starch wastes, wheat bran and rice bran.

Heden and Levin (1959) have grown photosynthetic bacteria in closed stainless steel fermentors by means of internal illumination. A light intensity of 100 lumens/litre in preliminary glass-flask cultivations has been found to be optimal. Light intensitities, higher than optimum, have been found to depress the growth.

The optimum temperature for the growth of most photosynthetic bacteria is approximately 37°C. Temperatures, lower than optimum may have harmful effects on growth. The pH and H_2S concentration are very important for the predominant development of many forms of purple photosynthetic bacteria. The optimum pH value of the medium for growing most of these bacteria is between 7.0 and 8.5. On the other hand, the optimum pH value of the culture medium for growing a limited number of photosynthetic bacteria is between 6.5 and 6.8 (Pfenning, 1967).

3 ACTINOMYCETOUS PROTEINS

Some actinomycetes have been reported as a potential single-cell protein source. These include *Nocardia paraffinica, Mycobacterium phlei* NRRL B-3423, *Thermomonospora fusca* and *Streptomyces* spp.

For the purpose of single-cell protein products, actinomycetes may be grown on media containing a suitable source of carbon, (e.g. methanol, gaseous *n*-alkanes, liquid *n*-alkanes or cellulosic wastes) depending upon the species or strain employed.

Other conditions, (e.g. medium, pH of medium, temperature for satisfactory growth) are almost similar to that of the bacteria discussed above.

4 YEAST PROTEINS

This has already been dealt with in Chapter 23.

5 FUNGAL PROTEINS

As a general rule, protein from fungi is deficient in methionine, and there could be a problem regarding excessive nucleic acid. Otherwise, fungi produce first-rate proteins with the advantage that they can be produced rapidly from any organic waste.

The ability of fungi to break down cellulose is of special interest. The most common natural waste product is cellulose. It is a cheap and abundant source of carbon and energy. Therefore, cellulose is a potential substrate for growing mixed fungal cultures, particularly those containing the cellulose-attacking fungus, *Trichoderma viride.* However, much of the work has centred on the amylolytic fungi. For example, these fungi, (e.g. *Aspergillus oryzae* or *Rhizopus arrhizus*) are used to inoculate barley mash. Of course, urea or ammonium sulphate is added as a source of nitrogen. An *Aspergillus* spp. is exploited in the production of Japanese soya based condiments, whereas *Rhizopus* spp. is used in the manufacture of Indonesian miso. The very first conversion that occurs is that of barley starch into glucose by the amylolytic enzymes from fungi. Thereafter, glucose and nitrogen are assimilated with the formation of fungal protein. The starch/protein conversion rate is 23 per cent for *Rhizopus* spp. whereas it is 40 per cent for *Aspergillus* spp. The factors affecting the growth rate of fungi are: the starch content and the nature of nitrogenous compound. In the case of *Aspergillus oryzae,* yield of 10.5 gms. mycelium (45% protein content) per litre of mash is obtained after an incubation period of 34 hours at 30°C. The product obtained in this way is readily accepted by pigs.

The present indication is that emphasis should be laid on the initial breakdown of cellulose-utilizing fungi. If this major raw material can be partly digested, it is easy to grow yeasts and productive bacteria on the hydrolyzed cellulose.

6 ALGAL PROTEINS

The protein content of algae is approximately 50 gms./100 gms. algal mass (dry weight). In addition to protein, they contain abundant amounts of β-carotene, vitamin K and a little quantity of vitamin C. But they are poor sources of vitamins of the B group. Moreover, protein contains low levels of sulphur-containing amino acids. Also, the digestibility of algae is low and the taste can be unpleasant.

The algae which appear to have raised most interest are the members of the following genera:

Scenedesmus	*Chlorella*
Chlamydomonas	*Coelestrum*
Spirulina	*Nostocales*
Anabaena	*Chrysophyta*

Most of the industrial interest is in *Scenedesmus* and *Chlorella, Scenedesmus* has an interesting composition of:

	Per cent
Protein:	50-60
Fatty acids:	12-14
Carbohydrates:	10-17

The remainder of *Scenedesmus* is 3-10% fibre and 4-8% water.

Sosa Texcoco S.A. in Mexico has started harvesting *Spirulina* species growing wild in Lake Texcoco (Figure 24.3). The algae is supplied to food-processing industries for addition to a level of 10% in chocolate-flavoured biscuits and up to 30% in other foods, notably in jams and soups. There is a parallel interest in wild algae from Lake Chad in Africa.

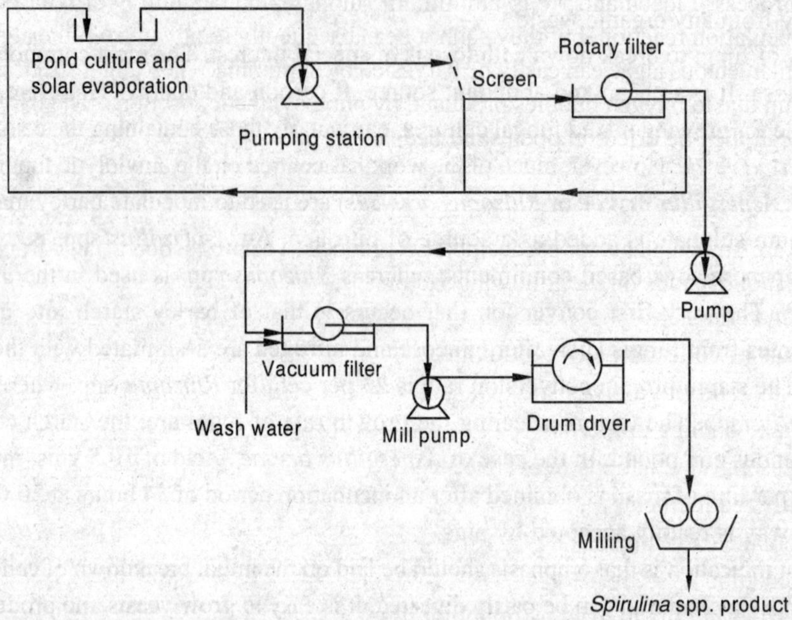

Fig. 24.3 *Flow-sheet of the Sosa Texacoco pilot plant for producing Spirulina species (Benemann et al., 1979).*

There have been various attempts to grow algae on a commercial scale. The Japanese method has used flat trays with artificial sunlight. The West German experiments used basins while the Czechoslovakians used sloping troughs. As a rule, industrial methods have produced at a rate of about 10 gms. dry algal protein/square metre/day.

Although the selection of species is yet far from clear, the production conditions are understood. The finer details of equipment for commercial production of algal biomass are determined by the medium and the selected organism. The algae have to be cultivated in shallow ponds with mechanical stirrers or in deeper ponds (not more than 20-30 cms. deep) with circulation pumps. The simplest and most suitable mixture of carbon and nitrogen for growth would be carbon dioxide, ammonia and a few trace minerals. For the continuous movement of the algal growth it is necessary to allow the maximum exposure of each organism. Thus, it was shown that *Scenedesmus* spp. grows twenty times faster in the optimum light provided than in natural conditions. The necessary temperature is 25-35°C depending upon the strain involved. The optimum temperature varies according to the intensity of the light. The optimum pH value for maximum growth is important. This depends upon whether the algae production is to be related to the industrial acid waste or to the agricultural waste, or whether the natural water concerned is acid or alkali. For example, *Spirulina* spp. appear to thrive at pH 9.5 and it is quite probable that strains can be found to thrive at pH 10.5. With open air ponds it is impossible to maintain culture purity.

Reported yields are not indicative of possibilities. Dortmund reported 28 gms./square metre/day, Bangkok claimed 35 gms. and Bulgaria is said to have reached 45 gms. It is possible that, commercial yields with suitable selection of the species will reach 70-170 tons/hectare with sewage as a growth medium.

There is a major problem for harvesting the algal biomass from the spent liquid medium. This is attributed to the low dry weight biomass concentrations, (i.e. 1-2 gms./litre) and the low sedimentation rates of this biomass. With unicellular algae, the harvesting of biomass involves flocculation or centrifugation process. Flocculants, (e.g. aluminium sulphate and calcium hydroxide) are effective in accelerating flocculation reaction, but they cannot be subsequently removed from the algal biomass. On the other hand, filamentous algae can easily be harvested by skimming. These algal species, (e.g. *Spirulina maxima*) come up due to oxygen bubbles produced by photosynthetic reactions. The harvested algae are dewatered and can then be dried on open sand beds.

7 ECONOMIC ASPECTS

At present there are limited plants for producing SCP products. The production of SCP from hydrocarbons, methanol or ethanol depends on the following factors:

1. Raw material
   - (a) ease of collection to a central area
   - (b) availability to a given site
   - (c) bulk handling properties
   - (d) seasonal fluctuations in availability.
2. Sterility requirements
   microbial encroachment.
3. Fermentation
   - (a) residence time in reactor (doubling time)
   - (b) cell concentrations attainable
   - (c) operating temperature (cooling water versus refrigeration)
   - (d) total oxygen requirements
   - (e) power requirements for mass transfer
   - (f) heat of reaction
   - (g) cell yields per pound of substrate consumed
   - (h) foaming tendency.
4. Cell harvesting techniques
   high speed centrifuges versus thickeners.
5. Washing and purification techniques for the removal of
   - (a) substrate residues
   - (b) raw material impurities
   - (c) nucleic acids
   - (d) metabolic by-products.
6. Product value
   - (a) percentage protein
   - (b) limiting amino acid
   - (c) digestibility.

Thus, the major factors that retard the present-day construction and operation of large-scale plants for producing feed-grade SCP products from hydrocarbons are:

(i) high costs of hydrocarbon feed-stocks
(ii) high construction and equipment costs

8 FUTURE

The prospect for profitable SCP production seems unlikely to improve in the present circumstances. And the technology will thrive only where special considerations make it worthwhile.

However, many small-scale plants producing animal feed material have been installed and are operating effectively. The safety and the nutritional value of SCP products have been demonstrated. The problems over their acceptance by regulating authorities for animal feeds have been for the most part overcome. On the other hand, serious commercial question marks hang over the large-scale plants designed for independent commercial viability and based on the use of hydrocarbon or petrochemical raw materials. It is probably true that the world is entering a period of increasing food shortage. SCP for animal feed will become increasingly scarce and costly.

At present the ICI process for SCP production is the only process that has been commercialized. It is certain that the technology developed for SCP production during the 1970s will almost certainly find a profitable application before the end of this century.

Whether SCP production comes to play a major role in food production or not, the contribution made by the development of these processes to the scientific understanding and technological development of the fermentation industry has been profound. By applying modern biochemical engineering concepts to the fermentation process, a new phase has been initiated in the development of the fermentation industry.

25 Microbial Insecticides

1 INTRODUCTION

Biologists have successfully tried to control insects harmful to crops with the help of pathogenic microbes. Such control is referred to as biological control. Fortunately, microbes useful as insecticidal agents are harmless to man and useful to animals. Many companies in the world produce such microbes to use them as biopesticides. The principal difficulty is in their economically competitive production technology as compared to chemical insecticides. Every day new microbial agents are successfully introduced as biopesticides. At present, bacteria, fungi and viruses have been introduced for the successful control of insects. Nematodes, protozoa and rickettsiae have also been reported as potential biopesticides.

2 HISTORY

The silkworm played a major role in the development of microbial insecticides. There was a disease of silkworms called muscardine, or calcino. This disease was causing serious losses to the Italian silk industry. Agostino Bussi took up the study of this disease. He was the first person to demonstrate that an insect disease could be caused by a micro-organism. In 1834, Bussi showed experimentally that the fungus *Beauveria bassiuna* was responsible for calcino. He also suggested that the micro-organism might be used for control purposes. Today spores of Bassi's fungus are produced by fermentation and are the subject of extensive field trials. There are some applications of these spores as microbiological control agents for a range of insect pests.

Another disease of the silkworm, flacherie, also contributed to the development of microbial insecticides. There were epidemics of this disease in Japanese silk farms. A Japanese bacteriologist, Ishiwatu, isolated *Bacillus sotto* from diseased caterpillars and showed them to be the cause of this disease (flacherie). Thereafter, Japanese bacteriologists found that only old sporulated cultures of *Bacillus sotto* were intensively infectious for healthy silkworms. On the contrary, young, non-sporulated cultures did not produce ill effects in the healthy silkworms. They, therefore, concluded that spores of *Bacillus sotto* were pathogenic for silkworms, but not the vegetative cells.

The isolation of *Bacillus sotto* marks the start of the development of microbiological control of insects but progress was slow at first. The Japanese isolate attracted little attention and the bacterium *(Bacillus sotto)* was deposited in a Japanese culture collection. In 1915, German microbiologist Berliner isolated a similar bacterium from the diseased caterpillars of the flour moth *Anagusta kuhniella.* In 1927, another German microbiologist, Mattes, also isolated a similar bacterium (*B. thuringiensis,* as he called it) from the same universal pest. Both these German bacteriologists observed that matured vegetative cells produced the Restekörper (or parasporal body). This was a structure in the cell differing from the typical refractile oval spores. The importance of this observation was not appreciated, though it was later seen to be critical for the development of our understanding of the pathogenicity of the organism. *B. thuringiensis* was widely disseminated amongst research laboratories and stimulated interest in the possibility of using the bacterium for control purposes.

Ultimately, scientists interested in controlling insect pests were attracted by the idea of using a bacterium for control. The 1930's saw fairly extensive field trials of spore dusts based on *B. thuringiensis* against the European corn borer (a serious pest of growing corn). The field trials were largely successful. Therefore, a commercial product called Sporiene was produced in France just before the Second World War. Sporiene consisted of a large number of spores of *B. thuringiensis* suspended in an inert carrier. The firm producing Sporiene was largely destroyed during the war. But an interesting study on Sporiene was carried out by Jacobs at Imperial College in London during the war. He used Sporiene for controlling the flour moth in stored flour. This study focused attention on *B. thuringiensis* as a potential control agent and set the scene for rapid progress in the subject.

3 USEFUL MICROBES

Insects are attacked by a multitude of pathogens. Many insect pathogens have been elaborately described (Table 25.2). A number of these are adaptable for mechanical dissemination as microbial insecticides for the inoculation of insect populations, soils, fields, orchards, forests with spores, microbial toxins, or virus suspension. Microbial insecticides are highly specific in action against a few closely related pests and are generally harmless to other animals.

3.1. Bacteria

Bacillus thuringiensis

Pathogenicity: There was a belief that the spore of *Bacillus thuringiensis* was responsible for pathogenicity. Subsequently a Canadian, Angus, separated the parasporal crystals (diamond-shaped) from the spores in aged cultures of *Bacillus sotto*. He also showed that pathogenicity for the silkworm was a function of the crystal alone. Spores freed from parasporal bodies were not capable of causing a disease when fed to silkworms.

At present, we recognize 12 groups of *B. thuringiensis.* All strains produce protein crystals. These crystals appear to play similar roles in pathogenicity. The crystals dissolve under alkaline reducing conditions and are insoluble in water under neutral or mildly acid conditions. Fortunately, the mid-gut contents of susceptible caterpillars (larvae) are generally alkaline. On ingestion the crystal dissolves in the mid-gut fluid. There also occurs partial hydrolysis of the dissolved crystal, since proteolytic enzymes are present in the mid-gut fluid. This digested protein crystal attacks the cementing substances which are present in the gut wall. Thus, loosening of the epithelial gut wall occurs. This helps in a consequent diffusion of liquid from the gut into the blood. As a result of this, the blood of the insect becomes highly alkaline. This leads to gut paralysis followed by general body paralysis often in as little as 60 min. In this way, mixtures of spores and crystals are pathogenic for a wide range of lepidopterous larvae and for a few sawflies. Fortunately, such mixtures are apparently non-toxic to other life forms, (e.g. humans animals and plants). This is one of the most desirable characteristic of a protein crystal in its use for pest control purposes.

Using electron microscopy and X-ray diffraction techniques, the basic sub-unit of the crystal has been characterized. It appears to be a rod-shaped or dumb-bell shaped structure of average length 12 nm. by 5 nm. (Norris 1969 and 1971). The two processes, namely crystal synthesis and spore formation, proceed simultaneously in *B. thuringiensis* cells. Moreover, both processes appear to be intimately related to one another. The amino acids required for the biosynthesis of a protein crystal are provided from the cellular protein, which undergoes breakdown at an early stage in sporulation.

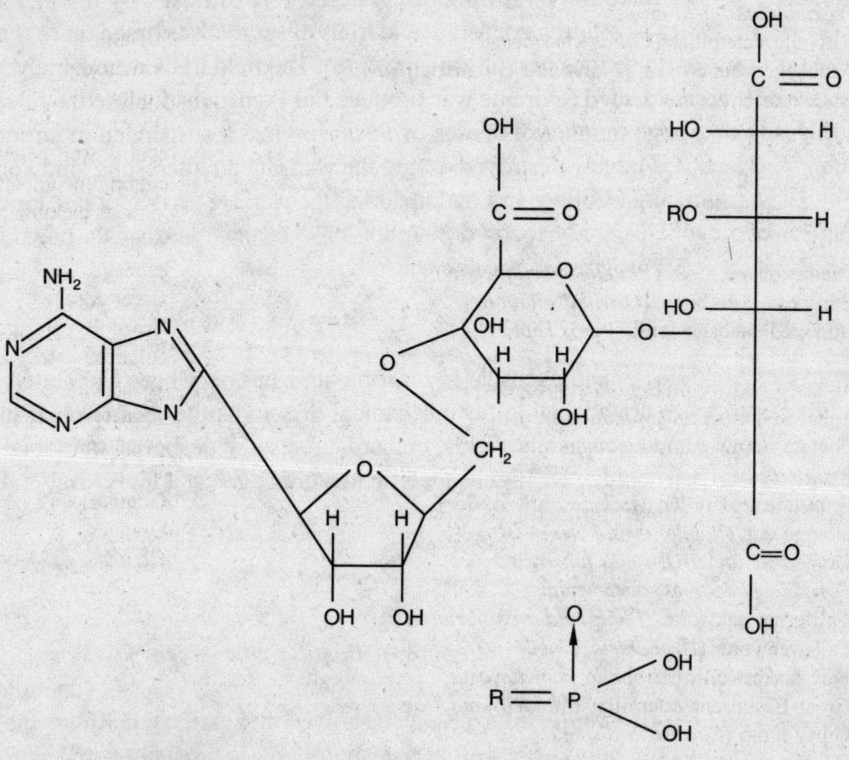

Fig. 25.1 *Structural formula of B. thuringiensis θ-exotoxin.*

The parasporal body is highly toxic for caterpillars. Some preparations of toxic protein give LD_{50}, (i.e. quantities killing 50 per cent of test larvae) values as low as 0.9 μg. per gm. of larvae. This value is comparable with some of the more widely used chemical insecticides. The spore/crystal mixtures of *B. thuringiensis* exert toxic effect for larvae of some 130 lepidopterous species, including many important pests. The symptoms following ingestion of a toxic dose vary greatly from one species to another. Often the potency of a spore/crystal mixture in the field is dependant on the presence of both components, but with some species this is not so.

Many strains of *B. thuringiensis* have been isolated and studied for their pathogenicity against different insect species.

The toxin production by a culture can be enhanced by controlling many factors, (e.g. use of efficient strain, fermentation conditions, etc.). Efficient strains are being isolated for the better production of toxins. Some strains of *B. thuringiensis* produce a second toxin which is called thermostable θ-exotoxin. This toxin is active against a wide range of insect species, including lepidoptera, diptera, hymenoptera, coleoptera and orthoptera. It is also toxic for other animals and this property does not permit us to use these strains as microbial insecticides in the field.

Table 25.1 *Some registered uses of Bacillus thuringiensis in the U.S.A.*

Pest Species	*Crop*
Vegetable and field crops:	
Alfa-alfa caterpillar (*Colias eurytheme*)	Alf-alfa.
Artichoke plume moth (*Platyptilia carcuidactyla*)	Artichokes.
Bollworm (*Heliothis zea*)	Cotton.
Cabbage looper (*Trichoplusia ni*)	Beans, broccoli, cabbage, cauliflower, celery, collards, cotton, cucumber, kale, lettuce, melons, potatoes, spinach, tobacco.
Diamondback moth (*Plutella maculipennis*)	cabbage.
European corn-borer (*Ostrinia nubilalis*)	sweet corn.
Imported cabbageworm (*Pieris rapae*)	Broccoli, cabbage, cauliflower, collards, kale.
Tobacco budworm (*Heliothis virescens*)	Tobacco.
Tobacco hornworm (*Manduca sexta*)	Tobacco.
Tomato hornworm (*M. quinquemaculata*)	Tomatoes.
Fruit crops:	
Fruit-tree leaf roller (*Archips argyrospilus*)	Oranges.
Orange dog (*Papilio cresphotes*)	Oranges.
Grape-leaf folder (*Desmia funeralis*)	Grapes.
Forests, shade-trees, ornamentals:	
California oakworm (*Phrygunidia californica*)	
Fall webworm (*Hyphantria cunea*)	
Fall cankerworm (*Alsophila pometaria*)	
Great Basin tent caterpillar (*Malacosoma fragile*)	
Gipsy moth (*Porthetria dispar*)	
Linden looper (*Erannis tiliaria*)	
Salt-marsh caterpillar (*Estigmene acrea*)	
Spring cankerworm (*Paleacrita vernata*)	
Winter moth (*Operophetera brumata*)	

The θ-exotoxin has a low molecular weight (800-900). It is a nucleotide containing one phosphate group per adenine molecule. In addition, an unusual sugar called allomucic acid is present in the molecule (Figure 25.1).

Bacillus thuringiensis in the field

Recently, commercial products containing *B. thuringiensis* have been produced by at least 12 manufacturers in five different countries. In the U.S.A., hundreds of tons are produced each year and the production seems to be rising. Commercial preparations of *B. thuringiensis* are currently registered in the U.S.A. for use on more than 20 agricultural crops, in addition to trees, ornamental shrubs and forests. They are used to control at least 23 insect pest species as shown in Table 25.1.

Insecticides based on *Bacillus thuringiensis* are used for short term control. They do not persist from year to year. Also, they do not spread extensively in the insect population. During crop spraying, it is essential to make efficient coverage of a leaf surface. Thus destruction of foliage is limited. The user of such insecticides must protect the sprayed toxin from inactivation by sunlight, since it shares a degree of ultraviolet sensitivity with other proteins. In order to overcome the above difficulties, emulsifying agents, stickers and ultraviolet protectants can be used in the *B. thuringiensis* insecticide formulation. Thus, the value of a product can be increased.

Insecticides based on *B. thuringiensis* are currently used in the areas of forestry and on green crops, (e.g. tomatoes, tobacco, etc.). A well-known microbial insecticide, called Dipel, is based on a potent isolate of *B. thuringiensis,* strain HD-1. Dipel was tested for its effectiveness against various pest species. Field-trials of Dipel, combined with small amounts of chemical insecticides, were also made and promising results have been secured. Therefore, this field is rapidly being explored at the present time. Dipel is calibrated for field use at 16,000 IU mg-1. A minimum viable spore count of 2.5×10^{10} viable spores g-1 is set.

Bacillus popilliae/Bacillus lentimorbus

The Japanese beetle (*Popillia japonica*) was introduced into the U.S.A. around 1915. It soon spread to become a serious problem throughout the eastern seaboard and into Canada. The various larval instars of the beetle cause extensive damage to lawns, pastures and shrubberies. Pupation takes place in the soil during May and adult beetles emerge about the second week in June and live for some 40 days during which time the female will lay up to 60 eggs. Following its introduction and the recognition that a serious pest situation existed, the U.S. Department of Agriculture set up a Japanese Beetle Control Laboratory in 1929. The objective of this laboratory was to limit and possibly eradicate the pest. Before long, workers from this laboratory found diseased larvae in field populations. These were characterized by a milky white appearance. The milky diseases are caused, principally, by two closely similar bacteria: *Bacillus popilliae* and *Bacillus lentimorbus.* Both of these have been developed for microbial control of the pest. But *B. popilliae* has been studied more extensively, since it has proved to be most potent in the field.

Following the ingestion of viable spores of the bacterium by larvae, germination takes place in the gut. The vegetative cells (resulting from germination of spores) penetrate the gut wall and enter the body cavity, where they multiply. Infected larvae usually live for a considerable time, and extensive growth of the micro-organisms occurs in the haemolymph. The mechanism of the pathogenicity of the bacterium is obscure. There is a refractile body called the parasporal body enclosed in the sporangium of *B. popilliae.* There is no evidence that this parasporal body plays any significant role in pathogenicity. The major pathogenic effect of the bacterium on the larvae may be attributed to the removal of nutrients and essential growth factors from the blood by the actively growing cells. Toxins are also involved, since cell-free filtrates of bacterial cultures exhibit toxicity if injected in small amounts into larvae.

The spore-containing material is applied to soil in 2 gram amounts at intervals of a few feet. There are various means for the dissemination of the disease in the insect population. Some of them are:

(1) the wandering of infected larvae
(2) dissemination by wind and water
(3) passive transfer by birds and animals
(4) by human activity

Such a treatment programme results in the dissemination of the infection throughout the Japanese beetle population in the treated area. The infection persists over a period of three seasons. Thus, the reduction of larval numbers occurs.

The milky disease bacteria have been found satisfactory in reducing the numbers of beetles in the treated area. Therefore, the Japanese beetle is no longer a serious pest over much of its original range. Some authorities in the U.S.A. have completely eradicated this pest by the use of milky disease bacteria.

The milky disease bacteria have also been employed in controlling pests other than the Japanese beetle. Several similar bacteria have been isolated from related larvae in other regions of the world. After a detailed study of these bacteria, it has been emphasized that *B. popilliae* and *B. lentimorbus* are only commercially important pest control agents in this large group of micro-organisms.

3.2. Microbes other than Bacteria

Viruses

At present the field of viral insecticides is the most rapidly expanding branch of microbial insecticides. We are likely to see the most significant developments in the next decade or so.

Unlike plant and animal viruses, many insect viruses have protein coats. These protein crystals are insoluble in water. The individual virus particles are encased, either singly or in large numbers, in these protein crystals. These protein crystals serve to protect the virus particles. Moreover, such insect viruses remain infective even under normal storage conditions for many years and this property of these viruses is due to existing protein crystals. In other words, the protective protein crystals give the viruses their remarkable potential as insecticides.

There are two major groups of useful insect viruses. They are:

(a) the polyhedroses—Many hundred of virus particles are embedded in each polyhedral crystal.
(b) the granuloses—Only one virus particle is contained in each protein crystal.

Both types have some potential as microbial insecticides due to their rapid dissemination in insect populations. In practice, much attention has been given to the nuclear polyhedrosis viruses. Following their ingestion, they develop within the nuclei of host cells. They cause damage to the skin, fat body and haemolymph of infected larvae. Infected tissues become packed with the characteristic polyhedral bodies. Subsequently, there occurs liberation of millions of these polyhedra when the insect dies and its skin ruptures. The liquefying body contents of a disintegrating cadaver leak from the ruptured skin. It is a highly infective fluid. It is readily disseminated amongst the healthy insect population. The spread takes place by wind, rain, and positive attraction of healthy insects.

At present Viron H (a viral insecticide) is registered for use in the U.S.A. Viron H is effective in the field against the Cotton bollworm (a pest of corn, cotton and tomatoes) and the closely related insect, the Tobacco budworm. This agent is safe for use. Products are also undergoing development for use against the Cabbage worm and the Army worm. Promising field trial results have also been obtained with a NPV against the Gipsy moth and for a similar virus active against the Spruce budworm. Many others are being assessed for their field potential.

In the field a disease may take some time to develop. Therefore, timing of the application is an important factor. This factor is important particularly in those cases where the young instar larvae are more susceptible to the virus than older, more mature pests, (e.g. NPV of *Heliothis)*. First, instar larvae will die very rapidly, often within a day or two of application. But it may take from four days to a week to kill a third instar larva. For the successful application of a viral insecticidal agent in the field, the following two features are important:

(1) Viruses must be ingested by the insect.

(2) Foliage should be effectively covered with the agent.

Repeated spraying during the growing season is required because infections are non-persistent.

Fungi: Most fungi potentially useful as insecticides are safe for human beings and domesticated animals. Some of them can be grown *in vitro*. Fungal spores are remarkably persistent. They are not disseminated by eggs but dispersion by air is often good. They normally infect insects through the cuticle.

Nematodes: They are safe for man and useful insects. Some of them grow *in vitro*. They persist by free resistant stages. Nematodes pathogenic to insects are under trial.

Protozoans: They are absolutely safe for man and useful insects. They can be produced only *in vivo*. Their transmission occurs by insect eggs. They are under experimentation as insecticides.

Rickettsiae: They are supposed to be safe for man and useful insects. They grow only *in vivo*. Their dissemination commonly takes place through eggs. They do not have any resistant stage. Many of them are very virulent. Their use as an insecticidal agent is possible.

4 BIOMASS PRODUCTION

A number of insect pathogens have been reported. Many of the insect pathogens have been and are being produced in mass. Table 25.2 provides information regarding their production technology.

Table 25.2 *Production technology in biopesticides.*

Insect pathogen	*Described*	*Mass Produced*	*Available*	*Production technology*		
				Whole-organism	*Fermen-tation*	*Tissue-Culture*
Fungi	530	030	02	07	27	0
Viruses	721	033	03	33	00	9
Protozoa	1000	015	00	11	03	02
Bacteria	100	014	04	04	10	0
Rickettsiae	012	00	00	00	00	1

Production of important insect pathogens on a commercial scale is dealt with in this chapter.

4.1. Bacterial Insecticides

Two processes are used commercially in the mass production of *Bacillus thuringiensis:*

(a) sub-merged fermentation

(b) semi-solid fermentation

(a) Sub-merged fermentation: Different contributing factors have been briefly dealt with here.

Preparation of inoculum: On the basis of Megna patent, inoculum is prepared by passing *B. thuringiensis* cells through two media in shake flasks and then to a seed fermentor. Nutrient broth can be used for the first shake flask. The following medium is advisable for use in the second flask as well as seed fermentor:

Name of component	Amount (%)
Beet molasses	1.00
Corn-steep solids	0.85
$CaCO_3$	1.00

Each flask is to be incubated for 24 hours. The seed tank is incubated until vigorous vegetative growth is obtained.

Production medium: A typical production medium is:

Name of component	Amount (%)
Beet molasses	1.86
Cotton seed flour (pharmamedia)	1.40
Corn-steep liquor solids	1.70
$CaCO_3$	0.10

The production medium is designed so that the fermentable carbohydrate and nitrogen available for growth present are exhausted at approximately the same time after the commencement of sporulation.

Dulmage (1970) reported three production media (Table 25.3).

Table 25.3 *Production media for Bacillus thuringiensis.*

Ingredient	LEVEL (I./ltr.) Tryptone medium	Proflo medium	Soya bean meal medium
Tryptone	10.00		
Proflo		10.00	
Soya bean meal			15.00
Bacto-peptone R		02.00	
Dextrose	05.00	15.00 (a)	05.00
Corn starch	05.00		05.00
Yeast extract	02.00	02.00	
K_2HPO_4	01.00		
KH_2PO_4	01.00		
$MgSO_4.7H_2O$		00.30	00.30
$FeSO_4.7H_2O$		00.02	00.02
$ZnSO_4.7H_2O$		00.02	00.02
$CaCO_3$		01.00	01.00

(a) In the latter experiments, level reduced to 10.00 gms.

N.B.: All media diluted to volume with distilled water.

The proflo and the soya bean meal media would be economical for mass production of *B. thuringiensis* var. Kurstaki, HD-1, the tryptone medium would not.

Recovery: Different steps involved in the recovery process may be schematically represented as under:

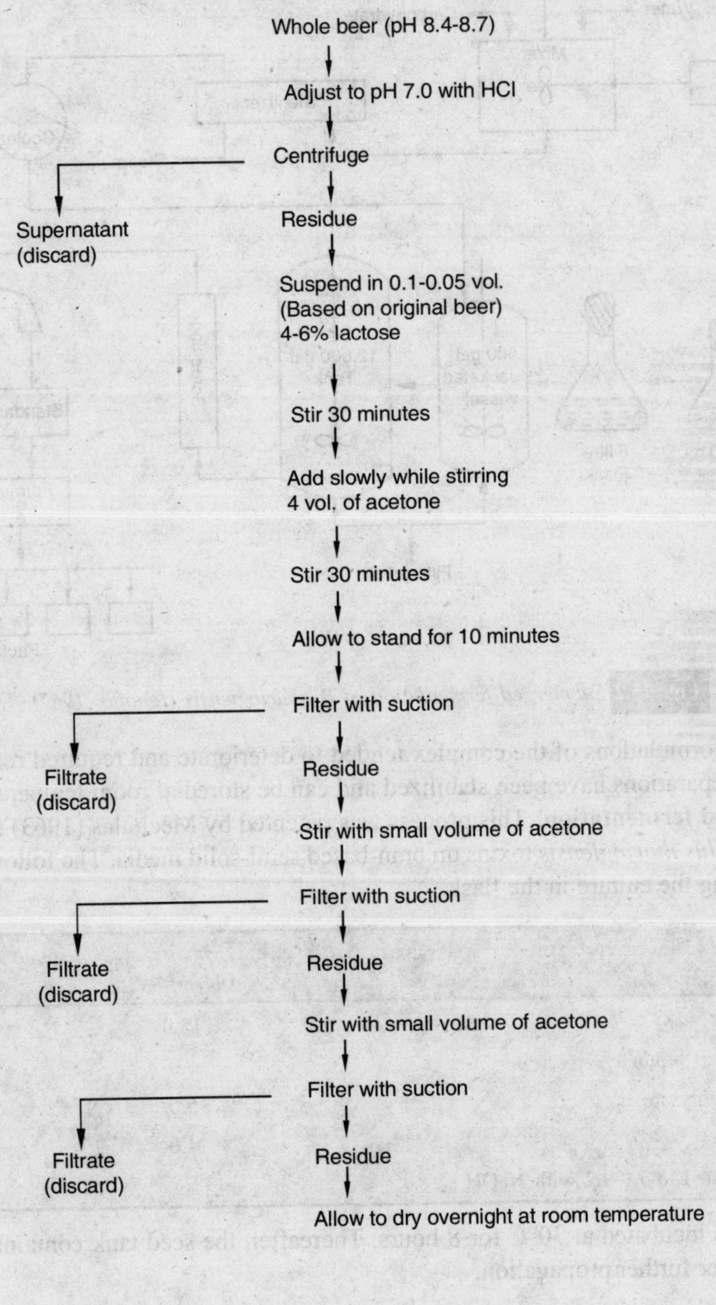

Flow-sheet: see Figure 25.2.

Submerged fermentations have been mainly used to produce flowable formulations (suspensions) of the spore-crystal complex. These can be used as sprays, but, of course, are not available as dusts.

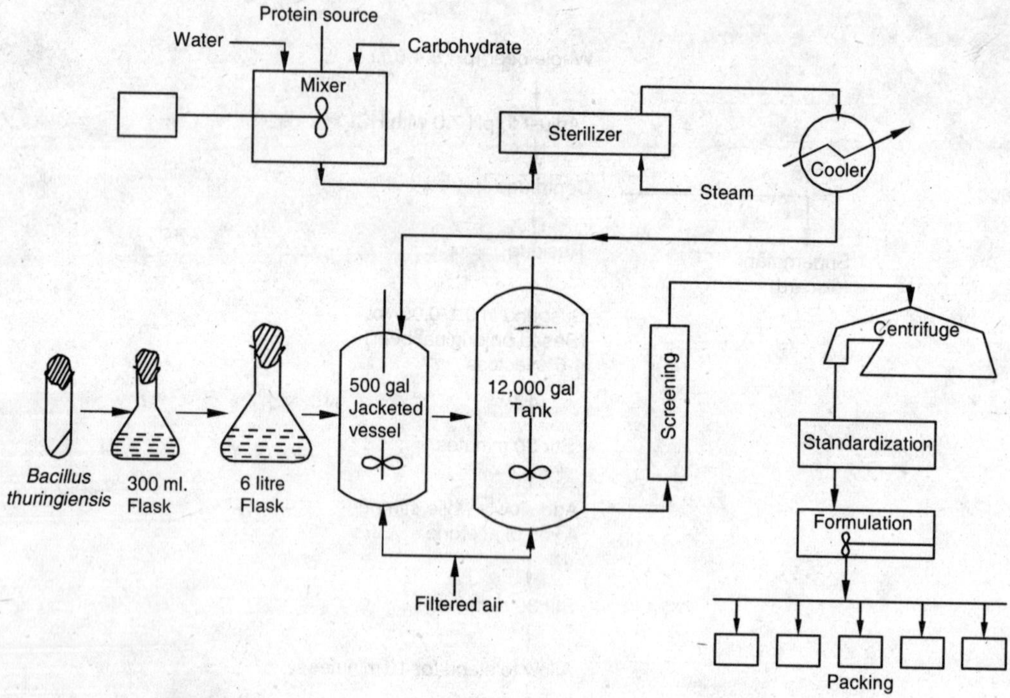

Fig. 25.2 *Submerged fermentation of B. thuringiensis (Ignoffo, 1967).*

In the past, liquid formulations of the complex tended to deteriorate and required refrigerated storage. Presently, these preparations have been stabilized and can be stored at room temperature.

(b) Semi-solid fermentation: This process was patented by Mechulas (1963) for the large-scale production of *Bacillus thuringiensis* toxins on bran-based semi-solid media. The following medium was used for propagating the culture in the flask.

Ingredient	*Amount (gms./litre)*
Dextrose	15.0
Corn-steep liquor	5.0
Yeast autolysate	5.0
K_2HPO_4	4.0
pH adjusted to 7.0-7.2 with NaOH	

The flask was incubated at 30°C for 8 hours. Thereafter, the seed tank containing the following medium was used for further propagation.

Ingredient	Amount (gms./litre)
Dextrose	10.0
Corn-steep liquor	4.5
Yeast autolysate	6.0
K_2HPO_4	3.5
NaOH	0.43
$CaCl_2$	0.10

The seed tank was inoculated with 0.1 per cent by volume of the shake flask seed and incubated at 30°C for 16 hours. Approximately 400 ml. of this tank seed was used to inoculate each kilogram of the semi-solid production medium. A production medium with the following composition was used to produce spores and toxins.

Ingredient	Amount (gms.)
Wheat bran	545
Expanded perlite	380
Soya bean meal	062
Dextrose	036
Lime	3.6
NaCl	0.9
$CaCl_2$	0.29
Water	160 ml.

After partial sterilization with flowing steam for 60 minutes, the mixture weighed about 1 kg. This was inoculated with 400 ml. of tank seed giving the final mixture a moisture content of 60 per cent by weight and pH 6.9. In an actual fermentation, about 250 kg. of the inoculated bran was placed in bins with perforated bottoms through which air at 30-34°C and 95-100 per cent R.H. was passed at about 0.4-0.6 vol. of air/vol. inoculated medium/minute for the first 3 hours and 1.0-1.2 vol. thereafter. The bins were incubated for about 36 hours, during which time the pH rose to 7.5 and the moisture content dropped to about 53 per cent. At this time, dry air at 50-55°C was passed through the bran for another 36 hours. The bran medium was then harvested and ground through 80 mesh screens. The final product had a moisture content of 4 per cent and pH of 7.0.

The semi-solid process gives a wettable powder or dust formulation containing spores, crystals and exotoxin by merely drying and grinding the final bran cake. The product being dry, is inherently stable. However, bran is a major constituent of the final formulation. Bran is difficult to wet and tends to swell when wetted. These characteristics make bran an unsatisfactory component of a wettable powder. Thus, the semi-solid product has been a satisfactory dust-base, it has been difficult to use as a wettable powder.

4.2. Fungal Insecticides

Beauveria bassiana, Meturrhizum anisopliae and *Acrostalagmus aphidum* can all be grown in semi-solid fermentation. The media and procedures are similar to those employed in the *Bacillus thuringiensis* semi-solid fermentation. Here, a liquid inoculum is used to inoculate the bran, and many media have been developed. Many fungi prefer corn meal to bran in the medium. A useful seed medium has the following formula:

Ingredient	Amount (gms./litre)
Soya bean meal	10.00
Corn meal	10.00
Glucose	15.00
NaCl	05.00
$CaCO_3$	02.00

Semi-solid production medium contains more corn meal than in the inoculum medium. In other words, it is necessary to replace about half the bran in the formulation with corn meal. Since the developing mycelium binds the bran medium tightly, it is usually necessary for maximum yields to mix the fermenting bran once or twice before sporulation begins. To obtain sporulation after the mycelium is well developed, drying of cultures should be done as quickly as possible.

The semi-solid fermentation is preferable for most of these fungi, since conidia need not be recovered from the bran base. The dry conidia-bran mixture is relatively stable, but cannot be stored for prolonged periods of time.

At present, submerged fermentation is not used in mass production of fungal pathogens. The reason behind this is the failure of many entomopathogenic fungi to produce true conidia.

Vezina *et al.* (1965) have reported the production of true conidia by an insect pathogenic fungus, *Aspergillus ochraceus.* Blastospores obtained in a submerged culture are difficult to preserve and tend to lose their viability easily. The submerged cultivation of the fungal insect pathogens would be more useful if true conidia could be obtained. Much research work needs to be done in this area.

4.3. Viral Insecticides

Since viruses are obligatory parasites, mass production of viruses is undoubtedly linked to the mass production of the host cells. Viruses have successfully been grown *in vitro.* But the information pertaining to large-scale production has not been available although ~20 per cent of the viruses studied were 'mass' propagated. Viruses were propagated mostly in caterpillars, however, sawflies and mites were also used. Artificial diets and continuously reared hosts are at present the best approach to systematic mass propagation of viral insecticides.

5 FORMULATION

To be a desirable microbial insecticide, the agent should not diminish in potency by inappropriate processing. The choice of formulation may be vitally important in practical application. For example, if heavy rainfall is expected, dust would be unsuitable and a spray containing a good sticker would be required.

There are two main types of formulation of microbial insecticides:

(1) liquid formulations

(2) dust formulations

(1) *Liquid formulations:* Water is generally used as a diluent. Suitability of water as a diluent is relatively higher for relatively resistant forms such as polyhedra and spores.

The use of oils as a carrier for microbial insecticides came, in part, from the need to fit such products into established procedures and to use existing equipment for their application in the field. Solubility, cheapness and good sticking and spreading properties are the desired characteristics of the oils used. The use of oils is sometimes limited by possible phytotoxicity. In some cases foliage damage can be tolerated, but where the foliage is used as a food, (e.g. lettuce), the oil residue might be unacceptable.

(2) *Dusty formulations:* The earliest microbial insecticides were formulated in a dry form and were used as dusts or wettable powders. Such formulations are still prepared for certain microbial insecticides, (e.g. viral insecticides).

Additives, namely spreaders, stickers and protectants, are used during formulation of microbial insecticides. Spreaders, (e.g. dried milk, powdered casein, dried blood, gelatin, saponins, soaps, sodium dodecyl sulphate) are added to the water diluent to ensure 'wetting' of the surface to be sprayed (Frear, 1968). Stickers or adhesives, (e.g. flour, dextrins, gums, amine-stearate solutions, polyviny! chloride) are used to ensure the formation of a weather-resistant film on foliage. Protectants are also added to minimize the effects of desiccation, sun-light and/or ultraviolet radiation.

6 ECONOMICS

There are many factors responsible for the success of any industry. One of the important factors is the cost of the product. Microbial insecticides are effective in the field against the target insect pests. They are also non-toxic to man, animals and plant species. Thus, these are some attractive features of microbial insecticides. Unfortunately, their cost of production is prohibitive. Therefore, at present, the establishment of industries for large-scale production of microbial insecticides has not become possible. The present effectiveness and future potential of microbial insecticides depend on a subtle interaction between cost/effectiveness as compared with chemical insecticides.

7 SAFETY

Different insect pathogens have different spectra of action on insects. For example, *Bacillus thuringiensis* exerts a pathogenic effect for a wide range of lepidopterous larvae, including many pest species. On the other hand, most of the insect viruses of interest for biological control are highly specific. They attack only insects of one species and sometimes of one type within a species. It is strictly desirable that any microbial insecticidal agent should not have a pathogenic effect for other forms of life, (e.g. humans animals and plants), since it is used in a viable form and is apt to multiply in the environment. Therefore, microbial insecticide formulations themselves and possible residues from them are subjected to stringent toxicological and metabolic studies. In the case of microbial insecticides, the US regulatory agencies have made clear that they require proof of safety before a preparation can be fully registered for use. Moreover a petitioner for registration of a microbial insecticide must also produce information about the method of production and demonstrate an adequate technique for the production of a uniform product and for the standardization of that product. He must also produce evidence that he can remove extraneous microbes from the preparation or control them to an acceptable level. At present, only *B. thuringiensis* based insecticides and products based on the nuclear polyhedrosis virus of *Heliothis* have been fully registered in the U.S.A. Thus, these insecticidal agents may be used without tolerance limitation right up to harvest on edible crops in the U.S.A.

From the foregoing discussion, it is clear that the production of a microbial insecticide is a more difficult process than the synthesis of a chemical insecticide. Therefore, microbial insecticides are expensive as compared to chemical agents. However, microbial insecticidal agents owe their present success and future potential primarily to their safety.

8 ADVANTAGES AND DISADVANTAGES

The various characteristics of the different groups of disease-producing microbes tend to regulate the utilization of many insect pathogens and have indicated certain advantages and disadvantages of microbial control methods.

The principal advantages of microbial control methods are:

(1) The harmless and non-toxic nature of insect pathogens for other forms of life with the resultant absence of toxic residues.

(2) The relatively high degree of specificity of most pathogens tends to protect beneficial insects.

(3) The compatibility of many pathogens with many insecticides to the degree that chemical and biological pesticides may be used concurrently and, in some cases at least, synergistically, since infection may cause the insects to be more susceptible to chemical poisoning.

(4) The ease and inexpensiveness with which some pathogens can be produced.

(5) The high versatility of microbial pathogens in so far as methods of applying them are concerned. Some microbes may be introduced and colonized with the result that the control brought about may be permanent. Other microbial agents may be used as sprays or dusts in the same way as insecticides.

(6) The apparent slowness by which a susceptible host develops resistance to a microbial pathogen.

(7) The low dosages required in some cases.

The disadvantages of microbial control methods may be listed as under:

(1) The necessity for careful and correct timing of the application of the pathogen with respect to the incubation period of the disease.

(2) The relatively marked specificity of most pathogens sometimes narrows the spectrum of effectiveness to only one insect species in cases where several pests are involved, all of which might be destroyed by a single chemical insecticide.

(3) The necessity of maintaining the pathogens in a viable state and at a high virulence and in a durable or resistant state until contact with the insect is made.

(4) The difficulty of producing some pathogens either in large quantities or inexpensively or both.

(5) The tendency of some diseases to cause insects, or parts of the insects, to remain attached to the foliage of the host plant. This may be particularly objectionable with food crops on which insect parts cannot be tolerated.

(6) The requirement of some pathogens for favourable climatic conditions in order to invade and infect their arthropod host.

9 FUTURE PROSPECTS

The effectiveness of *B. popilliae* is well demonstrated. Therefore, its use in the U.S.A. for the control of the Japanese beetle will undoubtedly continue. The future development of milky disease agents must be governed by the success of attempts to grow the organisms in the pathogenic form in nutrient media.

B. thuringiensis has also given fruitful results as an insecticide. For example, it is exploited for certain types of crop, particularly leafy vegetable crops (cabbage and lettuce) and for forestry applications. Significant pointers for future developments are to be seen in the use of such products against the Cabbage looper in Arizona and California. This is to be expected because this insect has long been resistant to many chemical insecticides. Moreover, chemicals that are effective against it may leave toxic residues on the edible plants. For the control of insects on tobacco, *B. thuringiensis* is as effective as chemical insecticides for the Tobacco hornworm and for the Tobacco budworm. Again, the attractiveness of *B. thuringiensis* is the freedom from chemical residues in the crop.

There are several insect viruses capable of producing persistent infection in forestry pests. Therefore, application of insect viruses as control agents will, undoubtedly continue to expand. It is more difficult to forecast the future potential for the non-persistent virus diseases. There has even been a recent registration of such a viral insecticide in the U.S.A. which has established a landmark in the developments of these control agents.

Microbial insecticides having a narrow-spectrum activity and high innate safety will certainly play a vital role in integrated pest control programmes. According to Falcon (1971), the future of bacterial control agents looks extremely bright in the coming era of ecological pest control. In addition to bacterial agents, the development of viral agents has been dramatic since 1971. It could well be that a thorough understanding of the mechanisms of microbial pathogenicity by illuminating the molecular basis of specificity could make a fundamental contribution to man's efforts to control insect pests.

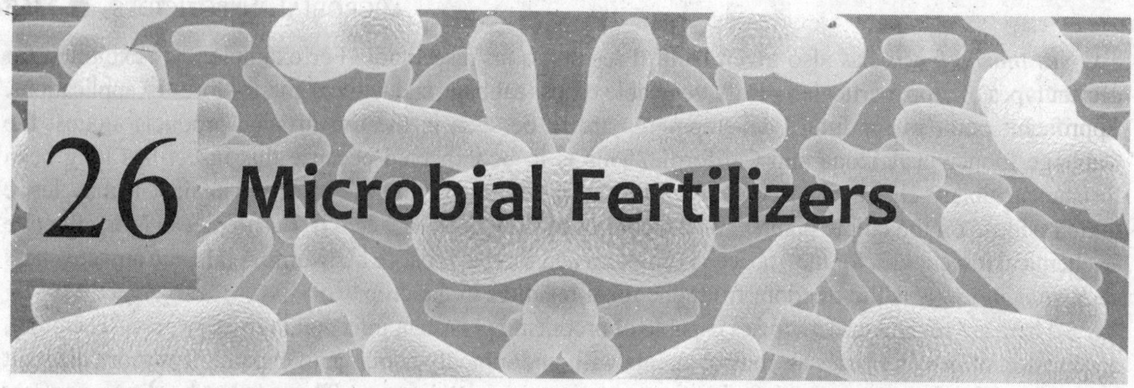

26 Microbial Fertilizers

1 INTRODUCTION

Fertilizers are compounds that promote the growth of plants. They supply major plant nutrients including nitrogen, phosphorus and potassium. These nutrients are often needed in quantities larger than the soil can naturally supply. Obtainable nutrients are the key to growth for all organisms. Without construction materials, building is impossible. Nitrogen is an abundantly available element in the earth's atmosphere, having a 79 per cent concentration in its N_2 gas form. Organisms use nitrogen in a variety of forms necessary to life like protein synthesis, in nucleic acids and for other Nitrogen compounds. Unfortunately, the N_2 gas in the atmosphere is very difficult for organisms to use due to the strong triple bond between the Nitrogen atoms. For organisms to access the vast amounts of environmental Nitrogen, it must be fixed, combined with Hydrogen. Nitrogen can be fixed as either Ammonium (NH_4) or Nitrate (NO_3). Once it is available, green plants can process it into proteins and it may travel along the food chain. All living things require Nitrogen, as it is a major component of amino acids, DNA, RNA and proteins. The key microbes responsible for fixing N_2 are *Azotobacter* spp. and *Rhizobium* spp. which use the enzyme Nitrogenase exclusively.

Chemical fertilizers have been in use for about 78 years. Unfortunately, their prolonged use has negative side effects for the environment as under:

- decrease in soil fertility
- hardening of soil
- pollution of air and water
- deteriorating plant quality that can lead to human health problems

Researchers in the United States, Japan, Germany, Australia, India and China have been looking into natural fertilizer alternatives that do not harm the environment.

2 HISTORY

During the nineteenth century, various individuals began to discover the steps of the Nitrogen cycle as well as the micro-organisms that ran or assisted the process. Reyset, who described the decomposition of organic material in soil and the resulting release of Nitrogen gas, was the first to do so in 1856. The Nitrification process was discovered by Schlossing and Muntz in 1877. In 1886, Gayon and Dupetit discovered Denitrification and isolated the first pure culture of bacteria to perform the reaction. The French scientist Jodin observed that unknown organisms in a nutrient solution under controlled conditions fix Nitrogen gas in 1862 and was observed again in 1888 by Martinus Bejerinck, a Dutch scientist. Bejerinck and Lipman isolated *Azotobacter* together in 1901. As the end of the ninteenth century approached, the idea that plants assisted the proliferation of micro-organisms capable of fixing Nitrogen was widely accepted.

3 THE NITROGEN CYCLE (See Figure 26.1)

There exists a cycle that converts Nitrogen gas into organic substances and then back into Nitrogen. This natural phenomenon is the Nitrogen cycle, and it is constantly occurring. The cycle is maintained by *Rhizobium* spp. and *Azotobacter* spp., both of which are decomposers and Nitrogen fixers. The cycle is composed of four different reaction steps as under:

1. nitrogen fixation
2. decomposition (of plant/animal matter) and ammonification
3. nitrification
4. denitrification

In these steps, atmospheric nitrogen is converted into organic Nitrogen, then Ammonia, then Nitrate and finally returned to the atmosphere as Nitrogen. Nitrogen fixation is very important for a number of reasons. Commercially, it is a key in agriculture where more naturally fixed Nitrogen is, the less artificial fertilizer needs to be used. In recent years, artificial fertilizers have greatly increased their cost, minimizing their need as well as their production's impact on the environment.

3.1. Nitrogen Fixation

The actual scheme of movement in the cycle starts with Nitrogen gas, found in abundance in the atmosphere. The N_2 gas is converted into Ammonia (NH_3) by combining atmospheric Nitrogen with hydrogen. This bioconversion is mediated by Nitrogen fixing micro-organisms in the presence of Nitrogenase. This reaction is a series of electron transfers which require a lot of energy in the form of ATP. The equation for such a reaction is:

$$N_2 + 16ATP + 8e^- + 8H^+ \xrightarrow{\text{Nitrogenase}} 2NH_3 + 16ADP + 16Pi + H_2$$

Such a reaction is energy intensive, requiring 16 ATP molecules to break the Nitrogen bonds to combine it with the Hydrogen. The Ammonia experiences Nitrification and becomes Nitrites (NO_2), and then is converted to Nitrates (NO_3) by *Nitrobacter*. The Nitrate is either assimilated into plants or denitrified back into Nitrites (NO_2) which may return to the atmosphere. This happens in the soil when the bacteria run out of Oxygen and need access to more for respiring organic matter. The Nitrates used by plants are held within them until plant death.

3.2. Decomposition and Ammonification

Plant and animal matter are decomposed by microbial decomposers naturally found in the *soil* (*Azotobacter* spp. and *Rhizobium* spp.) to form proteins. Then, proteins are hydrolyzed to yield amino acids. Microbes internalize the amino acids and convert them into ammonia.

When the plant dies, it decomposes and returns its Nitrogen through Ammonification into the form of Ammonia, which continues on in the cycle:

$$\text{Organic Nitrogen} \xrightarrow{\textit{ammonifying bacteria}} NH_3 + R$$

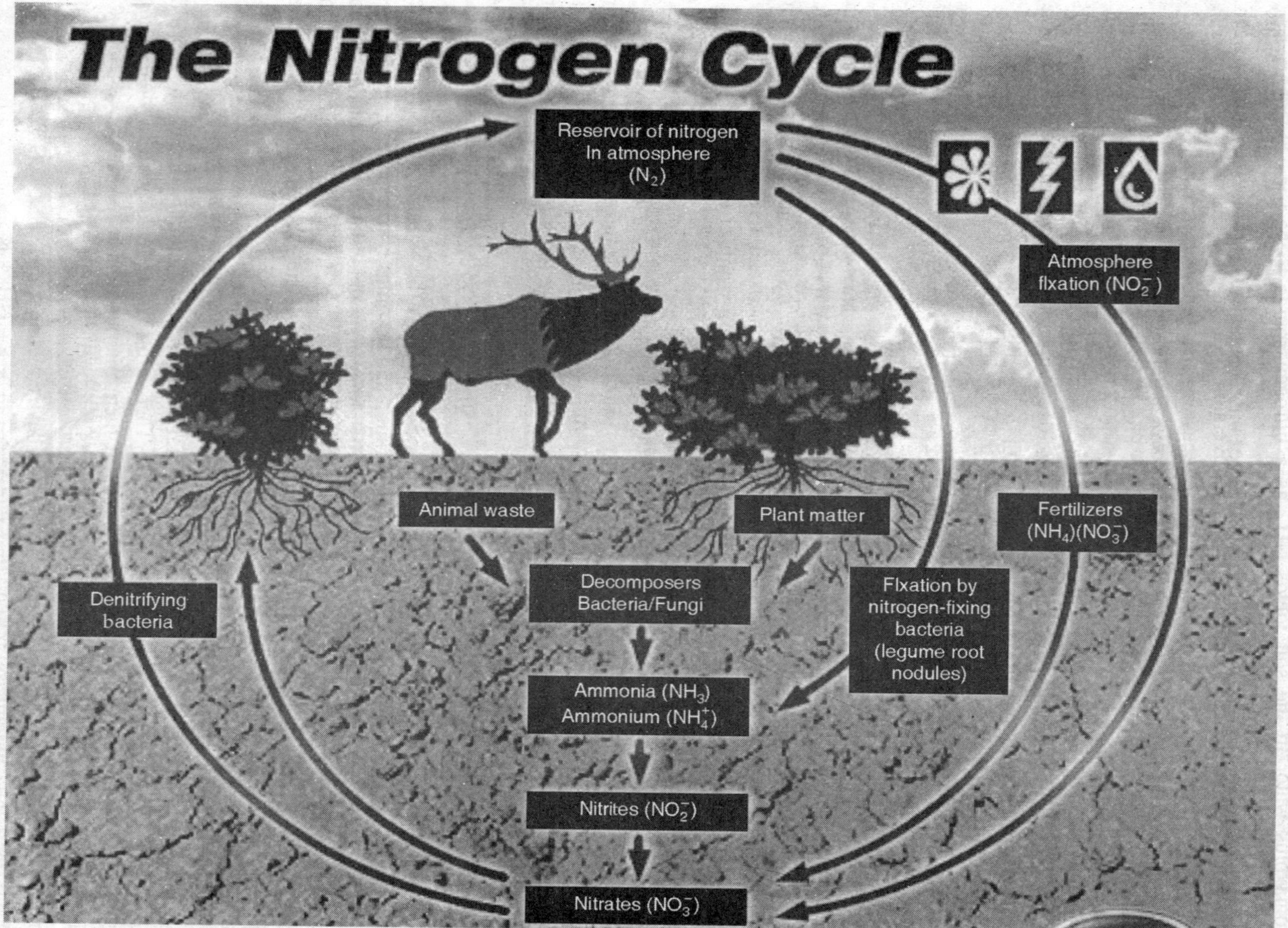

Fig. 26.1 *The nitrogen cycle.*

Ammonia plus water in the soil yields NH_4^+ as a nitrogen source to be used by plants as building blocks for proteins and nucleic acids.

The enzyme Nitrogenase that catalyses N_2 gas conversion into Ammonia is made up of two proteins that include iron and molybdenum-iron. Nitrogenase is also rendered inactive by Oxygen, which may be a problem for aerobic bacteria like *Azotobacter* spp. if not for its high respiration rate that maintains a low O_2 level. *Rhizobium* spp. also control this factor by binding oxygen in their nodules with leghaemoglobin.

3.3. Nitrification

Nitrification is the biological oxidation of ammonia with oxygen into nitrite (NO_2) followed by oxidation of these nitrites (NO_2) as under:

$$\text{(i)}\quad NH_3 + O_2 \xrightarrow{\textit{Nitrosomonas}} NO_2 + 3H^+ + 2e^-$$

$$\text{(ii)}\quad NO_2 + H_2O \xrightarrow{\textit{Nitrobacter}} NO_3 + 2H^+ + 2e^-$$

Plants tend to use nitrate as their main nitrogen source because it is more mobile in the soil and more likely than ammonium to come into contact with plant roots.

3.4. Denitrification

Since our atmosphere is full of oxygen making dentrification unfavourable, the reaction occurs only in the soil. Denitrification by microbes reduce nitrate in the soil to nitrous oxide and then back to Nitrogen gas. Bacteria such as *Pseudomonas* spp. use the nitrate as an electron acceptor when no oxygen is present in the soil. Thus, the Nitrogen gas returns to the atmosphere to complete the cycle. The process can be expressed as the following redox reaction:

$$2NO_3 + 10e^- + 12H^+ \rightarrow N_2 + 6H_2O$$

4 USEFUL MICROBES

Bacteria capable of fixing atmospheric nitrogen into a useable form are known as diazotrophs. These diazotrophs have iron-molybdenum nitrogenase systems, nitorgenase being the only known enzyme needed by organisms to fix the nitrogen. Although these bacteria will readily conduct the energy intensive reaction of nitrogen fixation, whenever N_2 is readily available in the environment, microbes will use those sources instead of producing their own. High atmospheric oxygen content also inhibits the nitrogen fixing process and must be circumvented through a variety of ways.

4.1. Bacteria

The *Azotobacter* spp., (*e.g. Azotobacter vinelandii*) are helpful in making nitrogen available to all organisms by creating Ammonia from Nitrogen through Nitrogen fixation. In the form of ammonia, Nitrogen can be turned into a variety of proteins used to synthesize important compounds in the life cycles of other organisms. *Azotobacter* spp. also keep soil healthy and increase crop production. The *Azotobacter* spp. are Gram-negative large rods measuring up to 2 microns in diameter. They exist singly, in chains or clumps. They are free-living aerobes and have a resting stage spent as a thick-walled cyst that protects them in harsh climates. Species of *Azotobacter* can be found in soil, aquatic environments and in plants with neutral to alkaline pH values. They are sensitive to acidic pH, high salt concentrations and temperatures above 35°C, thus classifying them as alkalinophile, mesophile, and halophobic respectively. *Azotobacter* spp. have more DNA than most bacteria and transfer genetic material via conjugation or transformation. They also possess the highest metabolic rate and synthesize the unique enzyme Nitrogenase.

Rhizobium spp., (*e.g. Rhizobium leguminosarum*) have the capability to fix Nitrogen from the inside of legume nodules. Like *Azotobacter* spp., rhizobia also synthesize Nitrogenase. For this, they require a symbiotic relationship with the plant host. They are motile, Gram-negative aerobic bacteria. The first *Rhizobium* species was identified in 1889, *R. leguminosarum,* as well as the rest of the species. There are 52 different rhizobia species spread amongst 12 different genera. They exist in the soil, where they encounter roots of legumes and can initiate their symbiotic relationship. Plant cells proliferate and form nodules while the bacteria differentiate and fix Nitrogen. The rhizobias' efforts are repaid with sugars, proteins and oxygen from the plant. Therefore, this process of N_2 fixation is called symbiotic N_2 fixation.

5 BIOMASS PRODUCTION

Medium: The medium designed to cultivate N_2-fixing bacteria contains the following ingredients:

(a) Coral reef granular powder: This granular powder has a grain size in the range of 10-200 mesh. It is an alkaline compound to buffer against acidity during cultivation. Also, it absorbs odours from animal excreta and sewage disposable waste.

(b) Domestic animal excreta and sewage disposal waste: This is an organic nutrition source for microbes. It is a very good source of N_2 for microbes being cultivated.

(c) Ground cellulose: This serves as a source of carbon.

All of these ingredients are mixed with one another in the presence of organic compounds and inorganic salts, (i.e. NaCl) in a storage tank. This mixture is subjected to cultivation for a predetermined period of time in a cultivating bath.

Microbial mixture: Microbes listed below are used to build up the microbial mixture:

(a) *Azotobacter vinelandii* and *Rhizobium leguminosarum:* for nitrogen fixation.

(b) *Bacillus megaterium:* enhances amount of nitrogen fixation.

(c) Green algae: symbiotic with *Azotobacter* providing it with carbohydrates and taking nitrogen.

(d) *Candida utilis:* acts as a yeast fungus to provide ATP for the energy consuming nitrogen fixation reaction.

(e) *Trichoderma viride:* decomposes cellulose into a carbon source for *Azotobacter* and *Rhizobium.* Also, it helps reduce contamination of plants by unwanted plant pathogens.

Cultivation conditions: Apart from sources of carbon and nitrogen, the following three factors play an important role in the cultivation of microbial mixture:

(a) temperature: optimum temperature between 25-35°C

(b) pH: optimum pH between 6-7

(c) time: Total optimal amount of cultivation time is 12 hours. One cultivating bath has a 10 ton capacity and cultivation time per bath is 4 hours. Using 12 cultivating baths, a total of 249 tons of product can be produced per day.

Process brief: The apparatus for biomass production was designed by Yosiaki Kimura in 1991. It has various component parts as outlined below:

(a) Storage tank: in which domestic animal excreta, sewage disposal waste and ground cellulose derived from plant fibre are stored and to which a properly determined amount of coral reef in the form of powder is delivered from a coral storage tank.

(b) Steam drum: for sterilizing raw material mixed in the storage tank under the effect of heat of steam.

(c) Temperature adjusting tank: in which the raw material subjected to sterilization in the steam drum is received to cool it at a properly adjusted temperature. It utilizes a refrigerator and compressor to blow cool air into the tank, thus bringing temperature back to 25 - 35°C.

(d) Master bacillus adding tank: in which the raw material is received from the temperature adjusting tank to add to the raw material master bacilli, including nitrogen fixing bacillus.

(e) Cultivating bath: in which the mixture with master bacilli added thereto is received from the master bacillus adding tank, cultivating liquid is received from a bacillus cultivation source mixing tank and auxiliary master bacillus liquid is received from a master bacillus cultivating tank. The received mixture and liquids are subjected to cultivation for a predetermined period of time.

(f) Cultivated material accumulating tank: in which the cultivated material mixture is received from the cultivating bath and it is then agitated and dried by a drier connected to the cultivated material accumulating tank.

(g) Dried material: is delivered to a storage tank in which packing or a similar operation is performed for the final product to be shipped to users.

6 ECONOMICS

With the world population continually increasing, it has become necessary to increase agricultural production to meet current demands. This has been made possible thanks to the use of fertilizers. Bio-fertilizers also provide farmers and growers autonomy and independence. Bio-fertilizers are excellent, cost effective options for acidic soil. Moreover, they lessen the adverse ecological effects of chemical fertilizers. They have negligible material cost and are easily handled/shipped. They can usually be produced at the site of their use.

Chemical fertilizers are better options for nutrient deprived soil where bio-fertilizers will not work without the addition of additional nutrients.

7 SAFETY

Improperly processed organic fertilizers sometimes contain pathogens that are harmful to humans and must be removed via proper composting. It is also possible to excessively employ microbial fertilizers. Fortunately, their nutrient content, solubility and release rates are usually lower than chemical fertilizers and also provide increased physical and biological storage mechanisms to soils, which tend to mitigate their risks.

8 ADVANTAGES AND DISADVANTAGES

Microbial fertilizers have some key advantages over chemical fertilizers as under:

1. Microbial fertilizers (bio-fertilizers) can operate in acidic conditions because they are calcareous products containing both CaO (lime) and $CaCO_3$ (calcium carbonate) from coral reef powder. Unfortunately, chemical fertilizers lower soil pH in addition to a shortage of

alkaline components in soil. Shortage of calcareous component in chemical fertilizers leads to malfunctions such as root decay and greening of ears.

2. They do not produce the run off toxins in the rain water, leaving biodegradable residues in the soil.
3. They can be applied less frequently, thus reducing their costs.
4. They lack concentrated chemicals that can burn or harm plants.
5. They allow the better retention of soil moisture and improve soil structure.
6. They mobilize existing soil nutrients already in the soil to achieve good growth rates without the addition of more nutrients.

On the other hand, microbial fertilizers have some disadvantages listed below:

1. They depend upon environmental conditions. For example, carbohydrates are necessary in the soil because they are needed for the metabolism of nitrogen fixing bacteria. In other words, the carbohydrate content of the soil is the limiting factor for the use of microbial fertilizers.
2. They can sometimes contain human pathogens if they are improperly processed.
3. There can be difficulty in attempting to establish non-native bacteria within an already existing ecosystem.
4. They produce a green house gas called nitrous oxide.

9 FUTURE PROSPECTS

As more and more nations tend towards green practices and greater demand, new methods will be developed to allow for healthier and more effective fertilizers. Worldwide use of fertilizers, whether natural or artificial, has increased tremendously over the last decade. With the growing importance of nations' agriculture, not just as a source of food for their people but as a biofuel alternative, the necessity of fertilizers becomes even greater. While the planet's population expands and newer technologies become available, research will help discover new resources for this field.

27 Fermented Foods

1 INTRODUCTION

The term 'food' includes all items consumed by man. Out of these, some food items are prepared by fermentation. In other words, desirable chemical changes are brought about by desirable micro-organisms, (e.g. sauerkraut and yoghurt). It is also essential to preserve the food properties if they are to be consumed after a long period.

2 SOME IMPORTANT FERMENTED FOODS

Sauerkraut

Introduction: Sauerkraut is a neatly shredded cabbage which has been allowed to undergo a natural lactic acid fermentation in the presence of a small amount of salt. It is produced commercially in most northern and eastern European countries. It is normally consumed after cooking as a main course vegetable. At present, it ranks fifth in the order of importance amongst canned vegetables in the United States. Sauerkraut is only preceded by peas, beans, corn and tomatoes (Pederson, 1960). It derives its keeping quality from the acid produced during fermentation. It is a relatively stable product if it is kept cool and out of contact with air.

Sauerkraut, which is properly prepared, can be quite a rich source of vitamin C. Pederson *et al.* (1956) examined 217 samples of canned sauerkraut from various sources. It was found that 67% of the samples had an ascorbic acid, (i.e. vitamin C) content in the range of 10 to 25 mg. per 100 gm., 20% from 0 to 10 mg. per 100 gm., and 13% from 25 to 35 mg. per 100 gm. Sauerkraut also has a mild laxative action, which is attributed to the dextran content of the product. Dextran is produced during fermentation by *Leuconostoc mesenteroides,* dominant during the first part of the curing period.

Manufacturing Procedure

It is axiomatic that good foods cannot be prepared from poor materials and sauerkraut is no exception to the rule. Consequently, the quality of the cabbage used has considerable bearing on the quality of the final product. Many varieties of cabbage are suitable for sauerkraut production. The heads should be solid, crisp, with a minimum of green leaf. Once the heads have been cut, it is necessary to store them for a few days prior to shredding. They are usually stored in well-ventilated sheds or barns. The reason for this is that a slightly wilted cabbage can be more neatly sliced, with less damage to the shreds than when it is first harvested. Another advantage of the period of wilting is that it allows the heads to come to even temperature, and to warm up during cold weather.

Just before shredding, the outer green leaves are trimmed away together with any damaged or diseased parts. The core is cut with a spiral knife without actually disconnecting the core from the head. The heads are then sliced by a machine. The shreds should not exceed 2 mm. in thickness. Campbell (1950) recommends 1.5 mm. thickness of cabbage shreds.

Good sauerkraut can be prepared in any size of the fermentation vessel. The size of the fermentation vessel, which may be used, ranges from a gallon jar to the 80 tons redwood tanks. These fermentation vessels can be made of wood. The tiled concrete tanks are better than wooden tanks so far as sanitary aspect is concerned. All the metal equipment coming into contact with the product must be of stainless steel, or be suitably coated. The reason for this is that iron causes discoloration, whereas copper causes certain loss of vitamin C. Copper is also responsible for off flavours.

The shreds are deposited in the fermentation vessel and are mixed with $2\frac{1}{4}\%$ by weight of salt. This is a crucial stage in the manufacture of sauerkraut. It is absolutely essential that exactly this amount of salt should be used and that it should be incorporated evenly throughout the mass of shreds (Pederson, 1946). This requires care and patience, and it is necessary to weigh both salt and shreds in relatively small batches. As each batch of shreds is added to the vessel, the approximate quantity of salt is scattered over it and rapidly forked in.

The height of the shreds in the vessel should be maintained at an even level as it gradually rises. This is the responsibility of the operator who should be an experienced man. The shreds must also be compacted down tightly. Once the vessel is full, a false head is placed in position and then weighed in some way. It is, usually, weighed by blocks of concrete fitted with handles. This is to compress the mass and thus, ensure that the shreds are forced under the juices extracted by the salt, and that all the air is taken out. On the other hand, such covers suffer from the disadvantage that the juices, which rise over the head are in contact with air. This permits the growth of yeasts and moulds at the surface. These may spoil the product by causing off-odours and flavours (Pederson and Kelly, 1932). Various ways of overcoming this difficulty have been tried in the past, and the modern practice is of replacing the wooden covers with plastic sheets weighed down with water or brine. These effectively seal off the whole surface from contact with air. Owing to the flexible nature of the plastic sheet, the gases formed during fermentation can escape freely past the seal. This method would appear to be ideal.

A variety of sugars, proteins and other nutrients are present in the juices extracted from the cabbage. These support the growth of bacteria. Consequently, in a matter of a day or so fermentation sets in, this being apparent by the evolution of gas and the formation of acid. The time required for the fermentation to run its course depends on temperature, particularly filling temperature (Pederson, 1956). If this is favourable, a period of between three to six weeks will normally suffice. At the end of this time, the following changes are marked:

(1) The product will have acquired its typical aroma.

(2) All the fermentable carbohydrates will have been consumed.

(3) The acid content will have risen to between 1.8% and 2.2%.

(4) The pH will approximate to the range 3.5 to 3.7.

The product is now stable, provided that it is kept cool and out of contact with air. Therefore, when fermentation is complete the vat or container must be properly sealed to exclude air.

If the temperature is too warm, darkening and general loss of flavour and quality will take place. The progress of fermentation should be followed by periodic checks on the development of acidity in the juice. This not only shows whether fermentation is proceeding normally, but also enables finished vats to be recognised and sealed, and for the order in which other vats will reach completion to be predicted. A record of vat temperatures should be kept, since it is also important for predicting finishing dates.

The Fermentation

Three acid-producing species of bacteria are mainly responsible for normal sauerkraut fermentation, each being dominant in succession (Pederson, 1930). The first species to appear in numbers is *Leuconostoc mesenteroides.* This coccus attacks the sugars leaching from the cabbage, with the formation of lactic and acetic acids, ethyl alcohol, mannitol, carbon-dioxide, etc.

The pleasant flavour and odour of sauerkraut emanates from the activity of this bacterium (Pederson, 1930). In addition to this, there is a copious evolution of carbon-dioxide arising from its activity. This is instrumental in quickly creating anaerobic conditions in the mass. Thus, growth of spoilage yeasts is inhibited. Since *Leuconostoc mesenteroides* is unable to withstand acidities approximately to 1%, as this level is reached in the product it dies out and is replaced by *Lactobacillus plantarum.* This also produces lactic acid, and in turn, it is similarly destroyed by the steadily increasing acidity. Meanwhile, the numbers of *Lactobacillus brevis* go on building up. Since this species is capable of withstanding any acidity attainable in the product by natural means, (i.e. its own acidity), it takes the fermentation to completion.

Both the bacteria in the first and final stages of the fermentation are producers of CO_2. *Lactobacillus plantarum,* however, is a homofermentative species and so does not produce gas. Therefore, the first stage is indicated by the rapid evolution of gas, followed by a period of relative quiescence (second stage). Again, a moderate increase in the gassing rate is an indication of the final stage of the fermentation process. The principal acids produced during fermentation are lactic and acetic acids and at the completion of the process, their ratio normally approximates 4 to 1 respectively.

Temperature: Temperature is important in sauerkraut production in relation to quality. The growth of *Leuconostoc mesenteroides* is not favoured at temperatures above 70°F. Since the flavour and odour are dependent on this species, high temperatures result in a poor quality product. Also, two other species of bacteria appear when high fermentation temperatures prevail. They are: (1) *Streptococcus faecalis* and (2) *Pediococcus cerevisiae.* These are homofermentative bacteria, producing only lactic acid. They do not produce the CO_2 necessary to create anaerobic conditions. Therefore, their presence is normally associated with a lessening of product quality (Pederson and Albany, 1954).

Although high temperatures act to the detriment of quality, they do, nevertheless, favour the growth of both *Lactobacillus plantarum* and *L. brevis.* These do not grow well under cold conditions. So many early investigators advocated the use of heat during commercial sauerkraut production. Generally, it is not advisable to allow the temperature to rise above 70°F or to drop below 45°F. The optimum range is probably 55°F to 65°F.

Salt: It is extremely important that the right amount of salt should be present during fermentation. The optimum range is 2% to $2\frac{1}{2}$%. High salt contents are associated with the development of pink sauerkraut. Also, high salt content (3.5%) favours the growth of two undesirable bacterial species, (i.e. *Streptococcus faecalis* and *Pediococcus cerevisiae*), whereas the more desirable species tend to be inhibited by it (Pederson and Albany, 1954). Low salt contents are detrimental to the texture of sauerkraut.

Starters

The use of bacterial cultures as starters has been investigated but generally with little success and sometimes with a positive disadvantage (Pederson, 1930). The reason is that the right bacteria are always present on cabbage and if given the proper conditions, (i.e. salt and temperature) they will always develop and in the correct sequence.

Canning and *Packaging*

Canning is best accomplished by practising the following operations:

(i) Sauerkraut is heated to 165°F in its juice. Heating is carried out in a stainless steel or glass–lined pan (steam-jacketed).

(ii) Heated sauerkraut is immediately packed into unlacquered cans.

(iii) Filled cans are quickly exhausted, sealed and cooled.

Canned sauerkraut has a less pleasant odour than the fresh one but the disagreeable quality disappears on cooking (Cruess, 1958).

Some sauerkraut is packed in glass and vacuum capping and cool storage are advisable. Pliofilm bags have also been tried. Sedky *et al.* (1952) showed that when ascorbic acid was added to protect the colour, and the bags sealed under vacuum and kept refrigerated, a shelf-life of six weeks was attained. Recently, a laminated polythene aluminium foil pouch has been tried with success (Anon., 1967). The pouches are pasteurised by heating at 170°F for 20 minutes. Freeze-dried sauerkraut has also been prepared (Anon., 1969).

Spoilage of sauerkraut

There are a number of forms of spoilage of sauerkraut most of which arise as a result of faulty procedure or utensils.

Dark or discoloured sauerkraut is essentially caused by oxidation arising in a number of ways. According to Pederson and Fisher (1944), the following conditions are associated with darkening of the product:

(i) excessive exposure of the product to air

(ii) the use of too much salt or too high temperature during fermentation

(iii) the use of loose-headed cabbage

Contamination with iron also gives rise to dark stains. Pink sauerkraut arises from the growth of certain yeasts, which produce a red pigment (Pederson and Kelly, 1938). A deficiency of manganese in the cabbage (which element is required for the growth of certain lactobacilli) is also suspected to be an indirect cause of pink sauerkraut.

Soft sauerkraut is directly associated with a low salt content. Any mould growth on the shreds, or in the vats between fermentations (which has not been properly washed away) could also give rise to areas of softened sauerkraut.

Slimy sauerkraut is caused by the excessive secretion of dextran by *Leuconostoc mesenteroides.* This may result from the presence of a particular strain of the bacteria prone to this habit, or possibly from a higher than normal sucrose content in the cabbage.

Olives

Introduction: Olives are valuable pickles in the United Kingdom. Comparatively large quantities are consumed in this way and something like an average of 775 tons per year are imported. Their keeping quality depends on the absence of suitable nutrients to support microbial growth, the presence of salt and of lactic acid and its attendant low pH. There is a difference between pickled olives and canned olives. The pH of canned olives approximates to neutral and its preservation necessitates a heat process of 240°F for 60 minutes.

The olive has been cultivated by man since ancient times. There are a number of varieties of olive but only two are of major importance for pickling. These are the large Sevillano or the Gordal (Spanish Queen) and the smaller Manzanillo olive. The former is considered to have the better flavour and texture but is more difficult to pickle.

Fresh olives have an insipid and intensely bitter flavour. One of the purposes of the pickling process is to remove this bitter taste. The first step in the pickling process consists of treating the fruit with a dilute alkali to remove the bitter taste.

Pickled olives fall into two categories—green and black. They differ in the following two respects:

(a) the time of harvesting
(b) the manner in which they are subsequently processed

Green Olives

The fruit is allowed to become full grown, but is picked before any change in the colour begins. Immediately after harvesting, the olives are placed in shallow vats. They are covered with a lye which consists of water with slightly less than 2% sodium hydroxide, (i.e. caustic soda). The olives remain here for about two days until the lye has penetrated approximately two-thirds of the way to the pits. The lye penetration is followed, in practice, by the simple process of halving a fruit, washing and applying phenolphthalein to the cut surface. Lye-treated olives are washed in a number of changes of water until they are mostly free of alkali. Care must be exercised during both the lye treatment and the subsequent washing to avoid undue aeration. The reason for this is to avoid blackening of the product.

Once the olives are free of alkali, they are drained and packed in large barrels. Then the barrels are headed up and a 45° salometer brine (11.9% salt) is introduced through the bung. After filling, the barrels are placed in the sun to warm the contents. Thereafter, a natural fermentation will commence, if all goes well. This is a critical period in the pickling process. If fermentation fails to start, spoilage micro-organisms may become established. Starters may be added to each barrel to ensure the rapid commencement of fermentation. Starters consist of a few pints of old brine from well fermented olives.

The first bacterial species which appears during fermentation is *Enterobacter aerogenes.* This bacterium produces carbon dioxide and hydrogen, giving rise to 'gassing' of the brine. Soon lactobacilli of both the homo- and hetero-fermentative species predominate when production of gas is much reduced. As a result of this, formation of lactic acid begins to take place. More lactic acid formation can be had by the addition of small amounts of sugar to the brine. This, however, must be added late in the process, otherwise *Enterobacter aerogenes* is unduly stimulated with resultant excessive gassing. Olives grown on nutritionally poor ground may be lacking in nitrogen-containing compounds. These compounds are required by the bacteria for their growth. Samish (1955) has shown that the fermentation of such olives can be improved by the addition of a source of nitrogen, (e.g. ammonium salts) to the brine. Etchells *et al.* (1966) have investigated the use of pure cultures to induce satisfactory fermentation.

After consumption of all the fermentable carbohydrates, fermentation ceases and the product is now stable, (i.e. fully cured), provided it is kept out of contact with air. At this stage the lactic acid content of the brine should not be less than 0.5% and the pH will have dropped below 4.

After curing, the olives are carefully size-graded and all the blemished fruits removed by hand. Then, they are recasked and are ready for shipment. The Manzanillo olives are pitted and their stones are removed. Thereafter, the cavities are stuffed in a variety of ways. The olives are then packed in casks with about 15% salt brine, preferably with 0.5% lactic acid added.

The most frequently used form of stuffing is strips of pimento. These strips, before use, are preserved in strong brine. The bright red of the pimento against the green of the olive is a most attractive colour combination. Such ingenuity in the presentation of olives does much to assist in their sale. These stuffed olives are packed in jars.

After filling into the jars, the olives should be rinsed with several changes of water to dislodge and remove all sediment and should then be drained. Then the jars are covered with a clear brine containing between 6 – 9% salt and 0.5% lactic acid (edible grade). Capping should follow immediately, usually under vacuum, without headspace formation.

Black Olives

Olives which are to be processed into the black form are not harvested until the fruit is straw coloured on the verge of turning to red. The essential difference between the processing of the green and black form lies in the amount of aeration during the lye treatment. Aeration is reduced to a minimum during the preparation of the green olives. On the other hand, black olives are treated to a series of alternating lye and air exposures over a period of 3 to 5 days. Catechol-like substances present in the olives are readily oxidized to a dark colour.

Spoilage problems with Olives

Pickled olives suffer from a number of spoilage difficulties, the majority of which are of microbiological origin. These are:

Scum formation: This is by far the most prevalent spoilage difficulty encountered with olives. It, undoubtedly, causes severe wastage, particularly at the domestic level. It arises from the growth of yeasts in the form of a white to cream coloured scum, floating on the brine surface. These are able to grow by utilizing lactic acid in the pack as a source of carbohydrate.

Zapatera spoilage: Cases of this form of spoilage are encountered rarely. The trouble consists of the presence of a peculiar and very penetrating and clinging odour, (e.g. sagey, cheesy or faecal-like, etc.). It occurs only when lactic acid formation during fermentation has been insufficient to drop the pH below 4.5. This condition allows certain undesirable bacteria to grow. A number of species of *Clostridium* (fortunately not *Cl. botulinum)* and *Propionibacterium* have been detected in such spoilage cases.

Butyric acid fermentation: This form of spoilage is closely related with zapatera spoilage. Here, sugars are directly converted into butyric acid through fermentation. This occurs in the early stages of brining. Vaughn has implicated various clostridia as the causative agents of spoilage. He also considers that olives spoiled in this way could be dangerous if eaten.

Yeast spots: These take the form of minute white spots just below the skin surface of the olives. Vaughn has shown that they consist of a tightly growing mass of the bacterium, *Lactobacillus plantarum.*

Softening: The firm structure of the olives is lost. There is little doubt that softening results from the secretion of pectolytic enzymes by moulds. Vaughn isolated 36 species of moulds from olive brines.

Discoloration: It has already been explained how olives are prone to blackening in the presence of oxygen. This can be prevented by maintaining 0.5% or more lactic acid content and the use of tight closures. However, the reaction is assisted by the presence of iron. Iron equipment, therefore, should never be used in the preparation of olives.

Yeasts, too, can cause blackening, since they lower the acidity, creating more favourable conditions for oxidation.

Fisheye spoilage: This consists of a partial collapse of one side of the fruit in a somewhat wrinkled fashion resembling the eye of a fish (Cruess, 1946). This is caused by the initial *Enterobacter* fermentation proceeding too rapidly and so causing gas blisters beneath the skin of the olives.

Pickled Cucumbers

Introduction: The class of pickles based upon cucumbers (either as the immature or mature fruit) is both large and important. In North America, the name 'pickles' is synonymous with cucumber pickles and these pickles are widely consumed. They are also very popular in Europe where numerous varieties are to be found.

The number of kinds of well established pickles prepared from this crop is considerable. Fabian and Switzer (1941) classified these pickles into three main headings: dill, sour and sweet pickles. The fermented dill pickle is the major type of pickled cucumber.

Preparation of salt-stock pickles

The brining of cucumbers: Cucumbers are fermented in large wooden tanks holding between 400 and 1,600 bushels of the fruit. A layer of 40° salometer brine is added to the tank to act as a cushion. This prevents bruising of the cucumbers. Then the cucumbers (after grading and sorting) are added together with 9 lb. of salt for each 100 lb. of cucumbers. 40° brine is added in sufficient quantity to keep the fruit submerged. When the tanks are full, the brine is pumped and circulated to ensure that the salt content is evenly distributed throughout the liquid phase and to dissolve any salt still in the solid state. False covers are fixed to the tanks to ensure that the cucumbers are kept fully submerged beneath the brine.

The salt content of the brine is raised by 2° salometer, at weekly intervals by the addition of fresh salt until it reaches 50°. Thereafter, it is raised by 1° each week until 60° salometer is finally attained. At this stage the cucumbers are fully cured. They will be preserved satisfactorily for several years since they are stable and relatively resistant to change or deterioration. The acidity will probably be between 0.5% and 1% and the pH will approximate to 3.8. If long storage is expected, a higher salt content (70°) may be advisable. It is advisable to pump the brine, especially during the early stages of brining, at regular intervals. Thus, effective curing is achieved.

Microbiological aspects of brining: The curing action arises directly from the activity of various micro-organisms. They bring about the complicated, biochemical changes in the brine and fruit. The following three groups of micro-organisms are important in the brine:

(i) acid producers belonging to the *Lactobacteriaceae*

(ii) members from the genus *Enterobacter* (producers of CO_2 and H_2)

(iii) the yeasts producing alcohol and CO_2

Their development is controlled by the salt concentration, the temperature and the availability of fermentable substances. This process is an example of a mixed natural type of fermentation.

The predominating species in the bacterial flora is *Enterobacter aerogenes* (Etchells *et al.* 1943). It is responsible for the vigorous evolution of CO_2 and H_2. It is of vital importance perhaps for the anaerobic conditions that it establishes in the mass. Its stay is short-lived. By the end of a week this bacterium will have almost disappeared. By this time, the acid-producing bacteria will be reaching their zenith. Subsequently, they are to be found in appreciable numbers for up to three weeks. The most important species of this group is *Lactobacillus plantarum* (1966). This is a homofermentative organism and so produces mainly lactic acid but no gas.

If temperatures are low, *Leuconostoc mesenteroides* may be the dominant acid producer. Both *Streptococcus faecalis* and *Lactobacillus brevis* may also be found in cucumber brines at temperatures above 60°F. Pederson and Albury (1950) concluded that 75° to 86°F was the optimum temperature range for brining cucumbers. This is not to say that perfectly successful curing cannot be carried out below this range.

Many species of yeasts are present throughout all the stages of fermentation, but they reach their peak population during the third week of brining. Their presence is thought to be of value with respect to the development of flavour. Costilow and Fabian (1953) found *Torulopsis holmii* and *T. rosei* to be the predominant species in cucumber brines.

The effect of adding live cultures (starters) to cucumber brines has been investigated on a number of occasions without much success.

When the brine fermentation has been completed and the salt content raised to 60° or more, the cured stock attains a microbiologically stable condition, with the exception of certain film-forming yeasts. Film yeasts utilize the lactic acid present as a source of carbohydrate. They form a thick white to grey coat or film on the brine surface. Owing to the oxidation of the lactic acid by yeasts, the pH of the brine rises to the value of 5 or above and putrefactive bacteria may become established. These bacteria produce foul odours in the product and discolour the cucumbers with ferric sulphide (particularly those lying close to the brine surface).

Film-forming yeasts are prevented from growing on the surface of cured stock. They are sensitive to ultra-violet light. Therefore, they may be inhibited by exposing the brine surface to direct sunlight. These yeasts can also be inhibited by the addition of 0.1 % sorbic acid to the brine (Phillips and Mundt, 1950). A layer of liquid paraffin of approximately $\frac{1}{8}$ inch depth has been found to be reasonably effective. This paraffin layer can be removed by skimming, and floating the remaining traces over the sides of the vat by flooding with water. Lastly, plastic covers for the tanks as a means of controlling yeast growth have been suggested by Canadian workers (Finely and Johnston, 1957).

Pasteurization of Cucumber Pickles: From the work of Etchells and his co-workers (Etchells *et al.*), it was established that the optimum heat treatment for all types of pickled cucumbers consisted of heating the liquid contents of the pack to 165°F for 15 minutes, and then cooling. This method was found to be fool proof. Besides effectively destroying all potential spoilage micro-organisms within the product, this process was later found to inhibit polygalacturonase. The presence of this enzyme in the product is harmful because it softens pickles. In addition to this, heat treatment noticeably lowers the peroxidase activity of the product. This enzyme produces certain flavour deteriorations, (i.e. the development of hay-like and musty odours). In other words, pasteurization assists in preserving a good flavour in cucumber pickles.

Canning Pickled Cucumbers: The cured salt-stock pickles are soaked in changes of water for removing the surplus salt. This operation is called processing. Thereafter, processed salt-stock pickles are used in preparing different types of pickles and this final operation is referred to as finishing. For example, sour pickles are prepared by the addition of vinegar into the processed salt-stock pickles. Sweet pickles, likewise, are prepared by treating them with vinegar and sugar.

The cans are filled with packing liquor which is previously heated to and maintained at 194°F. The composition of packing liquor is dependent on the destination of the product. For example, for export the packing liquor can have either of the following formulae:

(a) 0.6 – 0.7% acetic acid, 0.6 – 0.8% lactic acid, 2% salt, 0.5% sugar

(b) 0.4 – 0.5% acetic acid, 0.8 – 1.0% lactic acid, 2% salt, 1 – 2% sugar

The acetic acid is, in all cases, derived from spirit vinegar. Pickles are packed into 9-litre cans (Pijanowski, 1951).

As soon as the cans are filled, they are sunk up to the brim in hot water at 194°F for 10 to 15 minutes until bubbles cease to rise from the contents. The purpose of this is to exhaust the cans and to remove bubbles attached to the fruit. The contents should reach 158°F and some lowering of the liquid level results (about 0.5 cm.). The cans are then carefully sealed, and pasteurized in a water bath at 176°F to 185°F for 40 to 45 minutes. Alternatively, pasteurization may be carried out at 185°F to 194°F for 30 minutes. 5-litre cans require two-third of the time taken by the former process.

Cans used for pickles should be internally coated with an acid and heat-resistant lacquer. The thickness of the coating should be 0.32 mm. for the larger cans and 0.28 mm. for the smaller ones.

Spoilage of Cucumber Pickles

Softening: The softening of salt stock has always been a serious problem in the American pickle industry. A great deal of research went into this problem and it was firmly established that softening was associated with the presence of polygalacturonase (a pectin-splitting enzyme) in the brine. Etchells, Bell and Jones (1955) showed that this enzyme arose from moulds which grew on the dead flowers still attached to the fruit. The removal of these flowers from the fruit produces salt stock of low polygalacturonase activity and firm texture. Efforts are also being made to breed varieties of cucumbers which drop their flowers at an early stage.

Other possible sources of softening were suggested to be *Byssochlamys nivea* (Anon., 1963).

Off-flavours: The presence of peroxidase is responsible for the development of hay-like and musty odours in pickled cucumbers. Also, other enzyme systems may be associated with the development of off-flavours in pickles.

The use of the insecticide Lindane on the cucumbers during growth has at times been found to cause off-flavours. These off-flavours also have a musty character.

Bitterness is another flavour problem associated with fresh pack pickles (Chandler, 1965). The reason for this is unknown.

Loss of colour: The retention of colour in cucumber pickles has attracted some attention by research workers, but with little success. Little is known of the causes of loss of colour and even less is known of empirical methods whereby colour can be preserved. Exposure to light (either natural or artificial) causes the colour of pickles to fade. There are indications that oxygen is involved, to some extent, in the fading reaction.

Bloaters: Large cucumber pickles are prone to a trouble known as 'bloater' formation. The reason for this is the formation of gas within the fruit during the curing period. As a result of this, the cucumbers become distended to a marked extent. The cause of this gas formation has not yet been definitely proved. Factors increasing the incidence of bloaters have been investigated by Samish *et al.* (1957). They are: (i) addition of alum or sugar to the brine (ii) high salt contents and temperatures (iii) increasing size of fruit

Effect of Holding Time on the Quality of Fresh Pasteurized Pickles

It has been shown that the time between picking fruit and its processing is critical with respect to the quality of colour, texture and flavour of these pickles. This time should be as short as possible. It was found that even delays of 24 hours at 40°F were detectable in reduced quality. Cucumbers intended for fresh pasteurized pickles should be picked and processed on the same day if optimum quality is to be retained. Alternatively, chilled storage of the fruit is of considerable assistance in maintaining quality if delays between harvesting and processing are unavoidable.

Yoghurt or Bulgarian milk (Figure 27.1)

Introduction: Yoghurt is a dairy product prepared by the fermentation of milk. It is consumed in large quantities as a food item by the Bulgarian people. Nowadays, it is also consumed by the people of Europe and North America. According to the Russian bacteriologist Metchnikov, E. (1908), growth of *Lactobacillus bulgaricus* in the colon produced a high concentration of lactic acid. This highly acidic environment did not permit the growth of proteolytic flora. Therefore, disorders that were supposedly connected with absorption of proteolytic wastes from the alimentary tract would not occur.

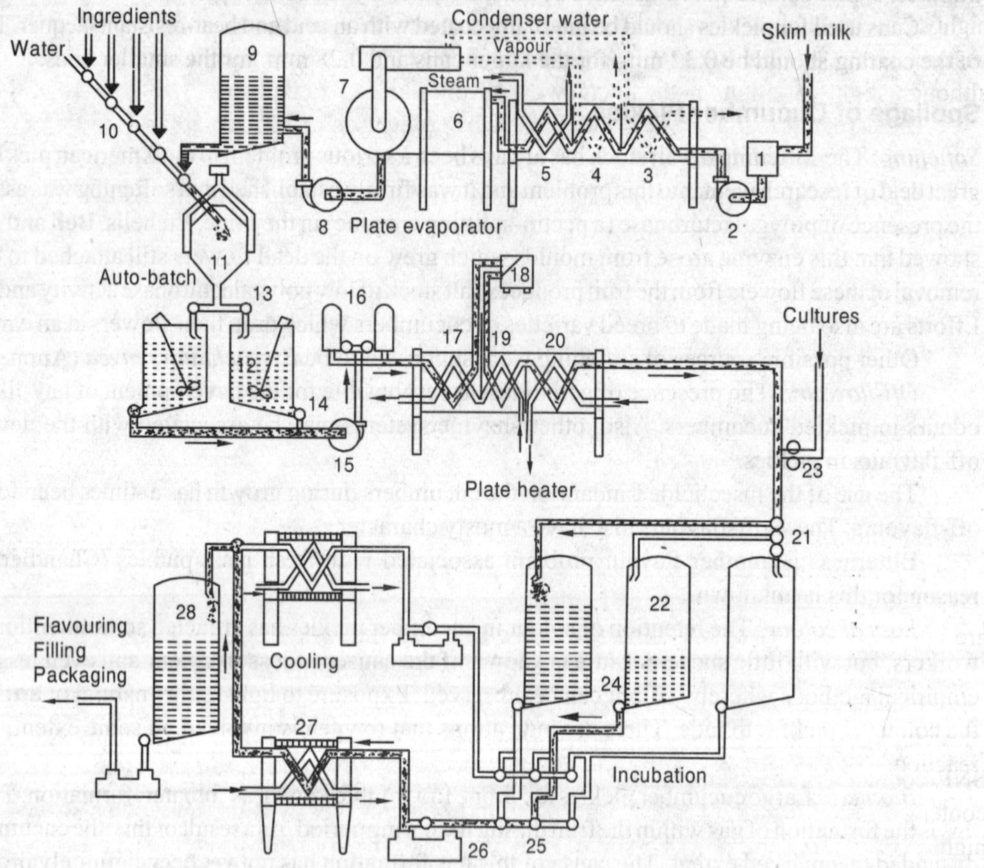

1. Feed balance tank
2. Feed pump
3. Paraflow preheater 1st stage
4. Paraflow preheater 2nd stage
5. Paraflow preheater 3rd stage
6. Plate evaporator
7. Separator
8. Concentrate pump
9. Concentrate tank
10. Ingredient inlet valves
11. Weigh hopper
12. Mix vat
13. Mix vat changeover valve
14. Mix vat outlet valve
15. Mix pump
16. Duplex filter
17. Paraflow 1st stage preheating
18. Homogeniser
19. Paraflow 2nd stage—U.H.T
20. Paraflow 3rd stage rapid cooling
21. Tank inlet valves
22. Incubation tank
23. Cultures metering pump
24. Tank outlet valve
25. Valve manifold
26. Positive pump
27. Paraflow cooler
28. Yoghurt tank

Fig. 27.1 *Flow diagram for yoghurt manufacture.*

It is believed that if a man regularly consumes 250 to 600 gms. of yoghurt daily from his childhood, he can expect to lead a healthy, buoyant and youthful life up to the age of 70 to 80 years. It decelerates the clock of aging, if not reverses or stops it. It also lowers the blood cholesterol level and, therefore, improves the functioning of the cardio-vascular system making the heart strong. Today, therapeutic, prophylactic and nutritional properties of *dahi* are widely accepted.

Manufacture of Yoghurt: Yoghurt is commercially manufactured by the fermentation of milk with two thermophilic bacteria, viz., *Streptococcus thermophilus* and *Lactobacillus bulgaricus* (Figure 27.1). The latter rapidly grows at optimum temperature through the process of symbiosis. The final product is highly viscous, slightly sour, with its characteristic 'curd' flavour and odour. It has consistency resembling custard. The manufacturing details of the production of yoghurt have been described in some publications, although slight variations in details may exist from industry to industry.

The essential features of all the methods of production of yoghurt are:

(1) selection of milk
(2) standardization (fat extracted)—milk powder of concentrated milk added
(3) sarming and homogenisation
(4) heat treatment (steaming for half an hour)
(5) cooling to about 50°C
(6) inoculation with yoghurt cultures
(7) packaging
(8) incubation in a warm room (42° to 44°C)
(9) cooling in the air
(10) cold storage

The alternate steps are:

(1) incubation (2) cooling (3) agitation (4) packaging

The variations in commercial practices in the manufacture are related to two aspects only. They are :

(1) standardization of milk in respect of fat and solids-not-fat (SNF)
(2) the stage of packaging the product

Yoghurt contains 2 – 3% of fat, but even a fat-free product is acceptable to many. Usually, the SNF content is raised to 14 – 16% to make it a more viscous product and to increase its protein content. In regard to the second aspect, the modification involves the marketing of the yoghurt as a highly viscous uniform product 'stirred curd' instead of the normal 'firm-set' or 'junket form'. As shown above, as an alternate step of manufacture, the milk after inoculation is held in the vat at 42° to 44°C. Until it coagulates, then it is cooled by circulating chilled water with gentle agitation and packed in suitable containers. The filled containers are then cold stored (at about 5° to 6°C) and held overnight to develop the characteristic 'Yoghurt' flavour.

The product so obtained will be slightly sour, but with a clean flavour. The acidity is about 0.9% but can go up to 1.1% (pH 4.0 to 4.2). Higher acidities may be tolerated if milk fortified with extra milk solids is used.

Fruit and Flavoured Yoghurts

A recent innovation which has enhanced the popularity of yoghurt is the practice of fortifying it with sweetening agents, fruits and fruit flavours. Fruit and flavoured yoghurts are sweet with the characteristic flavour of the fruits or the added flavours. They retain all the nutritional and curative properties of yoghurt. They are readily accepted by children, adolescents and aged people.

Common sugar is the most common form of sweetening agent added. Some people add honey which also increases the nutritive value of the product. The sugar level depends upon the consumer's preference and is in the range of 6 – 8% as sucrose.

Fruits and flavours can be added to yoghurt in a variety of forms. Some add synthetic flavours not to the liking of many. Quite a common practice is the addition of fruit extracts or concentrates. Fruit powders are also added in some Western Countries. However, the most popular practice is the addition of cut fruits or fruit pulps which retain the fresh flavour of the fruit. The types of fruits added are many and variable. The popular fruits are the berries, citrus and apples. Usually, fruits are added at the rate of 15 – 20% of the product. Fruits and sugar may be added separately or as 'fruits in syrup preparations'.

Fruit yoghurts can be preserved up to a week in a refrigerator and for longer periods at temperatures below 5°C. Deep freezing of the product is not desirable as it can cause 'wheying off' when taken out for thawing.

Nutritive Value of Yoghurt

During fermentation of milk, the composition of the minerals remains unchanged, while proteins, carbohydrates, vitamins, and to some extent, fat constituents are subjected to changes which produce special physiological effects. Dietary and therapeutic qualities of sour milk products are determined by micro-organisms and substances formed as a result of the biochemical process accompanying milk souring. These substances are lactic acids, alcohol, carbon dioxide, antibiotics and vitamins. The following processes make yoghurt more nutritive than milk:

Proteolysis in milk: Proteolysis in milk takes place by exo- or endo-peptidase of lactic acid bacteria. The biological value of protein increases significantly during yoghurt manufacture from a value of 85.4 – 90%. This increase is due to breakdown of protein into peptones, peptides and amino acids. The contents of essential amino acids such as leucine, isoleucine, methionine, phenylalanine, tyrosine, threonine, tryptophane and valine increase considerably, which offer special advantages not only to healthy people but also particularly to the physically weak.

Hydrolysis of lactose: Lactose in milk is hydrolysed by metabolic activity of bacteria. Approximately, 2.5% lactose, 0.8 – 1.3% galactose and 0.03% glucose are obtained from lactose hydrolysis. Lactose hydrolysis takes place due to β -galactosidase production by lactic acid bacteria. The importance of lactose is due to the lactic acid produced from the hydrolysis of lactose, which leads to a pH range in the bowel inhibiting the growth of putrefactants. In addition to this, lactic acid is important for organoleptic properties and calcium absorption.

Lipolysis: The homogenization process reduces the size of fat globules which become digestible. The production of free fatty acids as a consequence of lipolytic activity increases as compared to milk. This leads to some physiological effects.

Changes in Vitamins: There is more than a two-fold increase in vitamins of B-group, especially thiamine, riboflavin and nicotinamide as a result of the biosynthetic process during milk fermentation. Subsequently, vitamin B_6, vitamin C and vitamin B_1 (thiamine) decrease by approximately one half only, as they are utilized by the bacteria in milk.

Antibacterial activities: The bactericidal properties of sour products are determined by the antibiotic activity of the bacteria growing in the product. The antibiotic properties are generally associated with lactobacilli in yoghurt, and materials responsible for such antibacterial action are described as lactic acid, hydrogen peroxide and other substances such as lactobacilline, etc.

Therapeutic Importance: The main advantages of regular intake of yoghurt are:

(1) This product is easily absorbed and better assimilated than sweet whole milk. Assimilation of milk is 32% in one hour, while that of yoghurt is 91% in the same period.

(2) Yoghurt improves appetite due to its pleasant, refreshing and pungent taste. It is highly nourishing and invigorating not only to healthy persons but also to the ill and infirm.

(3) Gastric juice secreted by the action of yoghurt and the desirable ratio of calcium and phosphorus induced by it leads to a high digestive capability.

(4) Yoghurt consists of a sufficient amount of indispensable amino acid, methionine, which removes excessive fat from the liver and enhances bile secretion. It is, therefore, an important therapeutic adjuvant in gastrointestinal disturbances, hepatitis, nephritis, diarrhoea, colitis, anaemia and anorexia. It provides wonderful relief to patients of chronic diarrhoea, sprue and ulcerative colitis. Fat-free yoghurt is necessary for those who suffer from heart disease, arteriosclerosis, hypertension and chronic inflammation of the liver.

(5) Yoghurt possesses potent antitumor activity according to Shahani, professor of Food Sciences at the University of Nebraska, U.S.A. It may prevent or retard the onset of diseases. Pathogenic bacteria are not able to survive in it because of its low pH and other adverse factors for their growth. Consumption of less than one litre of yoghurt per day will not affect human health adversely, but more may involve the risk of acidosis. The nutritional and therapeutic effects of yoghurt do not depend on living bacteria but on their metabolites.

Cheese

The manufacture of cheese is a microbiological process. There are several hundred varieties of cheese. The particular combination of salt, incubation, temperature, pH and culture used, determines the kind of cheese manufactured. The production of most cheeses is made from cows' milk (either whole or skimmed). Some cheese varieties are manufactured from cream and whey or from the milk of other animals. The basic steps involved in cheese production are as under:

(1) *Curdling the milk:* Curd may be prepared either by adding starter culture (pure or mixed type) or the enzyme rennin to milk. The following organisms may be used as starter culture in the pure or mixed form: *Streptococcus cremoris, S. lactis, S. thermophilus, Lactobacillus lactis, L. helveticus, L. bulgaricus,* etc. The milk protein, casein, gets coagulated during curdling reaction. The mechanism of casein coagulation is an interesting phenomenon. The initial pH of the milk is about 6.8 at which casein does not undergo coagulation. But casein coagulation occurs when the pH reaches 4.78 to 4.64. Lactic acid produced by the starter culture lowers the pH of the milk to a sufficient extent to precipitate casein. Actually, casein is dispersed in fresh milk as calcium caseinate. Lactic acid reacts with the calcium with the formation of calcium lactate. The casein gets precipitated when the calcium content is decreased to a certain low level. Also, curdling of milk may be brought about by adding the enzyme rennin to milk. The added enzyme reacts with the calcium caseinate with the formation of calcium paracaseinate. Calcium paracaseinate, in turn, reacts with free calcium ions to form curd. Here, the pH of milk is not changed. In some varieties of cheese production, a combination of the starter culture and rennin is used.

(2) *Draining the curd to remove moisture:* The straw-coloured liquid called whey is produced during curdling of milk. It is separated from curd with or without the application of pressure. The kind of cheese to be produced determines how much moisture must be removed. The production of soft varieties of cheese involves draining without pressure. On the other hand, the manufacture of hard

varieties of cheese requires draining with pressure. Also, semi-hard varieties of cheese are produced, where a limited pressure is used during draining. After removing whey to the desired limit, the curd is moulded into various sizes and shapes according to the variety of cheese that is being produced.

(3) *Salting:* Sodium chloride is practically always added at some step of production to all varieties of cheese. Sodium chloride is added in either of the following two ways: (1) It may be applied to the surface of the pressed curd by rubbing with salt. (2) Salt may be mixed with the drained curd. Salting operation has three objectives:

(a) It serves as a preservative, and therefore, does not permit the development of unwanted micro-organisms.
(b) It imparts flavour to the cheese.
(c) It serves to control cheese moisture by withdrawal of water.

(4) *Ripening:* All varieties of cheese, except a few, require ripening of curd. Cottage cheese and cream cheese are examples of unripened varieties of cheese.

Usually, ripening is brought about either by bacterial or mould culture. In both cases, changes in the curd take place through enzymatic reactions. Importance of ripening operation can be understood by considering some important varieties of cheese:

Roquefort cheese: Roquefort cheese originated in southern France many years ago, and in its original form, was made only from sheep's milk. Today most Roquefort and all of its imitations are produced from cows' milk. The curd is inoculated with *Penicillium roqueforti* by mixing in bread crumbs on which the fungus has been growing. After inoculation, the inoculated curd is placed in a ripening room. The temperature and humidity of the ripening room are carefully controlled. The inoculated curd is left in the ripening room until the fungus has grown to the desired extent. The inoculated curd is then placed at lower temperature to inhibit the fungal growth, but fungal enzymes remain active. Since filamentous fungi, in general, are strongly aerobic organisms, the question naturally arises as to how *Penicillium roqueforti* is able to grow in the interior of the pressed curd. *Penicillium roqueforti* can grow in an atmosphere with oxygen concentration as low as 4.25% and for this reason holes are punched in the pressed curds to provide means of aeration deep in the curd.

Currie reported that during the ripening of Roquefort cheese, *Penicillium roqueforti* produces a water soluble lipase which catalyzes the hydrolysis of milk fats. He recovered acetic, butyric, capric, caprylic and caprolic acids by steam distillation. Currie stated that of these acids capric, caprylic and caproic and their salts have a 'peppery' taste. These compounds are responsible for the burning effect of Roquefort cheese on the tongue and palate.

Hammer and Bryant believed that the taste and odour of blue cheese (ripening by *Penicillium roqueforti)* are suggestive of 2-heptanone. 2-Heptanone is formed from caprylic acid through the action of *Penicillium roqueforti.* Using an ether extraction procedure, Patton recovered material from Blue cheese which contained a high concentration of methyl ketones. Fractional distillation gave relatively pure fractions of 2-pentanone, 2-heptanone and 2-nonanone. It was suggested that these methyl ketones are formed in Blue cheese as a result of beta-oxidation of fatty acids. Patton pointed out that a number of observers had noted the similarity in odour between these ketones (particularly 2-heptanone) and Blue varieties of cheese. Therefore, it was suggested that the unique flavour of these varieties of cheese was probably due to the minute quantities of methyl ketones.

Camembert cheese: Like Roquefort, Camembert is said to have originated in France and it has several imitations. Named cheese varieties which are basically of the Camembert type are: Brie, Thenay,

Troyes and Vendome. The principal and obvious action of the fungus in the ripening of Camembert type cheese is that of changing the texture of the cheese, although changes in the odour and the flavour also occur. Thom was of the opinion that certain characteristic flavours are due to the action of lactic acid bacteria.

During the ripening process (which ordinarily lasts three or four weeks) temperature and humidity are closely controlled. During this period, the texture of the curd is changed by the activity of the micro–organisms. They grow on the outside (rind) of the curd. Ordinarily, these micro-organisms are *Penicillium camemberti,* its white variety (var. *rogeri), Oidium lactis* and several species of bacteria. The curd is softened to a smooth butter consistency through the action of proteolytic enzymes produced by micro-organisms growing on the rind. It is obvious that in the ripening of Camembert cheese, a pure culture of a single organism is not involved. Therefore, in all probability, it really doesn't matter what species is involved, as long as it produces the proper extracellular protease in adequate quantity to bring about the desired texture change and does not give any off odours and flavours.

Swiss cheese: The ripening process in the manufacture of Swiss cheese is carried out by lactobacilli, streptococci, and propionibacteria. These bacteria are added to the milk prior to the curdling reaction. Lactose of the milk is attacked by lactobacilli and streptococci with the formation of lactic acid. Now lactic acid is further fermented by a species of *Propionibacterium.* The end products of this fermentation are: carbon dioxide, propionic and acetic acids and water. The presence of these two acids develops flavour in the cheese. The accumulated CO_2 forms the eyes (holes).

Cheddar cheese: It is the most commonly sold variety of American cheeses. It is a hard curd cheese without eyes. The organisms, primarily involved, are probably *Streptococcus lactis* and related streptococci and various lactobacilli.

Limburger cheese: It is produced by the action of the lactic acid streptococci and a number of other bacteria (especially *Brevibacterium linens*) and yeasts. *Brevibacterium linens* is actively proteolytic bacterium and is believed to be the main organism for imparting a characteristic flavour in Limburger cheese. High salt concentration, low pH, and temperature combine to produce the Limburger type of ripening.

The most important commercial varieties of cheese are given in Table 27.1.

Table 27.1 *Important commercial varieties of cheese.*

General group of cheese	*Variety of cheese*	*Primary ripening organisms*
Hard	Cheddar	Bacteria, not including eye-formers
	Swiss	Bacteria, including eye-formers
Semi-hard	Brick	Bacteria
	Roquefort	Mould
	Blue	Mould
	Gorgonzola	Mould
Soft	Limburger	Bacteria
	Camembert	Mould
	Cottage	Unripened
	Cream	Unripened
	Neufchatel	Unripened

Butter

Butter is manufactured by churning pasteurized sweet or sour cream. There is a better separation of butterfat globules from the other constituents in the churning process, if sour cream is used. Natural souring of raw cream may take place, since *Streptococcus lactis* is present. However, no control on flavour is possible because unwanted bacteria, other than *S. lactis*, may be present. Moreover, there may be absence of flavour and aroma producing micro-organisms. Nowadays, it is a common practice to inoculate pasteurized sweet cream (heated at 71°C for a period of 30 min.) with a butter culture (starter). The inoculation of pasteurized fresh cream with a butter culture is practised for two reasons: (i) to increase the yield of butter, and (ii) to develop a characteristic type of flavour and aroma in the butter.

Butter cultures consist of a mixture of two types of micro-organisms: (1) micro-organisms which can attack lactose (milk-sugar) with the formation of lactic acid. *Streptococcus lactis* and *S. cremoris* are the micro-organisms with this ability, and (2) micro-organisms which develop the desired aroma and flavour. Generally a mixture of *Leuconostoc dextranicum* and *L. citrovorum* is used. These micro-organisms produce small amounts of acetoin. Acetoin is spontaneously oxidized to diacetyl, the compound which imparts the flavour and aroma.

The pasteurized sweet cream is inoculated with a starter, and then held at a proper temperature until desired changes have taken place. Usually, 0.25 – 1% starter is used for the inoculation of pasteurized sweet cream. The resulting cream is called ripened cream, which is subjected to churning. The butterfat globules coalesce into granules during this operation. The liquid part (buttermilk) is drained off, and the granules are further processed.

3 PRESERVATION OF FOOD

There are various methods for the preservation of food. The various methods involve the use of: (a) heat (b) cold (c) dehydration (d) high osmotic pressure (e) preservatives (f) antibiotics (g) radiation.

Heat

Many food products, (e.g. fruits and vegetables) are preserved by the use of heat. It is one of the safest and most reliable methods of food preservation because heat is effective in destroying vegetative cells as well as spores. It is necessary to control the temperature during the thermal processing of the food products. The temperatures required for the thermal processing are different for different foods. Excessive heating may bring about changes in its appearance, flavour and composition. In the canning method, steam under pressure is used to kill all forms of life present in the food products.

Canning may be loosely defined as a process for preserving food by the application of a thermal sterilization procedure adopted for the products packed in hermetically sealed containers. The basic operations connected with the canning procedure are as follows:

(A) preparation of the food
(B) filling of the container
(C) exhausting or pre-heating
(D) sealing of container
(E) thermal processing
(F) cooling of the container and contents

(A) *Preparation of the food:* A variety of processes, such as grading, trimming, washing, blending, blanching, pre-cooking, etc. are used. These pre-canning operations should be carried out effectively but rapidly, since undue delay at this stage will permit the development of rapidly growing micro-organisms which may render the heat process inadequate.

The following three operations are of importance to the microbiologist: (1) washing (2) peeling (3) blanching. There occurs a reduction in the numbers of surface micro-organisms. If these operations are incorrectly carried out, or carried out in poorly designed equipment, this will lead to an increase in the initial bacterial load.

(1) *Washing:* There are various washing processes. Usually, the food material is subjected to high pressure water sprays, or is immersed in streams of water while it is moved on a conveyer belt. Various industrial washing machines are described by Slade (1967). The washing operation serves two purposes: (a) It decreases the load of spoilage micro-organisms. (b) It improves the appearance of the food material.

(2) *Peeling*: The peeling of fruits and vegetables is an important preliminary operation. It removes surface soiling and associated microbial contamination. Incorrect or inefficient peeling can lead to excessive wastage of sound raw material in addition to increasing the effluent problem. Various methods are employed, such as steam peeling, flame peeling, mechanical peeling, abrasive peeling and lye peeling or combinations thereof.

(3) *Blanching:* In this operation the raw fruits or vegetables are immersed for a short time in hot water at 87.5° to 95°C or are subjected to live steam. Blanching has various objectives:

(a) Expellation of respiratory gases which would reduce the ultimate vacuum in the can if they are liberated during processing.
(b) Inhibition of respiratory enzymes, especially those of oxidation. This prevents adverse effect on product quality and nutritive value.
(c) Either shrinkage or expansion of the food product takes place. This permits an adequate fill of the container.
(d) The adhering materials not removed by cold water may be removed.
(e) Fixation of the natural colour of certain food products which makes them attractive in appearance.
(f) Hydration of certain dried products which are not fully rehydrated before this operation.

The most common type of equipment used for water blanching is the rotary immersion blancher. The food product is conveyed by a screw on the inside of a rotating reel, the bottom part of which is immersed in hot water. Chain conveyers are also employed for transporting food materials through hot water zone. Other techniques of blanching have been suggested: namely, hot-gas blanching and IQB (Individual Quick Blanch), but neither has yet been widely applied on a commercial scale.

(B) *Filling of the container:* The prepared foods are next packed into containers. This is accomplished mechanically or manually. The containers should be filled with the correct weight of food material. It is also necessary to avoid the trapping of air in the filling of semi-solid products, such as meat and fish pastes.

(C) *Exhausting:* The filled cans are exhausted. This operation removes air from the container before it is sealed. Exhausting has four objectives as follows:

(a) Minimization of strain on the container through expansion of air during thermal processing operation.
(b) Removal of oxygen which accelerates the internal corrosion of the container.
(c) Development of a vacuum when the container is cooled. Creation of a vacuum is accepted as evidence of soundness of the canned food material.

(d) Prevention of oxidation of the food materials. For example, vitamin C content of the food material is preserved.

In commercial practice the methods employed for exhausting are: (1) heat-exhaust (2) mechanical exhaust (3) steam injection. In the case of the mechanical exhaust method, air is removed from the cans and then they are sealed simultaneously.

(D) *Sealing of container:* The filled containers after being exhausted are carefully sealed before thermal processing. The sealing of the container prevents contamination of the contents.

(E) *Thermal processing:* During this operation, all contaminating micro-organisms (vegetative as well as spore forms) are killed. This operation involves the use of steam under pressure. The containers must be heated for an accurately predetermined time to obtain specific temperature. The duration of heating and the temperature used depends upon various factors, such as the character and chemical composition of the food, the types and numbers of the micro-organisms present in the food product, the pH of the food product, size of can, etc. There is a slow penetration of heat in solid food materials, as opposed to liquid foods. If the food is more acidic than pH 4.5, it is usually sufficient to raise the temperature to 200°F at the centre of the container. The pH is important for two reasons: (1) killing of micro-organisms proceeds much more rapidly at a fairly acidic pH and (2) *Clostridium botulinum* does not grow at a pH more acidic than 4.5.

Non-acidic food products, (e.g. beans, meat, peas, etc.) must be heated to at least 240°F for safe processing. The spores of *Clostridium botulinum* must be destroyed as they are capable of growing in non-acidic food products. Spores are destroyed by holding them at a temperature of at least 240°F for 10 minutes.

Processing may be carried out in batch retorts (Figure 27.2) or in a continuous pressure cooker.

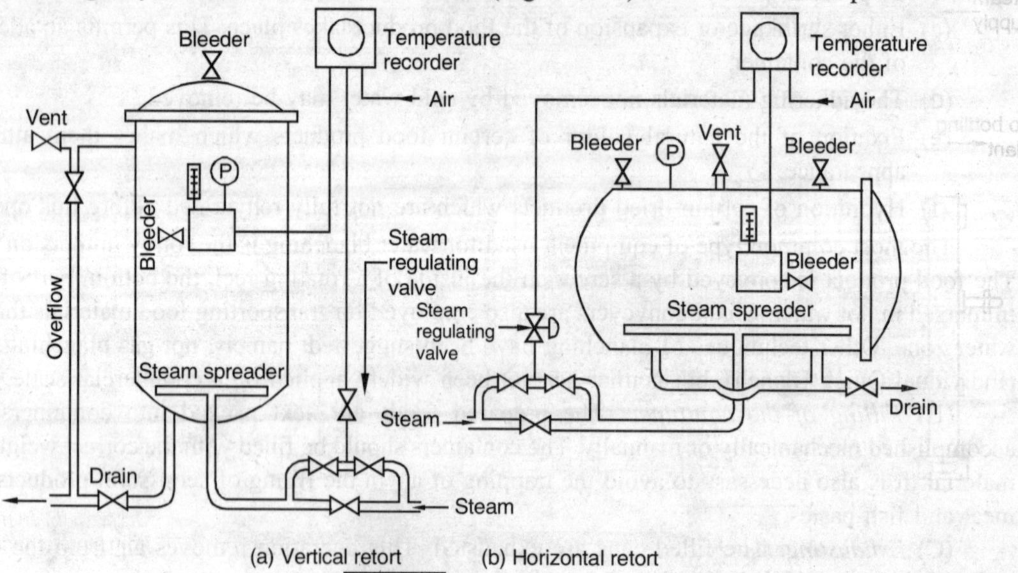

Fig. 27.2 *Batch retorts.*

(F) *Cooling of the container and contents:* This operation is necessary to check the action of the heat. Therefore, undue softening or alteration in the colour of the foods is prevented. Either air or water may be used for cooling purpose. The cans should be cooled to an average temperature of 37°C. This cooling temperature is helpful in retaining sufficient heat to ensure rapid drying and thus obviate rusting.

Pasteurization

The process of heating a product to improve its keeping qualities was applied to wine by Louis Pasteur (1860-64). The reason of this attempt was to save the breweries of France from ruin due to the wholesale spoilage of the wine. The process proved eminently successful. Therefore, the inventor has been honoured by having the process named after him, (i.e. pasteurization).

In 1886, Soxlet suggested that milk which was to be fed to infants should be heated. Soon after this, the dairy industry began the practice of pasteurizing market milk.

This process of pasteurization is applied to improve the keeping qualities of milk, fruit juices, (e.g. grape juice and apple juice) and many other food items. It is commonly practised in the beer and wine industries. It produces a marked reduction in the bacterial content. It may kill as many as 99% or even more of the mesophilic bacteria present. On the other hand, thermoduric bacteria and bacterial spores largely resist pasteurization. In addition to this, thermophilic bacteria may grow rather rapidly at pasteurization temperature.

There are two pasteurization procedures applied in milk industry. They are:

(a) the long time-low temperature (LTLT) or holding method. It consists of holding the milk in a vat at 145°F (62.8°C) for 30 minutes.

(b) the high temperature-short time (HTST) or continuous method. It consists of heating the milk at 161°F (71.7°C) for 15 seconds (Figure 27.3).

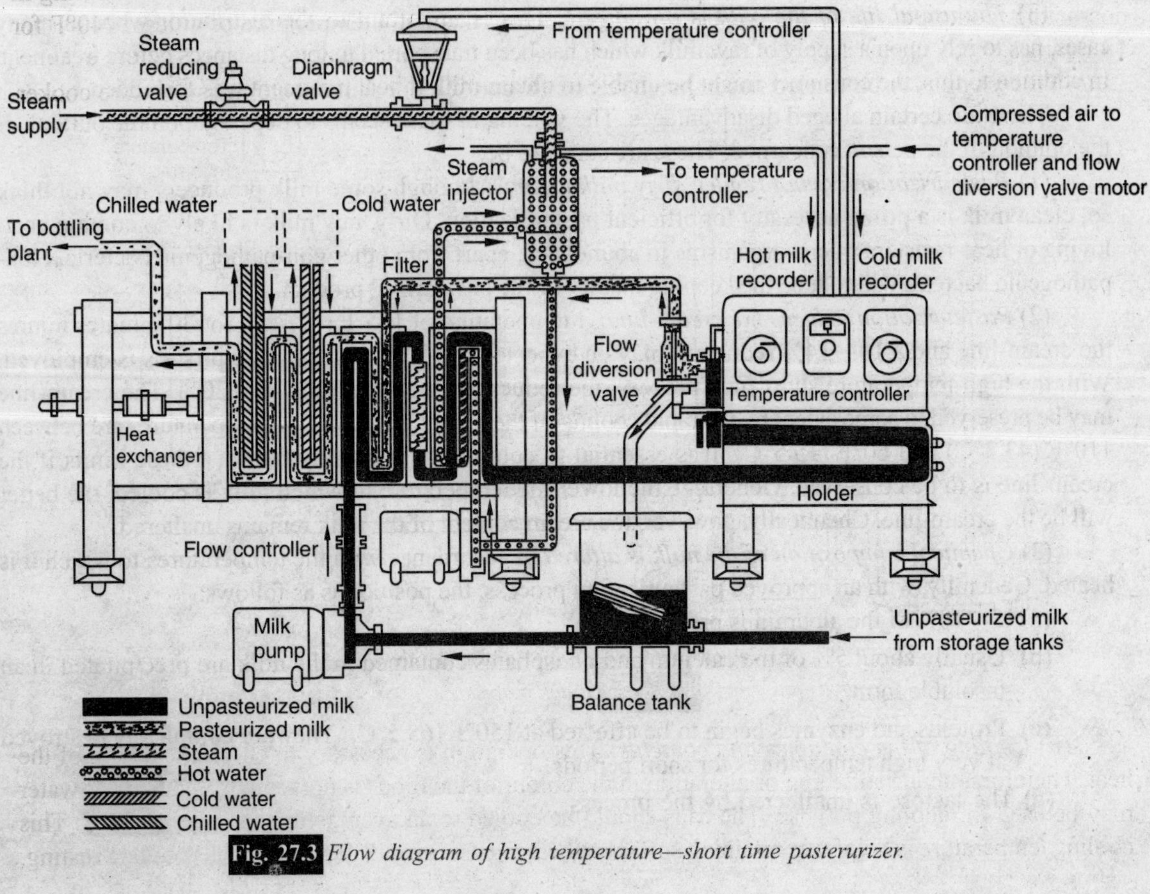

Fig. 27.3 *Flow diagram of high temperature—short time pasterurizer.*

In both methods, it is necessary to heat every particle of milk or milk product to the required temperature for a specified time. The pasteurized milk or milk product should be stored at a cold temperature.

Recently, ultra high temperature (UHT) pasteurization has been developed. The temperatures used are 200°F or higher for a brief period of time. This gives the milk much longer shelf-life and reduces storage costs. The milk even keeps for a period at room temperature if it has been heated high enough to destroy the natural milk enzymes which might impair the flavour.

Advantages and Disadvantages of Pasteurization

The advantages to be obtained from the efficient pasteurization of milk may be summarized as follows :

(1) *Milk is rendered 'safe' as all pathogens are destroyed:* It would not be possible to obtain a safe milk supply for the community at all times without using this process. This is the simplest way by which safety may be guaranteed.

(2) *The bulk of the non-pathogenic bacteria are also destroyed:* It is probably quite true to state that approximately only 1% of non-pathogenic bacteria remain after milk treatment by any officially approved method of pasteurization while, with some methods, a greater percentage is destroyed. The percentage which remains may be greatly increased when conditions are such as will allow large numbers of heat-loving and heat-resistant organisms to be present. Those micro-organisms which do survive are usually spore-formers. These do not affect the keeping quality of the liquid or the health of the consumer.

(3) *The useful life of the milk is prolonged:* This is important to the distributor who, in many cases, has to rely upon a supply of raw milk which has been transported to long distances before treatment. In addition to this, the consumer might be unable to obtain milk if heat treatment was forbidden.

There are certain alleged disadvantages. The voicing of these seems to depend upon the outlook or the opinion of the person concerned. These are stated to be:

(1) *Pasteurization cannot render dirty milk clean:* Although some milk producers may not think so, clean milk is a prime necessity for efficient pasteurization. Dirty, raw milk is likely to contain heat-loving or heat-resistant micro-organisms in abundance, apart from other non-pathogenic bacteria. Non-pathogenic bacteria will reduce the keeping quality of the pasteurized product.

(2) *Pasteurization reduces the cream-line: A* temperature of 145°F (62.7°C) for 30 minutes injures the cream-line and a 10 – 14% reduction may be expected if the low temperature process is employed. With the high temperature short time process, the reduction can vary from 10 – 20%. The cream-line may be preserved to some extent by the rapid cooling of the hot milk. The critical temperatures are between 110°F (43.3°C) and 60°F (15.5°C). It is essential to cool the milk quickly between these limits if the cream-line is to be conserved. Generally, the lower the temperature to which milk is cooled, the better will be the cream-line. Chemically, however, the cream content of the milk remains unaltered.

(3) *Chemical composition of the milk is altered* in accordance with the temperatures to which it is heated. Generally, with an approved pasteurization process, the position is as follows:

(a) 5 – 10% of the albumin is precipitated.

(b) Usually about 5% of the calcium and phosphates contained in the milk are precipitated in an insoluble form.

(c) Proteins and enzymes begin to be affected at 150°F (65.5°C). They are completely destroyed at very high temperatures for short periods.

(d) The lactose is unaffected by the process.

(4) *The vitamin content of the milk is altered* to a varying degree as follows:

Vitamin A — not affected by pasteurization

Vitamin B_1 — not heat tolerant, a reduction of approximately 10 – 25% occurring

Vitamin B-Complex — not affected

Vitamin C — heat sensitive and destroyed to a greater or lesser degree by the process according to the method used. With the low-temperature process, a reduction of up to 25% may be expected, but the short-time process is less harmful in this respect, not more than 5% being destroyed during the treatment. It would appear that the amount destroyed depends upon previous exposure of the milk to light, particularly sunlight and the degree of oxidation which occurs, this latter process being accelerated by the catalytic actions of metals. There is, however, little action of this kind with stainless steel or the other metals used in the construction of modern plants. In any case, milk is not a good source of this vitamin and it certainly cannot provide sufficient quantities of this vitamin for ones bodily needs even if it were unimpaired by the process. Other food-stuffs have to be consumed to supply the amount the body requires, irrespective of the type of milk consumed.

Vitamin D — absolutely heat tolerant at pasteurizing temperatures

Vitamin E — so far as is known, this vitamin is not affected by the process

(5) The flavour of pasteurized milk may be slightly different from that of raw milk due to the removal by filtration and during heating of the volatile substances derived from the cow and its surroundings. Overheating will, however, affect the flavour.

It would appear that a safe, long-keeping milk does not suffer from any really serious disadvantages as a result of pasteurization. Apart from a slight reduction in the amount of vitamin C, the nutritive value is not materially impaired.

Cold

Low (cold) temperatures are used in the preservation of foods. The following two methods make use of cold temperatures:

(a) refrigeration (chilling)

(b) freezing

(a) *Refrigeration:* In this method, the temperature is kept just above the freezing point. This is the situation encountered in the domestic refrigerator (temperature about 5°C), refrigerated trucks and railway cars and ships' storage vaults. Refrigeration owes its effectiveness to the marked reduction in the rate of micro-organisms. The generation time of micro-organisms may increase from 20 minutes at optimum temperature to several days or weeks at refrigerator temperature. Refrigeration, therefore, retards the growth and metabolic activities of micro–organisms. In other words, microbial growth is not prevented but slowed down. Moreover, the action of the food enzymes continues at a reduced rate. This brings about undesirable chemical changes in the food and food quality is impaired. Therefore, refrigerated foods can be kept for short periods.

Commercial refrigeration or chilling is employed in the preservation of many types of food, (e.g. milk, meat, eggs, fruits and vegetables). In dairy industries, milk after pasteurization is chilled (about 1°C) and is kept for many days. Nowadays it has, therefore, become possible to transport perishable foods like milk to distant places. Moreover, the physical state of the food is unaltered.

(b) *Freezing:* Foods must be completely frozen and stored at a low temperature if long term storage is required. Micro-organisms (which are the main spoilage agents) become inactive at about –10°C, while enzymes (which cause chemical spoilage and subsequent loss of quality) are largely inactivated below –18°C. Domestic freezers store foods at about –18°C, while commercially a temperature of –29°C is used to ensure high quality and a long storage life.

For successful preservation of foods by the freezing method, the following five points should be kept in mind:

(a) blanching
(b) proper packing
(c) freezing at –18°C or lower
(d) storage at –18°C with a minimum fluctuation
(e) pre-determined period for frozen food storage

It is desirable that the food to be frozen for preservation purposes should have a microbial load as low as possible. This may be accomplished by carefully washing the food and blanching it. For example, vegetables, and some fruits, (e.g. apples) are blanched with boiling water or steam before freezing. Blanching not only reduces the microbial load, but also destroys some of the enzymes which might otherwise cause slow deterioration in flavour and quality of the foods.

Blanched foods should be packed with superior packaging materials, (i.e. containers and wrappers). The ideal packaging material should possess the following characteristics:

(1) It should be easily and cheaply made.
(2) It should be clean and sanitary.
(3) It should be attractive in appearance.
(4) It should be free from any objectionable smell and taste. Also, it should be impenetrable to smells.
(5) It should offer protection against desiccation.
(6) It should be stainproof and greaseproof.
(7) It should offer resistance to air, vapour and moisture.
(8) It should be strong and not fragile at cold temperatures.
(9) One should be able to use it mechanically.
(10) It should be resistant to oxidation.

Packed foods should be subjected to freezing. The rate at which foods are frozen is important. Good quality is only obtained if freezing is quick. Quick freezing methods employ temperatures of –32°C or lower. Sharp freezing methods, using temperatures between –18°C and –29°C, are also satisfactory. With the quick freezing or sharp freezing method, the quality of the food is essentially unimpaired. During rapid freezing tiny ice crystals are formed and they do not rupture cell structures in the food. On the other hand, slow freezing methods are not satisfactory. Slow freezing refers to temperatures above –18°C. At such temperatures, the formation of large ice crystals takes place. These large ice crystals disrupt the cellular structure of the food. Therefore, slow freezing methods are not employed.

Frozen foods should be stored in the cold storage rooms. The temperature in the storage room should be –18°C with a minimum fluctuation. Fluctuation promotes dehydration and it is usually indicated by the presence of snow and ice on the inside of a package. There are two reasons for maintaining the temperature of the cold storage room at –18°C. They are: (1) to avoid rapid loss of vitamin content of foods and (2) to avoid more rapid development of rancidity in foods. Psychrophilic bacteria cannot grow at this temperature. But they liberate proteolytic and other enzymes at this temperature. And it is impossible to eliminate the enzymatic activity entirely, but it can be slowed by decreasing the temperature. The microbial count of most frozen foods goes on decreasing during storage. But many micro–organisms, (e.g. species of

Salmonella) survive for long periods of time at –9°C and –17°C. If the temperature rises above –18°C, food starts to deteriorate more rapidly. For example, although strawberries can be stored successfully at –20°C for over a year, at 10°C they show some taste deterioration after a few months.

Frozen foods cannot be stored for indefinite periods because they slowly undergo deterioration. Therefore, it is best to plan on a normal turnover with the season. For example, ground meats may be safely preserved for six months. Thus, this method for preservation of foods is the simplest, safest and surest.

Dehydration

Dehydration is an old method of preserving foods. In old days preservation of fish, meats, vegetables and fruits was done by dehydrating these foods by the sunlight.

Micro-organisms require water for their growth and multiplication. Therefore, water content of the food is reduced to below a certain crucial value, which varies from food to food. Thus, the growth of micro-organisms does not occur in dried foods. This method of preservation is now of great industrial importance and practically every type of food is prepared in a dehydrated form, (e.g. vegetables, fruits, eggs, milk, soups, nuts, etc.).

Nowadays many types of equipment are used for dehydrating food products. Drying is usually accomplished by passing air of carefully regulated temperature and humidity over or through the food in tray driers, tunnel type driers or rotating drum driers. Heated vacuum driers are also used. In these driers, the temperature required for drying under reduced pressure is much lower than would be required at ordinary pressures. Moreover, the atmosphere above the food product contains a much lower concentration of oxygen than in the normal methods of drying. This condition, therefore, reduces the extent to which oxidative changes take place. There is a modern development of vacuum drying called freeze-drying. Drying of the frozen food is carried out under high vacuum. Again *Accelerated Freeze Drying* (AFD) is a modern development of freeze drying. In this technique the time required for drying is reduced as compared to usual freeze drying. Therefore, this technique is now used commercially. Freeze drying is particularly attractive for dehydrating heat-sensitive food products. Dehydration takes place without discoloration and sensitive nutrients, (e.g. vitamins) remain unharmed. Freeze-dried food can be rapidly re-hydrated by adding cold water. A fairly wide range of freeze-dried foods is available commercially, (e.g. vegetables, coffee and sea foods).

Before vegetables are dried, they are scalded or 'blanched' by immersion in boiling water or by treatment with steam. This is practised for inactivating oxidative enzymes, (e.g. catalase and ascorbic acid oxidase). Thus, the stability of the dried product is improved. With coloured vegetables, the blanching operation also improves the colour of the product. Undesirably, there occurs some loss of water-soluble vitamins during water-blanching. But such loss can be minimized by allowing the concentration of water-soluble substances in the blanching water to build up. Sodium sulphite is the substance of choice for building up of concentration in the water used for blanching vegetables. This improves both the colour and ascorbic acid retention. In steam-blanching, vegetables may be treated with a sodium sulphite spray before steaming. Losses due to solution of water-soluble substances are much less with steam blanching than with water-blanching. Blanching also destroys a large number of micro-organisms present in a food. For example, the microbial count is reduced by a factor of 2,000 for peas and over 40,000 for potatoes.

Dried foods containing fats are prone to develop rancidity after a period. This happens particularly if the water content is reduced to too low a figure, (e.g. potatoes). But in the case of non-fatty vegetables, (e.g. cabbage), as much water as possible should be removed because this helps to conserve ascorbic acid. The absence of oxygen increases the storage life of dried food. In addition to this, loss of vitamin A and ascorbic acid is also decreased to a great extent. This can be accomplished by completely filling the container with compressed dried food. This practice reduces the amount of oxygen to a minimum. In some cases air in the container is replaced by nitrogen. Most dried foods can be stored for two years or more in sealed tins in which the air has been replaced by nitrogen, (e.g. whole milk powder).

Advantages:

(1) Dried foods may be easily preserved. They may be used over longer periods of time. For example, dried milk can be preserved for fairly long periods.

(2) Dried foods occupy very little space. For example, dehydrated potato in powder form has a volume of only 10% compared to that of an ordinary potato. This feature of dried foods is attractive, so far as transportation costs are concerned.

(3) Dried foods do not require sterilization or the maintenance of sterile conditions during their production. In this respect they differ from most canned foods.

(4) Dried foods are more economical to use because they are not wasted during utilization. For example, 'instant' milk powders can be reconstituted according to the requirement.

Disadvantages:

(1) Dried foods require a long soaking period to restore the water lost during dehydration. The soaking period varies from food to food.

(2) Dried foods may absorb moisture from the air. This happens, particularly in the hygroscopic foods. Such moistened foods favour the growth of micro-organisms and subsequently food spoilage occurs.

(3) Dried foods are not sterile. They may contain Gram-positive spore-formers, as well as mould spores. Also pathogenic *Salmonella* bacteria have been found to persist in occasional samples of powdered eggs. Therefore, dried foods are prone to undergo decomposition or develop unwanted flavours during the period of reconstitution.

High Osmotic Pressure

Sodium chloride (common salt) is one of the oldest preserving agents used by man. Its preserving action seems to be largely due to the high osmotic pressure present in strong salt solutions. Under this condition, water is withdrawn from micro-organisms. The microbial cells are plasmolyzed, and metabolism is arrested. They eventually die. Thus, the antimicrobial condition imposed by the increased osmotic pressure is related in principle to inhibition by dehydration. In salt concentrations of 25%, all microbes, except halophilic (salt-loving) bacteria, are practically inhibited. Therefore, the salt concentration in pickling brines (used for curing hams and similar types of meat) is about 25%. Sugar (10 – 15%) and sodium nitrate may also be added to the pickle.

At a salt concentration of 10%, there occurs a marked inhibition of the growth of the great majority of bacterial species. The saprophytic bacteria are more susceptible to strong brines than the pathogenic forms. *Clostridium botulinum* is generally inhibited by a salt concentration of 10%. Certain moulds are inhibited by 15% salt, although a few moulds and wild yeasts have been isolated, which can grow slowly in a saturated salt solution.

Occasionally, the halophiles may cause spoilage of salt-preserved foods. They usually come from the salt itself.

Cane sugar, likewise, exercises its preserving action because of the high osmotic pressure of strong sugar solutions. Sugar concentrations of 60 – 70% usually prevent growth of all types of micro-organisms. Owing to high sugar content (about 50%) and acidity, jellies and jams are protected from bacteria, yeasts, and most moulds. Some moulds, especially *Penicillium* and certain yeasts have been found, which can grow very slowly even in a 70% sugar concentration. Concentrated aqueous sugar solutions are shipped by tank cars. This is a more economical method of transporting than handling sacks of sugar. Occasionally, the tank cars are contaminated with a yeast capable of reproducing in the concentrated sugar solution. This produces marked bubbling and frothing, and spoilage of the sugar solution (syrup) results. Condensed milk is preserved in part by the increased concentration of lactose (milk-sugar) and supplemental sucrose (cane-sugar). It should be remembered that high osmotic pressure may inhibit microbial growth, but it cannot be relied upon to kill micro-organisms.

Preservatives

Substances which are added to foods to prevent microbial spoilage are called preservatives. This includes traditional 'natural' preservatives, (e.g. sugar, salt, etc.) and synthetic preservatives, as well as antibiotics. These preservatives exert their effect by poisoning the micro-organisms. In doing this, they inevitably tend to have, to a lesser or greater degree, a similar toxic effect on the cells of the body of the consumer. In certain circumstances preservatives may legitimately be used. For example, sulphurous acid and sulphites can be added to fruit pulp that is subsequently to be made into jam. These substances are almost entirely eliminated during the process of jam boiling.

The recommendation of preservatives for their use is made on the basis of the type of infecting organisms, and the prevalent manufacturing and marketing conditions. Preservatives are divided into two classes:

Class I: Includes those which can be added to any food, generally on a 'no limit' basis. The substances included in this group are acetic acid, acetate, vinegar, common salt, dextrose, edible oils, honey, spices and sugar.

Class II: Includes the following chemicals or their sodium, potassium or calcium salts, benzoic acid methyl and propyl parahydroxy benzoic acid, lactic acid, propionic acid, nitrates and nitrites, sodium diacetate, sorbic acid, sulphur dioxide, sulphites and acid calcium phosphate. Most of these preservatives are regarded as safe. Nitrites are readily absorbed from the intestine into the blood-stream. They may oxidize the haemoglobin to methaemoglobin. Methaemoglobin cannot function as oxygen carrier. Thus, nitrites exert their toxic effect on the consumer of food containing these substances. Even nitrates may be harmful to babies less than about six weeks old. This is because of insufficient HCl in their stomachs. Under such conditions, nitrate reducing bacteria are not inhibited, and hence, they convert nitrates to nitrites. Babies have, however, been poisoned in this way by drinking well water containing nitrates. There are certain limitations in the use of sulphur dioxide (sulphurous acid). It is volatile and tends to disappear from an open system or may be inactivated by combination with food components. It is somewhat corrosive and leads to the destruction of vitamin B_1, and has a taste which becomes apparent at concentrations above 500 p.p.m.

In Great Britain, the use of chemicals to preserve food is controlled by the preservatives in Food Regulations (1974/75). In these regulations, the word "preservative" means any substance which is capable of inhibiting, retarding or arresting the growth of micro-organisms or any deterioration of food due to micro-organisms or of masking the evidence of any such deterioration.

Preservatives, which exert their antimicrobial effect by virtue of their acidity include sulphur dioxide (the commonest preservative presently in use), benzoic acid, propionic acid and sorbic acid.

Diphenyl and o-phenylphenol are solids which suppress mould growth, but because of their strong smell and taste they are not suitable for general use as preservatives. The sodium salt of o-phenylphenol is used as a spray or a dip for citrus fruits. Diphenyl is used to impregnate the wrapping papers used for fruit.

The preservatives in Food Regulations lists the foods in which preservatives may be used and specifies the maximum permissible amount of preservative which may be present, its main provisions are summarized in Tables 27.2 and 27.3.

Table 27.2 *Names of Permitted Preservatives.*

Code No.	Name of Preservative	Alternative forms
1.	Sulphur dioxide	Sodium sulphite and metabisulphite, Potassium metabisulphite, calcium sulphite and bisulphite
2.	Benzoic acid	Sodium, potassium and calcium salt
3.	Propionic acid	Sodium and calcium salt
4.	Sorbic acid	Sodium, potassium and calcium salt
5.	Methyl, ethyl and propyl p-hydroxy benzoates	Corresponding sodium salts
6.	Diphenyl	
7.	Nisin	
8.	Sodium nitrate	Potassium salt
9.	Sodium nitrite	Potassium salt
10.	o-Phenyl phenol	Sodium salt
11.	Thiabendazole	
12.	Hexamine	

For the purposes of the regulations, the following substances are not regarded as preservatives:

(a) any permitted antioxidant
(b) any permitted artificial sweetener
(c) any permitted bleaching agent
(d) any permitted colouring matter
(e) any permitted emulsifier
(f) any permitted improving agent
(g) any permitted miscellaneous additive
(h) any permitted solvent
(i) any permitted stabilizer
(j) vinegar
(k) any soluble carbohydrate sweetening matter, potable spirits or wines
(l) herbs, spices, hop extract or essential oils when used for flavouring purposes
(m) common salt
(n) any substance added to food by the process of curing known as smoking

It should be noted that most of the traditional preservatives are included in this list and hence, from a legal point of view, are not regarded as preservatives at all.

Besides the permitted preservatives, the use of hydrogen peroxide (H_2O_2) in milk is considered harmless, as most of it evaporates during boiling. An amount of 3-4 ml. of 30 per cent H_2O_2 is suffcient for one litre of milk. The joint FAO/WHO Expert Committee on Food Additives (1962) was, however, of the view that use of H_2O_2 should be permitted only in exceptional circumstances. Neutralizers such as calcium hydroxide, sodium bicarbonate, sodium carbonate and sodium pyrophosphates are sometimes added to milk to prevent clotting, especially in summer. However, as per PFA specifications, neutralizers could be added only in butter manufactured from sour cream.

Table 27.3 *Use of Permitted Preservatives.*

Food	Code number of Preservatives	Max. permitted amount in mg. per kg. food
Beer	1, 2, 5	70
Bread	3	3,000
Cheese	4	1,000
Cheese, exc. Cheddar	8	100
Cheshire or soft cheese	9	10
Provolone cheese	12	5
Cider	1, 4	200
Coffee extract, solid	1	150
Colouring matter, exc. E1 50 caramel,	2, 5	2,000
as a solution	4	1,000
Colouring matter, exc. E1 50 caramel	1	1,000
Flavouring (including syrups)	1	350
	2, 5	800
Flour confectionery	3, 4	1,000
Fruit, dried exc. prunes	1	2,000
Fruit, fresh		
Bananas	11	3
Citrus fruit	6	70
	10	12
	11	10
Fruit juices	1	350
	2, 5	800
Gelatin	1	1,000
Glucose drinks	1	350
	2, 5	800
Jam, exc. diabetic	1	100
Jam, diabetic	1	100
	2, 5	500
Meat, cured	8	500
	9	200
Pickles	1	100
	2, 5	250
Potatoes, raw, peeled	1	50
Prunes	1	2,000
	4	1,000
Sauces, exc. horseradish	1	100
	2, 5	250
Sausages or sausage meat	1	450
Soft drinks requiring dilution	1	350
	2, 5	800
Sugar	1	70
liquid glucose syrups	1	450
Vegetables, dehydrated		
cabbage, sprouts	1	2,500
Potatoes	1	550
Others	1	2,000
Vinegar, malt	1	70
Yoghurt, fruit	1	60
	2, 5	120

Antioxidants

Certain food constituents such as fats, oils and oily flavour compounds become rancid when exposed to excessive air. As the name implies, antioxidants are substances which prevent or greatly retard the oxidation of easily oxidizable materials such as fats.

The antioxygenic action of antioxidants can be explained by chain reaction theory. According to this theory, the molecules in an auto-oxidisable material, capable of reacting with oxygen, consist, at any instant, of a limited number that have been activated by the absorption of energy. When reaction occurs, activating energy of these molecules is released and can be absorbed by another molecule of the oxidisable material. Thus the limited amount of energy will set up a chain reaction and may lead to the oxidation of large amounts of material unless the chain is broken by absorption of the activating energy in a side reaction. Antioxidants interrupt the chain propagation mechanism, that is, it is capable of receiving but not of passing on the activating energy. In the process, the antioxidant is oxidized and gradually destroyed.

Synergists, on the other hand, are substances which have little or no activity alone but which enhance the antioxygenic action of antioxidants.

Certain restrictions are imposed on the use of food antioxidants because they are incorporated into food products meant for human consumption. Thus an antioxidant should possess the following characteristics:

(a) The compound and oxidation product should be non-toxic,

(b) It should be easily incorporated into the foodstuff and should not impart any foreign flavour, odour or colour to the food even after prolonged storage,

(c) It should be easily available and cheap so that its addition does not increase the price of the food product.

(d) It should be active at low concentrations.

Antioxidants in food were known and used long before their function was appreciated. Several vegetable oils such as palm oil, soya bean oil, rice oil, peanut oil, olive oil, cottonseed oil, sesame oil, coffee oil, etc. have been used to improve the stability of animal fats. Several spices remove unpleasant tastes and odours by preventing peroxide development and thus increase the stability of fats toward oxidation. Flours of several grains, condiments and flavourings have also been found to possess antioxygenic properties.

No single compound is equally effective in all substrates under all conditions. Table 27.4 lists some of the antioxidants used for specific food items. The most commonly used antioxidants are propyl gallate and NDGA, the latter is preferred because of its lower cost and slightly higher solubility in fats. However, they are ineffective under alkaline conditions where BHT and BHA are used. A mixture of BHT and BHA is slightly synergistic and gives better results. Citric acid appears to be the most commonly used synergist. Commercially available antioxidants are sold in the form of mixtures, prepared specially for the product in which they are to be used. A mixture of propyl gallate-BHA-citric acid is the most common.

Incorporation of small amounts of antioxidants in large amounts of food may cause mixing difficulties and thus can damage the texture or appearance of the food. This can be minimized by the use of diluents in which the antioxidant can be dissolved or uniformly dispersed. Propylene glycol is the most widely used diluent or carrier. Some antioxidants, (e.g. BHT) are completely insoluble in propylene glycol and in that case some other diluent such as fat or fat derivatives should be used.

Addition of antioxidant to fats does not require diluent. In liquid foods, fine antioxidant powder or a dispersion in liquid carrier is added. The carrier solvent, if needed, can be removed by evaporation. In dehydrated products antioxidant is often added before the drying process. Food materials such as meat, fish or nuts are treated by spraying antioxidant suspension over them.

Table. 27.4 *Antioxidants for specific food items.*

Food item	*Antioxidant*
1. Dry fat	Oat extract, citric acid or tartaric acid.
2. Chicken fat	Mixture of BHA, propyl gallate, citric acid.
3. Ghee	Wheat germ oil.
4. Meats	Certain spices; mixture of NDGA, BHA and ascorbic acid.
5. Fish	Smoking, or a mixture of BHA and BHT.
6. Baked foods	Mixture of BHT and BHA.
7. Fried foods (applied in the form of antioxidant coated salt)	Mixture of BHA, popyl gallate and citric acid or NDGA and citric acid.
8. Milk powder	Ethyl gallate, NDGA, BHA.
9. Ice-cream mixes	Vanilla.
10. Wrappers and cartons	Mixture of BHA and BHT.

Antibiotics

Nisin is the principal antibiotic permitted as a food preservative. It is permitted as a preservative for cheese, clotted cream and canned foods. It is used to increase the shelf-life of sterilized milk for periods as long as 60 days. Nisin is considered safe for foods because it is readily destroyed by digestive enzymes. Therefore, it is harmless to man and no maximum permitted quantity is specified.

Nisin is effective only against certain species of Gram-positive bacteria. It is ineffective against Gram-negative bacteria, moulds or yeasts. For this reason, it is not suitable as a general-purpose food preservative. But it is attractive as a 'mopping-up' preservative for heat-processed foods, (e.g. canned foods), as heat-resistant spores are found only among the Gram-positive bacteria. The quantity of nisin required to prevent spoilage by the members of genus *Clostridium* is about 2 to 3 p.p.m. Nisin does not kill the bacterial spores, but prevents their development.

Apart from nisin, the only other antibiotic allowed for preserving food is *thiabendazole.* It is used to control rot on the skins of bananas and citrus fruit.

Broad-spectrum antibiotics, (e.g. tetracyclines) have been used to greatly increase the storage life of foods, (e.g. poultry, fish, etc.). The usual practice is to add the antibiotic to the ice for chilling or to the wash water. For example, very small amounts (5 p.p.m.) have been added to the ice used to preserve fresh fish or trawlers. The maximum amount of antibiotic which may be used is rigidly defined. Proper cooking is sufficient to destroy the antibiotic. Chlorotetracycline (CTC) has been extensively used in the U.S.A. and other countries. Narrow spectrum antibiotics have been found to be of no value. They fail to inhibit the Gram-negative psychrophilic bacteria, (e.g. *Pseudomonas* spp.) which are involved in the spoilage of foods stored under refrigeration.

Radiation (Cold sterilization)

The question of food preservation by ionising radiation has been the subject of extensive research in a number of countries for some years. Technologically a large number of applications of ionising radiations are advantageous; irradiation could substitute a number of alternative traditional methods of food preservation and could cope with a number of special problems which cannot be solved by older methods.

The absence or inadequacy of alternate techniques for food preservation on a large-scale confers special advantages in the adoption of food irradiation. The cost of putting up an irradiator and operating it would compare favourably with that of constructing a food freezer plant and its associated chain of frozen storage, transportation and retailing facilities extending from the processor to the consumer. Apart from the fact that not all foods are amenable to canning, the economics of the method is limited to the availability of the low cost tin plate containers. At present such packing material is scarce and expensive, and the cost of the container often forms a major part of the price of the product. Dehydration is also of limited applicability, especially as it severely restricts the end use of the product. It would, therefore, appear to be cheaper and easier to introduce food irradiation than any of the other established food preservation methods.

The radioisotopes emit, spontaneously, radiations of three types. They are: alpha, beta and gamma rays. Gamma rays have been used more than any of the others for the sterilization of foods. These rays have deep penetration and are effective to a depth of about 15 cm. in most food materials. The most commonly used radiation source has been cobalt-60. Radioisotopes nowadays are used in many fields of industry, agriculture, medicine, etc. Gamma radiation in large enough doses can destroy bacteria and enzymes in a material without significantly raising its temperature. The recent availability of large sources of irradiation has enhanced our interest in radiation sterilization and has increased the feasibility in irradiation treatment of foods and pharmaceuticals. Large-scale sterilization and pasteurization of perishable food-stuffs such as meals, potatoes and beverages hold interesting possibilities.

Radiation inhibition of the sprouting of soya beans is applicable for a new purpose. Irradiation of soya beans in the process of germination results in the reduction of the oligosaccharide content of the beans leading to a product which does not cause the well-known phenomenon of flatulence (excessive gas formation in the intestines after eating beans), at the same time, faster-cooking soya beans can be produced in this way.

Regarding potatoes, onions and garlic, the major benefits are three-fold—present day losses in cold storage (4°C potatoes and 0°C for onions) as well as in marketing channels and in homes can be reduced; the products have a better look, a better texture and taste better than the conventionally cold-stored products. Furthermore, the shelf-life at the green grocer's and at home is considerably extended as there is no sprouting. With citrus fruits the same fungicidal control can be attained as is currently obtained by wrapping the fruit with diphenyl paper.

Special importance should be assigned to the radiation disinfestation of fresh fruits and vegetables for quarantine purposes, (i.e. to facilitate inter-country trade of fruits without the risk of infesting the importing country with dangerous insects inhabiting the fruits). Elimination of insect damage in a number of food items has been successively achieved after many research programmes on food irradiation in laboratories of various countries. According to this confirmation, the radiation disinfestation is a highly practical proposition to control insects in grain, flour, peanuts, walnuts, dried fruits, etc.

In the case of starch irradiation, where pasteurization is the aim, the anticipated benefit is the elimination of taste, colour and nutritional changes occasioned in heat sterilization. In comparison with alternative technical methods of disinfestation, radiation offers the advantage of removing harmful residues. The microbicidal action of radiation was found to improve the quality of maize starch preparations remarkably, thus improving the hygienic characteristics of food prepared with starch. The same was observed for mixed seasonings and paprika powder.

Irradiation sterilization can be used successfully to store ham, bacon, pork, pork sausage, beef, canned beef, chicken, cod fish, cakes and shrimps for long periods under non-refrigerated conditions. The technical feasibility of using ionising radiations to preserve such highly perishable protein foods has been proved under U.S. Army Radiation Preservation of Foods Programme and the technology is advanced enough to carry this out.

In recent days the technology of radiation processing has advanced much in the aspects of selection of radiation source, irradiation in frozen states, use of 12-D dose (dose required to reduce the number of micro-organisms per gram to 10^{-12} of the original number present) for destruction of *Clostridium*. Meat products prepared with irradiated spices had to be heated less and kept longer than those contaminated by the high bacteria content of the non-treated spices. It has been observed that irradiation of Wiener sausages and of a Japanese fish paste product called Karuaboled, along with heat treatment, has prolonged storage stability considerably. High dose radiation sterilization of meat, meat products and poultry is a practical possibility, if it is combined with a mild treatment to inactivate radiation resistant proteolytic enzymes of the tissues. Products of practically unlimited shelf-life can be had without the necessity of using refrigeration.

There is no problem in the irradiation of tin plate at doses up to 7.5 mega radiations at temperatures as low as –90°C, provided the can enamels used are of the epoxyphenolic or phenolic types and the end sealing compounds are of the butadiene-styrene or neoprene types. Four food contacting plastic films are being used as components of flexible packaging in the form of pouch laminates with aluminium foil as the gas, water-vapour and light barrier. The most favourable balance of quality, cost and required irradiation dose appears at about –30° ± 10°C. Vitamin losses during preparation for eating will be negligible, regardless of the prior treatment.

Safety for consumption is the other main and important point, which has been thoroughly studied. Newer protocols, based on the most recent technological procedures are, therefore, currently being used by the U.S. Army to evaluate the safety of radiation sterilized foods. The question of toxicity is not unique to the irradiated foods only, but is one which should be considered for all substances including all processes to which food is subjected.

The possibilities of using radiation for food preservation as a way of making up food deficiency in the world are obvious.

4 FOOD SPOILAGE

Food items may undergo many different forms of spoilage. As a result of this, abnormal colours, flavours, odours, or other alterations, e.g. in consistency, develop in the foods. Food spoilage may take place due to one or more of the following causes:

(1) action of micro-organisms
(2) action of enzymes of animal or plant origin
(3) insects
(4) physical alterations, (e.g. caused by freezing)
(5) purely chemical reactions other than enzymatic reactions

Only microbial spoilage has been dealt with in some detail here. Micro-organisms involved in food spoilage are: bacteria, yeasts and moulds.

Food spoilage by bacteria is as varied as the capabilities of this large group of micro-organisms. Most food products are subject to bacterial spoilage. The types of bacteria involved in the spoilage depend on many factors, (e.g. processing). Some common types of bacterial spoilage include souring of milk, putrefaction of meat and so on. Lactic acid bacteria have been isolated from many spoiled food products, (e.g. milk, fresh meats, etc.). Heat-resistant spore-formers are found in case of spoiled canned food items. Non-spore-formers, (e.g. members of *Micrococcus*, *Lactobacillus, Microbacterium,* etc.) have also been reported in the spoilage of canned foods.

Food products rich in acid or sugar are subject to spoilage by yeasts and moulds. Yeasts, capable of growing under anaerobic conditions, may bring about spoilage of food either by the formation of slime or by fermentation of the entire food mass. Yeasts are responsible for the spoilage of milk and milk products, (e.g. Cheddar cheese, and butter), catsup, syrup, olives, pickles and mayonnaise. Moulds grow on the surface only, since they are strict aerobes. They occasionally present a spoilage problem in cream, eggs, butter, cheese and various other food items containing sugar.

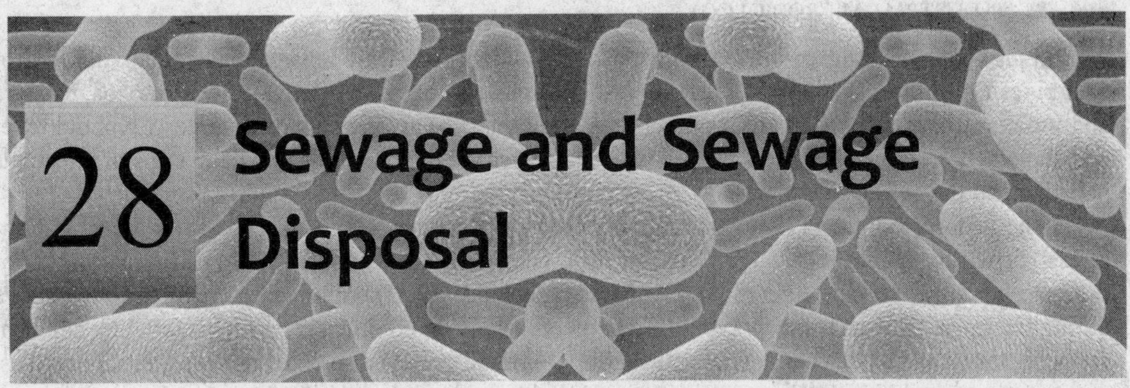

28 Sewage and Sewage Disposal

1 INTRODUCTION

Sewage is the waste water of a community. There are two classes of sewage, namely domestic and industrial. There is a considerable variation in the composition of domestic and industrial sewage, therefore, each type requires separate discussion.

Domestic Sewage: It is composed of human wastes (faeces and urine) and waste waters resulting from personal washing, food preparation, laundry and the washing of kitchen utensils. It is generally relatively low in solids, usually being more than 99% water. Fresh sewage is a grey, turbid liquid with an earthy odour. Approximate analyses of human faeces and urine are given in Table 28.1.

Table 28.1 *Composition of human faeces and urine.*

	Faeces	*Urine*
Quantity (wet) per person per day	135-270 gm.	1.0-1.3 Kg.
Quantity (dry solids) per person per day	35-70 gm.	50-70 gm.
Approximate composition (%)		
Moisture	66-80	93-96
Organic matter	88-97	65-85
Nitrogen	5.0-7.0	15-19
Phosphorous (as P_2O_5)	3.0-5.4	2.5-5.0
Potassium (as K_2O)	1.0-2.5	3.0-4.5
Carbon	44-55	11-17
Calcium (as CaO)	4.5	4.5-6.0

The composition of domestic sewage, in simpler form, is as shown in Figure 28.1.

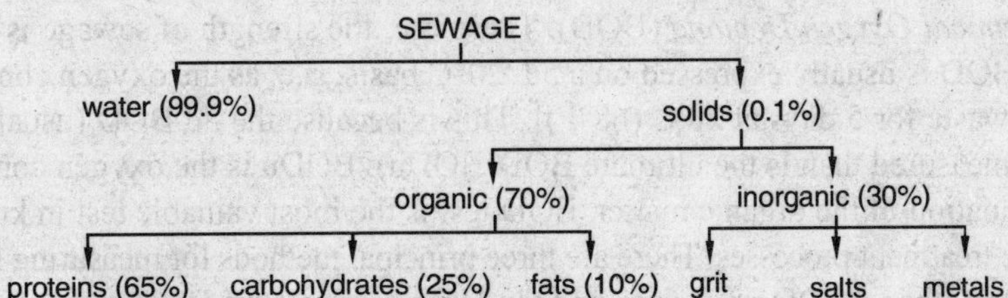

Fig. 28.1 *Composition of domestic sewage.*

It also contains fungi, protozoa, algae, bacteria and viruses. Raw sewage may contain millions of bacteria per millilitre. Domestic sewage represents a real risk because of pathogenic protozoa, bacteria and viruses it may carry. Human excrement may add the causative agents of dysentery, typhoid fever and cholera. Also faeces of infected individuals may carry the poliomyelitis virus, the virus of infectious hepatitis and the coxsackie viruses.

The widespread use of home garbage disposal units creates a marked increase in the amount of solids carried by domestic sewage. This, in turn, complicates the process of sewage treatment. For example, some synthetic detergents are resistant to biodegradation. Therefore, they are difficult to remove during treatment. Nowadays, new types of detergents which do not interfere with bacterial metabolism are being tested and made commercially available. Domestic sewage also contains soaps, fats, pesticides, greases of various kinds, kitchen wastes, (e.g. vegetable peelings, sour milk, tea leaves, etc.) and so on. In other words, the number of different chemical compounds in domestic sewage is too vast to list. Therefore, sanitary engineers use special parameters to characterize wash waters received from a domestic zone.

Industrial Sewage: It contains both organic and inorganic compounds. The nature and amount of these compounds depend upon the type of product that a factory produces, and also the nature of the process involved in it. For example, organic compounds are discharged by sugar factories, paper mills, creameries, breweries, slaughter houses, etc. On the other hand, inorganic substances are added by mines and metal industries, (e.g. acids and salts of metals). Organic matter which may readily be degraded by microbes, is called putrescible material. Its decomposability is due to a high content of sugars, proteins, urea, amines, amino acids, fats, etc. The amount produced by a single factory may be equivalent in terms of decomposable matter to that produced by a relatively large city. The following statistics indicate the great amount of waste water resulting from industrial processes. A ton of rayon requires the expenditure of a quarter of a million gallons of water, for every gallon of whiskey, eighty gallons of waste water are dumped into the sewer, for every gallon of gasoline, a thousand gallons of water are spent.

Characterization of Sewage: Organic matter in the sewage serves as food for micro-organisms. These micro-organisms oxidize the waste, and this requires oxygen supply. This may be represented by the following general equation:

$$\text{Organic matter of sewage + oxygen} \xrightarrow{\text{micro–organisms}}$$

$$\longrightarrow \text{Treated sewage + new mico–organisms.}$$

If the organic matter content of sewage is high, the oxygen required by micro-organisms to oxidize is also more. If the organic matter of sewage is in low concentration, the oxygen needed by micro-organisms during oxidization is also less. Between these two extremes, there exists a gradient for the requirement of oxygen, depending upon the amount of organic matter in the sewage. The amount of oxygen required for the oxidation of the organic matter of the sewage by micro-organisms is called *Biochemical Oxygen Demand* (BOD). Therefore, the strength of sewage is expressed in terms of BOD level. BOD is usually expressed on a 5d, 20°C basis, [i.e. as the oxygen consumed during the oxidation of the waste for 5 days at 20°C (68°F)]. This is because the 5d BOD (usually written 'BOD_5') is more easily measured than is the ultimate BOD(BODu). BODu is the oxygen consumed during the complete bio-oxidation of the organic matter. BOD test is the most valuable test in knowing the efficiency of the sewage treatment processes. There are three principal methods for measuring BOD. They are: the standard bottle test, the Δ COD procedure and the Hach manometric BOD meter. These methods are described in the Appendix I.

There is an alternate test for measuring the load of organic matter in sewage, i.e. *Chemical Oxygen Demand (COD)* test. COD is determined by oxidizing the waste (organic matter) with a boiling acid dichromate solution. Almost all organic compounds undergo oxidization with the formation of carbon dioxide and water. Usually, the reaction proceeds to more than 95% completion. COD measurements are made very quickly (within 3 hrs). But there are two disadvantages with this test, namely (i) such measurements do not give any information on the proportion of biodegradable waste, (ii) COD tests do not provide any information regarding the rate at which biooxidation may take place. A COD test is described in Appendix I.

2 OBJECTIVES OF SEWAGE TREATMENT

Sewage cannot be disposed of without proper treatment. However, many communities and municipalities dump improperly treated sewage into natural bodies of water, (e.g. lakes, rivers, etc.). It is believed that sewage gets diluted in a large body of water and thus prevents hazards to society. Actually, the dilution method for sewage disposal is not a reliable method. In other words, the solution for pollution is not dilution. Therefore, sewage should be adequately treated before its ultimate disposal in a receiving watercourse in order to:

(1) reduce the spread of pathogenic micro-organisms
(2) prevent the pollution of ground and surface waters
(3) prevent the contamination of oysters and other shell-fish which are used for human consumption
(4) maintain the oxygen balance of the receiving watercourse, permitting the aquatic life to survive
(5) avoid health and aesthetic hazards while swimming and boating in the water
(6) prevent the development of objectionable conditions, such as foul odours (of hydrogen sulfide and malodorous compounds) and objectionable colours and tastes
(7) prevent damage to the concrete structures by the chemicals present or produced at the site by microbial action. Also paint chipping and subsequent peeling through contact with H_2S are prevented

3 COLLECTION OF SEWAGE

Sewage is conveyed through pipes, known as sewers, from its place of production to its place of treatment and disposal. There are three types of sewerage systems. They are as under:

(a) *Sanitary sewers:* These sewers carry only domestic and industrial sewage.
(b) *Storm sewers:* These are meant for dumping storm and surface waters.
(c) *Combined sewers:* These are designed to carry all the sewage through a single system of sewers.

Large towns and cities may have a provision for a separate storm sewerage systems, under which the high value commercial buildings in the centre of the town are protected from floods by allowing the storm water falling in this area, to enter the storm sewers.

4 SEWAGE TREATMENT METHODS

There are many different methods available for the treatment of raw sewage. The selection of the method requires many factors to be considered. All these factors are peculiar to each situation. The undermentioned factors are kept in mind at the time of selecting a method:

(1) the quantity and nature of the raw sewage
(2) the cost of the plant and its operation
(3) sanitary requirements by state health department
(4) circumstances regarding the natural body of water for final disposal purposes

In this chapter sewage treatment methods applied at present will be dealt with. For better explanation, the methods used by the single dwelling units and municipalities have been dealt with separately.

Single Dwelling Units

Septic tanks: They are small, rectangular chambers. They are usually sited just below ground level. Two purposes are served by such tanks, namely sedimentation, as well as biodegradation of the sedimented sludge. The raw sewage is allowed to come into the tank, where the retention time is from 1 to 3 days. During this time the solids settle to the bottom of the tank, where they are digested under anaero' conditions. A thick crust of scum is formed at the surface, and this helps to maintain anaerobic conditions. The tank must be desludged at regular intervals, usually once in every 1-5 years. The effluent from septic tanks requires further treatment before disposal.

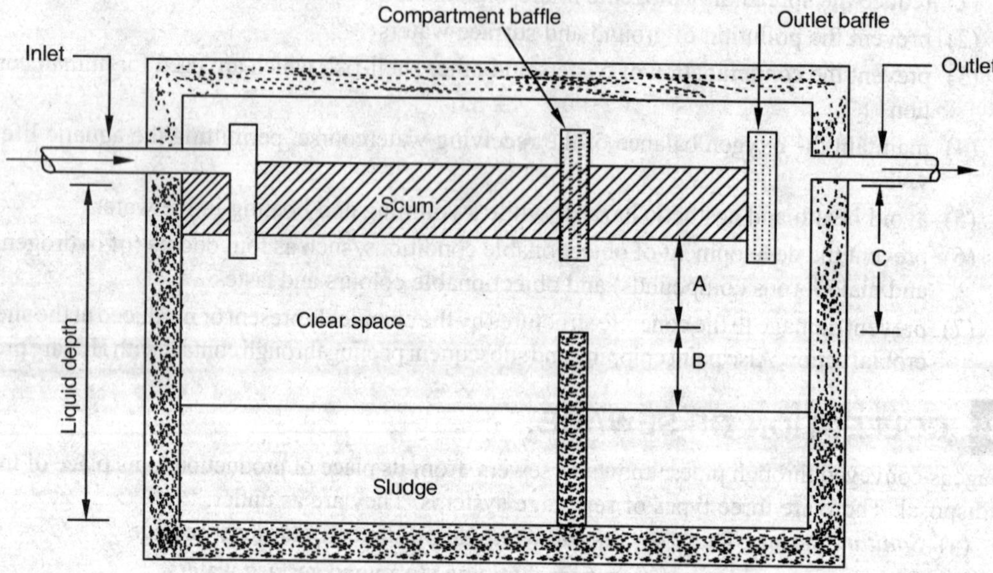

Fig. 28.2 *Two compartment septic tank: A—scum clear space (75 mm. minimum); B—sludge clear space (300 mm. minimum); C—40 per cent of liquid depth.*

A two-compartment septic tank (Figure 28.2) is now generally preferred to one with only a single compartment. The reason for selecting a two-compartment septic tank is of considerably low concentration of suspended solids in its effluent. The first compartment is usually twice the size of the second. The liquid depth is 1 to 2 meters, and the overall length to breadth ratio is 2-3 to 1.

This method can be used for small installations, (e.g. isolated rural buildings). This method is not reliable, as far as removal of pathogens is concerned. Therefore, the drainage from the tank is prevented from seeping into the water supply meant for drinking.

Municipal Sewage Treatment Methods

Municipal raw sewage passes through three major steps before its disposal. The nature of treatment in each step is a unique property of respective operations. The different serial operations in modern municipal sewage treatment plants are:

(A) *Primary treatment:* meant for removing the total solids in sewage by sedimentation

(B) *Secondary treatment:* designed to biooxidize putrescible materials of the sewage. In other words, BOD level of putrescible materials is reduced

(C) *Final treatment:* practised for disinfecting effluent, (e.g. chlorination) before it is dumped into a natural body of water, (e.g. rivers, lakes, etc.) and for disposing of sludge, (e.g. land-filling, fertilizer, incineration, etc.)

A. Primary Treatment

Raw sewage contains two types of solid materials: coarse solid materials and settleable solid materials. Coarse solid materials are removed by a variety of mechanical methods, (e.g. screening, grit chambers, grinding, etc.). Settleable solid materials are removed by holding the sewage in sedimentation units, (e.g. tanks and basins) for an appropriate time. The modification of simple sedimentation is sometimes used, where certain aluminium or iron salts are added to the sewage. This speeds up the rate of sedimentation for settleable solids and thus time is saved.

B. Secondary Treatment

The liquid portion, effluent, resulting from primary treatment operation still contains some small amount of solids and soluble putrescible matter. In other words, the effluent is highly unstable and, therefore, cannot be dumped into a natural body of water, (e.g. rivers, lakes, oceans, etc.) without stabilization. The effluent, therefore, undergoes secondary treatment in which the effluent is exposed to a massive culture of micro-organisms under aerobic conditions, (i.e. biodegradation). This method is biological in nature because stabilization or mineralization of putrescible matter takes place through the microbial agency:

$$\text{Unstable matter} \xrightarrow[\text{under aerobic conditions}]{\text{microbes}} \text{Stabilized matter}$$

There are different methods employed in the secondary treatment. They are:

(1) biofiltration, [i.e. (a) intermittent sand filters, (b) trickling filters, and (c) contact filters]

(2) aeration by the activated sludge process or by contact aerators

(3) lagoons

In addition to the effluent treatment facility, the secondary treatment unit is also equipped with large sludge digestion tanks. The solid portion, sludge, is digested under anaerobic conditions, and subsequently, digester gas (biogas) is produced.

(1.a) *Intermittent sand filters:* An intermittent sand filter consists of a bed of sand about $2\frac{1}{2}$ to 4 feet deep, with drains at the bottom. The effluent is applied to a depth of 3 to 4 inches over the whole area of a bed. The sand particles get coated with the suspended putrescible matter and large numbers of micro-organisms during sewage (effluent) filtration.

Micro-organisms, present in a gelatinous sand coating, oxidize sewage solids, as well as the dissolved matter during the seeping of sewage through the coated sand. An adequate air supply required to maintain aerobic conditions is available, since the bed is dosed only once or twice a day. In other words, sewage is applied intermittently over the sand bed, and this is the reason why this method is called the intermittent sand filter method. The readily oxidizable compounds are oxidized to more stable end-products. The effluent from a well operated intermittent sand filter is clear, and has a relatively low BOD level. Although this method gives a high grade effluent, it has the following two disadvantages. They are:

(i) This method requires large areas of land, since the capacity of any bed is small, and many acres of filter beds are required to handle the sewage of a sizeable city.

(ii) The resulting effluent has objectionable odours and, therefore, it cannot be dumped into receiving water courses like rivers, lakes, oceans, etc.

At present this method is not employed owing to its two above mentioned disadvantages.

(l.b) *Contact filters:* Contact filters consist simply of tanks filled with crushed gravel, slag or coke. The pieces employed vary in size from about $\frac{1}{2}$ to 3 inches. The depth of the bed is 4 to 6 feet. The tank is equipped with underdrains. The effluent from a sedimentation tank or Imhoff tank is applied over the contact bed. The sewage is permitted to stand for an hour or two. The suspended solids and putrescible materials are retained by the rocks. These rocks are coated after the first few loads with a film of stalked bacteria, other bacteria, water-moulds and protozoa. All these micro-organisms are actively involved in the degradation of the putrescible matter. The effluent is then drawn off, and air is drawn in.

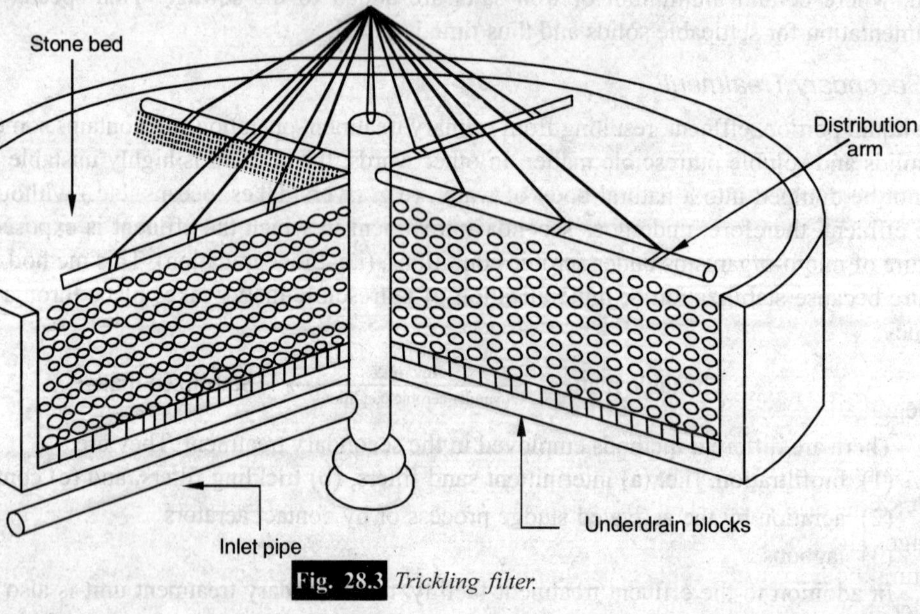

Fig. 28.3 *Trickling filter.*

This produces active aerobic degradation, and is permitted to continue for several hours. Then, the process is repeated. If properly handled, up to 90% of the solids and 80% of the BOD will be removed. This method is not widely employed at present because its capacity is relatively low, compared with some other methods.

(l.c) *Trickling filters:* The trickling filter (Figure 28.3) has been designed on the basis of experiments carried out at the Lawrence Experiment Station in 1894. It consists of a bed of coarse pieces of stone, gravel and slag or related material. The depth of the bed is 1 to 3 inches. This leaves a great many open spaces in the bed for the penetration of air. The tank is equipped with drains underneath. These filters have been described as 'a pile of rocks over which sewage or organic wastes slowly trickle'.

The liquid sewage is applied over the surface of the bed. The application of the effluent is made, either from fixed pipes or from rotating metal pipes with holes at intervals along the pipe. Such a practice of spraying saturates the effluent with oxygen. In addition to this, it is usual practice to operate the spray for a few minutes, and then turn off the spray. Then the effluent is allowed to trickle through. These operational techniques create aerobic conditions for biodegradation. After the initial operation the rocks build up a coating of microbial flora or zoogloeal film. A microbial film consists of the stalked bacteria forming an important part of the flora, protozoa., algae and fungi. Bacteria oxidize the organic compounds present in the seeping sewage (Figure 28.4). As the sewage effluent is subjected to

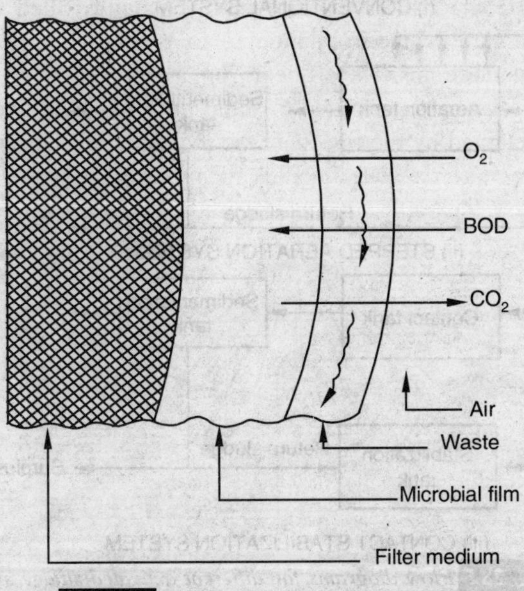

Fig. 28.4 *Formation of a microbial film.*

biodegradation; the microbial film grows with the formation of new bacterial cells:

$$\text{Settled sewage + Oxygen} \xrightarrow{\text{bacteria}} \text{Oxidized effluent + new bacterial cells.}$$

Some of the new cells so formed are washed away from the film by hydraulic action of the sewage. These cells exert a high BOD, and must be removed before the effluent is finally discharged to a natural body of water. This is accomplished in secondary sedimentation tanks. It is required to build up a coating of microbial flora on the rocks before it can work efficiently. This is achieved by operating the filter over a period of a few weeks.

Extensive growth of algae in the upper part is objectionable, because it impairs the operation of the trickling filter.

Many modifications have been made in the trickling filters for its greater effectiveness and higher capacity. They are: use of high rate filters or biofilters, recirculation of the sewage, use of plastic media, etc. Thus the resulting effluent is of high grade.

(2) *Activated Sludge Process:* This process was developed in 1912 at the Lawrence Experiment Station. The process is based on vigorous aeration of sewage. This practice of vigorous aeration causes the finer particles to clump into flocs. Such flocs are allowed to settle and then, added to fresh sewage that is vigorously aerated. This results in more rapid flocculation than before and thus, time is saved. Therefore, the process is repeated, (i.e. addition of settled flocs to fresh sewage, vigorous aeration, sedimentation, etc.) to bring about quick and complete flocculation of the fresh sewage. In other words, most of the sludge is recycled from the sedimentation tank to the aeration tank. There are three major activated sludge systems (Figure 28.5).

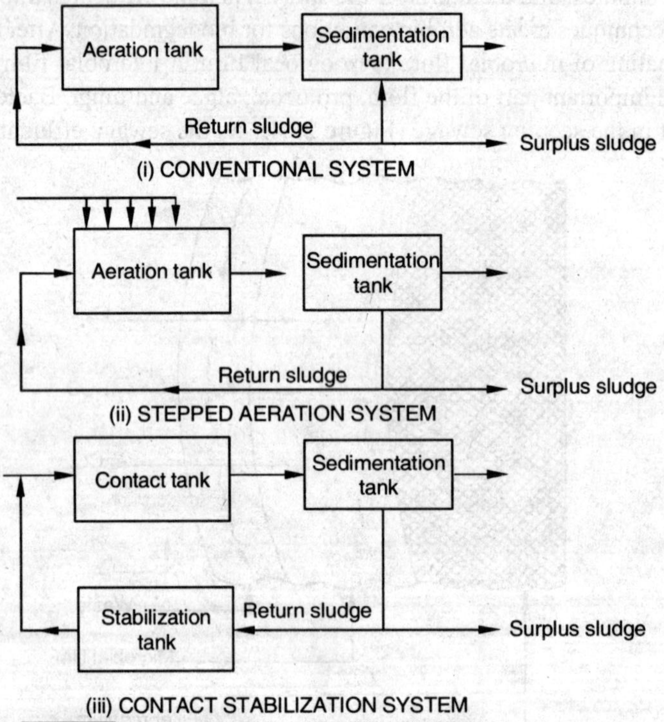

Fig. 28.5 *Flow diagrams for different activated sludge systems.*

These systems usually employ an aeration period of 4 to 8 hrs. The final sludge, thus formed, contains some inert solids and large numbers of living (actively metabolizing) bacteria and protozoa. Therefore, the process is referred to as the activated sludge process. These micro-organisms carry out appreciative oxidation of organic matter, and thus, BOD is reduced. The presence of filamentous micro-organisms in the activated sludge is highly objectionable, because sedimentation of activated sludge flocs occurs slowly. Such inefficient sedimentation creates a problem in a sewage treatment plant owing to the large volume of sewage. Some filamentous species of micro-organisms isolated from voluminous sludge are: *Sphaerotilus natans, Streptothrix hyalina, Microthrix parvicella,* etc.

Advantages: (1) This sewage treatment method requires a small land area. (2) Effluent resulting from an activated sludge unit has a low BOD and, therefore, does not require high dilution for its disposal. Also, there is great reduction of suspended solids.

Disadvantages: (1) Detergents containing alkylbenzene sulfonate (ABS) as their active component create a major problem during this process. ABS is only about 40% biodegradable when ABS-containing sewage passes through a modern sewage treatment plant. When the effluent from such ABS-containing sewage is discharged into a lake or a river, a great deal of objectionable foaming may occur. If the river or lake water is used as a domestic supply, the detergent molecules will pass unchanged through the water purification plant. This may cause undesirable foaming, and occasionally objectionable tastes in the water collected from the tap. Fortunately, ABS compounds are completely non-toxic. (2) Also foam formation takes place during activated sludge process, because it needs air supply. This is objectionable and attempts are made to suppress foam by mechanical and chemical methods.

$$\text{Organic-C} + O_2 \xrightarrow{\text{bacteria}} CO_2$$

$$\text{Organic-H} + O_2 \xrightarrow{\text{bacteria}} H_2O$$

$$\text{Organic-N} + O_2 \xrightarrow{\text{bacteria}} NO_3^-$$

$$\text{Organic-S} + O_2 \xrightarrow{\text{bacteria}} SO_4^{2-}$$

$$\text{Organic-P} + O_2 \xrightarrow{\text{bacteria}} PO_4^{3-}$$

Biooxidation is the bacterial conversion of elements from organic forms to their most highly oxidized inorganic forms. This process is called mineralization.

(3) *Lagoons:* Lagoons are also known as oxidation ponds or stabilization ponds. Lagoonization is now being practised by a large number of smaller communities in the Middle West. It is, without doubt, the most important method of sewage treatment in hot climates where sufficient land is normally available. The preliminary treatment to the sewage is given to remove most of the solids. The sewage effluent resulting from the preliminary treatment stage is dumped into a large pond having a depth of 2 to 4 ft. Lagoonization is a natural process of sewage purification because oxidation of putrescible matter occurs through symbiotic action of algae and aerobic bacteria present in the effluent (Figure 28.6).

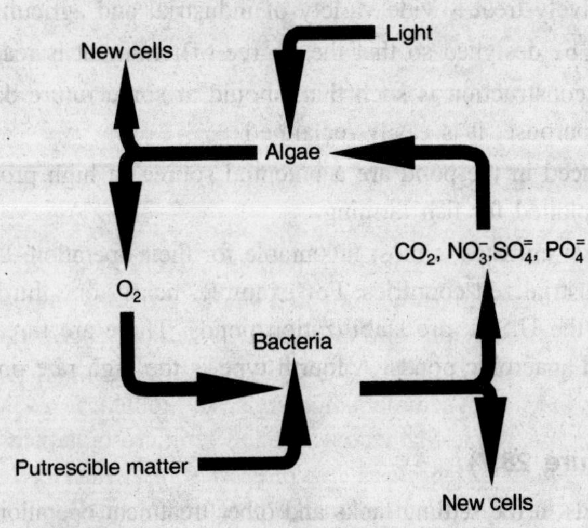

Fig. 28.6 *Symbiosis of algae and bacteria in lagoons.*

Aerobic bacteria stabilize the putrescible matter, and thus BOD of effluent is reduced. Aerobes, therefore, require oxygen which is supplied from the air and algal photosynthetic reaction, (e.g. *Chlorella pyrenoidosa*). Algae use carbon dioxide as a source of carbon for the synthesis of new cells, and evolve oxygen from water by the classic mechanism of plant photosynthesis:

$$6CO_2 + 12H_2O \xrightarrow[\text{Algae}]{\text{Solar energy}} \underset{\text{Glucose}}{C_6H_{12}O_6} + 6H_2O + 6O_2$$

Moreover, algal growth is greatly stimulated by nitrates, sulfates, and phosphates which are usually present in sewage effluents. Thus, mineral salt content of the effluent is reduced through assimilatory reactions, accompanied with extensive algal growth, (i.e. an algal bloom). Sometimes, inorganic nitrogenous fertilizers are added to the effluent to stimulate the algal growth. Effluent treatment by this method requires 30 days or more. This method has some *disadvantages:* They are:

(1) Lagoonization requires large land space.

(2) It is an inefficient method for removing faecal bacteria.

(3) Algae may die, and if not removed from the lagoon they themselves would become organic matter of sewage effluent. Dead algae, therefore, should be removed from lagoon for healthy lagoonization.

However, this method has several *advantages* too.

(1) They can achieve any required degree of purification at the lowest cost and with the minimum of maintenance by unskilled operators.

(2) The removal of pathogens is considerably greater than in other methods of sewage treatment.

(3) They are able to withstand both organic and hydraulic shock loads.

(4) They can effectively treat a wide variety of industrial and agricultural wastes.

(5) They can easily be designed so that the degree of treatment is readily altered.

(6) The method of construction is such that, should at some future date the land be required for some other purpose, it is easily reclaimed.

(7) The algae produced in the pond are a potential source of high protein food, which can be conveniently exploited for fish farming.

The temperature in hot climates is most favourable for their operation. It is an important method of treatment in many industrialized countries. For example, nearly one-third of all municipal waste water treatment plants in the U.S.A. are stabilization ponds. There are three major types of ponds: facultative, maturation and anaerobic ponds. A fourth type is the high rate pond which is still largely experimental.

Sludge Digestion (Figure 28.7)

Sludge is produced as solids in the settling tanks and other treatment operations in a sewage treatment plant. The Imhoff tank (designed by Karl Imhoff in 1907) combines a settling basin and a sludge digestion tank. It is arranged in such a way that the sludge settles to the bottom, and the sewage is allowed to flow through slowly. The septic tank resembles the Imhoff tank, in that the sewage is

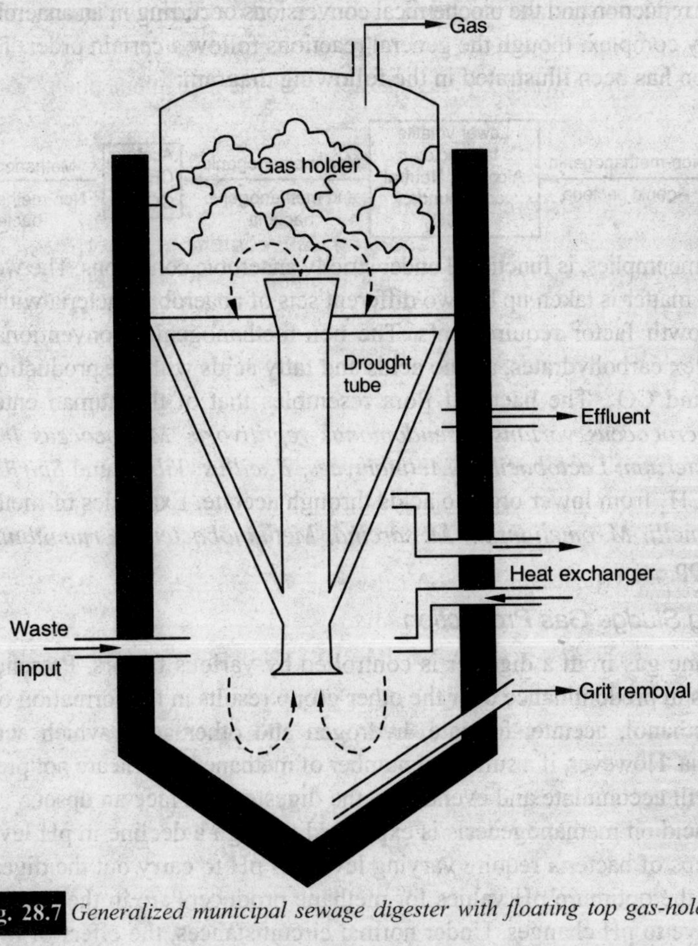

Fig. 28.7 *Generalized municipal sewage digester with floating top gas-holder.*

allowed to flow through slowly, settling a portion of its solids in the process. Raw sludge, activated sludge and secondary sludge are the common types of sludge possessing varying characteristics. Sludge also has to be disposed of safely, and the right disposal option has to be chosen. In course of technological progress, sludge stabilization before disposal through anaerobic digestion has been viewed as a beneficial process, in view of the inherent advantages.

Anaerobic digestion of raw sewage sludge carries with it the twin benefits of generating energy and stabilizing sludge. Whereas raw sludge presents disadvantages from highly putrescible organic matter and heavy loads of pathogens in the disposal system, the digested sludge poses no such problem. Moreover, anaerobic digestion produces fuel gas. The gas contains about 60 to 65% methane (CH_4) and 30 to 35% carbon dioxide (CO_2) and very low levels of nitrogen (N_2), hydrogen (H_2) and hydrogen sulphide (H_2S). The fuel value of digester gas is governed by methane content of the gas. The level of methane in fuel gas shifts with the change in the condition of digestion. For example, increased loading rate enhances the level of CO_2 in the digester gas. This digester gas is used as fuel to supply the energy needs of the plant processes and to generate electricity.

Mechanism of Anaerobic Sludge Digestion

Mechanisms of solid reduction and the biochemical conversions occurring in an anaerobic sludge digestion system are extremely complex, though the general reactions follow a certain order. The route of organic matter decomposition has been illustrated in the following diagram:

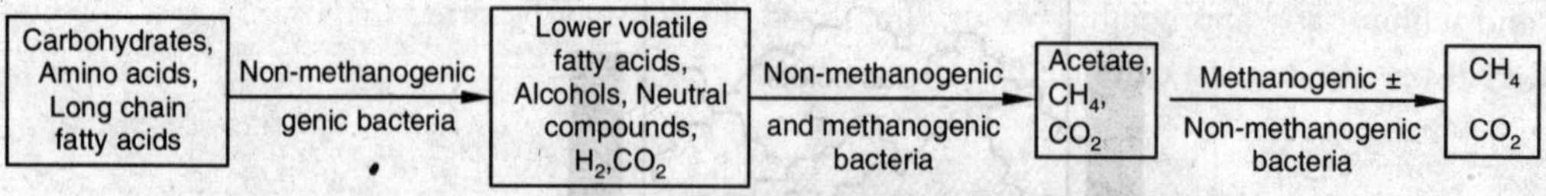

The system, as its name implies, is functional under strictly anaerobic conditions. The work of decomposing the complex organic matter is taken up by two different sets of anaerobic bacteria with different substrate specificities and growth factor requirements. The non-methanogenic, conventional bacteria tend to hydrolyse the complex carbohydrates, amino acids and fatty acids with the production of lower volatile acids, ethanol, H_2 and CO_2. The bacterial flora resembles that of the human enteric tract and may contain *E. coli*, *Micrococcus varians, Pseudomonas reptilivora, Micrococcus luteus, Clostridium, Peptococcus, Eubacterium, Lactobacillus, Actinomyces, Bacillus, Vibrio* and *Spirillum.* The next step is the formation of CH_4 from lower organic acids through acetate. Examples of methane producers are *Methanococcus vannelli, M. omelianskii, M. sarcina, Methanobacterium ruminatum, M. mobilis* and *Methanospirillum* spp.

Factors Controlling Sludge Gas Production

The yields of methane gas from a digester is controlled by various factors. Presence of acid-forming bacterial populations in predominance over the other group results in the formation of a combination of compounds like methanol, acetate, formate, hydrogen and other acids which act as substrates for methanogenic bacteria. However, if a sufficient number of methane bacteria are not present in appropriate proportions, acids will accumulate and eventually, the digester will face an upset.

The effect of acid on methanogenesis is expressed through a decline in pH levels of the digested slurry. The two groups of bacteria require varying levels of pH to carry out the digestion process. Past research shows that the optimum pH values for methane producers are in the range of 6.4 to 7.5, and they are very sensitive to pH changes. Under normal circumstances, the effect of low pH is countered by the destruction of acids by methane bacteria and the subsequent production of bicarbonate. Addition of alkaline compounds containing sodium, magnesium or calcium has been shown to be a valid method to control pH in the digester.

The presence of nutrients necessary for anaerobic bacteria in sufficient amounts is an important factor in deciding gas production. Generally, nitrogen and phosphorus are required in higher amounts. Research findings indicate a minimum ratio of carbon, nitrogen and phosphorus at 100:15 :1 for digestion purposes. Addition of certain trace materials like iron and sulphur is also beneficial.

The other factors responsible for the rate of methane formation are the temperature (around 30°C) the conversion of acetate to CH_4, and CO_2 and even the removal of methane produced during formation. The presence of toxic heavy metals in the solution can cause serious problems. The heavy metal contents are more in the raw sludge, as most of the elements present in sewage tend to concentrate in the solid sludge during settling. The toxicity of heavy metals can be removed through their precipitation as sulphides, but this has to be practised with adequate care as sulphides may lead to the production of H_2S gas or sulphuric acid. Depending upon sludge quality, the sludge loading rates and the sludge retention time, gas production will vary. The volume of gas produced, and its methane content per unit mass of solids destroyed, are essential considerations, because all cost analyses are based on them only. Complete digestion requires as many as 30 to 40 days.

C. Final Treatment

The organic matter remaining after sludge digestion is stabilized and is not easily decomposed further. It is black and without any appreciable odour. Stabilized sludge may be dried (Figure 28.8) and sold as fertilizer, or it may be burned or buried.

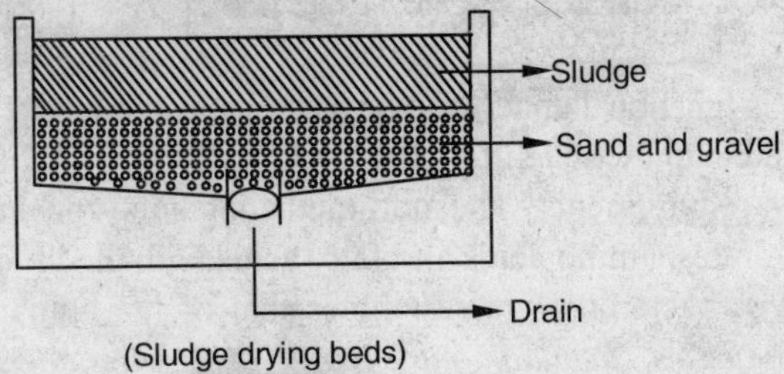

Fig. 28.8 *Sludge drying bed.*

The effluent is chlorinated (5 to 10 p.p.m.), and is generally dumped into some water channel for final disposal. Satisfactory chlorination can kill the bacteria more than 99% of the original bacterial population. Flow diagram of the hyperion activated sludge plant (city of Los Angeles is presented) in Figure 28.9.

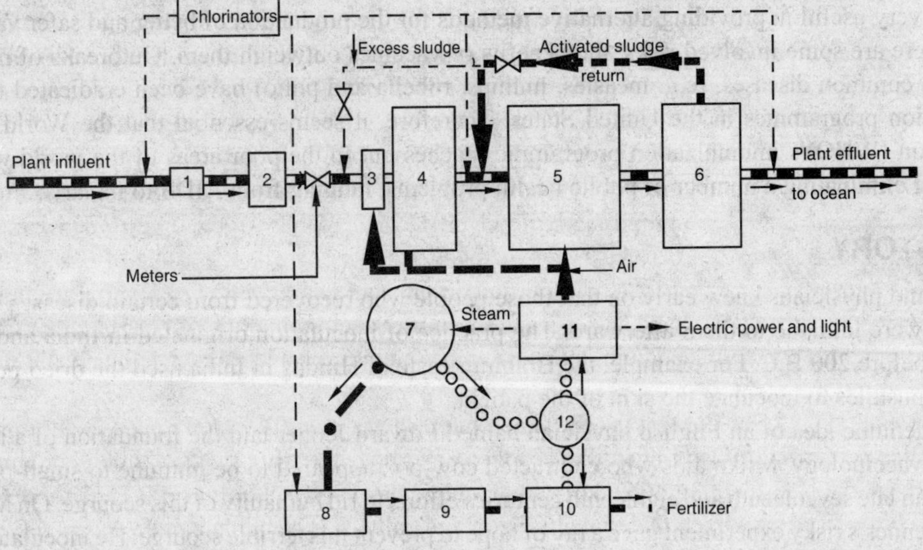

1. Bar screens
2. Grit removal
3. Preaeration
4. Primary settling
5. Aeration
6. Final settling
7. Sludge digestion
8. Elutriation
9. Filters
10. Dryers
11. Blowers and power generation
12. Gas

Fig. 28.9 *Diagram showing the flow of sewage through the hyperion activated sludge plant, and the various steps involved in the treatment process (Courtesy of Los Angeles).*

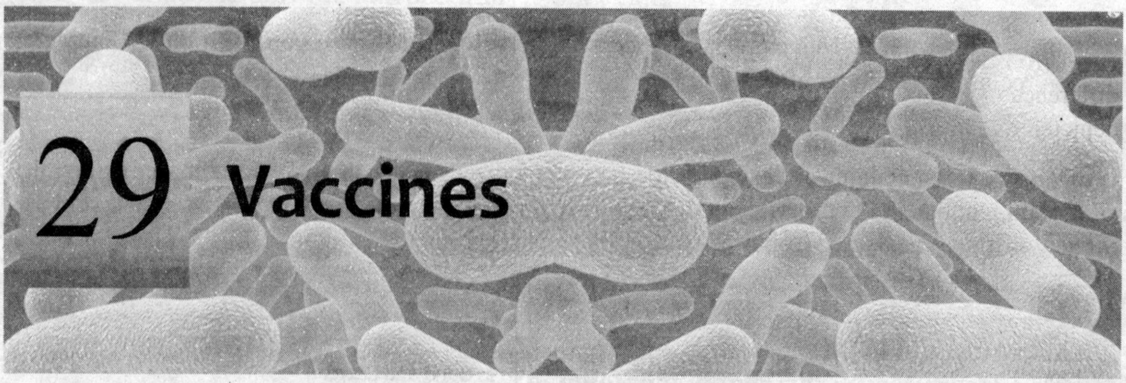

29 Vaccines

1 INTRODUCTION

Biological materials capable of developing an immune response are called vaccines or immunizing agents. The reader will find the use of the terms vaccination and immunization interchangeably throughout the chapter. With the successful development of the first vaccine against small-pox, a new era of immunization had sprouted. This prompted the development of several other vaccines which saved the lives of many people during the years of World War I and World War II. Today, whole virus agents (live or killed) constitute the vast majority of vaccines in use. Moreover, rapid advances in biotechnology have been very useful in providing alternative methods for the production of better and safer vaccines. Though there are some involved risks, the benefits of vaccines outweigh them. Outbreaks of many of those once common diseases, (e.g. measles, mumps, rubella and polio) have been eradicated through immunization programmes in the United States. Therefore, it seems essential that the World Health Organization (WHO)'s immunization programme reaches out to the poor areas of the world with the intention of eliminating a number of public health problems, ranging from AIDS to malaria.

2 HISTORY

Scientists and physicians knew early on that those people who recovered from certain diseases such as small-pox were immune to them afterward. The practice of inoculation originated in India and China sometime before 200 B.C. For example, the Brahmin caste of Hindus in India used the dried pus from small-pox pustules to inoculate the skin of the patient.

The genuine idea of an English physician named Edward Jenner laid the foundation of a modern science of vaccinology. Milkmaids who contracted cow-pox appeared to be immune to small-pox. Dr. Jenner began late seventeenth and eighteenth centuries efforts to rid humanity of this scourge. On May 14, 1796, Dr. Jenner's risky experiment gave a ray of hope to prevent this terrible scourge. He inoculated both arms of a young boy named James Phipps with a matter collected from the milkmaid's pustular cow-pox lesions. Six weeks later, he challenged the success of the first inoculation (variolation) by injecting the boy with the small-pox virus. The boy survived without developing small-pox. Unfortunately, it took more than two centuries to prove Jenner correct. Since the first vaccine was derived from a virus affecting cows, (i.e. cow-pox virus), Jenner coined the term "vaccine" from the Latin word *vacca* meaning cow.

Louis Pasteur further adapted Edward Jenner's pioneering work in microbiology. In 1885, Pasteur used a hypodermic needle to inject a boy named Joseph Meister who had been bitten by a rabid dog. Since the boy survived, Pasteur's discovery was another stepping stone in the field of vaccinology.

From 1870 to 1910, most of the discoveries of the basic elements of immunology made a large difference in the development of several vaccines. The outline of the development of human vaccines is depicted in Table 29.1.

Table 29.1 *Outline of the development of human vaccines.*

Century	*Vaccines Developed*
18th	Small-pox
19th	Rabies, Typhoid, Cholera, Plague
20th–First Half	Diphtheria toxoid, Tetanus toxoid, Pertussis, Tuberculosis (Bacille Calmette-Guerin), Yellow fever, Influenza, Typhus
20th–Second Half	Polio (oral), Polio (injected), Measles, Mumps, Rubella, Adenovirus, Typhoid (*Salmonella* Ty21a), Varicella, Rotavirus reassortants, Cholera, Rabies (cell culture), Japanese encephalitis, Tick-borne encephalitis, Hepatitis A, *Pneumococcus* polysaccharide, *Meningococcus* polysaccharide, *Haemophilus influenza* type b polysaccharide, Meningococcal conjugate, *H.influenza* conjugate, Hepatitis B (plasma derived), Typhoid (VI), Polysaccharide, Acellular pertussis, Anthrax secreted proteins, Hepatitis B surface antigen recombinant, Cholera (recombinant Toxin B)
21st	Cold Adapted Influenza (CAIV), Rotavirus (attenuated and new reassortants), Zoster, Pneumococcal conjugate, Meningococcal quadrivalent conjugate, Human papillomavirus recombinant

3 DIFFERENT TYPES OF VACCINES

Both live and killed vaccines are the major vaccines used today, however new types of vaccines are emerging through the recent advances in biotechnology and related scientific disciplines. Hopefully, advanced biotechnology will bring less expensive and safer vaccines in the future. Different types of vaccines are described below:

3.1. Attenuated, live vaccines

These vaccines are nothing but weakened strains of microbes. The basic principle of such vaccines is to induce a protective immune response without causing any harmful effects to the host. These attenuated microbial agents provide continuous antigenic stimulation over a period of time, thus avoiding the necessity of booster doses. While these vaccines are 95 per cent effective and provide a life-long immunity, the microbial agents can back mutate to virulent forms. This can cause a disease, and also limits administration to immunocompromised and pregnant patients. Several methods have been developed to attenuate the strains intended for use in the production of vaccines as listed below:

1. use of related viruses from other animals, e.g. cow-pox virus to prevent small-pox
2. repeated passages of the virus in an unnatural host or host cell, e.g. 17D strain of yellow fever, polio and measles
3. administration of a pathogenic or partially attenuated virus by an unnatural route, e.g. adenovirus type 4,7 (used in military recruits against adult respiratory distress syndrome)

4. temperature-selected mutants, e.g. influenza
5. reassortant technology, e.g. rotavirus and influenza virus

3.2. Inactivated whole-agent vaccines

These vaccines are also called killed vaccines. Such vaccines have two key advantages over live vaccines: (1) generally more technically feasible to produce and (2) cannot multiply or revert to pathogenicity. The disadvantage, however, is the necessity of multiple doses to attain long-term protective immunity. The antigenicity of these vaccines can be enhanced by the administration with an adjuvant or delivery system. Inactivation of microbial agents is carried out by the use of physical agents, (e.g. heat) or chemical agents, (e.g. formalin, phenol and beta-propiolactone). Formalin is the traditional chemical agent for inactivation of these agents. Excessive treatment can destroy antigenicity, whereas insufficient treatment can leave an infectious virus capable of causing a disease, e.g. incomplete inactivation of polio virus with formaldehyde in 1955 by Cutter Laboratory of Berkely, California from which 40,000, developed mild polio, 200 had permanent paralysis and 10 were killed. Examples of inactivated vaccines include bacterial, (e.g. pertussis, cholera and enterotoxigenic *Escherichia coli*) and viral, (e.g. Salk polio virus, influenza virus, rabies virus and Japanese encephalitis).

The differences between attenuated, live and inactivated vaccines are summarized in Table 29.2.

Table 29.2 *The differences between attenuated, live vaccines and inactivated vaccines.*

Feature	*Attenuated, live vaccines*	*Inactivated vaccines*
Dose	low	high
Number of doses	single	multiple
Need for adjuvant	no	yes
Duration of immunity	many years	less
Antibody response	IgG	IgA, IgG
Cell-mediated immunity	good	poor
Reversion to virulence	possible	not possible

3.3. Toxoids

Toxoids are nothing but detoxified toxins. Inactivation of toxins is carried out by using formaldehyde at 37°C. Vaccines are directed at toxins produced by a specific pathogen. Toxoids require a series of injections for full immunity and then boosters every 10 years, e.g. tetanus and diphtheria.

3.4. Subunit vaccines

Subunit vaccines consist of highly purified preparations of only the antigenic segments of whole organisms and not the whole organisms themselves. As a result, these vaccines contain little or no extraneous matter, making them less reactogenic. Therefore, they can be used safely for immunosuppressed individuals. The key disadvantage is that these vaccines may not retain the native conformation, resulting in their failure to develop the defensive antibody. Like inactivated vaccines, these vaccines may be used with adjuvants, (e.g. aluminum salts) to enhance the immunogenicity. Examples of subunit vaccines are: the HA vaccines for influenza A and B and acellular pertussis vaccine. Advanced recombinant DNA technology provides an alternative way to produce subunit protein vaccines. These vaccines are produced by cloning the specific gene in other organisms, e.g. HBsAg gene in *Saccharomyces cerevisiae.*

3.5. Conjugated vaccines

Vaccines that consist of protein-linked polysaccharides, whereby a protein molecule serves as a carrier molecule, are called conjugated vaccines. Conjugation of polysaccharides with a carrier protein increases not only the immunogenicity, but also elicits the immunological memory in infants and young children below 2 years of age. The principle of conjugated vaccines is that proteins are T-cell-dependant immunogens, whereas polysaccharides are T-cell-independent immunogens. Owing to the conjugation process, a carrier protein converts polysaccharides from a T-cell-independent (TI) immunogen to a T-cell-dependant immunogen (TD), allowing conjugated vaccines to develop an immune response in children that are under 2 years of age. The best example of this kind of vaccine is *Haemophilus influenzae* type b (administered starting at 2 months old).

3.6. DNA vaccines

DNA vaccines are vaccines that involve the plasmids of naked DNA encoding a foreign protein, (i.e. a vaccine antigen) following the intramuscular injection into the host. The secreted foreign protein then elicits humoral or cellular immune responses in the host. These vaccines seem to be extremely safe, free from any adverse effects, inexpensive and easier to produce. The ill consequences following the integration of plasmid DNA into host chromosomes remains questionable. Vaccines produced by this technology are still under clinical trials for HIV, influenza, herpes simplex virus and malaria.

4 MANUFACTURING OF VACCINES

The reader should note that growth conditions optimized for a potential vaccine in the laboratory may not be amenable to a 'scale up' situation.

Major steps in the manufacturing of vaccines are briefly described below:

The 'seed': The beginning of the production of vaccines needs a starter culture called the 'seed'. Depending upon the specificity of the vaccine, the 'seed' will consist of 5 or 10 cubic centimeters of a highly purified viral or bacterial agent. The "seed" must contain only one specific strain of a microbe (viral or bacterial), excluding other similar microbial agents and even variants of the same type of a microbial agent. To keep the microbial agents stable, the "seed" is stored under "ideal" conditions, usually in a frozen state. Freezers used for the storage purpose are maintained at specified temperatures and charts and/or dials outside of the freezer keep a continuous record of the temperature. In case of any changes in temperature from a normal range, sensors will set off audible alarm signals and/or computer signals which help take the immediate corrective action to bring the temperature within the normal range.

Generation of the antigen: First, the "seed" of the viral agent is defrosted and warmed under specified conditions. Then, a small amount of the viral agent is placed into a "cell factory". A "cell factory" is a small device equipped with valves, tubes, and sensors that is used for the growth of the viral agent. A suitable medium is added to the "cell factory" in order to support the growth of primary cells, (e.g. chicken fibroblasts) or continuous cell lines, (e.g. MRC-5). The principal component of such media is protein from mammals in one form or another, (e.g. purified protein from cow blood). Factors such as temperature, pH and time are also significant in order to support the growth of the viral agent. The next step is to transfer the viral material from a "cell factory" to a container similar to the "cell factory". The container is larger and equipped with accessories similar to that of a "cell factory".

Finally, microscopic beads are added to the container, allowing a much greater area for viral agents to attach. Thus, beads help to produce more viral agents.

Bacterial agents are grown in bioreactors which use specific media that stimulate the formation of antigens without any alterations. Bioreactors are much more complex in their design. They are equipped with many accessories meant for different purposes. Key factors such as temperature, pH, time, chemical composition of a growth medium, etc. are carefully controlled.

Separation: There are several separation techniques employed for the isolation of viral agents from the beads. The most common separation technique is filtration. In this method, the broth is forced through a filter with a pore size large enough to allow the viral agents to pass through, but small enough to prevent beads from passing. Ultracentrifugation is another way to separate the viral agents.

Purification: Purification of the vaccine is extremely important as impurities can cause adverse effects, e.g. pyrogens. There are several purification techniques available as under:

(a) chromatography, (e.g. column, ion-exchange)
(b) anion-exchange
(c) filtration, (e.g. gel, density gradient)
(d) centrifugation
(e) precipitation, (e.g. ammonium sulphate, alcohol, detergent)
(f) diafiltration

Selection of a purification technique depends on the nature of the vaccine to be purified.

Formulation: Formulation is a process by which necessary ingredients and the vaccine itself are blended. The following ingredients are used in formulating the vaccines:

(a) pure water in order to decrease the potency of a vaccine
(b) adjuvants (e.g. alum) to enhance the immune response
(c) preservatives in order to preserve the vaccine
(d) stabilizers to prolong the shelf-life of a vaccine
(e) other vaccines in order to develop immunity against multiple diseases

Quality control: Since vaccine manufacturers must meet strict licensing guidelines established by law, all transfers of the antigens are carried out aseptically. Workers wear protective clothing and all equipment used is sterilized and highly controlled. Laboratory and manufacturing rooms used are also specially cleaned and air-conditioned.

Finally, the vaccine is tested for safety, efficacy, potency, sterility and purity as a part of the approval process.

5 EXAMPLES OF VACCINE PRODUCTION

5.1. Hepatitis B

The Hepatitis B vaccine is produced by recombinant DNA technology. In this methodology, a segment of viral DNA that expresses the Hepatitis B surface antigen (HBsAg) is cloned into a baker's yeast, *Saccharomyces cerevisiae.* Then, yeast cells are grown in a complex fermentation medium which consists of yeast extract, soy, peptone, dextrose, amino acids and mineral salts. The HBsAg protein is released from the yeast cells by cell disruption. Finally, the HBsAg protein is purified by a series of physicochemical steps. The purified protein is treated in a phosphate buffer with formaldehyde and then coprecipitated with alum to form bulk vaccine. The adjuvant used is amorphous aluminum hydroxyphosphate sulfate. The vaccine may contain yeast protein (1-5%) and it is a safe vaccine.

5.2. Cell culture rabies vaccines (CCV)

The attenuation of the rabies virus is carried out by using one of the following kinds of cells:

(i) vero cell (a continuous African green monkey kidney cell line)
(ii) human diploid cells
(iii) hamster kidney
(iv) chick embryo cells
(v) duck embryo
(vi) fetal rhesus lung

Chick embryo cells available from specialized forms are cheap and readily accessible.

The purification process involves the use of liquid and solid phases. The crude vaccine is separated into liquid and solid phases by centrifuging the crude vaccine from about 30,000 to about 80,000 G for about 15-45 mins. Then, suspension is made by connecting and dispensing the solid phase in a diluent. Once the suspension is made, the process then calls for the solid and liquid phases to separate again by centrifuging from about 500 to about 2500 G for about 2 to 10 mins.

The formulation of the rabies virus involves not only the rabies virus itself but also other ingredients as well. RabAdvert's formulation for the rabies vaccine is: the inactivated virus itself, Flury LEP, neomycin, chlortetracycline, amphotericin B, potassium glutamate, sucrose, human albumin, bovine gelatin and serum. Another common formulation is: the inactivated rabies virus, beta-propiolactone, aluminum phosphate, thimerosal, phenol and red rhesus monkey fetal lung cells. Lastly, quality control of the rabies vaccine is extremely important. Closely regulating the production of the rabies vaccine assures that everything is effective and done correctly.

6 ECONOMICS

The number of vaccines being produced and administered has risen dramatically in recent decades. The main reason for the increase in the production and administration of vaccines is that they are among the most cost-effective interventions in healthcare. Vaccines are cost-effective because they help prevent epidemics. The prevention of epidemics, in turn, aids the general population in survival since epidemics hinder progress and growth. Disease also requires the healthcare system to utilize exorbitant funds, resulting in needless debt. Thus, immunization programmes are a pragmatic way to spend money and protect the population against disease outbreaks. Economic analysis reveals that the costs of immunization programmes against the treatment of vaccine-preventable infectious diseases are surprisingly low. This justifies how immunization programmes benefit a nation's economy. Therefore, it is of utmost importance that each nation's immunization programmes work efficiently and effectively.

7 SAFETY

Vaccines are the best weapons of the defense mechanism, however, no vaccine is 100 per cent effective or safe. While vaccines have an immense benefit to mankind, they have also brought their risks along with them. Therefore, it is important to have every vaccine closely scrutinized and proven to be safe without adverse effects prior to its release to the general public. Furthermore, monitoring health problems after vaccination is essential to ensure that a safe and effective supply of vaccines is available to each and every community.

VAERS

VACCINE ADVERSE EVENT REPORTING SYSTEM
24 Hour Toll-free information line 1-800-822-7967
P.O. Box 1100, Rockville, MD 20849-1100
PATIENT IDENTITY KEPT CONFIDENTIAL

For CDC/FDA Use only
VAERS Number
Date Received

Patient Name:
Last First M.I.
Address
City State Zip
Telephone no. ( )

Vaccine administered by (name):
Responsible Physician
Facillity Name/Address
City State Zip
Telephone no. ( )

Form completed by (Name):
Relation to Patient ☐ Vaccine Provider ☐ Patient/Parent ☐ Manufacturer ☐ Other
Address (if different from patient or provider)
Telephone no.

1. State	2. County where administered	3. Date of birth (mm / dd / yy)	4. Patient age	5. Sex ☐ M ☐ F	6. Date form completed (mm / dd / yy)

7. Describe adverse event(s) (symptoms, signs, time course) and treatment, if any

8. Check all appropriate:
☐ patient died (date mm / dd / yy)
☐ Life threatening liness
☐ Required emergency room/doctor visit
☐ Required hospitalization (________days)
☐ Resulted in prolongation of hospitalization
☐ Resulted in permanent disability
☐ None of the above

9. Patient recovered ☐ YES ☐ NO ☐ UNKNOWN

12. Relevant diagnostic tests/laboratory data

13. Enter all vaccines given on date listed in no.10

	Vaccine (type)	Manufacture	Lot number	Route/Site	No. of Previous doses
a.					
b.					
c.					
d.					

14. Any other vaccinations within 4 weeks prior to the date listed in no. 10

	Vaccne (type)	Manufacturer	Lot number	Route/Site	No. of Previous doses	Date given
a.						
b.						

15. Vaccinated at:
☐ Private doctor's office/hospital ☐ Military clinic/hospital
☐ Public health clinic/hospital ☐ Other/unknown

16. Vaccine purchased with:
☐ Private funds ☐ Military funds
☐ Public funds ☐ Other/unknown

17. Other medications

18. Illness at time of vaccination (specify)

19. Pre-existing physician-diagnosed allergies, birth defects, medical conditions (specify)

20. Have you reported this adverse event previously? ☐ No ☐ To health department ☐ To doctor ☐ To manufacturer

Only for children 5 and under
2. Birth weight lb. oz.
23. No. of brothers and sisters

21. Adverse event following prior vaccination (check all applicable, specify)

	Adverse Event	Onset Age	Type Vaccine	Dose no. in series
☐ In patient				
☐ In brother or sister				

Only for reports submitted by manufacturer/ immunization project
24. (Mfr./imm. proj. report no.)
25. Date received by mfr./imm. proj.
26. 15 day report? ☐ Yes ☐ No
27. Report type ☐ Initial ☐ Follow-Up

Health care providers and manufactures are required by law (42 USC 300aa-25) to report recations to vaccines listed in the Table of reportable Events Following Immunization. Reports for reactions to other vaccines are voluntary except when required as a condition for immunization grant awards.

Fig. 29.1 *The Vaccine Adverse Event Reporting System (VAERS) form.*

As vaccine usage increases and the incidence of vaccine-preventable infectious diseases is reduced, vaccine-related side effects become more frequent and prominent. Thus, the importance of vaccine safety concerns will continue to grow in the future. Some common questions that people raise include:

1. How are vaccines manufactured?
2. Are vaccines effective?
3. Are vaccines safe?

The information that people receive is sometimes complex and at times misleading. Therefore, it is required that the healthcare community and the public get accurate and clear information which will enable them to make the best possible rational decisions.

The United States has been the leading country in the world's campaign for vaccine safety, and has remained the leading country due to the many assessment tests and procedures in place to ensure vaccine safety. The continued post-market assessments of data is best exemplified by the Vaccine Adverse Event Reporting System (VAERS) shown in Figure 29.1. Additionally, inspections of manufacturing facilities are very helpful in evaluating the safety of vaccines.

It is important for us and future generations to see that the sword which vaccines have put in our hands is never allowed to dull due to human error.

8 ADVANTAGES AND DISADVANTAGES

Undoubtedly, vaccines have many proven advantages over disadvantages. They help to:

(a) eradicate deadly contagious diseases (e.g. small-pox)

(b) reduce the morbidity rate of a disease (e.g. reduction of poliomyelitis by 99 per cent)

(c) reduce the mortality rate of a disease (e.g. reduction in the estimated number of neonatal tetanus deaths per year in 2002 to 180,000)

(d) take prophylactic measures (e.g. prevention of influenza in children, all older adults and healthcare professionals)

Although many accomplishments have been achieved in this arena, there remains much to be done with the use of new biotechnological techniques. Immunization programmes are effective in keeping the nation healthy and growing, thus leading to prosperity.

Looking to the dark side of vaccines, some important disadvantages of common vaccines should not go unnoticed. Very little is known from developing countries about harmful effects following the vaccination. While discussing the disadvantages of vaccines, it would be appropriate to include the factors associated with the rates of adverse reactions following the vaccination.

A. Vaccines: (1) type of a strain of the micro-organism (2) titer of an antigen (3) medium used to grow the micro-organism (4) method of attenuation (5) adjuvants (6) stabilizers or preservatives

B. Recipients: (1) age (2) sex (3) pregnancy (4) previous doses of the vaccine (5) prior illness by the offending agent (6) passively acquired immunoglobulins, (e.g. maternal) (7) coincidental illness (8) immune deficiency

C. Administration: (1) use of injecting device (use of jet gun, needle and syringe) (2) injection site (3) tissue injected (intradermal, subcutaneous and intramuscular)

Therefore, pre-licensure evaluations as well as post-licensure evaluations of vaccine safety help avoid or minimize vaccine-related adverse effects.

Vaccine-associated adverse effects of some commonly used vaccines are exemplified below:

(a) *MMR (Measles, Mumps and Rubella) vaccine:*

Mild problems (within 7-12 days):

- fever, mild rash, swelling of glands in cheeks or neck

Moderate problems:

- seizure (jerking or starring), pain and stiffness of joints, low platelet count that can cause a bleeding disorder

Severe problems:

- serious allergic reaction, deafness, coma, or lowered consciousness, permanent brain damage

(b) *TD vaccine*:

Mild problems (within hours to a day):

- soreness, redness, pain, tenderness, swelling, hard lump at the site of vaccination, mild fever, joint pain, body aches, mild drowsiness, or mild vomiting

Severe problems (within 2 days to 4 weeks):

- extreme drowsiness, fainting, severe headache or vomiting, confusion, seizure, or high fever

Influenza vaccine:

Mild problems:

- soreness, redness, swelling at the site of the injection, uneasiness, fever, muscle aches

Severe problems:

- high fever or behaviour changes, signs of a serious allergic reaction can include difficulty in breathing, hoarseness or wheezing, hives, paleness, weakness, an accelerated heart rate or dizziness

(c) *Rabies vaccine:* Mild problems:

- soreness, redness, swelling, or itching at the injection site (30 - 74%), headache, nausea, abdominal pain, muscle aches, dizziness (5 - 40%), diarrhea, insomnia

Moderate problems:

- hives, joint pain, fever (about 6 % of booster doses), illness resembling Guillain-Barre Syndrome (GBS) with complete recovery (very rare)

Severe problems:

- difficulty in breathing, hoarseness or wheezing, paleness, weakness or high fever

The reader should note that several brands of the rabies vaccine are available in the market and reactions may vary among brands.

9 FUTURE PROSPECTS

Ongoing advances of molecular biology and other scientific disciplines seem to explore promising results in the field of vaccines. In other words, biotechnological achievements will enhance the prospects of developing new vaccines that are less expensive and safe.

The international community needs to be reminded that the potential bioterrorist threats cannot be overlooked as the world faces nuclear threats today. Vaccines such as small-pox and anthrax can help prepare for responding to an emergency. Other biodefence pathogens that may be adapted as biological weapons include plague, tularemia, brucellosis, Q-fever and botulism toxin. Therefore, we foresee serious challenges ahead in developing and testing counter bioterrorism vaccines against these agents successfully.

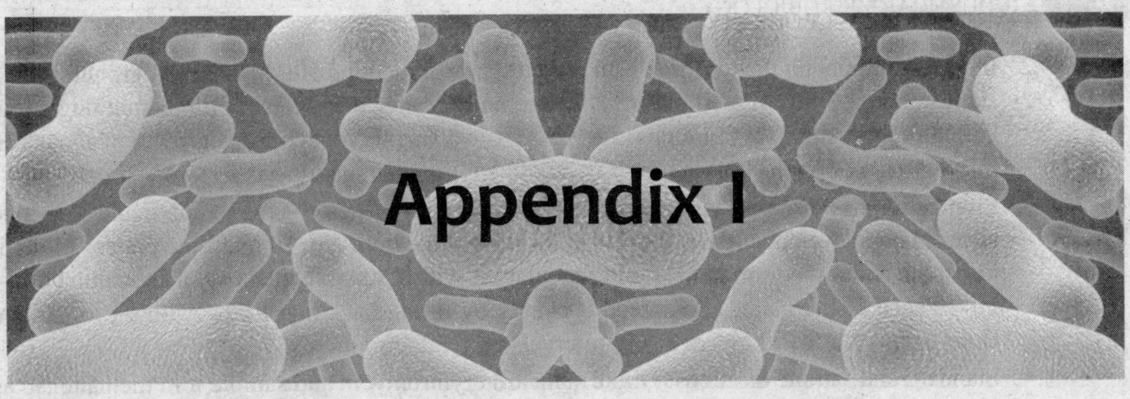

Appendix I

1 MICROBIOLOGICAL ASSAY OF ANTIBIOTICS

Streptomycin Assay

1. Test organism: *Bacillus subtilis* ATCC 6633.
2. Inoculum: Maintain the test organism on the slants of medium No. 8. Wash off growth from a slant in about 3 ml. of sterile normal saline and transfer it to Roux bottle containing about 250 ml. of medium No. 9. Incubate the bottle in a horizontal condition at 37°C for 7 days. Wash off this growth, which consists mainly of spores in 100 ml. sterile normal saline. Store it in a cold environment.
3. Preparation of Petri Dishes: Put 4 flat-bottomed petri dishes on a leveled surface and melt medium No. 10 and put 21 ml. in each of the plates. Allow it to harden. Melt medium No. 11 and inoculate it with a suitable quantity of inoculum. Put 4 ml. in each dish and spread it evenly by tilting back and forth. Allow it to harden. Bore 4 cups with sterile borer, so that they are at 90° intervals on a 2.8 cm. radius.
4. Standard and sample dilutions: Dilute standard streptomycin sulphate and sample suitably in sterile distilled water to finally contain 5 mcg. Standard Low (SL) and Test Low (TL) and 20 mcg. Standard High (SH) and Test High (TH) of streptomycin base in each ml.
5. Assay procedure: Fill each of the four cups with a different dilution, so that each one of the dishes, all four dilutions are filled. Transfer the dishes to incubator (37°C) after about two hours. After the incubation of about 18 hrs measure the diameters of zones of inhibition.

   Total the diameters of the zones for each dilution and calculate the content of streptomycin base in the sample by the following formula:

$$a = \frac{(TH + TL) - (SH + SL)}{(TH - TL) + (SH - SL)}$$

   I = ratio of dilutions = 4 : 1 = 4/1 = 4

   Per cent potency = Antilog (2.0 + a log I)

   Multiply it by suitable dilution factors to final content of streptomycin base in the sample.

Erythromycin Assay

1. Test organism: *Micrococcus luteus* ATCC 9341.

2. Inoculum: Maintain the test organism on the slants of medium No. 8. Wash off growth from a slant in about 3 ml. of sterile normal saline and transfer it to Roux bottle containing about 250 ml. of medium No. 8. Incubate at 32–35°C overnight. Wash off this growth in about 100 ml. of sterile normal saline. Store it in a cold environment and use it up to one week.
3. Preparation of Petri Dishes: Put 15 flat-bottomed petri dishes (12 for standard and 3 for one sample) on a leveied surface. Put 21 ml. of melted medium No. 10 in all the plates. Allow it to harden. Melt medium No. 11 and inoculate it with a suitable quantity of inoculum. Put 4 ml. in each of the plates and spread it evenly by tilting back and forth. Allow it to harden. Bore 6 cups, so that they are at 60° intervals on a 2.8 cm. radius.
4. Standard and sample dilutions: Dilute standard erythromycin and sample it in methanol so as to contain 10 mg. of erythromycin in each ml. Dilute it further with solution No. 19, so as to contain 10 mcg. of erythromycin in each ml. Further dilute the standard to contain 0.64, 0.80, 1.0, 1.25 and 1.56 (a to e) mcg. of erythromycin in each ml. and sample it to contain 1.0 mcg. (U) of erythromycin in each ml.
5. Assay procedure: To derive standard curve, fill the alternate cups on each of the 3 dishes with the median dilution (c) of the standard and each of the remaining nine cups with standard dilution *a*. Repeat the process with standard dilutions *b, d* and *e*. For the sample, fill the alternate cups with median dilution (c) of the standard and each of the remaining nine cups with sample dilution (U). Transfer the petri dishes to the incubator (32–35°C) after about 2 hrs. Incubate it overnight and measure the diameters of zones of inhibition.

Find the average of the 9 diameters of the zones of median standard dilution (c) on a set of 3 dishes and also the average of other standard dilution in that set. Similarly, find the averages for all the sets of the standard. Also, find the average of all the 36 diameters of median standard dilution. Correct the average diameter obtained for *a, b, d* & *e* by adding the difference algebraically between the average of 36 diameters and the average of 9 diameters of a particular set. Using these corrected values, find the calculated zone diameters for the lowest (L) and the highest (H) dilutions of the standard by the following formulae:

$$L = \frac{3a + 2b + c - e}{5}$$

$$\text{and } H = \frac{3e + 2d + c - a}{5}$$

Plot these values on 2-cycle semi-log paper on the abscissa against the concentrations of the standard on the ordinate (the logarithmic scale). Connect the points with a straight line.

Find the average of the 9 diameters of the zones of median standard concentration on the set of 3 plates of the sample and also the average of the 9 diameters of the zones of the sample dilution. Add the difference algebraically between these averages to the diameter of zone of median concentration (c) as seen on the standard curve. Read off this value on the curve and multiply it by suitable dilution factors to obtain the content of erythromycin in the sample.

Tetracycline Assay

1. Test organism: *Staphylococcus aureus* ATCC 6538P.
2. Inoculum: Prepare inoculum as shown under erythromycin assay incubating slant and Roux bottle at 37°C.

3. Standard and sample dilutions: Dilute the standard tetracycline hydrochloride and the sample in 0.1N HCl, so as to contain 1 mg. of tetracycline hydrochloride in each ml. Further, dilute the standard in sterile distilled water so as to finally contain 0.192, 0.215, 0.240, 0.268 and 0.300 mcg (*a* to *e*) tetracycline hydrochloride in each ml. and dilute the sample in sterile distilled water to finally contain 0.240 mcg. of tetracycline hydrochloride in each ml.
4. Assay Procedure: Add 1.0 ml. of each of the doses to each of the 3 test tubes (19 × 150 mm). Add 9.0 ml. of the previously inoculated medium No. 12. Place all the tubes, including one tube containing 1.0 ml. of sterile distilled water and 9.0 ml. of the uninoculated medium, in water bath maintained at 37°C. Arrest the growth after incubation of 3–4 hrs. and measure the absorbance in a suitable photoelectric colorimeter. Average the optical density of each set of doses. Find the calculated absorbance values for the lowest (L) and the highest (H) dilutions of the standard by the formulae given under erythromycin assay.

Plot these values on a semi-log graph as shown under erythromycin assay. Average the absorbance values of the sample dilution and determine the antibiotic concentration from the standard curve and multiply it by suitable dilution factors to obtain the content of tetracycline hydrochloride in the sample.

2 MICROBIOLOGICAL ASSAY OF VITAMINS AND AMINO ACIDS

Vitamin B_{12} Activity Assay

1. Test Organism: *Lactobacillus leichmannii* ATCC 7830.
2. Inoculum: Maintain the test organism as stab culture in medium No. 1. Inoculate a tube of broth (medium No. 2) with a test organism from stab culture and incubate it overnight at 37°C. Transfer a suitable aliquot from this to 50 ml. sterile single strength assay medium and incubate it at 37°C for 5 hrs. Store this stock inoculum to cold storage and use it for one week. For the preparation of inoculum, dilute the stock inoculum suitably with sterile normal saline.
3. Assay Medium: Dissolve L-cystine and DL-tryptophan in dil. HCl and vitamin-free casamino acids and anhydrous sodium acetate in distilled water. Mix and add stock solutions except the Tween 80 solution. Adjust the pH to 6.0, autoclave for 10 minutes at 15 p.s.i., cool and filter it. Add solutions of ascorbic acid, dextrose and Tween 80 to make the desired volume.
4. Standard and Sample Dilutions: Dilute standard cyanocobalamin and sample it suitably with distilled water to finally contain 20 pg. of the vitamin B_{12} activity in each ml.
5. Assay Procedure: Use 25 mm. × 200 mm. lipless tubes with loose-fitting aluminium caps for the assay. Prepare a series of tubes containing different aliquots between 0.5 and 5 ml. of the standard and sample dilutions, raise the volume in all the tubes to 5 ml. with distilled water, add 5 ml. of a double strength assay medium, autoclave at 15 p.s.i. for 5 mins, cool, inoculate and incubate it at 37°C overnight, arrest the growth of the test organism with formalin, measure turbidities (or per cent transmittances) using a suitable photoelectric colorimeter. Draw a standard curve of per cent transmittances against the concentration of vitamin on an arithmatic graph paper. Find the equivalent values of the standard for the responses of the sample by interpolation and compute the vitamin B_{12} activity in the sample.

Folic Acid Assay

1. Test Organism: *Streptococcus faecalis* ATCC 8043.
2. Inoculum: Maintain the test organism and prepare inoculum as shown for the vitamin B_{12} activity assay using medium No. 3 for the stab and No. 4 for the broth.
3. Assay Medium: Dissolve L-cystine in dilute HCl and other solid ingredients in distilled water. Add stock solutions, adjust the pH to 6.8 and make the desired volume.
4. Standard and Sample Dilutions: Dilute standard folic acid and the sample suitably to finally contain 1.0 mmcg. of folic acid in each ml.
5. Assay Procedure: Proceed as shown under the vitamin B_{12} activity assay.

Panthenol (Pantothenyl Alcohol) Assay

1. Test organism: *Leuconostoc mesenteroides* ATCC 8042.
2. Inoculum: Maintain the test organism as stab culture in medium No. 5. Inoculate a tube of broth (medium No. 6) with a test organism from the stab culture and incubate it overnight at 37°C. Dilute an aliquot with sterile normal saline to prepare the inoculum for the assay.
3. Assay Medium: Dissolve ingredients in distilled water and add stock solutions. Adjust the pH to 6.8 and make the desired volume with distilled water.
4. Standard and Sample Dilutions: Dilute standard D-panthenol and sample it suitably with distilled water to finally contain 1 mcg. of D-panthenol in each ml.
5. Assay Procedure: Proceed as shown under the vitamin B_{12} activity assay.

Total vitamin B_6 Assay

1. Test Organism: *Saccharomyces carlsbergensis* ATCC 9080.
2. Inoculum: Maintain the test organism on the slants of medium No. 7. To prepare inoculum, transfer a loopful of growth from the fresh slant to sterile normal saline.
3. Assay Medium: Dissolve solid ingredients (except dextrose) in distilled water and add stock solutions. Adjust the pH to 5.5, heat on waterbath for about 30 mins., filter, cool and add dextrose (previously dissolved in distilled water). Mix and make the desired volume.
4. Standard and Sample Dilutions: Dilute standard pyridoxine hydrochloride and sample it suitably with distilled water to finally contain 10 mmcg. of pyridoxine hydrochloride in each ml.
5. Assay Procedure: Use a 50 ml. conical flasks for the assay. Proceed further as directed under the vitamin B_{12} activity assay.

L-Lysine Monohydrocbloride Assay

1. Test Organism: *Leuconostoc mesenteroides* ATCC 8042.
2. Inoculum: Maintain the test organism and prepare inoculum as shown under the panthenol assay.
3. Assay Medium: Dissolve solid ingredients No. 1 to 11 in distilled water and No. 12 to 23 in 1N HCl. Mix and then add the stock solutions. Adjust the pH to 6.8 and make up the desired volume with distilled water.

4. Standard and Sample Dilutions: Dilute standard L-lysine monohydrochloride and sample it suitably with distilled water to finally contain 18 mcg. of L-lysine monohydrochloride in each ml.
5. Assay Procedure: Proceed as shown under the vitamin B_{12} activity assay.

Formulae for different media and solutions

1. Medium No. 1:

Yeast extract	..	0.75 g.
Peptone	..	0.75 g.
Dextrose	..	1.00 g.
Potassium dihydrogen phosphate	..	0.20 g.
Manganese sulphate (0.1 per cent solution in distilled water)		1.00 ml.
Tween 80 (10 per cent solution in ethanol)	..	1.00 ml.

Dissolve it in distilled water, adjust the pH to 6.8 and make the volume to 100 ml. Add 2.5g agar, dissolve it by heating on waterbath. Distribute it in test tubes and autoclave it at 15 p.s.i. for 15 mins.

2. Medium No. 2
Same as medium No. 1 but without agar.
3. Medium No. 3

Peptone	..	0.5 g
Dextrose	..	1.0 g
Anhydrous sodium acetate	..	0.6 g
Potassium dihydrogen phosphate	..	0.2 g

Dissolve it in distilled water, adjust the pH to 6.8 and make the volume to 100 ml. Add 2.5 g. agar, dissolve it by heating on waterbath. Distribute it in test tubes and autoclave at 15 p.s.i. for 15 mins.

4. Medium No. 4 :
Same as medium No. 3 but without agar.
5. Medium No. 5 :

Yeast extract	..	2.00 g.
Peptone	..	0.50 g.
Dextrose	..	1.00 g.
Potassium dihydrogen phosphate	..	0.20 g.
Tween 80 (10 per cent solution in enthanol)	..	0.05 ml.

Dissolve in distilled water, adjust the pH to 6.7 and make the volume to 100 ml. Add 2.5 g. of agar, dissolve it by heating on waterbath. Distribute it in test tubes and autoclave at 15 p.s.i. for 15 mins.

6. Medium No. 6:
Same as medium No. 5 but without agar.

7. Medium No. 7:
Dissolve 3.0 g. malt extract in distilled water, adjust the pH to 5.5 and make the volume to 100 ml. Add 2.0 g. agar, dissolve it by heating on water bath. Distribute it in test tubes, autoclave for 15 mins. at 15 p.s.i. and allow it to cool in a slanting position.
8. Stock solution No. 1:
Dissolve 1.0 g. of L-aspargine in distilled water with heating and make up to 100 ml. Store it under a layer of toluene.
9. Stock solution No. 2:
Dissolve 100 mg. each of adenine sulphate, guanine hydrochloride and uracil in dilute HCl by heating. Make the total volume to 100 ml. with distilled water and store it under a layer of toluene.
10. Stock solution No. 3:
Suspend 100 mg of xanthine in 20 ml. distilled water. Heat on waterbath and dissolve xanthine by the addition of ammonia solution and by stirring. Make up to 100 ml. with distilled water and store it under toluene.
11. Stock solution No. 4:
Dissolve 10.0 mg. of D-biotin in 200 ml. of 50 per cent ethanol. Store it in a cool place.
12. Stock solution No. 5:
(A) Dissolve 5.0 mg. of riboflavin in 50 ml. of 0.02N acetic acid by warming.
(B) Dissolve 5.0 mg. of thiamine hydrochloride and 10.0 mg. of niacin in 50 ml. of 0.02N acetic acid. Mix (A) and (B), add 1 ml. of stock solution No. 4 and make up the volume to 200 ml. with 0.02N acetic acid. Store it in a cool place under toluene.
13. Stock solution No. 6:
Dissolve 10.0 mg. of P-amino benzoic acid and 44.0 mg. of pyridoxine hydrochloride in 25 per cent ethanol. Store it in a cool place.
14. Stock solution No. 7:
(A) Dissolve 5.0 mg. of folic acid in dilute ammonia solution.
(B) Dissolve 25.0 mg. calcium pantothenate in 25 per cent ethanol. Mix (A) and (B) and make up to 500 ml. with 25 per cent ethanol. Store it in a cool place.
15. Stock solution No. 8 :
Dissolve 10.0 g. each of potassium dihydrogen phosphate and dipotassium hydrogen phosphate in 200 ml. of distilled water. Add a drop of concentrated HCl. Store it under toluene.
16. Stock solution No. 9:
Dissolve 4.0 g. of magnesium sulphate and 0.2 g. each of manganese sulphate, ferrous sulphate and sodium chloride in 200 ml. distilled water. Store it under toluene.
17. Stock solution No. 10:
Dissolve 10.0 g. of Tween-80 in 100 ml. of ethanol. Store it in a cool place.
18. Stock solution No. 11:
Dissolve 2.0 g. of DL-tryptophan in the minimum quantity of dilute HCl and make up to 500 ml. with distilled water. Store it. under toluene.
19. Stock solution No. 12:
(A) Dissolve 10.0 mgs. of riboflavin in 50 ml. of 0.02N acetic acid by warming.
(B) Dissolve 10.0 mgs. of P-amino benzoic acid, 4.0 mg. of thiamine hydrochloride, 40.0 mgs. of pyridoxine hydrochloride, 8.0 mg. of calcium pantothenate and 8.0 mg. of niacinamide in 50 ml. of 0.02N acetic acid.

(C) Dissolve 1.9 g. of anhydrous sodium acetate in 50 ml. of 0.02N acetic acid. Mix (A), (B) and (C), add 4.0 ml. of stock solution No. 4 and make it up to 200 ml. with 0.02N acetic acid. Store it under toluene in a cool place.

20. Stock solution No. 13:

Dissolve 2.0 g. of L-cystine and 1.0 g. of DL-tryptophan in the minimum quantity of dilute HCl and make it up to 500 ml. with distilled water. Store it under toluene.

21. Stock solution No. 14:

(A) Dissolve 10.0 mg. of riboflavin in 100 ml. of 0.02N acetic acid by warming.

(B) Dissolve 5.0 mg. of thiamine hydrochloride in 100 ml. of 0.02N acetic acid. Mix (A) & (B) and 0.4 ml. of stock solution No. 4. Make it up to 500 ml. with 0.02N acetic acid. Store it under toluene in a cool place.

22. Stock solution No. 15:

Dissolve 2.5 mg. of P-amino benzoic acid, 12.5 mg. of niacin and 10.0 mg. of pyridoxine hydrochloride in 250 ml. of 25 per cent ethanol. Store it in a cool place.

23. Stock solution No. 16:

(A) Dissolve 5.0 mg. of riboflavin in about 50 ml. of 0.02N acetic acid by warming.

(B) Dissolve 5.0 mg. of thiamine hydrochloride and 10.0 mg. of niacin in 25 ml. of 0.02N acetic acid. Mix (A) and (B), add 0.5 ml. of stock solution No. 4 and make it up to 100 ml. with 0.02N acetic acid. Store it under toluene in a cool place.

24. Stock solution No. 17:

Dissolve 5.0 mg. of P-amino benzoic acid and 100.0 mg. of pyridoxine hydrochloride in 500 ml. of 25 per cent ethanol. Store it in a cool place.

25. Stock solution No. 18:

(A) Dissolve 5.0 mg. of folic acid in 10 ml. of dilute ammonia solution.

(B) Dissolve 10.0 mg. of calcium pantothenate in 25 per cent ethanol, add 0.2 ml. of solution (A) and make it up to 200 ml. with 25 per cent ethanol. Store it in a cool place.

26. Medium No. 8:

Peptone	:	0.6 g.
Casitone	:	0.4 g.
Yeast extract	:	0.3 g
Beef extract	:	0.15 g.
Dextrose	:	0.1 g.
Distilled water	:	to 100 ml.
pH	:	6.5 : 6.6
Agar	:	2.5 g.

Distribute it in test tubes and sterilize.

27. Medium No. 9:

Perparation is same as that of medium No. 8, but containing 0.1 mg. of manganese sulphate. Distribute it in Roux bottle and sterilize.

28. Medium No. 10:

Beef extract	.. 0.15 g.
Yeast extract	.. 0.3 g.
Peptone	.. 0.6 g.
Water	.. to 100 ml.
pH	.. 7.8 – 8.0
Agar	.. 2.5 g.

Distribute it in conical flasks and sterilize.

29. Medium No. 11:

Same as medium No. 10, but with 1.1 g. of agar.

30. Medium No. 12:

Peptone	0.5 g.
Yeast extract	0.15 g.
Beef extract	0.15 g.
Sodium chloride	0.35 g.
Dextrose	0.1 g.
Dipotassium hydrogen phosphate	0.368 g.
Potassium dihydrogen phosphate	0.132 g.
Water	to 100 ml.
pH	7.0

Distribute it in conical flasks and sterilize.

31. Solution No. 19 :

Dipotassium hydrogen phosphate	.. 16.73 gms.
Potassium dihydrogen phosphate	.. 0.523 gms.
Distilled water	100 ml.

Adjust the pH with 18N phosphoric acid or 10N potassium hydroxide to yield pH of 8.0 after sterilization.

Composition of 250 ml. double strength basal media for vitamin assay.

Dry ingredients/stock solutions	*Vitamin B_{12} Activity Assay*			*Folic Acid Assay*			*D-panthenol Assay*			*Vitamin B_6 Assay*		
	gms.	*mgs.*	*ml.*	*gms.*	*mgs.*	*ml.*	*gms.*	*mgs.*	*ml.*	*gms.*	*mgs.*	*ml.*
1. Vitamin-free casamino acids	2.5	–	–	2.5	–	–	2.5	–	–	2.0	–	–
2. Anhydrous sodium acetate	5.0	–	–	–	–	–	5.0	–	–	–	–	–
3. Sodium citrate	–	–	–	13.0	–	–	–	–	–	–	–	–
4. Dextrose	10.0	–	–	10.0	–	–	10.0	–	–	25.0	–	–
5. Ascorbic acid	1.0	–	–	–	–	–	–	–	–	–	–	–
6. L-cystine	–	100.0	–	–	190.0	–	–	–	–	–	–	–
7. DL-tryptophan	–	100.0	–	–	–	–	–	–	–	–	–	–
8. Dipotassium hydrogen phosphate	–	–	–	1.6	–	–	–	–	–	–	–	–
9. Stock solution No. 1	–	–	5.0	–	–	15.0	–	–	–	–	–	–
10. Stock solution No. 2	–	–	5.0	–	–	2.5	–	–	5.0	–	–	–
11. Stock solution No. 3	–	–	5.0	–	–	5.0	–	–	–	–	–	–

Contd.

12. Stock solution No. 5	–	–	10.0	–	–	–	–	–	–	–	–	–
13. Stock solution No. 6	–	–	10.0	–	–	–	–	–	–	–	–	–
14. Stock solution No. 7	–	–	5.0	–	–	–	–	–	–	–	–	–
15. Stock solution No. 8	–	–	5.0	–	–	–	–	–	5.0	–	–	–
16. Stock solution No. 9	–	–	5.0	–	–	5.0	–	–	5.0	–	–	–
17. Stock solution No. 10	–	–	5.0	–	–	0.25	–	–	0.25	–	–	–
18. Stock solution No. 11	–	–	25.0	–	–	–	–	–	–	–	–	–
19. Stock solution No. 12	–	–	5.0	–	–	–	–	–	–	–	–	–
20. Stock solution No. 13	–	–	–	–	–	–	–	–	25.0	–	–	–
21. Stock solution No. 14	–	–	–	–	–	–	–	–	5.0	–	–	–
22. Stock solution No. 15	–	–	–	–	–	–	–	–	5.0	–	–	–

Composition of 250 ml. double strength basal medium for lysine monohydrochloride assay.

Dry ingredients/stock solutions	*gms.*	*mgs.*	*ml.*
1. Dextrose	10.0	–	–
2. Anhydrous sodium acetate	6.0	–	–
3. L-alanine	–	50.0	–
4. L-arginine hydrochloride	–	100.0	–
5. L-aspargine	–	50.0	–
6. Glycine	–	100.0	–
7. L-histidine hydrochloride	–	100.0	–
8. L-methionine	–	50.0	–
9. L-proline	–	50.0	–
10. L-serine	–	50.0	–
11. L-theronine	–	50.0	–
12. L-aspartic acid	–	50.0	–
13. L-cystine	–	25.0	–
14. L-glutamic acid	–	250.0	–
15. L-isoleucine	–	50.0	–
16. L-leucine	–	50.0	–
17. L-phenylalanine	–	50.0	–
18. L-tryptophan	–	50.0	–
19. L-tyrosine	–	50.0	–
20. L-valine	–	50.0	–
21. L-cysteine hydrochloride	–	25.0	–
22. L-hydroxyproline	–	50.0	–
23. L-norleucine	–	50.0	–
24. Stock solution No. 5	–	–	5.0
25. Stock solution No. 11	–	–	2.5
26. Stock solution No. 12	–	–	2.5
27. Stock solution No. 16	–	–	5.0
28. Stock solution No. 17	–	–	5.0
29. Stock solution No. 18	–	–	5.0

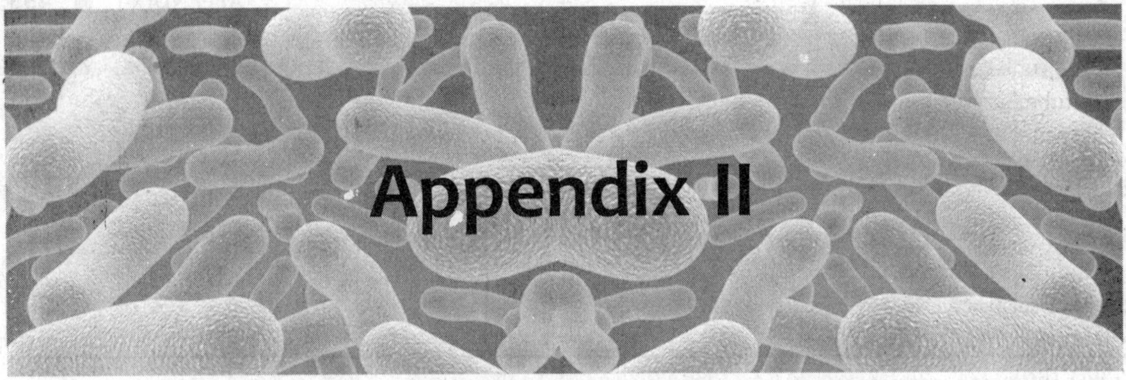

Appendix II

1 METHYLENE-BLUE REDUCTION TEST

This test was first employed by Barthel and Orla-Jensen in 1900. The original test was modified by Wilson and his colleagues. It is the official test for both 'Untreated' and 'Pasteurized' milks. The majority of organisms in milk are capable of altering the oxidation potential of the liquid to a point, at which the colour of methylene blue is discharged. The rate at which this change takes place varies considerably. This is related to the species of organisms present, their rate of growth, and their activity at any one temperature. Generally speaking, it may be stated that the greater the number of organisms in a given quantity of milk and the greater their activity, the shorter the time will be required for the dye to be reduced.

The apparatus required to carry out the test consists of:

(1) A thermostatically controlled water bath with temperature maintained at 37°C ± 0.5°C.

(2) Sterile test tubes possessing an etched mark at the 10 ml. level. They should be plugged with cotton wool or covered with closely fitting aluminium caps.

(3) Glassware and rubber stoppers for sealing the tubes which must be sterilized immediately prior to use.

(4) One-millilitre, straight-sided, blow out pipettes which should be plugged with cotton wool at their upper ends.

(5) Methylene blue solution is prepared from standard methylene blue tablets, one tablet being dissolved in 200 ml. of cold glass distilled water in a sterile stoppered flask. The flask and its contents should be well shaken to dissolve the tablet, afterwards adding sufficient sterile distilled water to bring the contents up to 800 ml. The flask should be sealed with a rubber stopper to prevent evaporation and stored in a cool, dark place. Solutions over two months old, or those which have been exposed to sunlight must not be used. The amount of solution required for use each day must be poured from the stock bottle into a suitable glass container.

The technique of the test is as follows: The sample to be examined is well mixed by inversion and shaking. The stopper and the mouth of the sample bottle are flamed along with the mouth of the test tube. Well mixed milk is poured rapidly into the tube until the liquid reaches the 10 ml. mark, one side of the tube being left dry. One millilitre of the methylene blue solution is then added by means of a pipette. The pipette should not touch the milk or the wet interior of the tube. After three seconds have elapsed, any solution remaining in the tip of the pipette is blown out. This pipette should never be used a second time to withdraw the dye solution from its container. The tube is aseptically sealed with a sterile rubber stopper. Care is to be taken to ensure that the fingers do not touch the mouth of the tube. The tube and its contents are slowly inverted twice and within 5 minutes it is placed in the water bath. The water in the bath must be maintained at a higher level than that of the milk in the tube. The bath must be kept in dark and the temperature maintained between 37°C and 38°C. Control tubes are required for each batch of samples tested, each of these containing 10 ml. of mixed milk of a type similar to that under test, plus 1 ml. of tap water. These control tubes are stoppered and immersed in boiling water for 3 minutes in order to destroy the reducing properties of the milk. Control tubes are important as they denote the beginning of the decolourisation and the completion of the change. The tubes containing the samples are examined at the end of 30 minutes, a common method being to inspect a whole rackful *en bloc.* If this is the normal practice, slight changes of colour might not be readily observed, especially if the inspections are made under artificial light. For this reason, it is preferable to lift each tube separately from the rack for inspection, along side the control tube in a good light. The sample is regarded as being decolourised when the whole column of liquid has lost its colour or is decolourised within 5 mm. of the surface. If no colour change has occurred within the prescribed time, the sample complies with the test.

By prolonging the incubation of the sample in the water bath, the test may be employed for the rough grading of milk as shown in the Table below:

Result of Test	*Regarded as*	*Grade*
Reduced in 15 minutes		Very bad
Reduced in ½ to 1 hour	Unsatisfactory	Bad
Reduced in 1½ to 2 hours		Poor
Reduced in 3 to 4 hours	Doubtful	Doubtful
Reduced in 4½ to 6 hours	Satisfactory	Good
Not reduced in 6 hours		Very Good

The tubes should be inspected at 30 minute intervals. Care is to be taken to ensure that they are not exposed to direct sunlight while any tube which has lost its colour should be removed. Tubes which appear to be losing their colour should be allowed to remain, while all others should be inverted once and replaced. Inversion is necessary, otherwise the cream content would not be distributed as evenly as possible throughout the liquid. If cream was allowed to collect on the surface of the milk, the reduction period would be delayed.

Certain *precautions* must be strictly observed in carrying out the test. These are:

(1) The interior of the water bath must be kept dark.

(2) Its temperature should be checked at frequent intervals during testing.

(3) Thermometers used on baths should be checked at least once a month.

(4) Water bath should be cleaned out weekly and fresh water added.

Advantages:

(1) Cheap material and labour.

(2) The quality of any official sample is obtained within half an hour and, if the longer test is carried out, within a few hours.

(3) It can be carried out by any intelligent person who has received little training.

(4) Where laboratory facilities are limited the control of milk supplies can be exercised at low cost.

(5) It is fairly foolproof if due precautions are taken.

(6) If the test is carried out at 55°C, it may be used as a means of detecting thermophilic micro-organisms.

In spite of the official approval given to the test, it is by no means perfect, and has certain *Disadvantages:*

(1) The temperature at which the milk is held, i.e. 37°C is not the ideal temperature for the growth of all types of micro-organisms present in milk. Therefore, the reduction period may not give a true indication of their activity.

(2) It cannot be used as a test for freshly pasteurized milk, having been originally devised as a raw milk test. If the pasteurized milk has been aged, however, as is the case prior to delivery to the consumer, it is useful for assessing the quality of the liquid.

(3) The test is not sufficiently sensitive to allow any close differentiation between different grades of milk. Also it does not provide a test for the milk of individual animals.

(4) It is an 'all-or-nothing' test. The dye is either reduced completely, proving the test satisfactory or, if no reduction takes place, the milk is considered satisfactory. No latitude is allowed for the laboratory to assess the position in the light of conditions which prevail at the point of production.

(5) The presence of any antibiotic in the milk is likely to prolong the reduction period.

Biochemical Oxygen Demand

Outline of the method: Biochemical oxygen demand (BOD) is the quantity of oxygen required by a definite volume of the liquid effluent for oxidizing the organic matter contained in it by micro-organisms under specified conditions. For its determination, the dissolved oxygen content of the sample, with or without dilution, is measured before and after incubation at 20°C for 5 days.

Apparatus

Glass-stoppered bottles: Narrow-neck bottles of about 250 ml. capacity with suitable water sealing.

Reagents

Sodium Hydroxide Solution: Approximately 1N *Hydrochloric Acid:* Approximately 1N.

Sodium Sulphite Solution: Dissolve 1.5 g., of anhydrous sodium sulphite in l litre of water. Prepare fresh solution daily for use.

Dilution Water: Distilled water of good quality, free from metals, particularly copper, and aerated. *Phosphate Buffer Solution:* Dissolve 8.5 g. of potassium dihydrogen phosphate (KH_2PO_4), 21.75 g. of dipotassium hydrogen phosphate (K_2HPO_4), 33.4 g. of disodium hydrogen phosphate (Na_2HPO_4, $7H_2O$) and 1.7 g. of ammonium chloride in about 500 ml. of water and dilute to 1 litre. The pH of this solution should be 7.2.

Magnesium Sulphate Solution: Dissolve 22.5 g. of magnesium sulphate ($MgSO_4$, $7H_2O$) in water and dilute to 1 litre.

Calcium Chloride Solution: Dissolve 27.5 g. of anhydrous calcium chloride in water and dilute to 1 litre.

Ferric Chloride Solution: Dissolve 0.25 g. of ferric chloride ($FeCl_3$, $6H_2O$) in water and dilute to 1 litre.

Seeding Material: Supernatant liquor of domestic sewage stored for 24 to 36 hours at 20°C. In the case of industrial effluents containing organic compounds which are not easily oxidized by sewage seed, the receiving water collected about 3.5 km. below the discharge point may be used.

Procedure

Samples containing acid or caustic alkalinity should be neutralized to pH about 7.0 with sodium hydroxide solution or hydrochloric acid respectively by adding a predetermined quantity.

Samples containing residual chlorine or chloramines should be dechlorinated if chlorine is not dissipated on standing for 2 hours. To dechlorinate, first determine the quantity of sodium sulphite solution required for a known aliquot of the sample by titration to starch-iodide end point after acidifying the sample with acetic acid (1 : 1) or sulphuric acid (1 : 50) followed by 10 ml. of 10 per cent potassium iodide solution. Then add to the requisite volume of the sample the predetermined quantity of sodium sulphite, avoiding any excess, and check for the absence of chlorine after 20 minutes.

Samples containing toxic substances in a large amounts would require special treatment. However, the effect of small amounts may be overcome by using the proper dilution so that toxicity is removed and the maximum BOD value is obtained. If increasing dilutions show increasing BOD, the dilution should be increased to a level where BOD levels off at a maximum.

To check the quality of the dilution water and the effectiveness of the seed, determine the BOD of a standard solution of 300 mg./*l* of either glucose or glutamic acid in the dilution water. Standard glucose solution should show a BOD of 224 ± 10 mg./*l*. and glutamic acid 217 ± 10 mg./*l*.

Store the dilution water at 20°C and use it when near that temperature. Take the desired volume of dilution water required for the test sample and add, for every 1 litre of water, 1 ml. each of phosphate buffer solution, magnesium sulphate, calcium chloride solution and ferric chloride solution. Seed the dilution with seeding material. The quantity of seeding material (0.1 to 1% of settled sewage or 1 to 5 % receiving water) added should be such that oxygen depletion in the dilution water control is between 0.2 and 0.8 mg./ltr. after incubation at 20°C for 5 days.

Prepare, as follows, several dilutions of the sample (usually 0.1 to 1.0 % for strong industrial effluents and 5 to 25% for treated effluents) so as to obtain a depletion of at least 2 mg./*l* of dissolved oxygen, after incubation for 5 days. In the case of dilutions greater than 1 : 100, prepare a 10 per cent primary dilution in a volumetric flask and from this make the final dilutions.

Carefully siphon the prepared seeded dilution water into a graduated 1000-ml. measuring cylinder and fill it to the 500-m1. mark. Add the requisite quantity of the carefully well mixed sample to make the particular dilution and fill it with dilution water to 1 litre. Mix it thoroughly but gently with a plunger type of rod without allowing air to enter. Siphon the dilution into two glass-stoppered bottles, fill completely and stopper. Prepare succeeding dilutions of lower concentrations in the same manner. Determine the initial dissolved oxygen concentration in one of the two bottles of each dilution by the appropriate method. Methods of determining dissolved oxygen are: (a) the Winkler method, (b) the Alsterberg (Sodium azide) modification. Water-seal the other bottles and incubate at 20°C for 5 days: At the same time, siphon the dilution water alone into two glass-stoppered bottles and determine the blank in one and incubate the other at 20°C for 5 days. After incubation for 5 days, determine the dissolved oxygen in the dilutions and the blank in the same manner as the initial dissolved oxygen content.

Calculation

$$\text{Biochemical Oxygen Demand (5 days at 20°C) mg./}l = \frac{(D1 - D2) - (C1 - C2)F}{P}$$

where

D1 = initial dissolved oxygen content of the diluted sample.

D2 = dissolved oxygen content of the diluted sample after incubation.

C1 = initial dissolved oxygen content of the seeded dilution water.

C2 = dissolved oxygen content of the seeded dilution water after incubation.

F = ratio of the seed in the sample to that in the control, that is, per cent seed in D1 divided by per cent seed in C1, and,

P = decimal fraction of the sample used.

Express the result to the nearest 0.2 mg./*l*.

Chemical Oxygen Demand

(i) *Principle:* A sample is refluxed with a known amount of potassium dichromate and sulphuric acid, and the excess dichromate is titrated with ferrous ammonium sulphate. The amount of oxidizable matter, measured as an oxygen equivalent is proportional to the potassium dichromate consumed. The straight chain compounds are more effectively oxidized in the presence of silver sulphate catalyst. However, silver sulphate reacts with chlorides, bromides and iodides to produce precipitates which are only partially oxidized by the procedure.

The oxidization and other difficulties caused by the presence of chlorides in the sample are overcome by adding mercuric sulphate to the samples before refluxing. This ties up the chloride complex, which greatly reduces its ability to react further.

(ii) *Apparatus:* Reflux apparatus 250 ml. or Erlenmeyer flasks with ground glass 24/40 neck and 300 mm. jacketed Liebig Werstor equivalent condenser with 24/400 ground glass joint. A hot plate having sufficient power to produce at least 9 watts. inc. of heating surface or equivalent to ensure adequate boiling of the contents of the refluxing flask. Bunsen burner can be used instead of a hot plate.

(iii) *Reagents:*

(a) Std. potassium dichromate solution 0.250N: Dissolve 12.259 g. $K_2Cr_2O_7$ standard grade previously dried at 103°C. for 2 hours in distilled water and dilute to 1000 ml.

(b) Sulphuric acid reagent: concentrated H_2SO_4 containing 22 g. of Ag_2SO_4/91b. bottle (1 to 2 days required for dissolution).

(c) Std. ferrous ammonium sulphate titrant analytical grade crystal, 0.10N: Dissolve 39 gms. of $Fe(NH_4)_2 (SO_4)_2 \cdot 6H_2O$ in distilled water. Add 20 ml. conc. H_2SO_4. Cool and dilute to 1000. This solution must be standardized against std. $K_2Cr_2O_7$ solution daily.

(iv) *Standardization:* Dilute 10.0 ml. std. $K_2Cr_2O_7$ solution to about 100 ml. Add 30 ml. conc. H_2SO_4 A.R. and allow it to cool. Titrate with ferrous ammonium sulphate titrant using 2 or 3 drops (0.10 to 0.15 ml.) of ferrion indicator.

$$\text{Normality} = \frac{\text{ml. of } K_2Cr_2O_7 \times 0.25}{\text{ml. of } Fe\,(NH_4)_2\,(SO_4)_2}$$

(d) Ferrion indicator solution—dissolve 1.485 g/*l*,10-phenolphthalein monohydrate, together with 695 mg., $FeSO_4 \cdot 7H_2O$ in water and dilute to 100 ml. This indicator solution may be purchased.

(e) Mercuric sulphate—analytical grade.

(f) Silver sulphate—reagent powder for sulphuric acid reagent.

(v) *Procedure:* The volume of the effluent sample taken for the determination of chemical oxygen demand (COD) is called sample size. If the COD of the sample is in the range of 60,000 to 1,00,000 the volume of the sample for the analysis should be 0.1 ml. For this an initial dilution should be made in a volumetric flask by taking 5 ml. of original sample and making it up to 100 ml. *2* ml. of this should be taken for analysis. The volume of the effluent sample should be taken as per the volume recommended below, with respect to the range and also the dilution given:

	Volume ml.			
60,000 to 1,00,000	0.1	5	100	2
30,000 to 60,000	0.25	5	100	5
15,000 to 30,000	0.5	10	100	5
5,000 to 15,000	0.1	10	100	10
1,000 to 5,000	2.0	20	100	10
500 to 1,000	5.0	Not necessary. Can be taken directly.		

Place 0.4 gm. of H_2SO_4 in the reflux flask. Pipette out suitable volume of the sample as mentioned above. Pipette out 10 ml. of std. potassium dichromate and add several glass beads which are previously heated to 600°C for 1 hour. Slowly add 30 ml. of conc. H_2SO_4 containing Ag_2SO_4 by a cylinder along the sides of the conical flask, and mix it thoroughly during addition. Add 20 to 30 ml. of water slowly along the sides.

Mix reflux mixture thoroughly before heat is applied. If this is not done, local heating occurs at the bottom of the flask or the mixture may be blown out of the condenser.

Use 0.4 gm. H_2SO_4 to complex 40 mg. chloride ion or 4000 mg./litre chloride ion conc. when 0.1 ml. of original effluent is used. If more chloride is present, add more H_2SO_4 to maintain a H_2SO_4: Cl ratio of 10:1. If it precipitates slightly, it does not adversely affect the determination. Connect the flask to the condenser and reflux the mixture for *2* hours. Cool and wash down the condenser with distilled water. Make the final volume to 150 ml. approximately in the same conical flask before titration. Titrate the excess dichromate with the std. ferrous ammonium sulphate using 2 to 3 drops of ferroin indicator. Consider as the end point when the sharp colour changes from blue green to reddish brown, even though the blue green may reappear within minutes.

Reflux in the same way a blank containing distilled water in place of a sample, together with the reagents.

(vi) *Calculation:*

$$\text{mg./litre, COD} \;—\; \frac{(a-b)\text{N} \times 8{,}000}{\text{ml. of sample}}$$

where, COD — chemical oxygen demand

a — ml. of $Fe\,(NH_4)_2\,(SO_4)_2$

b — ml. of $Fe\,(NH_4)_2\,(SO_4)_2$

N — normality of $Fe(NH_4)_2\,(SO_4)_2$

8000 — Factor for COD.

(vii) *Precautions:*

(1) The complete glass equipment should be washed with chromic acid and then with distilled water before use.

(2) As this is a very sensitive test, care should be taken during the analysis.

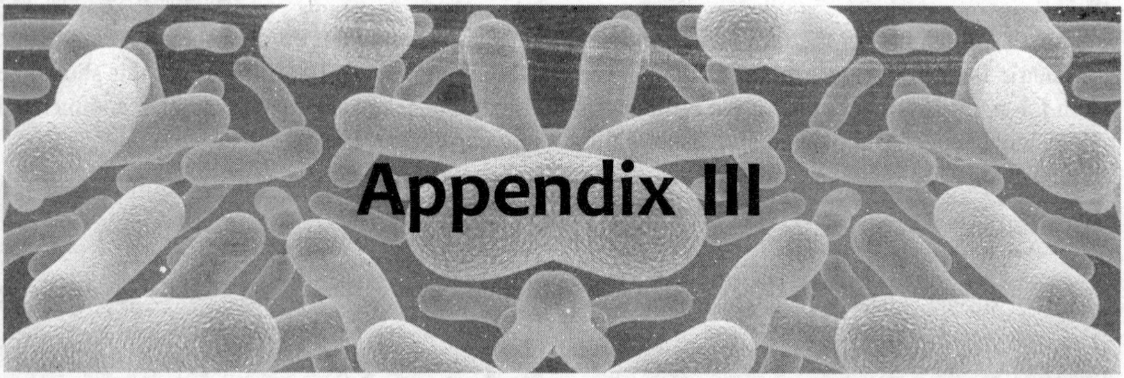

Appendix III

DATA INDEX 1*

ENZYME GROUPINGS

Enzyme groupings	*Data Index reference*
Amylases—General	A
Bacterial	B
Fungal	C
Plant	D
Amyloglucosidases (debranching)	E
Analytical/Pharmaceutical/Research	F
Cellulases	G
Dextranase and invertase	H
Galactosidases and inulases	I
Glucanases	J
Isomerases	K
Lipases and esterases	L
Oxidases and catalases	M
Pectinases	N
Pentosanases	O
Proteases—General	P
Animal	Q
Microbial	R
Plant	S
Rennets	T
Immobilized—General	U
Mixed activity preparations	V

*Godfrey and Reichelt (1983), 'Industrial Enzymology' published by Macmillan Publishers Ltd., New York.

DATA INDEX 2*

ENZYME SUPPLIER COMPANIES

No.	Company	Country	Comments	Enzyme types (see Data Index 1)
1.	A.B.M,C. Food Division Poleacre Lane Woodley Stockport Cheshire SK6 1PQ	UK	Bulk industrial	A, B, C, E, J, P, R, S
2.	Akzo Chemie Statwinstraat 48 PO Box 247	Holland		Range not specified
3.	Aktieboluget Montoil Fack S 100 55 Stockholm	Sweden		Range not specified
4.	Alltech Inc. 271 Gold Rush Road Lexington Kentucky 40503	USA	Specialists for alcohol production	A, B, E
5.	Alpha Color S.P.A. Via Valtellina 48 20159 Milan	Italy	Textile applications	B
6.	Amano Pharmaceutical Co. Ltd. 1–21–chome Nishiki Naka-Ku Nagoya	Japan	Bulk industrial	A, B, C, E, G, L, P, R S
7.	B.D.H. Chemicals Ltd. Broome Road Poole Doiset BHl2 4NN	UK	Agents for Merck	H, S
8.	Biddle Sawyer and Co. Ltd. PO Box 170 3 Lovat Lane London EC3P 3EX	UK	Brewing suppliers	S
9.	Biochemie GmbH A-6250 Kundi Tirol	Austria	Pharmaceutical specialities	F
10.	Biocon Ltd. Kilnagleary Carrigaline Co. Cork	Eire	Bulk industrial	A, B, C, D, E, G, H, J, N, P, R, S, V
11.	Biozyme Laboratories Ltd. Unit 6 Gilchrist-Thomas Estate Blaenavon Gwent NP4 9RL	UK	Analytical and research Wide range Agents in USA, Italy, Switzerland and France	F
12.	Boehringer Mannheim GmbH PO Box 51 D-6800 Mannheim-31	W. Germany	Analytical and research Wide range Worldwide agents	F

*Godfrey and Reichelt (1983), 'Industrial Enzymology', published by Macmillan Publishers Ltd., New York.

Contd.

No.	Company	Country	Comments	Enzyme types (see Data Index 1)
13.	Calbiochem PO Box 54282 Los Angeles California 90054	USA	Analytical and research Wide range Worldwide agents	F
14.	Chr. Hansen Laboratories A/S 3 Sankt Annae Plads DK-1250 Copenhagen	Denmark	Dairy specialities	L, T, V
15.	Corning Biosystems Corning Glass Works Corning New York 14830	USA	Immobilized enzyme systems Agents for Rohm	A, B, I, N, P, R, U
16.	Dairyland Food Laboratories Inc. Progress Avenue Waukesha Wisconsin 53187	USA	Dairy specialities	L, M, T, V
17.	Daiwa Kasei KK 3-11 Vehonmachi-5-chome Tennoji-ku Osaka	Japan	Bulk industrial	A, B, C, P
18.	Diamalt Friedrichstrasse 18 D-8000 München 40 Postfach 400 469	W. Germany	Textile and leather industries	A, P, Q, R
19.	Diamond Shamrock 620 Progress Avenue Waukesha Wisconsin 53186	USA		Range not specified
20.	Enzyme Development Corporation 2 Penn Plaza New York New York 10121	USA	Bulk industrial Research and development	A, B, C, E, G, H, I, J, K, L, M, N, O, P, Q, R, T, U
21.	Fermco Biochemics Inc. 2638 Delta Lane Elk Grove Village Illinois 60007	USA	Beverage industry Wine and juice industry Analytical	A, B, D, E, F, G, H, K. M, N
22.	Genzyme Biochemicals Ltd. Springfield Mill Maidstone Kent ME14 2LE	UK	Analytical/diagnostic Division of Genzyme Corp. Connecticut, USA	F
23.	Gist Brocades NV PO Box 1 Wateringseweg 1 Delft 2600MA	Holland	Bulk industrial	A, B, C, E, G, H, I, J, K, L, M, N, P, R, S, T, V
24.	Glaxo Operations UK Ltd. Ulverston Cumbria LA12 9DR	UK	Bulk industrial	C, E, J
25.	Godo Shusei Co. Ltd. 6-2-10, Ginza, Chuo-ku Tokyo	Japan	Bulk industrial	A, B, I, K, R
26.	Grindestedvaerket A/S 38 Edwin Rahrs Vej 8220 Braband	Denmark	Bulk industrial Wine/juice industries	J, M, N

Contd.

No.	Company	Country	Comments	Enzyme types (see Data Index 1)
27.	Hankyu Kyoei Bussan Co. Ltd. 5/6 chome Tenjin Bashi-suji Oyodo-Ku Osaka	Japan	Bulk industrial	Range not specified
28.	Hayashibara Shoji Co. Ltd. 198 Shimoishii Okayama City	Japan	Bulk industrial	Range not specified
29.	Henkel KG a.A D-4000 Düsseldorf 1 Postfach 1100	W. Germany	Bulk industrial	Range not specified
30.	Hoechst, D-6000 Frankfurt/M80 Postfach 800 320	W. Germany	Bulk industrial Textile industry	A, B, C
31.	Hughes and Hughes (Enz.) Ltd. Elms Industrial Estate Church Road Harold Wood Romford Essex RM3 0HR	UK	Analytical/research	F, G, L, M, S
32.	Koch-Light Laboratories Ltd. Colnbrook Buckinghamshire SL3 0BZ	UK	Analytical/research	F, U
33.	Kingsbridge Industrial Inc. PO Box 24 200 Tapei	Taiwan	Bulk (bromelain)	S
34.	Kyowa Hakko Kogyo Co. Ltd. Ohtemachi Building 6-1, Ohtemachi 1-chome Chiyoda-ku Tokyo	Japan	Bulk industrial	G, T
35.	Laboratories Sanders SA 47-51 rue Henri Wafelaerts Brussels 6	Belgium		Range not specified
36.	Dr. Madis Laboratories Inc. 375 Huyler Street S. Hackensack New Jersey 07606	USA	Bulk industrial	P
37.	Miles Kali-Chemie GmbH 3 Hannover-Kleefeld Hans-Buckler Allee 20 Postfach 690 307	W. Germany	Bulk industrial Analytical and research	A, B, C, E, F, G, J, K, L, M, N, P, Q, R, S, T, U
38.	Miles Laboratories Inc. Elkhart Indiana 46514	USA	Bulk industrial Analytical and research	A, B, C, E, F, G, J, K, L, M, N, O, P, Q, R, S, T, U, V
39.	Mitsui and Co. Ltd. PO Box 822 Tokyo Central	Japan	Bulk industrial	Range not specified

Contd.

No.	Company	Country	Comments	Enzyme types (see Data Index 1)*
40.	Munton and Fison Ltd: Cedars Factory Stowmarket Suffolk IP14 2AG	UK	Malt extract specialities	D, S, V
41.	Murphy and Son Ltd. Wheathampstead St Albans Hertfordshire	UK	Bulk (papain)	S
42.	Naarden International NV Postbus 2 Naarden-Bussum	Holland	Bulk industrial	Range not specified
43.	Nagase and Co. Ltd. Konishi Building 2, 2-chome Honcho Nihonbashi Chuo-ku Tokyo	Japan	Bulk industrial Analytical/pharmaceutical	A, B, D, E, F, G, K. P, R
44.	Novo Industri A/S Novo Alle DK-2880 Bagsvaerd	Denmark	Bulk industrial Analytical/research/pharmaceutical	A, B, C, E, F, G, H, I, J, K, L, N, O, P, Q, R, T, U, V
45.	Okasa Industria e Comercia de Diastase Ltd. Rue Araraquara 41-Diadema-SP Brazil CEP 09900 Caixa Postal 352	Brazil	Bulk industrial	Range not specified (A)
46.	Oriental Yeast Co. Ltd. Enzyme Development Centre 4-1 Minamisuita 4-chome Suita, Osaka 564	Japan		Range not specified
47.	P. L. Biochemicals Inc. 1037 West McKinley Avenue Milwaukee Wisconsin 53205	USA	Analytical/research	Wide range
48.	Premier Malt Products Inc. 1137 North 8th Street Milwaukee Wisconsin 53201	USA		D, S
49.	Pfizer Inc. World Headquarters 235 East 42nd Street New York New York 10017	USA	Bulk industrial	A, B, C, G, O, P, Q, R, T
50.	Pharmachim State Economic Trust 15 Iliensko Chaussee Sofia	Bulgaria		Range not specified
51.	Powell and Scholefield Ltd. 38 Queensland Street Liverpool L7 3JG	UK	Brewing Industry	A, B, C, J, P, S

Contd.

No.	Company	Country	Comments	Enzyme types (see Data Index 1)
52.	Rohm GmbH Kirschenallee Postfach 4242 D-6100 Darmstadt 1	W. Germany	Bulk industrial	A, B, C, G, N, O, P, Q, R, S, V
53.	Rohm and Hass Co. Independence Mall West Philadelphia Pennsylvania 19105	USA	Bulk industrial	A, B, C, G, N, P, Q, R, S
54.	Sigma Chemical Company Inc. St. Louis Morrisville Missouri	USA	Analytical/research	F
55.	Société Rapidase 15 rue des Comtesses 59113 Seclin	France	Gist-Brocades associate company	*See* no. 23
56.	John and E. Sturge Ltd. Denison Road Selby North Yorkshire Y08 8EF	UK	Bulk industrial	C, D, E, F, G, I, L, M, N, O, V
57.	Sturge Enzymes /Henley and Co. Inc. 750 Third Avenue New York New York 10017	USA	Analytical *See also* no. 56	A, B, D, F, H
58.	Sumitomo Shoji Kaisha Ltd. 2 Nishikicho Building 24-1 Kandanishikicho 3-chome Chiyoda-ku Tokyo	Japan	Bulk industrial	A, B, C, E, F, G, I, N, O, P, R, S
59.	Swiss Ferment Co. Vogesenstrasse 132 4056 Basel 13	Switzerland	Bulk industrial Novo Industri subsidiary	A, C, H, L, N, O, V *Also see* no. 44
60.	Tanabe Seiyaku Co. Ltd. via Siber Hegner Benelux BV Postbus 414 Rotterdam Westersingel 107	Japan and Holland	Bilk industrial Analytical/pharmaceutical	A, F, L, N, P
61.	Ubichem Ltd. 281 Hithermoor Road Stanwell Moor Staines TW19 6AZ	UK	Industrial/analytical	A, G, P, S, V
62.	Vifor SA 48 route de Drize 1227 Carouge Geneva	Switzerland		Range not specified
63.	W.B.E. Ltd. Sandyford Industrial Estate Foxrock Dublin 18	Eire	Bulk industrial	Mixed systems

Contd.

No.	Company	Country	Comments	Enzyme types (see Data Index 1)
64.	Worthington Biochemical Co. New Jersey 07728	USA	Analytical/research	Range not specified
65.	Windsor Laboratories Ltd. Bedford Avenue Slough Berkshire	UK	Analytical/research	Wide range
66.	Yakult Biochemical Co. Ltd. 8-21 Jingikan-Machi Nishinomiya-shi Hyogo	Japan	Industrial/pharmaceutical	B, C, F, G, N, O, T

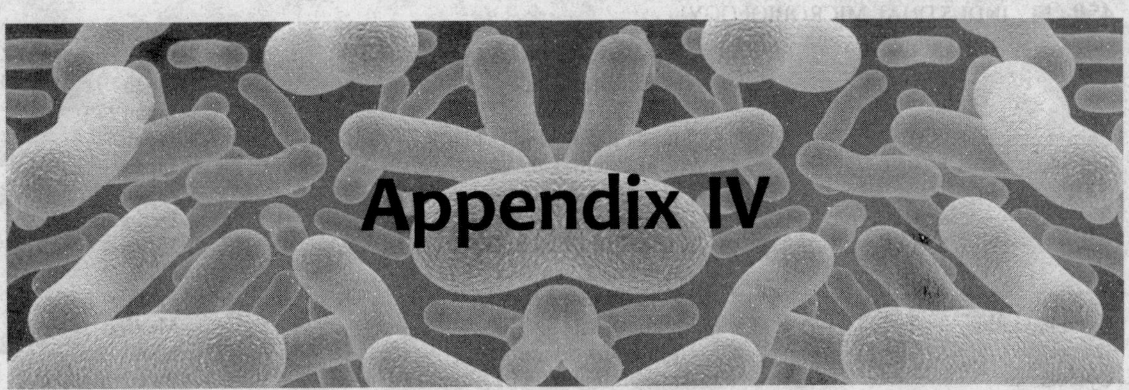

Appendix IV

LIST OF FERMENTATION PRODUCTS AND COMPANIES (AFTER PERLMAN 1977)

Product	Company	Country
Antibiotics		
Adriamycin	Farmitalia SpA	Italy
Amphomycin	H. Lundbeck & Co.	Denmark
Amphotericin B	Standard Brands Inc.	USA
Avoparcin	American Cyanamid	USA
Azalomycin F	Sankyo Co.	Japan
Bacitracin	Apothokernes Labs. A/S	Norway
	Asahi Chem. Ind. Co.	Japan
	Dumex Ltd.	Denmark
	IMC Chemicals Corp.	USA
	H. Lundbeck & Co.	Denmark
	NOVO Industri A/S	Denmark
	S.P. Penick & Co.	USA
	Pfizer Inc.	USA
	Publicker Industries Inc.	USA
	Societe Chimique Pointet Girard	France
Bambermycin	Hoechst AG	FRG
Bicyclomycin	Fujisawa Pharm. Co.	Japan
Blasticidin S.	Kaken Chem. Co.	Japan
Bleomycin	Nihon Kanyaku Co.	Japan
Cactinomycin	Farbenfabriken Bayer AG	FRG
Candicidin	Dumex Ltd.	Denmark
	H. Lundbeck & Co.	Denmark
	S.B. Penick & Co.	USA
Candidin	H. Lundbeck & Co.	Denmark
Capreomycin	Dista Products Ltd.	UK
Cephalosporins	Anheuser Busch Inc.	USA
	Bristol Myers Co. Inc.	USA
	China National Chemicals	PR China
	Companhia Industrial (CIPAN)	Portugal
	Farmitalia SpA	Italy
	Fervet SpA (CIBA-Geigy)	Italy
	Fujisawa Pharmaceutical Co.	Japan
	Glaxo Labs. Ltd.	UK
	Instituto Biochimico Italiano	Italy
	Eli Lilly & Co.	USA
	Leo Pharmaceutical Products	Denmark
	Novo Industri A/S	Denmark
	Pierrel SpA	Italy

	Proter SpA	Italy
	Takeda Chem. Ind.	Japan
	Toyo Jozo Co.	Japan
Chromomycin	Takeda Chem. Ind. Ltd.	Japan
Colistin	Asahi Chem. Co.	Japan
	Banyu Pharmaceutical Co.	Japan
	Dumex Ltd	Denmark
	Kayaku Antibiotica Res. Co.	Japan
	American Cyanimid	USA
	Rhone Poulenc SA	France
Cycloheximide	Fermic SA	Mexico
	Upjohn Co.	USA
Cycloserine	Aktiebiologet KABI	Sweden
	IMC Chemicals Group	USA
Dactinomycin	Merck & Co.	USA
	Reanal	Hungary
Daunorubicin	Farmitalia SpA	Italy
Destomycin	Meiji Seika Kaisha Ltd.	Japan
Enduracidin	Takeda Chem. Ind. Ltd.	Japan
Erythromycin	Abbott Labs.	USA
	Alembic Chemical Works	India
	Anheuser-Busch Inc.	USA
	Archifar ApA	Italy
	Chemibiotic Ltd.	Ireland
	China National Chemicals	PR China
	Companhia Ind. CIPAN	Portugal
	Comp. Esp. Penicillina Antibiotica	Spain
	Dista Products Ltd.	UK
	Fermic SA	Mexico
	Inst. Biochimico Italiano	Italy
	Eli Lilly & Co.	USA
	Pfizer Inc.	USA
	Pierrel SpA	Italy
	Proter SpA	Italy
	Roussell-UCLAF	France
	'Polfa'	Poland
	Upjohn Co.	USA
Fortimycins	Kyowa Hakko Kogyo	Japan
Fumagillin	Abbott Labs.	USA
	Chinoin	Hungary
Fungimycin	H. Lundbeck & Co.	Denmark
Fusidic acid	Leo Pharmaceutical Products	Denmark
Tentamicins	China National Chemicals	PR China
	Chinoin	Hungary
	Schering Corp.	USA
Gramicidin A	H. Lundbeck & Co.	Denmark
	S.B. Penick & Co. Ltd.	USA
	Wallerstein Labs. Inc.	USA
Griseofulvin	China National Chemicals	PR China
	Glaxo Laboratories	UK
	ICI	UK
	Leo Pharmaceutical Products	Denmark
	Nikon Kayaku Co.	Japan
	Takeda Chem. Ind.	Japan
Hygromycin B	Biogal	Hungary
	Eli Lilly & Co.	USA

	Takeda Chem. Ind.	Japan
Josamycin	Sanraku Ocean Co.	Japan
Kanamycins	Alembic Chemical Works	India
	Bristol-Myers Co.	USA
	China National Chemicals	PR China
	Dong Myung Industrial Co.	S. Korea
	Meiji Seika Kaisha Ltd.	Japan
	Nihon Kayaku Co.	Japan
	Pierrel/SpA	Italy
Kasugamycin	Banyu Pharmaceutical Co.	Japan
	Hokko Kangaku Kogyo Co.	Japan
	Meiji Seika Kaisha Ltd.	Japan
	Sanraku Ocean Co. Ltd.	Japan
Kitasatamycin	Toyo Jozo Co. Ltd.	Japan
Lasalocid	Hoffman-La Roche Inc.	USA
Lincomycin	Upjohn Co.	USA
Lividomycin	Kowa Co.	Japan
	Rhone-Poulenc SA	France
Macarbomycin	Meiji Seika Kaisha Ltd.	Japan
Mepartricin	Societa Prodotti Antibiotici SpA	Italy
Midecamycin	Meiji Seika Kaisha Ltd.	Japan
Mikamycins	Banyu Pharmaceutical Co.	Japan
	Kanegafuchi Chemical Ind.	Japan
Mithramycin	Pfizer Inc.	USA
Mitomycin C	Bristol-Myers Co.	USA
	Kyowa Hakko Hogyo Co.	Japan
Mocimycin	Gist-Brocades	Holland
Mononsin	Eli Lilly & Co.	USA
Myxin	Hoffman-La Roche Inc.	USA
Neomycins	Antibioticos SA	Spain
	Apothokernes Laboratorium	Norway
	Biogal	Hungary
	China National Chemicals	PR China
	Farmitalia SpA	Italy
	H. Lundbeck & Co.	Denmark
	Nihon Kayaku Co.	Japan
	S.B. Penick & Co.	USA
	Pfizer Inc.	USA
	Rhone-Poulenc SA	France
	Roussel-UCLAF	France
	E. R. Squibb & Sons Inc.	USA
	Takeda Chem. Ind. Co.	Japan
	Upjohn Co.	USA
Novobiocin	Meiji Seika Kaisha Ltd.	Japan
	Rhone-Poulenc SA	France
	Upjohn Co.	USA
Nystatin	American Cyanamid	USA
	China National Chemicals	PR China
	Chinoin	Hungary
	Pierrel SpA	Italy
	Rhone-Poulenc SA	France
	E. R. Squibb & Sons Inc.	USA
Oleandomycin	Pfizer Inc.	USA
Oligomycin	Reana	Hungary

Antibiotic	Manufacturer	Country
Paromycins	Parke Havis & Co.	USA
Penicillin G	Aktiebolaget Astra	Sweden
	Aktiebolaget Fermenta	Sweden
	Aktiebolaget KABI	Sweden
	Alembic Chemical Works	India
	Antibioticos SA	Spain
	Banyu Pharmaceutical Co.	Japan
	Beecham Pharmaceutical Co. Ltd.	UK
	Biochemie GmbH	Austria
	China National Chemicals	PR China
	Farbenfabriken Bayer AG	FRG
	Fermion Oy.	Finland
	Gist-Brocades	Holland
	Glaxo Laboratories	UK
	Hindustan Antibiotics Ltd.	India
	Hoechst AG	FRG
	ISF. SpA	Italy
	Eli Lilly & Co.	USA
	Meiji Seika Kaisha Ltd.	Japan
	Merck & Co.	USA
	NOVO Industri A/S	Denmark
	Pfizer Inc.	USA
	Pierrel SpA	Italy
	Rhone-Poulenc SA	France
	Sociedade Produtora de Leveduras	Portugal
	E. R. Squibb & Sons Inc.	USA
	Sarabhai Chemicals Ltd.	India
	Toyo Jozo Co. Ltd.	Japan
	Wyeth Laboratories	USA
Penicillin V	Abbott Laboratories	USA
	Aktiebolaget Astra	Sweden
	Aktiebolaget Fermenta	Sweden
	Aktiebolaget KABI	Sweden
	Banyu Pharmaceutical Co.	Japan
	Biochemie GmbH	Austria
	Biogal	Hungary
	Bristol-Myers Co.	USA
	Dista Products Ltd.	UK
	Fermion Oy	Finland
	Gist-Brocades	Holland
	Glaxo Laboratories Ltd.	UK
	Hoechst AG	FRG
	Eli Lilly & Co.	USA
	Meiji Seika Keisha Ltd.	Japan
	NOVO Industrie A/S	Denmark
	Pfizer Inc.	USA
	Rhone-Poulenc SA	France
	E. R. Squibb & Sons	USA
	Toyo Jozo Co. Ltd.	Japan
	Wyeth Laboratories	USA
Penicillins (semisynthetic)	Aktiebolaget Astra	Sweden
	Anheuser-Busch Inc.	USA

	Antibioticos SpA	Italy
	Banyu Pharmaceutical Co.	Japan
	Beecham Pharmaceutical Co.	UK
	Biochemie GmbH	Austria
	Bristol-Myers Co.	USA
	China National Chemicals	PR China
	CIPAN	Portugal
	Farbenfabriken Bayer AG	FRG
	Fermion Oy	Finland
	Fujisawa Chemical Co.	Japan
	Gist-Brocades	Holland
	ISF SpA	Italy
	Inst. Biochimico Italiano	Italy
	Leo Pharmaceutical Products	Denmark
	Meiji Seika Kaisha Ltd.	Japan
	Merck & Co.	USA
	Pfizer Inc.	USA
	Pierrel SpA	Italy
	Quimasa SA	Brazil
	E. R. Squibb & Sons Inc.	USA
	Takeda Chemical Industries	Japan
	Toyo Jozo Co. Ltd.	Japan
	Wyeth Laboratories	USA
Pentamycin	Nikken Chemicals Ltd.	Japan
Pimaricin	Gist-Brocades	Holland
Polymyxins	Dumex Ltd.	Denmark
	NOVO Industri A/S	Denmark
	Pfizer Inc.	USA
Polyoxins	Kaken Chemical Co.	Japan
	Sanraku Ocean Co. Ltd.	Japan
Pristinamycins	Rhone-Poulenc SA	France
Quebemycin	Kyowa Hakko Kogyo Co.	Japan
Ribostamycin	Meiji Seika Kaisha Ltd.	Japan
Rifamycins	Archifar ApA	Italy
	Fevet SpA	Italy
	Gruppo Lepetit SpA	Italy
Sagamycin	Kyowa Hakko Kogyo Co.	Japan
Salnomycin	Kaken Chemical Co.	Japan
Siccanin	Sankyo Co. Ltd.	Japan
Sisomycin	Farbenfabriken Bayer AG	FRG
	Sohering Corpn.	USA
Spectinomycin	Abbott Laboratories	USA
	Kaken Chemical Co.	Japan
	Upjohn Co.	USA
Streptomycins	Alembic Chemical Works	India
	Antibioticos SA	Spain
	China National Products	PR China
	Gist-Brocades	Holland
	Glaxo Laboratories Ltd.	UK
	Meiji Seika Kaisha Ltd.	Japan
	Merck & Co.	USA
	NOVO Industri A/S	Denmark

	Pfizer Inc.	USA
	Sarabhai Chemicals Ltd.	India
	Wyeth Laboratories	USA
Tetracyclines		
Chlortetracycline	American Cyanamid	USA
	China National Chemicals	PR China
	Diaspa SpA	Italy
	Farmitalia SpA	Italy
	Hindustan Antibiotics Ltd.	India
	ISF SpA	Italy
	Lohmann & Co.	FRG
	Proter SpA	Italy
	Rachelle Laboratories Inc.	USA
	Shionogi & Co. Ltd.	Japan
	SPA	Italy
	Takeda Chemical Industries	Japan
Demecocycline	American Cyanamid	USA
	CIPAN	Portugal
	Pierrel SpA	Italy
	Proter SpA	Italy
	Rhone-Poulenc SA	France
	Takeda Chemical Industries	Japan
Oxytetracycline	Ankerfarm SpA	Italy
	Biogal	Hungary
	China National Chemicals	PR China
	Chinoin	Hungary
	CIPAN	Portugal
	Fermic SA de SV	Mexico
	Gist-Brocades	Holland
	ICI	UK
	Instituto Biochimico Italiano	Italy
	'Polfa'	Poland
	KRKA Pharmaceutical Works	Yugoslavia
	Lark SpA	Italy
	Pfizer Inc.	USA
	Pliva Pharmaceutical Works	Yugoslavia
	Proter SpA	Italy
	Rachelle Laboratories Inc.	USA
	Societe Cimique Pointet Girard	France
	Tarchomin Pharmaceutical Works	Poland
Tetracycline	Aktiebolaget Astra	Sweden
	American Cynamid	USA
	Ankerfarm SpA	Italy
	Antibioticas SA	Spain
	Archifar ApA	Italy
	Banyu Pharmaceutical Co.	Japan
	Bristol-Myers Co.	USA
	Chemibiotic Ltd.	Ireland
	China National Chemicals	PR China
	Chong-Kun-Dong Corpn.	S. Korea
	CIPAN	Portugal
	Com. Esp. de la Penicillina y Antibioticos	Spain

	Diaspa SPA	Italy
	Farmitalia SpA	Italy
	Fermentfarma SpA	Italy
	Fermic SA de SV	Mexico
	Gruppo Lepeti SpA	Italy
	Hindustan Antibiotics Ltd.	India
	Hoechst AG	FRG
	Hokko Kagakyu Kogyo Co.	Japan
	ICN-Chimica SpA	Italy
	ISF SpA	Italy
	Instituto Biochimica Italiano	Italy
	Linson Ltd.	Ireland
	NOVO Industri A/S	Denmark
	Pfizer Inc.	USA
	Pharmacin Antibiotic Works	Bulgaria
	Proter SpA	Italy
	Quimasa SA	Brazil
	Rachelle Laboratories Inc.	USA
	Roussel-UCLAF	France
	Societa Prodotti Antibiotici	Italy
	Societe Chimique Pointet Girard	France
	E. R. Squibb & Sons Inc.	USA
	Takeda Chemical Industries	Japan
	'Polfa' Tarchomin	Poland
	Upjohn Co.	USA
Tetranactin	Chugai Pharmaceutical Co.	Japan
Thiopeptin	Fujisawa Pharmaceutical Co.	Japan
	Kaken Chemical Co.	Japan
Thiostrepton	E. R. Squibb & Sons Inc.	USA
Tobramycin	Biogal	Hungary
	Eli Lilly & Co.	USA
Trichomycin	Fujisawa Pharmaceutical Co.	Japan
Tylosin	Dista Products Ltd.	UK
	Eli Lilly & Co.	USA
Tyrothricin	Biochemie GmbH	Austria
	H. Lundbeck & Co.	Denmark
	S. B. Penick & Co.	USA
	Wallerstein Laboratories Inc.	USA
Tyrocidine	H. Lundbeck & Co.	Denmark
Uromycin	Reanal	Hungary
Validomycin	Takeda Chemical Industries	Japan
Vancomycin	Eli Lilly & Co.	USA
Variotin	Leo Pharmaceutical Products	Denmark
	Nihon Kayaku Co.	Japan
Viomycin	Toyo Jozo Co. Ltd.	Japan
Virginiamycin	Recherche et Industrie Therapeutique	Belgium
Enzymes		
Amylases	Biogal	Hungary
	C.H. Boehringer Sohn	FRG
	Clinton Corn Processing Co.	USA
	CIPAN	Portugal
	Comp. Espanola de la Penicillina	Spain

	Gist-Brocades	Holland
	Hailsun Chemical Co. Ltd.	Taiwan
	Meiji Seika Kaisha Ltd.	Japan
	Miles Laboratories Inc.	USA
	Nagase & Co. Ltd.	Japan
	NOVO Industri A/S	Denmark
	Premier Malt Products Inc.	USA
	Publicker Industries Inc.	USA
	Rohm GmbH	FRG
	Rohm & Haas	USA
	Sankyo Co. Ltd.	Japan
	Standard Brands Inc.	USA
	Wallerstein Laboratories Inc.	USA
Amyloglucosidase	Clinton Corn Processing Co.	USA
	CPC International Inc.	USA
	Finnish State Alcohol Monopoly	Finland
	Gist-Brocades	Holland
	Glaxo Laboratories Ltd.	UK
	Miles Laboratories InW.	USA
	NOVO Industri A/S	Denmark
	Wallerstein Laboratories Inc.	USA
Anticyanase	Tanabe Seiyaku Co. Ltd.	Japan
1-Asparaginase	Farbenfabriken Bayer AG	FRG
	Kyowa Hakko Kogyo Co.	Japan
Catalase	C. H. Boehringer Sohn	FRG
	Dawe's Laboratories Inc.	USA
	Kyowa Hakko Kogyo Co.	Japan
	Toyo Jozo Co. Ltd.	Japan
Cellulase	Aktiebolaget Astra	Sweden
	C. H. Boehringer Sohn	FRG
	Gist-Brocades	Holland
	Meiji Seika Kaisha Ltd.	Japan
	E. Merck	FRG
	Miles Laboratories Inc.	USA
	NOVO Industri A/S	Denmark
	Rohm GmbH	FRG
	Rohm & Haas	USA
	Wallerstein Laboratories Inc.	USA
Dextranase	NOVO Industri A/S	Denmark
	Sankyo Co. Ltd.	Japan
'Diagnostic enzymes'	Kyowa Hakko Kogyo Co.	Japan
	Miles Laboratories Inc.	USA
	Toyo Jozo Co. Ltd.	Japan
Esterase-lipase	Wallerstein Laboratories Inc.	USA
Glucanase	NOVO Industri A/S	Denmark
	Wallerstein Laboratories Inc.	USA
Glucose dehydrogenase	E. Merck	FRG
Glucose isomerase	Anheuser-Busch Inc.	USA
	Clinton Corn Processing Co.	USA
	CPC International Inc.	USA
	Gist-Brocades	Holland

	ICI Ltd.	UK
	Miles Laboratories Inc.	USA
	Nagase & Co. Ltd.	Japan
	NOVO Industri A/S	Denmark
	Standard Brands Inc.	USA
Glucose oxidase	C. H. Boehringer Sohn	FRG
	Dawe's Laboratories Inc.	USA
	Gist-Brocades	Holland
	Miles Laboratories Inc.	USA
	Nagase & Co. Ltd.	Japan
Glutamic decarboxylase	Kyowa Hakko Kogyo Ltd.	Japan
Gumase	Premier Malt Products Inc.	USA
Hemicellulase	Miles Laboratories Inc.	USA
	NOVO Industri A/S	Denmark
	Rohm & Haas	USA
	Wallerstein Laboratories Inc.	USA
Hespiriginase	Tanabe Seiyaku Co. Ltd.	Japan
Invertase	Gist-Brocades	Holland
	Lohmann & Co. AG	FRG
	E. Merck	FRG
	Sankyo Co. Ltd.	Japan
	Standard Brands Inc.	USA
	Universal Foods Corporation	USA
	Wallerstein Laboratories Inc.	USA
Lactase	Gist-Brocades	Holland
	NOVO Industri A/S	Denmark
	Wallerstein Laboratories Inc.	USA
Lipase	C. H. Boehringer Sohn	FRG
	Rohm & Haas	USA
	Tanabe Seiyaku Co.	Japan
	Toyo Jozo Co. Ltd.	Japan
	Wallerstein Laboratories Inc.	USA
Microbial rennet	Dairyland Food Labs. Inc.	USA
	Meito Sangyo Co. Ltd.	Japan
	NOVO Industri A/S	Denmark
	Pfizer Inc.	USA
	Wallerstein Labs. Inc.	USA
Naringinase	Tanabe Seiyaku Co.	Japan
	Wallerstein Labs. Inc.	USA
Pectinase	C. H. Boehringer Sohn	FRG
	Miles Labs. Inc.	USA
	NOVO Industri A/S	Denmark
	Rohm GmbH	FRG
	Rohm & Haas	UK
	Sankyo Co.	Japan
	Tanabe keiyaku	Japan
	Toyo Jozo Co.	Japan
	Wallerstein Labs. & Inc.	USA
Pentosanase	Rohm & Haas	USA
	Wallerstein Labs. Inc.	USA

Proteases	Biogal	Hungary
	C. H. Boehringer Sohn	FRG
	Chinoin	Hungary
	Gist-Brocades	Holland
	Kaken Chem. Co.	Japan
	Kyowa Hakko Kogyo Co	Japan
	Meiji Seika Kaisha Ltd.	Japan
	E. Merck	FRG
	Miles Labs. Inc.	USA
	NOVO Industri A/S	Denmark
	Pfizer Inc.	USA
	Premier Malt Products Inc.	USA
	Rohm GmbH	FRG
	Rohm & Haas	USA
	Sankyo Co.	Japan
	Takeda Chem. Ind.	Japan
	Toyo Jozo Co.	Japan
	Wallerstein Labs. Inc.	USA
Streptokinase	American Cyanamid	USA
Streptodornase		
Unicase	NOVO Industri A/S	Denmark
Organic acids		
Citric	J. A. Benekhiser GmbH	Austria
	C. H. Boehringer Sohn	FRG
	Citrique Belge	Belgium
	Miles Labs. Inc.	USA
	Pfizer Inc.	USA
	Rhone-Poulenc SA	France
	Santu Chem. Co.	Taiwan
	J. & E. Sturge Ltd.	UK
	Tai Nan Fermentation	Taiwan
Comenic	Pfizer Inc.	USA
Erythorbic	Fujisawa Pharmaceutical Co.	Japan
Gluconic	J. H. Benckhiser GmbH	FRG
	Bristol-Myers Co.	USA
	Fujisawa Pharm. Co.	Japan
	Kyowa Hakko Kogyo Co.	Japan
	E. Merck	FRG
	Pfizer Inc.	USA
	Premier Malt Products	USA
	World Champion Co.	Taiwan
Itaconic	Pfizer Inc.	USA
	Rhone-Poulenc SA	France
2-Keto-D-gluconic	Kyowa Hakko Kogyo	Japan
	Merck & Co. Inc.	USA
	Pfizer Inc.	USA
α-Ketoglutaric	Ajinomoto Co.	Japan
	Kyowa Hakko Kogyo Co.	Japan
Lactic	C.H. Boehringer	FRG
	Clinton Corn Processing Co.	USA
	Croda Bowmans Chemicals	UK

	Rhone-Poulenc SA	France
Malic	Ajinomoto Co.	Japan
	Kyowa Hakko Kogyo Co.	Japan
	Tanabe Seiyaku Co.	Japan
Urocanic	Ajinomoto Co.	Japan
	Tanabe Seiyaku Co.	Japan
Solvents		
Ethanol	Finnish State Alcohol Monopoly	Finland
	Gist-Brocades	Holland
	Grain Processing Corporation	USA
	Kyowa Hakko Kogyo Co.	Japan
	Publicker Industries	USA
	Sanraku Ocean Co.	Japan
	Taiwan Sugar Corp.	Taiwan
2; 3 Butanediol	Gist-Brocades	Holland
Vitamins and growth factors		
Giberellins	Abbott Laboratories	USA
	ICI UK	UK
	Kyowa Hakko Kogyo Co.	Japan
	Eli Lilly & Co.	USA
	Merck & Co. Inc.	USA
	Pierrel SpA	Italy
	Takeda Chem. Ind.	Japan
Riboflavin	Merck & Co. Inc.	USA
Vitamin B12	Alembic Chem. Works	India
	Chinoin	Hungary
	Farmitalia, SpA	Italy
	G1axo Labs. Ltd.	UK
	Merck & Co. Inc.	USA
	Rhone-Poulenc SA	France
	G. Richter	Hungary
	Roussel-UCLAF	France
Zearalanol	IMC Chemicals Inc.	USA
Nucleosides and Nucleotides		
5^1-Ribonucleotides and nucleosides	Ajinomoto Co.	Japan
	Asahi Chem. Co.	Japan
	Kyowa Hakko Kogyo Co.	Japan
	Takeda Chem. Ind.	Japan
	Yamasa Shoyu Co.	Japan
Orotic acid	Kyowa Hakko Kogyo Co.	Japan
Ara-A-(9-(-n-ara-binofurosanyl) adenine	Parke, Davis & Co.	USA
b-azauridine	Kyowa Hakko Kogyo Co.	Japan
Amino acids		
L-Alanine	Ajinomoto Co.	Japan
	Kyowa Hakko Kogyo	Japan
	Tanabe Seiyaku Co.	Japan
L-Arginine	Ajinomoto Co.	Japan

	Kyowa Hakko Kogyo	Japan
L-Aspartic acid	Ajinomoto	Japan
	Carlo Erba SpA	Italy
	Kyowa Hakko Kogyo	Japan
	Orsan SA	France
	Rhone-Poulenc SA	France
	Tanabe Seiyaku	Japan
L-Citruline	Ajinomotc Co.	Japan
	Kyowa Hakko Kogyo	Japan
	Tanabe Seiyaku	Japan
L-Glutamic acid	Ajinomoto	Japan
	Asahi Chem. Co.	Japan
	Kyowa Hakko Kogyo Co.	Japan
	Mi-Won	S. Korea
	Orsan SA	France
	Sanraku Ocean Co.	Japan
	Stauffer Chem. Co.	USA
	Takeda Chem. Ind.	Japan
	Tsin Tsin Foods Co.	Taiwan
	Tung Hai Fermentation Co.	Taiwan
	Wei Chuan Foods Corp.	Taiwan
	Wei Wang Fermentation Co.	Taiwan
	World Champion Co.	Taiwan
L-Glutamine	Ajinomoto Co.	Japan
	Kyowa Hakko Kogyo Co.	Japan
L-Glutathione	Kyowa Hakko Kogyo Co.	Japan
	Toyo Jozo Co.	Japan
L-Histidine	Ajinomoto Co.	Japan
	Kyowa Hakko Kogyo Co.	Japan
L-Homoserine	Kyowa Hakko Kogyo Co.	Japan
L-Isoleucine	Ajinomoto Co.	Japan
	Kyowa Hakko Kogyo Co.	Japan
	Toyo Jozo Co.	Japan
L-Leucine	Ajinomoto Co.	Japan
	Kyowa Hakko Kogyo Co.	Japan
L-Lysine	Ajinomoto Co.	Japan
	Eurolysine Co.	France
	Kyowa Hakko Kogyo Co.	Japan
	Rhone-Poulenc SA	France
L-Methionine	Kyowa Hakko Kogyo Co.	Japan
	Tanabe Seiyaku Co.	Japan
L-Ornithine	Ajinomoto Co.	Japan
	Kyowa Hakko Kogyo Co.	Japan
L-Phenylalanine	Kyowa Hakko Kogyo Co.	Japan
	Tanabe Seiyaku Co.	Japan
L-Proline	Ajinomoto Co.	Japan
	Kyowa Hakko Kogyo Co.	Japan
	Tanabe Seiyaku	Japan
L-Serine	Ajinomoto Co.	Japan
	Kyowa Hakko Kogyo Co.	Japan
L-Threonine	Ajinomoto	Japan
	Kyowa Hakko Kogyo Co.	Japan

L-Tryptophan	Ajinomoto	Japan
	Tanabe Seiyaku	Japan
L-Tyrosine	Kyowa Hakko Kogyo Co.	Japan
L-Valine	Ajinomoto Co.	Japan
	Kanegafuchi Chem. Ind.	Japan
	Tanabe Seiyaku Co.	Japan
Miscellaneous		
Acetoin	Gist-Brocades	Holland
Acyloin	Knoll GmbH	FRG
	Merck & Co. Inc.	USA
Anka-red	Taiwan Tobacco & Wine Monopoly	Taiwan
Blue cheese flavour	Dairyland Food Labs. Inc.	USA
Desferroxiamine	Fervet SpA (Div. of CIBA Geigy)	Italy
Dihydroxyacetone	E. Merck	FRG
	Pfizer Inc.	USA
	Roussel UCLAF	France
	Schering Corpn.	USA
	Tanabe Seiyaku Co.	Japan
	Wallerstein Labs.	USA
Dextran	Aktiebiologet Astra	Sweden
	Biogal	Hungary
	Meito Sangyo Co.	Japan
	Pharmacosmos	Denmark
Ergocornine	Biochemie GmbH	Austria
	Farmitalia SpA	Italy
	G. Richter	Hungary
	Sandoz-Wander AG	Switzerland
Ergocristine	Biochemie GmbH	Austria
	Farmitalia SpA	Italy
	Sandoz-Wander AG	Switzerland
Ergocryptine	Biochemie GmbH	Austria
	Farmitalia SpA	Italy
	G. Richter	Hungary
	Sandoz-Wander AG	Switzerland
Ergmetrine	G: Richter	Hungary
Ergotamine	Farmitalia SpA	Italy
	G. Richter	Hungary
Insecticide	Abbott Labs.	USA
	Kyowa Hakko Kogyo Co.	Japan
	Rhone-Poulenc SA	France
	Sandoz Inc.	USA
Lysergic acid	Farmitalia SpA	Italy
Paspalic acid	Biochomie GmbH	Austria
	Sandoz-Wander AG	Switzerland
Picibanil	Chugai Pharmaceutical Co.	Japan
Robose	Ajinomoto Co.	Japan
Scleroglucan	Orsan SA	France
Sorbose	Hoffman-la Roche NJ	USA
	Hoaffman-la Roche	Switzerland
	E. Merck. Darmstadt	FRG

	Pao Yeh Chem. Co.	Taiwan
	Pfizer Inc. NY	USA
	Takeda Chem. Ind.	Japan
Xanthan	Biogal	Hungary
	Lohmann & Co.	FRG
	Merck & Co.	USA
	Orsan SA	France
	Pfizer Inc.	USA
	Rhone-Poulenc SA	France
	Taiwan Sugar Corp.	Taiwan
	Tate & Lyle	UK

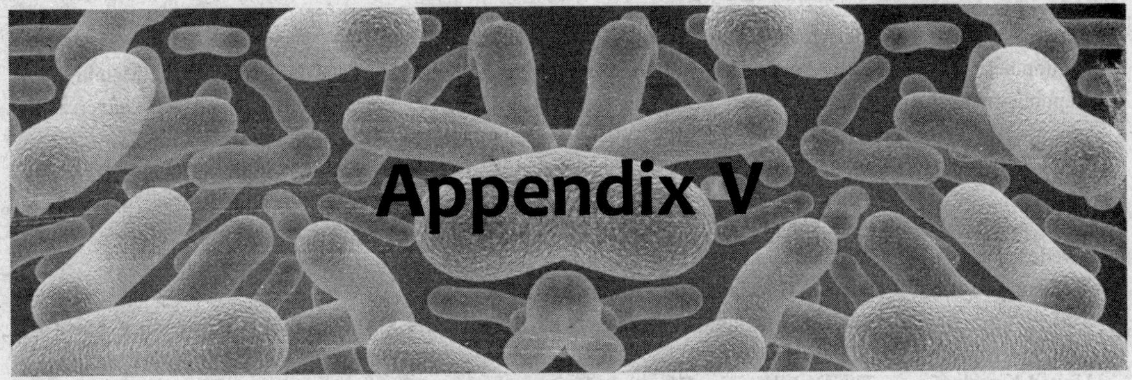

Appendix V

SOME COMPANIES GROWING YEAST FOR FOOD AND FEED

Bakers' Yeast

Argentina: Compania Argentina de Levaduras S.A.Jc, Buenous Aires; Levaduras de Pan S.A.I.C.A Buenous Aires

Australia: Fermentation Industries PTY, Ltd., Granville, N.S.W.; Mauri Brothers and Thomson, Ltd., Sydney

Austria: Vereinte Maunter Markhofsch Presshefe Fabriken, Vienna

Belgium: Gist & Spiritusfabrieken Bruggeman, Ghent

Brazil: Usina Itaiquara de Acucar e Alcool S.A., Sao Paulo

Canada: The Fleischmann Company, Montreal; Lallemand Inc., Montreal

Chile: Hoffman Prado y Co., Ltd., Santiago; Legrand Cia, Ltd., Santiago; Levaduras y Fermentos S.A., Santiago; Levaduras Collico Limited, Valdivia

Columbia: Compania Nacional de Levaduras Levapan Ltd., Bogota

Denmark: Danish Fermentation Industry, Ltd., Copenhagen

Egypt: Egypt Starch Products and Yeast Company, Alexandria

Finland: The Finnish State Alcohol Monopoly (ALKA), Helsinki; Lahden Polthimo Oy, Lahti

France: Establishments Fould-Springer, Maisons-Alfort (Seine); Societe Industrielle Lasaffre, Marcq-en-Baroeul; Societe Lesaffre, Lille; Syndicat des Producteurs de Levure Aliment de France, Paris

FRG: Asbeck, Hamm; Asmussen, Elshorn; Giegold, Schwartzenbeck; Lindenmeyer & Co., Heilbronn; Norddeutsche Hefe Industrie G.m.b.H., Hamburg Wandsbeck; Ostfriesische Hefefabriken, Leer; Pliser G.m.b.H., Darmstadt; Uniferm G.m.b.H., Werne; Weineger, Rettsteig

Guatemala: Central Productos Alimenticios Universal & Cia, Ltd., Guatemala City

Holland: Gist-Brocades, N.V., Delft

India: Dhampur Sugar Mills; Indian Yeast Company; Mohan Meakins

Iran: Iran Meyeh Company, Teheran

Japan: Chuetsu Yeast Industries, Ltd., Nagoaka; Kanegafuchi Chemical Industry Co., Ltd., Osaka; Kyowa Hakko Kogyo Co., Ltd., Tokyo; Nihon Tensaitoh, Hokkaido; Oriental Yeast Company Ltd., Tokyo; Sankyo Company, Ltd., Tokyo; Toyo Jozo Fermentation Company, Shizuoka

Mexico: Acidos Organicos S.A., Mexico D.F.; Industria Mexicana de Alimentos S.A. de C.V., Mexico D.F.; Levadura Aztica S.A., Mexico D.F.; Leviatan y Flor S.A., Mcxico D.F.

Peru:	Fleischmann Peruana Inc., Lima; Red Star del Peru S.A., Lima
Philippines;	Basic Foods Corporation, Manila; Standard Brands of the Philippines Inc., Manila
Republic of Korea:	Chun Ho Yeast Company, Seoul; Cho Hung Chemical Works, Seoul; Jeil-Universal Ltd., Seoul
South Africa:	Anchor Yeast (Pty) Ltd., Johannesburg; Union Yeast Products Ltd.; Johannesburg
Spain:	Compania de Azucares y Alcoholes S.A., Barcelona; Compania de Industrias Agricolas S.A., Valladolid; Union Alcoholera Espanola S.A., Madrid
Sweden:	Aktiebolaget S.J.A., Stockholm
Tunisia:	Societa Tunisianne de L'Industrie Laitiere, Tunis
Turkey:	Mayadag, Istanbul; Pak Gida Sinayii ve Ticaret S.A., Istanbul; Uludag Fenni Maya Ltd., Sti., Istanbul
United Kingdom:	The Distillers Co. (Yeast) Ltd., Surrey
United States:	American Yeast Company, Baltimore, Md.; Anheuser-Busch, Inc., St. Louis, Missouri; Federal Yeast Company, Baltimore, Md.; Fleischmann Laboratories division of Standard Brands, Inc., New York; Universal Foods Corporation, Milwaukee, Wisc.
Uruguay:	Levadura Uruguaya S.C.I., Montevideo
Zaire:	Leveudrie du Zaire (Leza), Kinhasa
Zambia:	Lee Yeast Company, Lusaka

Food and feed yeast

Belgium:	Raffinerie Tirelemontoise, Tienen
Finland:	Metaslüton Teollisuus Oy, Äänekosi; Ynthneet Paperitehtaat OY, Jämsänkosko
France:	Societe Franciase de Petroles B.P., Lavera
United States:	Amber Laboratories, Inc., Juneau, Wisc.; Amoco Foods Company, Chicago, Ill.; Boise-Cascade, Inc., Portland, Oregon; Diamond V Mills, Inc., Cedar Rapids, Iowa; Lake Stages Yeast Company, Rhinelander, Wisc., Stauffer Chemical Company, Westport, Conn.; Standard Brands, Inc., New York

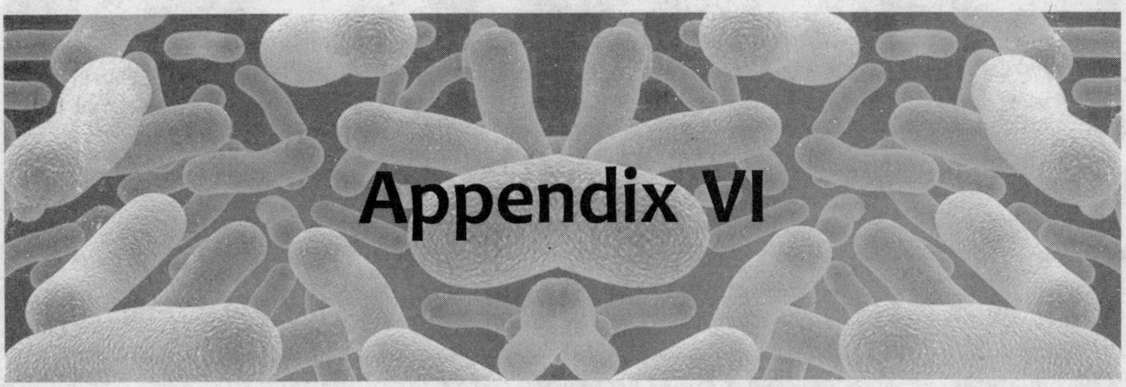

Appendix VI

FEED-GRADE SINGLE-CELL PROTEIN PLANTS [SCHWARTZ AND LEATHEN (1976)]

Plant class	*Company*	*Plant locations*	*Substrate*	*Type of microorganism*	*Plant size (ton yr^{-1})*
Demonstration	British Petroelum	United Kingdom	*n*-Alkane	Yeast	4000
	Chinese Petroleum	Taiwan	*n*-Alkane	Yeast	1000
	Dianippon	Japan	*n*-Alkane	Yeast	Unknown
	Imperial Chemical Industries	United Kingdom	Methanol	Bacteria	1000
	Kanegafuchi	Japan	n-Alkane	Yeast	5000
	Kohjin	Japan	Unknown	Yeast	2400
	Kyowa Hakko	Japan	n-Alkane	Yeast	1500
	Milbrew	United States	Whey	Yeast	5000
	Shell	Netherlands	Methane	Bacteria	1000
	Svenska-Socker	Sweden	Potato starch	Yeast	2000
Semi-commercial	British Petroleum	France	Gas oil	Yeast	20 000
	Imperial Chemical Industries	United Kingdom	Methanol	Bacteria	50 000
	United Paper Mills	Finland	Sulphite waste	Yeast	10 000
	USSR State	USSR	Unknown	Yeast	20 000
Commercial	British Petroleum	Italy	n-Alkane	Yeast	100 000
	Liquichimica	Italy	n-Alkane	Yeast	100 000
Other systems	LSU-Bechtel		Cellulose waste	Bacteria	
	Tate and Lyle		Critic acid waste	Fungi	
	ICAITI (Guatemala)		Coffee waste	Fungi	
	IFP		Carbon dioxide, sunlight	Algae	
	General Electric		Feed lot waste	Bacteria	
	Mitsubishi		Methanol	Yeast	
	Finnish Pulp and Paper		Paper pulp waste	Fungi	

Appendix VII

Name of the Biopesticides	Manufacturers/Distributors
(A) ***Bacteria***	
I. *Bacillus thuringiensis*	
1. Dipel	Abbot Laboratories, U.S.A.
2. Cajrab	BAFE Laboratories, Poona, India.
3. Thuricide HP	Sandoz (I) Ltd., India.
4. Agritrol	Merck & Co., U.S.A.
5. Bactospeine	Panchiney Progil Lab., Roger Bellon, France.
6. Bakthane L69	Rohm and Haas Co., U.S.A.
7. Baktukol	Serum Zavod, Katimovica, Yogoslavia.
8. Bathurin	Chemapol, Biokrma, Czechoslovakia.
9. Biospor 2802	Farbwerke Hoechst, Germany.
10. Biotrol	Nutrilite Products, Inc., U.S.A.
11. Dendrobacillin	Moskovs, Zavod, bakt, preparatory Atov. U.S.S.R.
12. Entobacterin 3	All Union Institute Plant Protection, U.S.S.R.
13. Parasporin	Grain Processing Corp., U.S.A.
14. Plantibae	Procida, France.
15. Sporeine	Laboratoire L.I.B.E.G., France.
16. Thuricide	Sandoz Wander and International Mineral & Chemical Corp., U.S.A.
17. *Bacillus thuringiensis* (Beta exotoxin)	
(i) Eksotoksin	Glavmikrobioprom, Russia.
(ii) Toxobakterin	Glavmikrobioprom, Russia.
(iii) Biotoksybacillin	All Union Inst. Agr. Microbiol., Russia.
II. *Bacillus popilliae*	
Milky spore bacteria	Reuter Laboratories, Inc., U.S.A.
III. *Bacillus moritai*	Rahirusu, Sumitomo Chem. and Seihu Chem., Japan.
(B) ***Fungi***	
I. *Beauveria bassiang*	
Biotrol FBB	International Minerals and Chem. Corp., Bioferm Division, U.S.A.
Beauveria bassiang	At Krasnodar, U.S.S.R.
Beaveria Spores	Nutrilite Products Inc., U.S.A.
Boverin	Glavmikrobiprom, Russia.
II. *Metarrhizium anisopliae*	
(i) *Metarrhizium anisopliae*	International Minerals and Chemicals Corp., Bioferm Division, U.S.A.
(ii) Biotrol FMA	Nutrilite Products Inc., U.S.A.

Contd.

Name of the Biopesticides	Manufacturers/Distributors
III. *Aschersonia aleyrodis*	
Aseronija	All Union Inst. Agr. Microbiol, Russia.
(C) ***Viruses***	
I. Nuclear Polyhedrosis Virus	
1. Heliothis	
(i) Bitrol VH_2	Nutrilite Products, Inc., U.S.A.
(ii) Heliothis Virus	International Minerals and Chemicals Corp. Bioferm Division, U.S.A.
(iii) Virion / H	Sandoz Wander, U.S.A.
(iv) Virex	Hayes, Sammons, U.S.A.
2. Prodenia (Spodoptera)	
(i) Biotrol VPo	Nutrilite Products Inc., U.S.A.
(ii) Biotrol VSE	Nutrilite Products Inc., U.S.A.
(iii) Viron / P	International Minerals and Chemicals Corporation, Bioferm Division, U.S.A.
(iv) *Spodoptera Virus* Viron/s	— do —
3. Trichoplusia (INI)	
(i) Trichoplusia Virus, Viron/T	Biological Control Suppliers, U.S.A.
(ii) Biotrol VTN	Nutrilite Products, Inc., U.S.A.
4. Neodiprion	Indiana Farm Bureau Co-op., Assoc., U.S.A.
5. Pieris	
Viron GKB	Latuian Agr. Acad., Russia.
II. Cytoplasmic Polyhedrosis Virus	
Dendrolimus CPV	Matsukemin, Chugai-Kumiai Chem., Japan.
(D) ***Nematodes*** :	
Neoaplectana carpocapsae (*DD* 136)	
Biotrol NCS	Nutrilite Products Inc., U.S.A.

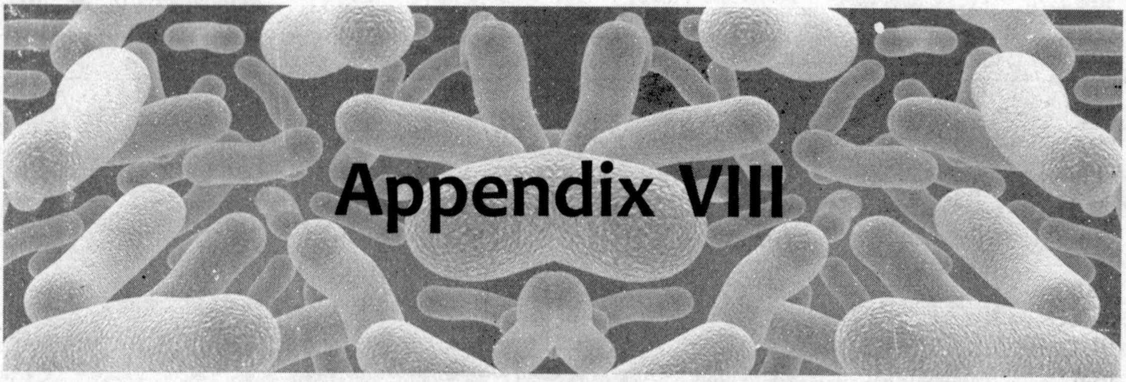

Appendix VIII

SOURCES OF INDUSTRIALLY IMPORTANT MICRO-ORGANISMS

ACC	Akers Culture Collection of Imperial Chemical Industries, Ltd., Alderley Park, Macclesfield, Cheshire, UK
AMC	Walter Reed Army Medical Center (formerly Army Medical School and Army Medical Department Research and Graduate School), Washington DC, USA
AMIF	American Meat Institute Foundation, Chicago, Illinois, USA
AMNH	American Museum of Natural History, New York, New York, USA
ARL	See ACC
ATCC	American Type Culture Collection, Rockville, Maryland, USA
ATU	See FAT
BBL	Baltimore Biological Laboratory, Baltimore, Maryland, USA
BKM	All-Union Collection of Micro-organisms, USSR
BRL	See ACC
BU	See BUCSAV
BUCSAV	Biological Ustav, Ceskoslovenska Akademie Ved, Prague, Czechoslovakia
CASE	Case Laboratories, Chicago, Illinois, USA
CBS	Centraalbureau voor Schimmelcultures, Baarn, Netherlands
CCAP	Cambridge Collection of Algae and Protozoa, Cambridge, UK
CCEB	Culture Collection of Entomogenous Bacteria, Prague, Czechoslovakia
CCF	Culture Collection of Fungi, Department of Botany, Charles University, Prague, Czechoslovakia
CCM	Czechoslovak Collection of Micro-organisms, J.E. Purkyne University, Brno, Czechoslovakia (same as MDB)
CCTM	Centre de Collection de Type Microbienne, Lille, France
CDA	Canadian Department of Agriculture, Ottawa, Canada
CDC	Center for Disease Control, Atlanta, Georgia, USA
CI	Carnegie Institute, Cold Spring Harbor, Long Island, New York, USA
CIP	Collection of the Institute Pasteur, Paris, France
CLMR	Central Laboratory, South Manchurian Railway Co., Ltd.
CMI	Commonwealth Mycological Institute, Kew, UK [strain numbers formerly prefixed 1MI (Imperial Mycological Institute) are included here]
CSIRO	Cornmonwealth Science and Industrial Research Organization, Sydney, Australia
CU	See CCAP
DAOM	Plant Research Institute, Department of Agriculture, Mycology, Ottawa, Canada
DIFCO	Difco Laboratories, Detroit, Michigan, USA
DMUR	Department of Mycology, University of Recife, Brazil
DSM	Deutsche Sammlung von Mikroorganismen, Gottingen, West Germany

EPA	Environmental Protection Agency, Washington DC
ETH	Eidgenosische Technische Hochschule, Zurich, Switzerland
FAT	Faculty of Agriculture, Tokyo University, Japan
FDA	Food and Drug Administration, Washington DC, USA
FERM	Fermentation Research Institute, Agency of Industrial Science and Technology, Chiba, Japan
FGSC	Fungal Genetic Stock Center, Humboldt State College, Arcata, California
FI	Farmitalia SpA, Milan, Italy
FRR	Division of Food Research, Food Research Laboratory, CSIRO, Sydney, Australia
GRIF	Government Research Institute of Formosa
HACC	Hindustan Antibiotics Ltd., Pimpri, Poona, India
HUT	Hiroshima University, Faculty of Engineering, Hiroshima, Japan
IAM	Institute of Applied Microbiology, University of Tokyo, Japan
IAUR	Instituto de Antibioticos da Universidade de Recife, Brazil
IAW	Institute of Antibiotics, Warsaw, Poland
ICI	Imperial Chemical Industries Ltd., Butterwick Research Laboratories, Welwyn, UK
ICPB	International Collection of Phytopathogenic Bacteria, University of California Davis, California, USA
IEM	Institute of Epidemiology and Microbiology, Prague, Czechoslovakia
IFM	Institute of Food Microbiology, Chiba University, Chiba, Japan
IFO	Institute for Fermentation, Osaka, Japan
IHM	Instituto de Higiene Experimental, Montevideo, Uruguay
IMCAS	Institute of Microbiology, Czechoslovak Academy of Science, Prague, Czechoslovakia
IMI	See CMI
IMRU	Institute of Microbiology, Rutgers-The State University, New Brunswick, New Jersey, USA
IMUR	Instituto de Micologia, Universidade de Recife, Brazil
INA	Institute for New Antibiotics, Moscow, USSR
INMI	Institute for Microbiology, USSR Academy of Sciences, Moscow
IOC	Instituto Oswaldo Cruz; Rio de Janeiro, Brazil
IP	Institute Pasteur, Paris, France
IPV	Instituto de Patologia Vegetale, Milan, Italy
ISC	International Salmonella Center
ISL	International Subcommittee on Lactobacilli and Closely Related Organisms
ISP	International Streptomyces Project
ITCC	Indian Type Culture Collection, New Delhi, India
ITCCF	Indian Type Culture Collection of Fungi, New Delhi, India
IU	See UTEX
KCC	Kaken Chemical Company, Ltd., Tokyo, Japan
MDB	See CCM
MTHU	See MTU
MTU	Faculty of Medicine, University of Tokyo, Japan
NADC	National Animal Disease Center, Ames, Iowa, USA
NADL	See NADC
NBL	Naval Biological Laboratory, Oakland, California, USA
NCA	National Canners' Association, Washington DC, USA
NCAIA	National Center for Antibiotics and Insulin Analysis, FDA, Washington DC, USA
NCDC	See CDC
NCDO	National Collection of Dairy Organisms, Reading, UK
NCIB	National Collection of Industrial Bacteria, Aberdeen, UK
NCIM	National Collection of Industrial Micro-organisms, National Chemical Laboratory, Poona, India
NCMB	National Collection of Marine Bacteria, Aberdeen, UK
NCPPB	National Collection of Plant Pathogenic Bacteria, Harpenden, UK
NCTC	National Collection of Type Cultures, London, UK
NCYC	National Collection of Yeast Cultures, Food Research Institute, Norwich, Norfolk, UK

NDRC	National Defense Research Committee, Washington DC, USA
NEA	Nobel Explosives Co., Ardeer, UK
NI	Nagao Institute, Tokyo, Japan
NIH	National Institutes of Health, Bethesda, Maryland, USA
NIHJ	National Institutes of Health, Japan
NIRD	National Institute for Research in Dairying, Reading, UK
NRC	See NRCC
NRCC	National Research Council, Ottawa, Canada
NRRL	Northern Utilization Research and Development Division, US Department of Agriulture, Peoria, Illinois, USA
OUT	Osaka University, Faculty of Engineering, Osaka, Japan
PD	Plantenziektenkundige Dienst, Wageningen, Netherlands
PRL	Praire Regional Laboratory, Saskatoon, Canada
PSA	Progetto Sistematica Actinomiceti, Instituto "P. Stazzi," Milan, Italy
QM	Quartermaster Research and Development Center, U.S. Army, Natick, Massachusetts, USA
RIA	See USS RIA
SAUG	Sammlung was Algenkulturen, University of Gottingen, West Germany
SMG	See DSM
TMC	Trudeau Mycobacterial Culture Collection, Trudeau Institute, Saranac Lake, New York, USA.
TRTC	Department of Botany, University of Toronto, Toronto, Canada
UAMII	University of Alberta Mold Herbarium and Culture Collection, Alberta, Canada
UC	Upjohn Company, Kalamazoo, Michigan, USA
U. Md.	University of Maryland, College Park, Maryland, USA
USDA	United States Department of Agriculture
USDI	United States Department of the Interior
USPHS	United States Public Health Service
USSR RIA	USSR Research Institute for Antibiotics, Moscow, USSR
UTEX	University of Texas, Culture Collection of Algae, Austin, Texas, USA
UTMC	University of Texas, Myxomycete Collection, Austin, Texas, USA
VPI	Virginia Polytechnic Institute and State University, Blacksburg, Virginia, USA
WB	University of Wisconsin, Bacteriology Department, Madison, Wisconsin, USA
WHO	World Health Organization
WLRI	Warner Lambert Research Institute, Morris Plains, New Jersey, USA
WRAIR	Walter Reed Army Institute of Research, Washington DC, USA
WRRL	Western Utilization Research and Development Division, U.S. Department of Agriculture, Albany, California, USA

Bibliography

A

Aiba, S.; Humphrey, A. E. and Mills, N., F. (1965), *Biochemical Engineering*, Academic Press, New York.

Ambasta, S. P. (Editor), (March '83) *Science reporter*, Council of Scientific and Industrial Research, New Delhi, India.

Ambasta, S. P. (Editor), (April '82) *Science reporter*, Council of Scientific and Industrial Research, New Delhi, India.

Ambasta, S. P. (Editor), (March '82) *Science reporter*, Council of Scientific and Industrial Research, New Delhi, India.

Ashley, M. H. J. (February '82), *The chemical engineer*, London.

Atkinson, B. and Mavituna, F. (1983), *Biochemical engineering and biotechnology handbook*, Macmillan Publisher Ltd., New York.

Atkinson, B. (1974), *Biochemical reactors*, Pion Ltd., London.

B

Bailey, J. E. and Oillis, D. F. (1977), *Biochemical engineering fundamentals*, McGraw-Hill book company, London.

Barton-Wright, E. C. (1952), *The microbiological assay of the vitamin B-complex and amino acids*, Pitman books Ltd., London.

Bender, A. E. (1975), *Dictionary of nutrition and food technology*, Butterworths, London.

Bhattacharjee, J. K. (1970), *Adv. appl. microbial.*, 13-139-159.

Binsted, R.; Devey, J D. and Dakin, J. C. (1971), *Pickle and sauce making*, Food trade press Ltd., London.

Bioprocessing in the eighties: Symposium, Southampton (1982), Institution of chemical engineers.

Biotechnology: report of a joint working party; By the advisory council for applied research and development, the advisory board for the research councils and the Royal society, Chairman A. Spinks (1980).

Biotechnology and education: report of a working group (1981), Royal society, London.

Birch, G. G. and other (Editors), (1981), *Enzymes and food processing*, Applied science Ltd., London.

Birch G. G.; Spencer, M. and Cameron, A. G. (1977), *Food Science*, Pergamon dress, New York.

Blakebrough, N. (Editor), (1967, 1968), *Biochemical and biological engineering Science*, Vols. 1 and 2, Academic Press, London.

Board, R. G. and Lovelock, D. W. (Editors) (1975), *Some methods for Microbiological Assay*, Academic press, London and New York.

Borkar, P. S. (Editor) (1982), *Hindustan Antibiotics bulletin* Vol. 24, H. A. Ltd., India.

Brenner, S.; Hartley, B. S. and Rodgers, P. J. (Editors), (1980), *New Horizons in industrial microbiology*, The Royal Society, London.

Brock, T. (Editor) (1961), *Milestones in Microbiology*, American Society for Microbiology, Washington.

Bryan, A. H. and Bryan, C. G. (1953), *Bacteriology*, Barnes & Noble, Inc., New York.

Buchanan, R. E. and Buchanan, E. D. (1951), *Bacteriology*, The Macmillan Company, New York.
Bull, A. T. and Meadow, P. M. (Editors) (1978). *Companion to Microbiology*, Longman group Ltd., London.
Bull, A. T. and other (Editors) (1979), *Microbial technology*, University of Cambridge, Cambridge.
Bulla, L. A. and Costilow, R. N. (1978), *Adv. Appl. microbiol.*, 23, 1-16.
Bu'Lock, J. D.; Nisbet, L. J. and Winstanely, D. J. (Editors)(1982), *Bioactive microbial Products*: search and discovery, Academic Press, London.
Bushell, M. E. and Slater, J.H. (Editors)(1981), *Mixed culture fermentations*, Academic Press, London and New York.

C

Canale, R. P. (Editor) (1971), *Biological waste treatment* (Biotechnology and bioengineering symposium No. 2) John Wiley & Sons, New York.
Carpenter, P. L. (1961), *Microbiology*, W. B. Saunders Company, London.
Carr, J. G.; Cutting, C. V and Whiting, G. C. (Editors) (1975), *Luctic acid bacteria in beverages and food*, Academic Press, New York.
Chakrabarty A.M. (1982), *Biodegradation and detoxification of Environmental Pollutants,* CRC Press.
Charney, W. and Herzog, H. L. (1967), *Microbial transformation of sterioids*, Academic Press, New York and London.
Clem-Fus (Editor) (1982), *Encyclopedia of science and technology*, Vols. 51 and 5, McGraw-Hill Book Company, New York.
Casida, L. E. (1968), *Industrial microbiology*, John Wiley & Sons Ltd., New York.
Clifton, C. E. (1950), *Introduction to the bacteria*, McGraw-Hill Book Company, Inc., New York.
Clifton, C. E. (1957), *Introduction to the bacterial Physiology*, McGraw-Hill Book Company, Inc., New York.
Codd, L. W.; Dijkhodd, K. and Others (Editors) (1975), *Materials and technology*, Vol. 5, 7 and 8, Longman Group Ltd., London.
Collard, P. (1976), *The developments in Microbiology*, Cambridge University Press, London.
Collins, C. H. (Editor) (1967), *Progress in microbiological techniques*, Butterworths, London.
Computer applications in fermentation technology (1982), Society of Chemical Industry, London.
Conn, H. W. (1912), *Bacteria, yeasts and molds in home*, Ginn & Company, New York & London.
Conney, C. L. (1979), *Computer application in fermentation technology:* a perspective; Biotechnology Bioengineering symposium, 9 1-11.
Conney, C. L. and Acevedo, F. (1977), *Biotechnol. Bioeng*. **19**, 1449-1462.
Cooper, P. f. and Atkinson, b. (Editors) (1981), *Biological fluidised bed treatement of water and waste water*, Ellis Horwood (Publisher) Limited Chichester.
Corum, C. J. (Editor) (1970), *Developments in industrial microbiology*, Vol.**10**, American Institute of Biological Sciences Washington, U.S.A.

D

Davies, C. N. (1973), *Air filtration*, Academic Press, London and New York.
Daies, R. (Editor) (1982), *Development in food microbiology*-1, Applied Science Publishers Ltd., London.
Davis, P. (Editor) (1974), *Symposium on SCP*, Academic Press, London and New York.
Dawes, B. (1952), *Hundred years of biology*, Gerald Duckworth & Co. Ltd., London.
Debach, P. (1974), *Biological control by natural enemies*, Cambridge University Press, London.
Dean, A. C. R., Ellwood, D. C.; Evans, C. G. T. and Metlling, J. (Editors) (1976), *Continuous culture* 6: *Applicaitonc and New Fields*, Ellis Horwood Ltd., Chichester.
Decker, H. M. *et al.* (1954), *Appl. microbiol.* 2 33.
Dellweg, H. (Editor) (1983), *Biotechnology*, Vol. 3, Verlag Chemie, Florida.

Dubos, Rene J. (1950) Louis Pasteur—*Free Lance of Science* Little, Brown and Company, Boston.

E

Escritt, L. B. (1972), *Sewerage and sewage disposal* Vol. II, Macdonald & Evans Ltd., Plymouth.

F

Ferguson, Andrew (1980), *Biochemical systematic and evolution*, Blackie, Glasgow and London.

Fogarty, W. M. (Editor) (1983), *Microbial enzymes and biotechnology*, Applied Science Publishers, London and New York.

Ford, B. J. (1970), *Microbiology and food*, Northwood Publications Ltd., London.

Fox, Brian A. (1970), *Food Science*—a chemical approach, Hodder and Stoughton, London.

Frazier, W. C. (1967), *Food Microbiology*, McGraw-Hill Book company, London & New York.

Freitas, Y. M. and Fernandes, F. (Editors) (1971), *Global impacts of applied microbiology*, University of Bombay, India.

Fried, J.; Thoma, R. W.; Gerke, J. R.; Herz, J. F.; Donin, M. N. and Perlman, D. (1952), *J. Am. Chem. Soc.*, 74, 3262.

Frobisher, M. (1968), *Fundamental principles of Microbiology*, W. B. Saunders Company, London.

G

Gardner, Eldon, J. (1972), *History of biology*, Wiley Eastern Limited, New Delhi, India.

Gebhardt, L. P. (1970), *Microbiology*, The C. V. Mosby Company, Japan.

Godfrey, T. and Reichelt, J. (1983), *Industrial enzymology*, Macmillan Publisher Ltd., New York.

Gordman, J. J.; May, M. and Smith, L. L. (1980), *J. Biol. chem.*, 235 965.

Gray, W. D. (1959), *The relation of fungi to human affairs*, Henry Holt and Company, Inc., New York.

Gray, W. D. (1970), *The use of fungi as food and in food processing*, CRS Press, Ohio.

Gutcho, S. J. (1973), *Chemical by fermentation*, Noyes Data Corporation, London and New York.

H

Hagino, H. Kobayashi, S.; Araki, K. and Nakayama, K. (1981), Biotechnology letters, Vol. 3, No. 8, 425-430.

Hammer, B. W. (1938), *Diary bacteriology*, Wiley, London.

Han, Y. M. (1978), *Adv. appl. microbiol.* **23**, 119-149.

Hanka, L. J. (1972), *Adv. appl. microbiol.*, **15**, 147-156.

Harrison, D. E.; Haggins, I. J. and Watkinson, R. (Editors) (1980), *Hyrdocarbons in biotechnology*, Heyden & Son Ltd., London.

Harrigan, W. F. And McCance, M. E. (1976), *Laboratory methods in food and Dairy Microbiology*, Academic Press, London and New York.

Hartman, P. A. (1968), *Miniaturized Microbiological methods* (Adv. app. microbiol. suppliment 1), Academic Press, New York and London.

Harvey, W. C. and Hill, H. (1967), *Milk production and control*, H. K. Lewis & Co. Ltd., London.

Hersom, A. C. and Hulland, E. D. (1980), *Canned foods*, Churchill Livingstone, London and New York.

Hewitt, W. (1977), *Microbiological assay*, Academic Press, London.

Hobson, P. N.' Bousfield, S.; and Summers, R. (1981), *Methane production from agricultural and domestic wastes*, Applied Science Publisher Ltd., London.

Hockenhull, D. J. D. (Editor) (1972), *Progress in Industrial Microbiology*, Vols **10** & **11**, Churchill Livingstone, London.

Hopwood, R. (1975), *Advanced food science*, G. Bell & Sons Ltd., London.

Hopwood, D. A. (1970), *Methods in microbiol.*, 3A, 363.

Howard, J. R. (Editor) (1983), *Fluidised beds*, Applied Science Publisher, London and New York.

Hugo, W. B. and Russell, A. D. (1980), *Pharmaceutical Microbiology*, Blackwell Scientific Publications, London.

I

Ingle, M. B. and Erickson (1978), *Adv. appl. microbiol*. 24, 257-276.

Iyengar, M. R. S. (1969), *Indian Chem.* J. 4, 55-59.

J

Jackson, J. M. and Shinn, B. M. (Editors) (1979), Fundamentals of food canning technology, AVI publishing Company, Inc., Connecticut.

Jay, J. M. (1970), *Modern Food Microbiology*, D. Van Nostrand Company, London and New York.

Jones, A. (1974), *World Protein resource*, Medical and Technical Publishing Co. Ltd., England.

K

Kalle, G. P. *et al*. (1978), *Perspectives in Industrial Microbiology*, Association of Microbiologists of India, Bombay, India.

Kapsiotis, G. D. (1978), *Single cell-protein* (*Food and nutrition* Vol. 4), Food and Agriculture Organization of the United Nations.

Kavanagh, F. (1963), *Analytical Microbiology*, Vol. 1. Academic Press, London and New York.

Kavanagh, F. (1972), *Analytical Microbiology*, Vol. 2. Academic Press, London and New York.

Kent, N. L. (1975), Technology of cereals, Pergamon Press, England.

Kharatyan, S. G. (1978), Ann. rev. microbiol., **32**, 301-327.

Kondo, E. and Masau, E. (1961), *J. Gen. Appl. Microbiol.*, 7, 113.

Kulhanek, M. (1970), *Ad. appl. microbiol.*, 12, 11-28.

Kieslich, Klaus (1976), *Microbial transformations of non-sterioid cyclic compounds*, John Wiley & Sons George Thieme Publishers, London.

Klocker, Alb. (1903), *Fermentation organisms*, Londmans, Gren and Co., new York.

L

Lafar, Franz (1898), *Technical mycology*, Vol. **I**, Charles Griffin and Company Limited, London.

M

Macdonald, K. D. (Editor) (1976), *Genetics of Industrial Micro-organisms*, Academic Press, London.

Madhok, M. R. (1942), *Bacteria in every-day life*, Lyallpur, India.

Mahanta, K. C. (1984), *Dairy Microbiology*, R. Kumar, Delhi, India.

Mara, D. (1978), *Sewage treatement in hot climates*, John wiley & Sons, New York.

Mark, H. F.; Othmer, D. F. and Others (Editors), (Third Edition) Encyclopedia of Chemical Technology, Vols. **2**, **9**, **11**, **12** & **15**.

Marstrand, P. K. (1981), *Pattens of change in biotechnology*, (SPRU Occasional Papers 15), University of Sussex, U. K.

Meynell, P. J. (1982), *Methane—Planning a digester*, Prism Press, London.

Miller, B. M. and Litsky, W. (1976), *Industrial Microbiology*, McGraw-Hill Book Company, New York.

Monot, F. and Engasser (1983), *Biotechnology letter*, Vol. 5, No. 4, 213-218.

Moss, M. O. and Smith, J. E. (Editors) (1977), *Industrial applications of Microbiology*, Surrey University Press,

London.

Mudambi, S. R. and Rajagopal, M. V. (1983), *Fundamentals of foods and nutrition*, Wiley Eastern Ltd., India.

N

Nicholls D.G. (1983), *Bioenergetics: An Introduction to the Chemiosmotic Theory, Academic Press.*

Norris, J. R. and Richmond, M. H. (Editors) (1978), *Essays in applied microbiology*, John Wiley & Sons, New York.

Norris, J. R. and Ribbons (Editors) (1975), *Methods in Microbiology*, Vols. **3A, 1, 2, 3B & 4,** Academic Press, London.

O

Onions, A. H. S.; Allsopp, D. and Eggins, H. O. W. (1981), *Smith's introduction to industrial mycology*, Edward Arnold, London.

P

Palmer, T. C. (1981), *Understanding enzymes*, Ellis horwood Ltd., England.

Patel, J. D.; Venkataramu, K. and Subba Rao, M. S., *Indian food Packer*, **31(6)**, 5-8, 1977.

Parker, A. (1950), *Aseptic Technique in industrial scale fermentations. Report on Recent Advances in the Fermentation Industries* (Royal Institute of Chemistry, 1950).

Patricia Cleveland-Peck (1980), *Making cheeses, Butters, cream and yoghurt at home*, Thorsons Publisher Ltd., Northamptonshire.

Pearson, D. (1976), *The chemical analysis of foods*, Livingstone, London and New York.

Pederson, C. S. (1971), *Microbiology of food fermentations*, Avi Publishing Co., Inc., Westport, Conn.

Pelczar, M. J.; Reid and Chan, E. C. S., (1977), *Microbiology*, McGraw-Hill Publishing Company, New Delhi.

Penicillin fifty years after Fleming (1979, 1980), a Royal Society discussion, Royal Society, London.

Peppler, H. J.; Pertman, D. (editors) (1979), *Microbial technology* Vols. **1** and **2**, Academic Press, New York.

Perlman, D. (Editor) (1969), *Fermentation advances*, Academic Press, New York and London.

Perlman, D. (Editor) (1977), *The fermentation industry*, ASM News **43** (2), 82-89.

Perlman, D. and Ogawa, Y. (1976), J. Antibiot. 29, 1112.

Perlman, D. (Editor) (1977 and 1978), *Advances in applied microbiology*, Vols. **22** and **23**, Academic Press, London and New York.

Pirie, N. W. (Editor) (1975), *Food protein sources*, Cambridge University Press, London and New York.

Pirt, S. J. and Righelato, R. C. (1967), *Appl. microbiol.*, **15**, 1284-1290.

Plofkin S.A, Orenstein W.A., Offit P.A., *Vaccines* (Fifth Edition), Sauders Elsevier.

Primrose, S. B. and Wardlaw, A. C. (Editors) (1982), *Sources book of experiments for the teaching of microbiology*, Academic Press, London and New York.

Postgate, J. (1975), *Microbesand man*, Penguin Books Ltd., London.

Pye, E. K.(1974), *Present and future trends in enzyme biotechnology and its applications*.

Pyke, M. (1981), *Food science and technology*, John Murray, London.

R

Radley, J. A. (Editor) (1976) *Examination and analysis of starch and starch products*, Applied science Publishers Ltd., London.

Radley, J. A. (Editor), (1976), *Industrial uses of starch and its derivatives*, Applied Science Publishers Ltd., London.

Rainbow, C. and Rose, A. H. (Editors), (1963), *Biochemistry of industrial micro-organisms*, London and New

York.

Ramakrishnan, C. V., Parekh, L. J.; Akolkar, P. N.; Rao, G. S. and Bhandari, S. D. (1976), *Plant foods for man*, 2, 12.

Reed, G. (Editor), (1982), *Present & Dunn's industrial microbiology*, Macmillan Publishers, Ltd., U. K.

Rance, Patrick (1982), *The Great British cheese book*, Macmillan, London.

Rehm, H. J. and Reed, G. (editors), (1981), *Microbial fundamentals* (Biotechnology, Vol. **1**), Verlag Chemie, Florida.

Reid, R. (1974), *Microbes and men*, The British Broadcasting Corporation, London.

Reusser, F.; Koepsell, J. and Savage, G. M. (1961), *Appl. Microbiol*., 9, 346.

Rhodes, A. and Fletcher, D. L. (1966), *Principles of Industrial Microbiology*, Pergamon Press Ltd., Oxford.

Rockwell, P. J. (1976), *Single cell proteins from cellulose and hdyrocarbons*, Noyes data Corporation, New Jersey.

Rose, A. H. (1961), *Industrial microbiology*, Butterworth & Co. (Publishers) Ltd., London.

Rose, A. H. (Editor), (1982), *Fermented foods* (Economics Microbiology 7), Academic Press, London.

Robinson, R. K. (Editor), (1981), *Dairy Microbiology*; Vol. 2, Applied Science Publishers, New Jersey and London.

Rose, A. H. (Editor), (1980), *Microbial enzymes and bio-conversions* (Economic Microbiology, 5), Academic Press, London.

Rose, A. H. (Editor), (1978), *Primary Products* (Economic Microbiology; 1) Academic Press, London.

Rose, A. H. and Harrison, J. S. (Editors), (1971), *Yeast technology* (The Yeast 3), Academic Press, London.

Rosebury, Theodor (1962), *Micro-organisms indigenous to man*, McGraw-Hill Book Company, New York.

Rosen B.P. (1978), *Bacterial Transport,* Dekker.

S

Salle, A. J. (1973), *Fundamental principles of bacteriology*, McGraw-Hill, Inc., New York.

Schild, H. O. (1980), *Applied pharmacology*, Churchill Livingstone, London and New York.

Scientific American (September '81), Vol **245**, No. 3.

Shirasuka, M.; Ozaki, M. and Sugawara, S. (1961), *J. Ferment. Assoc., (Japan)*, **19**, 335.

Sikyta, B.; Prokop, A. and Novak, B. (Editors), (1974), *Advances in microbial engineering, Part* **2**, Butterworths, London.

Sir Wilson G.S. (1967), *The Hazards of Immunization,* The Atholone Press, University of London.

Smith, J. E. (1981), *Biotechnology*, Edward Arnold (Publishers) Limited, London.

Smith, J. E. and Berry, D. R. (Editors), (1975), *Industrial mycology,* (Filamentous Fungi, Vol. **1**), Edward Arnold (Publishers) Ltd., London.

Solomons, G. L. (1969), *Materials and methods in fermentation*, Academic Press, London and New York.

Stanier, R. Y.; Edward, A. A. and John, L. I. (1977), *General microbiology*, The Macmillan Press Ltd., London.

Starr, M. P. (1964), *Global impacts of applied microbiology*, John Wiley & Sons, Inc., London.

Steel, R. (Editor), (1958), *Biochemical engineering*, Heywood & Co. Ltd., London.

Steel, R. and Miller, T. L. (1970), *Ad. appl. microbiol*., **12**.

Stodola, F. H. (1958) *Chemical transformations by micro-organisms*, John Wiley & Sons, Inc., New York.

Stryer L. (1981) *Biochemistry, 2nd* edition, Freeman.

Subba Rao, N. S. (1982), *Advances in agricultural microbiology*, Butterworths, London.

Swartz, R. W. (1979), *Ann. repts. ferm. processes*, **3**:75-110.

T

Taylor, M. J. and Richardson, T. (1979), *Adv. appl. microbiol*., **25**, 7-31.

Taylor, R. J. (1980), *Food additives*, John Wiley & Sons Ltd., Chichester.

Thatcher, F. S. and Clark, D. S. (Editors), (1968), *Micro-organisms in foods*, University of Toronto press, Canada.
Thoma, R. W. (Editor), (1977), *Industrial Microbiology*, Dowden Hutchinson & Ross Inc., Pennsylvania.
Tsao, G. T. (Editor), (1982), *Annual reports on fermentation processes*, Vol. **5**, Academic Press, London and New York.

U

Umbreit, W. W. (Editor), (1964), *Advances in applied Microbiology*, Vol. **6**, Academic Press, New York.
Underkofler, L. A. and Hickey, R. J. (Editors), (1954), *Industrial fermentations*, Vol. I, Chemical Publishing Co. Inc. new York.
Underkofler, L. (Editor), (1975), *Developments in Industrial Microbiology*, Vol. 16, American Institute of Biological Sciences, Washington.

V

Vandamme, E. J. (Editor), (1984), *Biotechnology of Industrial Antibiotics*, Marcel Dekker, Inc., New York.
Voznaya, N. F. (1981), *Chemistry of water and Microbiology*, Mir Publishers, Moscow.
Vanek, Zdenko; hostalek, Zdenek and Cudlin, Josef (Editors), (1973), Elsevier Publishing Company, London
Vezing, Claude and Singh, Kartar (Editors), (1980), *Advances in biotechnology*, Vol. III, Pergamon Press, London.

W

Wasley, G. D. and Warner, R. W. (1974), *Microbiology*, St. Paul's House, Warnik Lane, London.
Weaver, E. A.; Kenney, H. E. and Wall, M. E. (1960), *Appl. microbiol.*, **8**, 345.
Whitamarsh, J. M. (1954), *J. Appl. Bact.* 17, 27.
William, D. G. (1970), *The use of fungi as food and in food processing*, CRC Press, Ohio.
Winkler, M. A. (1981), *Biological treatment of waste-water*, Ellis Horwood Ltd., Chichester.
Wiseman, A. (Editor), (1975), *Handbook of enzyme biotechnology*, Ellis Horwood Ltd., Chichester.
Wiseman, A. (Editor), (1983), *Principles of biotechnology*, Surrey University Press, London.
Wiseman, A. (Editor), (1978), *Topics in enzyme and fermentation biotechnology* 2, Ellis Horwood Ltd., Chichester.

Y

Yamada, K. (1977), *Japan's most advanced industrial fermentation technology and industry academic review*, The International Technical Information Institute, Japan.

Z

Zaborsky, O. R. and Kennedy, J. F (Editors), (November 82), *Enzyme and microbial technology*, Vol. 4, No. 6, Butterworths, London.

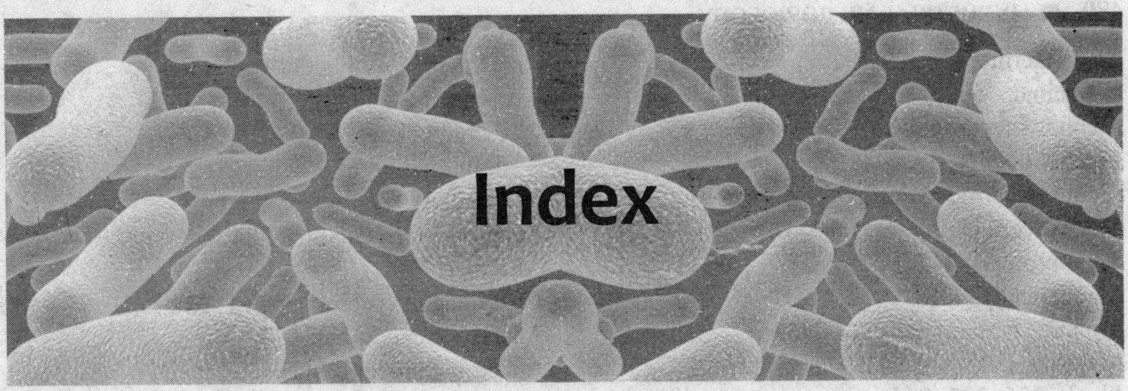

Index

A

C

G

N

O

P

Q

R

S

Sulfur-free toluene, 256